Handbook of
Full-Field Optical Coherence Microscopy

Handbook of Full-Field Optical Coherence Microscopy
Technology and Applications

edited by
Arnaud Dubois

PAN STANFORD PUBLISHING

Published by

Pan Stanford Publishing Pte. Ltd.
Penthouse Level, Suntec Tower 3
8 Temasek Boulevard
Singapore 038988

Email: editorial@panstanford.com
Web: www.panstanford.com

British Library Cataloguing-in-Publication Data
A catalogue record for this book is available from the British Library.

Handbook of Full-Field Optical Coherence Microscopy: Technology and Applications

Copyright © 2016 Pan Stanford Publishing Pte. Ltd.

All rights reserved. This book, or parts thereof, may not be reproduced in any form or by any means, electronic or mechanical, including photocopying, recording or any information storage and retrieval system now known or to be invented, without written permission from the publisher.

For photocopying of material in this volume, please pay a copying fee through the Copyright Clearance Center, Inc., 222 Rosewood Drive, Danvers, MA 01923, USA. In this case permission to photocopy is not required from the publisher.

ISBN 978-981-4669-16-0 (Hardcover)
ISBN 978-981-4669-17-7 (eBook)

Printed in the USA

Contents

Preface   xix

1 **Introduction to Full-Field Optical Coherence Microscopy**   1
  Arnaud Dubois
  1.1 Introduction   1
  1.2 Experimental Setup   4
    1.2.1 General Principle   4
    1.2.2 Interference Microscope Configuration   5
    1.2.3 Light Source   6
    1.2.4 Image Sensor   11
  1.3 Image Acquisition   13
    1.3.1 Axial Scan   13
    1.3.2 Phase-Shifting Interferometry   15
    1.3.3 Sinusoidal Phase Shifting and Four Integrating Buckets   18
  1.4 Performance   22
    1.4.1 Detection Sensitivity   22
    1.4.2 Axial Resolution   25
    1.4.3 Lateral Resolution   29
    1.4.4 Sample Motion Artifacts   30
  1.5 Clinical Applications   31
    1.5.1 Ophthalmic Tissue Imaging   32
    1.5.2 Skin Tissue Imaging   33
    1.5.3 Breast Tissue Imaging   34
    1.5.4 Other Tissue Imaging   36
  1.6 Applications in Biology   37
    1.6.1 Developmental Biology   37
    1.6.2 Cell Biology   38
    1.6.3 Plant Studies   39

1.7 Material Characterization 39
1.8 Conclusion 40

PART I THEORETICAL ASPECTS

2 **Theory of Imaging and Coherence Effects in Full-Field Optical Coherence Microscopy** 53
Anton A. Grebenyuk and Vladimir P. Ryabukho
2.1 Introduction 54
2.2 Signal Formation in FF-OCM 55
   2.2.1 Principle of the Analysis 56
   2.2.2 Optical Transmission Functions of the Sample and Reference Arms 61
   2.2.3 Resultant Signal of FF-OCM 66
2.3 Coherence Effects in FF-OCM 69
   2.3.1 Principle of Coherence Gating 70
      2.3.1.1 Sample with uniform interfaces 70
      2.3.1.2 Sample with transversal structure 73
   2.3.2 Impulse Response and Illumination Aperture Function 76
   2.3.3 Influence of the Sample Refractive Index 80
2.4 Conclusions 83

3 **Spatiotemporal Coherence Effects in Full-Field Optical Coherence Tomography** 91
Ibrahim Abdulhalim
3.1 Introduction 91
3.2 Basic Coherence Concepts 94
3.3 Finite Temporal and Infinite Longitudinal Spatial Coherence Lengths 97
3.4 Finite Longitudinal Spatial and Infinite Temporal Coherence Lengths 101
3.5 Lateral Mismatch as a Source of Lateral Decoherence 105
3.6 Finite Longitudinal and Finite Temporal Coherence Lengths 107
3.7 Effects on Multilayer Sample Measurement 111
   3.7.1 LSC Effect on the Fringe Size 111

        3.7.2   LSC Effect on the Thickness Determination
               When Imaging a Multilayered Sample   112
        3.7.3   LSC Effect on the Imaging Depth   115
  3.8 Conclusions and Future Trends   119

4 Cross Talk in Full-Field Optical Coherence Tomography   131
Boris Karamata, Marcel Leutenegger, and Theo Lasser
  4.1 Optical Cross Talk in FF-OCT   135
        4.1.1   Origin and Definition of Cross-Talk Noise   136
        4.1.2   Qualitative Analysis of the Cross-Talk
               Contribution   137
               4.1.2.1   Spatial coherence of cross talk light   138
               4.1.2.2   Interference with multiply scattered
                       light   139
               4.1.2.3   Questions raised by our analyses   140
        4.1.3   Cross-Talk Noise Suppression with Spatial
               Coherence Gating   141
  4.2 Theory and Model   143
        4.2.1   Mathematical Description   144
        4.2.2   Calculation of the Mean Signal   148
        4.2.3   Integration of Monte Carlo Results   151
  4.3 Method for Cross-Talk Investigation   151
        4.3.1   Setup   152
        4.3.2   Sample   154
        4.3.3   Presentation of Experimental Results   155
        4.3.4   Monte Carlo Simulation Details   157
  4.4 Experimental and Theoretical Results for Cross-Talk
       Noise   158
        4.4.1   Dependence on Sample Properties   158
        4.4.2   Cross-Talk Contribution Relative to the Useful
               Signal   161
        4.4.3   Additional Results and Conclusions   164
  4.5 Cross-Talk Suppression   164
        4.5.1   Setup and Sample   165
        4.5.2   Results   168
  4.6 Conclusions and Discussions   170
        4.6.1   Full-Field OCT   170
        4.6.2   Modeling Multiple Scattering in OCT   173

5 Signal Processing Methods in Full-Field Optical Coherence Microscopy 183
Igor Gurov
5.1 Introduction 183
5.2 Basics of Fringe Signal Registration 186
5.3 TD-OCT and FD-OCT Information Capacity 189
    5.3.1 Information Channel and Capacity 190
    5.3.2 2D and 3D OCT Imaging 193
5.4 Signal Downsampling in Full-Field OCT 194
    5.4.1 Spectral Characteristics of Downsampled Signals 197
5.5 Dynamic Fringe Signal Evaluation in State Space 203
    5.5.1 Recurrence Computational Algorithms Applied to Fringe Processing 206
    5.5.2 Application of Recurrence Processing Algorithm in State Space to Low-Coherence Profilometry 207
    5.5.3 Application of Recurrence Processing Algorithm in State Space When Evaluating Random Tissues 207
    5.5.4 Combined Use of Signal Downsampling and Kalman Filtering 210
    5.5.5 Recurrence Filtering Modifications and Processing Speed 211
5.6 Recurrence Processing of Swept-Source Spectral Interference Fringes 212
5.7 Discussion and Conclusions 217

PART II TECHNOLOGICAL DEVELOPMENTS

6 High-Speed Image Acquisition Techniques of Full-Field Optical Coherence Tomography 223
Byeong Ha Lee, Woo June Choi, Kwan Seob Park, and Gihyeon Min
6.1 Introduction 223
6.2 High-Speed 2D Imaging of Full-Field Optical Coherence Tomography 224

  6.2.1 Phase-Stepping Method Based on Hilbert Transformation 224
  6.2.2 Integrating Bucket Method Based on Hilbert Transformation 228
  6.2.3 Quadrature Fringes Wide-Field Optical Coherence Tomography 230
  6.2.4 Single-Shot Phase-Stepping Methods 234
 6.3 High-Speed 3D Imaging of Full-Field Optical Tomography 242
  6.3.1 2D Heterodyne Detection with a Pair of CCD Cameras 242
  6.3.2 2D Heterodyne Detection with a Single CCD Camera 248
  6.3.3 2D Heterodyne Detection with a Single CCD Camera and an Active Feedback Loop 255
 6.4 Summary 261

7 Toward Single-Shot Imaging in Full-Field Optical Coherence Tomography 267
Bettina Heise
 7.1 Introduction 267
 7.2 From Phase Shifting toward Single-Shot Techniques 269
  7.2.1 Temporal Phase-Shifting Techniques 270
  7.2.2 Simultaneous Phase-Shifting Techniques 273
   7.2.2.1 Dual-shot image acquisition 273
   7.2.2.2 Single-shot image acquisition 274
   7.2.2.3 Polarization-based schemes 275
   7.2.2.4 Unpolarized instantaneous phase-shifting schemes 276
   7.2.2.5 Heterodyne detection schemes 277
   7.2.2.6 SLM-based schemes 278
  7.2.3 Hyperspectral Single-Shot Techniques 279
 7.3 Signal Processing Methods 281
  7.3.1 Phase-Stepping Approaches 281
  7.3.2 The Subtraction Method 282
  7.3.3 Analytic Signal–Based Reconstruction 283
 7.4 Applications 289
 7.5 Summary and Outlook 295

8 Frequency Domain Full-Field Optical Coherence Tomography 303
Rainer A. Leitgeb, Abhishek Kumar, and Wolfgang Drexler
8.1 Introduction 303
8.2 General Concept 305
    8.2.1 Setup and System Components 305
    8.2.2 Standard SS-OCT Signal Processing 308
    8.2.3 Sensitivity 311
8.3 Digital Refocusing 313
8.4 Outlook 317

9 Full-Field OCM for Endoscopy 323
Anne Latrive and Claude Boccara
9.1 Introduction 324
    9.1.1 Overview of Endoscopy 324
    9.1.2 Endoscopic OCT 325
        9.1.2.1 OCT for flexible endoscopy 325
        9.1.2.2 OCT for rigid endoscopy 326
    9.1.3 Problematic of Endoscopic Full-Field OCM 326
9.2 Theory of Full-Field OCM with Two Coupled Interferometers 327
    9.2.1 Dual-Interferometer OCT Signal 327
    9.2.2 Performances 330
        9.2.2.1 Sensitivity 330
        9.2.2.2 Transversal resolution 331
        9.2.2.3 Axial resolution 332
    9.2.3 Conclusion 333
9.3 Problematics of Endoscopic Imaging with a Flexible Fiber Bundle 334
    9.3.1 Pixelation Effect Removal 334
        9.3.1.1 Fourier domain filters 334
        9.3.1.2 Deconvolution 335
        9.3.1.3 Interpolation between the cores 336
    9.3.2 Cross-Talk Interferences 337
        9.3.2.1 Core-to-core coupling 337
        9.3.2.2 Mode-to-mode coupling 340
    9.3.3 Conclusion 341
9.4 Full-Field OCM with a Flexible Endoscope 341

        9.4.1   FF-OCM System with Simple Imaging
Interferometer and SLD Source   342
        9.4.2   FF-OCM System with Complex Imaging
Interferometer and Xenon Arc Source   343
        9.4.3   FF-OCM System with Simple Imaging
Interferometer and Xenon Arc Source   345
        9.4.4   Discussion   346
  9.5   Full-Field OCM with a Rigid Endoscope   346
        9.5.1   Choice of the FF-OCM Setup   347
        9.5.2   FF-OCM Setup with a Rigid Probe   347
                9.5.2.1   Setup and performances   347
                9.5.2.2   Imaging results   348
        9.5.3   Discussion   349
  9.6   Conclusion   351

**10 Full-Field Optical Coherence Tomography and Microscopy
Using Spatially Incoherent Monochromatic Light   357**
*Dalip Singh Mehta, Vishal Srivastava, Sreyankar Nandy,
Azeem Ahmad, and Vishesh Dubey*
  10.1   Introduction   357
  10.2   Principle of Low Coherence Interferometry Based
on Temporal Coherence of the Light Source   360
        10.2.1   The Wiener–Khintchine Theorem of
Temporal Coherence   362
        10.2.2   Resolution of OCT Systems Based on
Temporally Low Coherent Light   365
  10.3   Principle of Spatial Coherence Gated Tomography
with a Monochromatic Light Source   368
        10.3.1   Transverse (Lateral) Spatial Coherence of
Light Vibrations   370
        10.3.2   Longitudinal Spatial Coherence of Light
Vibrations   373
  10.4   Principle of Spatial Coherence Gated Microscopy
with Monochromatic Light   376
  10.5   Experimental Details for Spatial Coherence Gated
Tomography with Monochromatic Light   377
        10.5.1   Determination of Longitudinal Spatial
Coherence Length   379

10.5.2 Application of Longitudinal Spatial
Coherence in Profilometry 381
10.5.3 High-Resolution Longitudinal Spatial
Coherence Gated Tomography 382
10.6 Coherence Holography 385
10.7 Conclusion 386

11 Real-Time and High-Quality Online 4D FF-OCT Using Continuous Fringe Scanning with a High-Speed Camera and FPGA Image Processing 393
P. C. Montgomery, F. Anstotz, D. Montaner, and F. Salzenstein
11.1 Introduction 394
11.2 Theory 398
11.2.1 Interference Microscopy 398
11.2.2 Modeling of the Optical Probe in
Interference Microscopy 400
11.3 Envelope Detection Algorithms 402
11.3.1 Peak Fringe Detection Algorithm 403
11.3.2 The Fringe Modulation Algorithm 404
11.4 Experimental 407
11.5 System Performance 412
11.6 Applications 415
11.6.1 Laterally Moving Microfluxgate Surface
Measured with the PFSM Algorithm 415
11.6.2 Laterally Moving Microfluxgate Surface
Measured with the FSA Algorithm 417
11.6.3 Laterally Moving GaN Surface Measured
with the FSA Algorithm 417
11.6.4 Drying Drop of Liquid Correction Whitener
Measured with the FSA Algorithm 419
11.7 Future Developments and Potential Applications in
FF-OCT 420
11.8 Conclusions 423

12 Digital Interference Holography for Tomographic Imaging 429
Lingfeng Yu, Mariana C. Potcoava, and Myung K. Kim
12.1 Introduction 429
12.2 Principle of Digital Interference Holography 432

		12.2.1	Basic Description of DIH	432

 12.2.1 Basic Description of DIH 432
 12.2.2 Phase Correction in DIH 433
 12.2.3 Spectral Shaping in DIH 434
 12.3 Techniques of DIH 437
 12.3.1 Detailed Description of DIH Experiments 437
 12.3.2 DIH by the Angular Spectrum Algorithm 439
 12.3.3 Variable Tomographic Scanning 442
 12.3.4 DIH Based on Spectral Interferometry 445
 12.4 Applications of DIH 450
 12.4.1 Animal Tissue 450
 12.4.2 Human Retina 451
 12.4.3 DIH for Biometric Application 454
 12.4.4 Submicron Tomography of Cells with DIH 457
 12.5 Discussions 457

PART III ADDITIONAL IMAGING MODALITIES

13 Technological Extensions of Full-Field Optical Coherence
 Microscopy for Multicontrast Imaging 467
 Arnaud Dubois
 13.1 Introduction 467
 13.2 Polarization-Sensitive FF-OCM 469
 13.2.1 Experimental Setup and Principle 470
 13.2.2 Image Calculation 472
 13.2.3 Validation 476
 13.3 Spectroscopic FF-OCM 479
 13.3.1 Experimental Setup 479
 13.3.2 Intensity-Based Tomographic Imaging 480
 13.3.3 Spectroscopic Measurements 481
 13.3.4 Interpretation of Spectroscopic
   Measurements 482
 13.3.5 Validation of the Spectroscopic
   Measurements 484
 13.3.6 Demonstration of Imaging Contrast
   Enhancement 486
 13.3.7 Spectroscopic Polarization-Sensitive
   FF-OCM 487
 13.4 Multispectral FF-OCM 490

		13.4.1	Dual-Band FF-OCM	491

        13.4.1 Dual-Band FF-OCM   491
              13.4.1.1 Method   491
              13.4.1.2 Simultaneous dual-band imaging   494
        13.4.2 Three-Band FF-OCM   498
              13.4.2.1 Materials and methods   498
              13.4.2.2 System characteristics and image results   501
  13.5 Combination of FF-OCM with Fluorescence Microscopy   505
        13.5.1 Materials and Methods   506
        13.5.2 System Performance and Image Results   508
  13.6 Conclusion   510

14 Spectroscopic Full-Field Optical Coherence Tomography   519
Julien Moreau
  14.1 FF-OCT Principle   520
  14.2 Spectroscopic FF-OCT   524
  14.3 Limitations of Spectroscopic FF-OCT   530

15 Multiwavelength Full-Field Optical Coherence Tomography   533
Mariana C. Potcoava, Nilanthi Warnasooriya, Lingfeng Yu, and Myung K. Kim
  15.1 Introduction   533
  15.2 Basic Single-Wavelength, Low-Coherence Interferography   536
        15.2.1 Digital Focusing in Low-Coherence Interferometry   538
  15.3 Multiwavelength, Optical-Phase-Unwrapping, Low-Coherence Interferography   544
        15.3.1 Two-Wavelength Optical Phase Unwrapping   547
        15.3.2 Three-Wavelength Optical Phase Unwrapping   548
        15.3.3 Application for Two-Wavelength Optical Phase Unwrapping   549
        15.3.4 Application for Three-Wavelength Optical Phase Unwrapping   551
  15.4 Full-Color FF-OCT   553

        15.4.1   Results of Full-Color FF-OCT   556
   15.5 Discussion   559

16 **Dual-Modality Full-Field Optical Coherence and
   Fluorescence Sectioning Microscopy: Toward All Optical
   Digital Pathology on Freshly Excised Tissue**   565
   Fabrice Harms
   16.1 Introduction: Clinical Context   565
   16.2 Optical Coherence Tomography in Pathology: Early
         Multimodal Approaches and Limitations   568
         16.2.1 Applying OCT to Pathology Assessment of
                 Biological Tissue   568
         16.2.2 High-Resolution OCT for Pathology:
                 Full-Field Optical Coherence Tomography   569
         16.2.3 Multimodal OCT Approaches   573
   16.3 Multimodal Full-Field Optical Coherence
         Tomography for Pathology Applications   574
         16.3.1 Translating FF-OCT to Pathology Diagnosis   574
         16.3.2 Combined FF-OCT and Fluorescence
                 Sectioning Microscopy   576
                 16.3.2.1 Setup   576
                 16.3.2.2 Full-field optical coherence
                           tomography: resolution   578
                 16.3.2.3 Structured illumination
                           fluorescence microscopy   579
         16.3.3 Multimodal FF-OCT/Fluorescence Images
                 of Healthy and Cancerous Tissue Samples   582
                 16.3.3.1 Sample selection and preparation   582
                 16.3.3.2 Imaging protocol   583
                 16.3.3.3 Results   584
   16.4 Discussion and Conclusion   589

PART IV APPLICATIONS

17 **Full-Field Optical Coherence Tomography for Rapid
   Histological Evaluation of ex vivo Tissues**   595
   Manu Jain and Sushmita Mukherjee
   17.1 Introduction   595

17.2 Current Practice for Intraoperative Diagnosis and
     Its Limitations   596
17.3 Basic Principles and Instrumentation of FF-OCT   597
17.4 FF-OCT-Generated Histology Atlas of Rat Organs   598
     17.4.1 Skin   599
     17.4.2 Stomach   600
     17.4.3 Liver   600
     17.4.4 Heart   601
     17.4.5 Kidney   602
     17.4.6 Prostate   602
     17.4.7 Lung   602
     17.4.8 Urinary Bladder   603
17.5 FF-OCT to Identify Spermatogenesis in Rat Testis   605
17.6 FF-OCT for the Analysis of Human Lobectomy
     Specimens   607
17.7 Potential Clinical Applications of FF-OCT   611
17.8 Current Limitations of FF-OCT and Possible
     Solutions   612

**18 FF-OCT Imaging: A Tool for Human Breast and Brain Tissue
   Characterization   617**
   Osnath Assayag
   18.1 Introduction   617
   18.2 Breast Tissue Features Recognition and
        Pathological Modifications   618
   18.3 Distinction between Benign/Normal and Malignant
        Tissue   622
        18.3.1 Malignant Tissue Characterization   622
        18.3.2 Benign Lesions Identification   624
   18.4 Breast Tissue Classification Using FF-OCT   625
        18.4.1 Diagnostic Accuracy of FF-OCT Images   627
   18.5 FF-OCT Imaging of Healthy and Tumorous Human
        Brain Parenchyma   627
        18.5.1 Human Brain Parenchyma Morphological
               Structure Recognition   629
        18.5.2 Benign Lesions Identification   630
        18.5.3 Malignant Brain Lesion Imaging   635
   18.6 Discussion and Conclusion   636

19 Full-Field Optical Coherence Microscopy in Ophthalmology   641
G. Latour, K. Grieve, G. Georges, L. Siozade, M. Paques,
V. Borderie, L. Hoffart, and C. Deumié
19.1   Introduction   641
19.2   Cornea   643
    19.2.1   Morphology and Characterization   643
        19.2.1.1   Description of the tissue   643
        19.2.1.2   Conventional corneal imaging techniques   646
    19.2.2   FF-OCM Imaging of the Healthy Cornea   648
    19.2.3   Pathological Corneas   651
        19.2.3.1   Monitoring the evolution of edema   651
        19.2.3.2   Imaging of common pathologies   658
    19.2.4   Characterization of Laser Ablations   664
    19.2.5   Perspectives   667
        19.2.5.1   Eye banking   667
        19.2.5.2   Characterization of artificial corneas   669
19.3   Lens   669
    19.3.1   Anatomy   669
    19.3.2   Imaging the Lens   669
    19.3.3   FF-OCM Lens Imaging   670
    19.3.4   Conclusion and Perspectives   670
19.4   Retina   671
    19.4.1   Anatomy   671
    19.4.2   Retinal Imaging Techniques   672
    19.4.3   FF-OCM Retinal Imaging   673
    19.4.4   Perspectives in Retinal Imaging   675
19.5   Future Developments in Ophthalmology   677
    19.5.1   In vivo FF-OCM Imaging   677
    19.5.2   Coupling with Other Modalities: A Way to Increase Information   677

20 Investigation of Spindle Structure and Embryo Development for Preimplantation Genetic Diagnosis by Subcellular Live Imaging with FF-OCT   689
Ping Xue and Jing-gao Zheng
20.1   Introduction   689

20.2	Experimental Design	691
20.3	Extraction of the Signals	693
	20.3.1 Phase Shifting	693
	20.3.2 Extract the Signals with Two-Phase Shifting	694
	20.3.3 Comparison between Two and Four-Phase Shifting	696
	20.3.4 3D Reconstruction Image Display	697
20.4	System Characteristics	698
	20.4.1 Axial Resolution	698
	20.4.2 Transverse Resolution	699
	20.4.3 Depth of Focus/Field	700
	20.4.4 Sensitivity	701
20.5	Applications in Embryology and Developmental Biology	704
	20.5.1 Imaging the Structures of Spindles	705
	20.5.2 Imaging Early Patterning and Polarity	710
	20.5.2.1 Static studies of early patterning and polarity	711
	20.5.2.2 Dynamic studies of early patterning and polarity	716
20.6	Conclusion	720

21 FF-OCT for Nondestructive Material Characterization and Evaluation — **727**
David Stifter

21.1	Introduction	727
21.2	Methods	729
21.3	Applications	734
	21.3.1 Surface Metrology	734
	21.3.2 Layer Thickness Determination	737
	21.3.3 3D Structural and Functional Evaluation	738
21.4	Conclusions and Outlook	745

Index — 755

Preface

Microscopic imaging for the visualization of the internal structure of objects, nondestructively, has been a subject of active research and developments for applications in the fields of medicine, biology, and materials science. In particular, the capability to noninvasively explore the microstructure of biological tissues with a spatial resolution similar to that of histology has prompted significant work in recent years.

Confocal microscopy is an optical imaging method to generate high-resolution (≤ 1 μm) cross-sectional views of semitransparent objects by rejecting light coming from out-of-focus regions. The imaging penetration depth of confocal microscopy in most biological tissues, however, is relatively weak (~ 200 μm). Using ultrashort pulse lasers, nonlinear optical effects such as harmonic generation and multiphoton absorption have been successfully applied in optical microscopy, offering deeper imaging penetration in highly scattering tissues compared with confocal microscopy. Optical coherence tomography (OCT) is another technique for noninvasive imaging in semitransparent objects. OCT allows deeper imaging (~ 1 mm), but with a lower spatial resolution. In OCT, light backscattered within the object is detected using low-coherence interferometry. Cross-sectional images are generated by performing axial measurements (A-scans) at different transverse positions. OCT has revolutionized the clinical practice of ophthalmology for the diagnosis and management of retinal diseases. OCT has been clinically demonstrated in a diverse set of other medical and surgical applications, including cardiology, gastroenterology, and dermatology, allowing the visualization of the microstructure of tissues without excision and processing.

Full-field OCT (FF-OCT), also termed full-field optical coherence microscopy (FF-OCM), is a particular version of OCT that uses full-field illumination and an array detector to acquire en face tomographic images without the need for transverse scanning. FF-OCM benefits from the transverse imaging resolution of optical microscopy along with the capacity of optical axial sectioning at micrometer-scale resolution. Since the introduction of FF-OCM in the early 2000s, significant progress in the technology has been achieved. FF-OCM has been demonstrated in a variety of applications and is now commercially available.

The diversity of scientific publications on FF-OCM has led to a need for a comprehensive handbook describing FF-OCM in terms of technology, performance, and applications. The aim of this book is to address this need by serving as a guide for engineers and scientists involved in the technological developments of FF-OCM. This book may also constitute a reference for users of FF-OCM who wish to understand the fundamentals of this imaging modality and to know its capabilities and limitations.

This handbook comprises 21 self-contained chapters, organized into four parts, written by internationally recognized experts and leaders in the field of FF-OCM. The first chapter provides a general introduction to FF-OCM. The fundamental characteristics of the technology are analyzed and discussed theoretically in the following four chapters, constituting the first part of the book. The recent main technological developments of FF-OCM, in particular for increasing the image acquisition speed, are presented in the second part. Extensions of FF-OCM for image contrast enhancement and functional imaging are reported in the third part. An overview of applications of FF-OCM in medicine, biology, and materials science is provided in the last part.

This book would not have been possible without the invaluable contribution of many people. I would like to express my sincere appreciation to all of my colleagues who have dedicated many hours to writing the chapters. I am grateful to Stanford Chong, director and publisher at Pan Stanford Publishing, for having initiated this work. I also thank Sarabjeet Garcha for the editing work.

I and all the coauthors hope you will find this handbook interesting, stimulating, and useful.

Arnaud Dubois
Palaiseau, France
May 2016

Chapter 1

Introduction to Full-Field Optical Coherence Microscopy

Arnaud Dubois
*Laboratoire Charles Fabry, CNRS, Institut d'Optique Graduate School,
Univ. Paris-Saclay, 2 av. Augustin Fresnel, 91127 Palaiseau, France*
arnaud.dubois@institutoptique.fr

This introductory chapter provides an overview of full-field optical coherence microscopy. The different technological approaches that have been developed are reported, specifying the interferometer configuration, the light source, the image sensor, and the method used to obtain images from the acquired data. Theoretical analyses of the system characteristics, including the detection sensitivity and spatial resolution, are presented. Challenges, advantages, and drawbacks of the technique are discussed. A review of the main applications of full-field optical coherence microscopy is provided.

1.1 Introduction

Optical microscopy has been a standard tool in life sciences as well as material sciences for over the last century and a half.

Handbook of Full-Field Optical Coherence Microscopy: Technology and Applications
Edited by Arnaud Dubois
Copyright © 2016 Pan Stanford Publishing Pte. Ltd.
ISBN 978-981-4669-16-0 (Hardcover), 978-981-4669-17-7 (eBook)
www.panstanford.com

Features of an object that are normally not visible to the naked eye can be accessed using an optical microscope by increasing the spatial resolution and providing image contrast. Three-dimensional microscopic imaging for the visualization of the internal microstructures of semitransparent objects, mainly in biomedical applications, has prompted intense technological developments. Diagnosis and study of many diseases, especially cancers, depend heavily upon biopsy and histopathological analysis of tissue at the cellular-level scale using an optical microscope. A large variety of stains and fluorescent dyes can be employed to enhance the contrast of the image and differentiate individual components within the tissue. However, as the penetration depth of visible light is very limited, only thin and transparent samples can be visualized with conventional optical microscopy. The imaged sample must be cut into slices a few micrometers thick after being frozen or embedded in paraffin, which is time consuming and may introduce structural deformation or damage. Moreover, some biological components can be deteriorated or removed during tissue processing. There is therefore great interest for an optical technique capable of noninvasive optical biopsy, that is, capable of real-time in situ imaging of tissue microstructures with a resolution similar to that of histology without the need for tissue excision and processing, or at least for a technique capable of immediate tomographic imaging of excised specimens at high resolution, while avoiding the long and delicate histological preparations.

The introduction of confocal microscopy was a significant advance in optical microscopy. By rejecting light coming from out-of-focus regions, optical sections can be produced using a confocal microscope without requiring physical sectioning. By labeling with fluorophores, specimens can be visualized in three dimensions at high resolution. However, the imaging penetration depth of confocal microscopy is limited to ~ 200 µm due to tissue scattering. Besides, the use of fluorophores may affect the viability of cells.

With the great advances in the technology of ultrashort pulsed lasers, nonlinear optical effects such as harmonic generation and two-photon absorption have been applied successfully in the field of microscopy. The advantage of two-photon excitation fluorescence microscopy over linear (one-photon) fluorescence microscopy

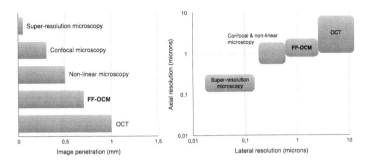

Figure 1.1 Comparison of typical resolution and imaging depth achieved with different optical imaging techniques.

includes efficient background rejection, reduced photobleaching and photodamage [1], and deeper imaging penetration because of a longer excitation wavelength. However, the penetration is limited to several hundreds of micrometers in highly scattering tissues [2].

Optical coherence tomography (OCT) is another technique for three-dimensional imaging that can penetrate deeper than confocal microscopy and nonlinear microscopy (see Fig. 1.1). OCT relies on low-coherence interferometry to measure the amplitude of light backscattered by the sample being imaged [3–6]. The most significant impact of OCT is in ophthalmology for in situ examination of the pathologic changes of the retina [7–10] and measurement of the dimensions of the anterior chamber [11, 12]. OCT has also been applied successfully to imaging of various highly scattering tissues [13, 14]. Ultrahigh axial resolution of ~ 1 μm can be achieved by using laser-based ultrabroad-bandwidth light sources [15–17]. However, OCT usually suffers from a limited transverse (lateral) resolution because relatively low-numerical-aperture lenses have to be used to preserve a sufficient depth of field. Several solutions have been implemented to improve the transverse resolution of OCT, including adjustment of focus while the depth is scanned [15–18], illumination of the sample with a Bessel beam [19, 20], or image postprocessing based on inverse scattering theory [21]. The most efficient approach for achieving fine enough transverse resolution for cellular-level imaging in scattering tissues is to acquire en face (parallel to the sample surface) images rather than cross-sectional (perpendicular to the sample surface) images. In this configuration, there is no limitation to the depth of field. Optical coherence microscopy, or OCM, is a version of OCT that produces en face tomographic images. High-numerical-aperture optics can be used in OCM to achieve higher transverse resolution. Two general approaches for OCM have been reported to date. The first approach is based on the combination of confocal microscopy with low-coherence interferometry, as demonstrated in Refs. [22, 23]. This combination was revisited later using modern technologies, leading to the scanning OCM technique [24, 25]. Broadband coherence gating was shown to significantly enhance the imaging depth of conventional confocal microscopy [24, 26]. Scanning OCM was applied successfully to cellular-level resolution imaging deep below the surface of various human tissues such as skin [27], oral mucosa [28], and colonic mucosa [26, 29, 30]. The second approach of OCM involves full-field illumination and detection. Also sometimes termed "full-field optical coherence tomography" (FF-OCT), full-field optical coherence microscopy (FF-OCM) is an alternative technique to scanning OCM, based on low-coherence (white-light) interference microscopy [31–34]. FF-OCM produces tomographic images in the en face orientation by arithmetic combination of several interferometric images acquired with an area camera and by illuminating the whole field to be imaged with low-coherence light. The major interest for FF-OCM lies in its high imaging resolution in both transverse and axial directions using a simple and robust experimental arrangement [35, 36].

This chapter gives an overview of FF-OCM. The key technological elements of the technology are discussed, showing their influence on the system performance. A model of the detection sensitivity and spatial resolution of FF-OCM is presented. Finally, the main applications of FF-OCM are described.

1.2 Experimental Setup

1.2.1 *General Principle*

FF-OCM is based on the combination of a microscope with a Michelson-type interferometer. The conventional experimental

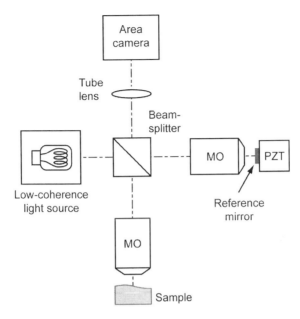

Figure 1.2 Diagram of conventional FF-OCM. MO: microscope objective; PZT: piezoelectric transducer.

setup is represented schematically in Fig. 1.2. The beam emitted by the broadband light source is split into two beams by a beam splitter, one of these beams being directed onto the sample and the other onto a flat low-reflectivity reference mirror. Light backscattered from the sample is combined at the beam splitter with light reflected by the reference mirror. The resulting interferometric image is formed on the image sensor of an area camera. A phase modulation is introduced in the interferometer, usually by varying the position of the reference mirror using a piezoelectric actuator. An arithmetic combination of several phase-shifted interferometric images yields an en face tomographic image of the sample.

1.2.2 Interference Microscope Configuration

Several configurations of interference microscopes have been used in FF-OCM, referred to as Michelson, Mirau, and Linnik interference microscopes (see Fig. 1.3).

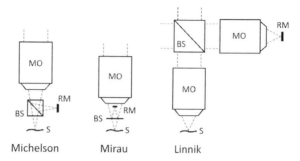

Figure 1.3 Interference microscope configurations used in FF-OCM. MO: microscope objective; RM: reference mirror; S: sample; BS: beam splitter.

In the Michelson interference microscope, a single microscope objective images both the sample and the reference mirror. The Mirau interferometer is a variant consisting of a beam-splitting surface and a reference mirror, arranged in such a way that both optical elements share the same axis with the microscope objective. The compactness of the Michelson and Mirau interferometers allows for incorporation into the microscope objective. The presence of the beam splitter and the reference mirror between the objective and the sample requires a relatively long working distance and prevents the use of a high numerical aperture.

Alternatively, the Linnik configuration requires two identical microscope objectives, one placed in each arm on the interferometer. Compared to the Michelson and Mirau interference microscopes, the Linnik configuration is less compact but offers more flexibility in adjustments. The optical length of the interferometer arms and the focus can be adjusted independently. Moreover, immersion objectives can be used, which is more difficult in the Michelson and Mirau configurations, although this has been demonstrated [37–39]. There is no limitation in the numerical aperture of the objectives of the Linnik microscope, which makes this configuration preferable when high lateral imaging resolution is desired [31, 35].

1.2.3 Light Source

The light source is a key element in FF-OCM. Various features of the light source, including spectral properties, noise, power, and spatial coherence, have a direct impact on the imaging performance of FF-OCM.

The spectral properties of light used for illumination determine the imaging spatial resolution in both transverse and axial directions. The transverse (lateral) resolution is related to the mean optical wavelength, whereas the axial resolution (the sectioning ability) is usually governed by the temporal coherence function, which is the Fourier transform of the optical power spectral density (Wiener–Khinchin theorem). The optical wavelength also determines the depth of penetration in the imaged sample. In biological tissues, the penetration is maximized in the near-infrared window (also known as optical window or therapeutic window), between 650 nm and 1350 nm typically, where the optical absorption is low. For the reasons mentioned above, an appropriate light source for FF-OCM should therefore emit broadband light in the near infrared with a spectral power distribution as close as possible to a Gaussian function (when expressed versus wavenumber or frequency).

The noise of the light source has an impact on the detection sensitivity, whereas the optical power plays a role on the imaging speed.

The spatial coherence of the light source is another essential parameter. High spatial coherence is required in point-scanning imaging systems. Spatially coherent light sources like superluminescent diodes (SLDs), femtosecond lasers, and supercontinuum laser sources are appropriate sources for scanning OCT. In full-field illumination imaging systems, spatially coherent light is not required. Light sources with low spatial coherence are then preferable to provide uniform illumination of the imaged field and avoid cross-talk effects [40].

Spontaneously emitting light sources are valuable sources for FF-OCM. Thermal light, in particular, has the advantage of extremely low temporal coherence. Examples of thermal light sources used in FF-OCM are the tungsten halogen lamp [32, 33] and the arc lamp [36, 41]. As a first approximation, thermal light sources emit blackbody radiation with a spectral power distribution given by Planck's law for cavity radiation. Many thermal light sources, however, do not emit true blackbody radiation. Besides, their radiation is usually filtered by the surrounding media and by optical

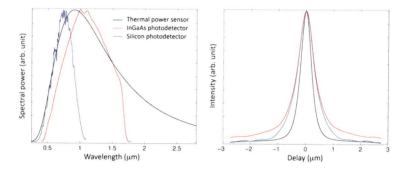

Figure 1.4 Left: Spectral power distribution of blackbody radiation at a temperature of 3200 K, measured with different detectors. Right: The corresponding temporal coherence functions.

components transmitting the light. Furthermore, the responsivity of the photodetector greatly determines the effective spectrum. For example (see Fig. 1.4), the spectral power distribution of ideal thermal light (blackbody emission) at a temperature of 3200 K peaks at a wavelength of 906 nm and has a full-width at half-maximum (FWHM) of 1090 nm. The corresponding temporal coherence function has a width of 0.46 μm. Using a silicon-based complementary metal-oxide semiconductor (CMOS) camera as a photodetector (Photonfocus HD1-D1312, for example), the effective spectrum is centered at 730 nm and the corresponding temporal coherence length is 0.70 μm. Using an indium-gallium-arsenide (InGaAs) camera with extended sensitivity in the visible (OWL SW1.7 CL-HSVIS-SWIR manufactured by Raptor Photonics, for example), the coherence length has a similar value of 0.73 μm. But the effective spectrum is shifted further in the infrared at a 1110 nm center wavelength.

Tungsten halogen lamps are the conventional light sources in FF-OCM due to many advantages [34]. They present a very broad and continuous spectrum that covers the spectral response of any conventional image sensor. Because of the high thermal inertia of the tungsten filament, they have excellent power stability at the scale of the image acquisition time. They are easy and inexpensive to operate. Halogen lamps are usually incorporated in a Köhler illumination system in order to obtain a homogeneous light source at the image plane and to control its spatial coherence to a limited degree. They can also be used in fiber optic illuminators.

A tungsten filament lamp itself has very low spatial coherence due to the large size of the light emitter. In contrast, arc lamps possess higher spatial coherence unless a large area of the plasma is utilized as the source. Arc lamps are therefore brighter light sources than halogen lamps. Light emitted by the arc can be efficiently injected in a multimode fiber to provide a bright broadband light source for FF-OCM [36, 41, 42]. The spectrum of xenon arc lamps is the combination of a thermal spectrum and spectral lines. Xenon arc lamps produce broadband, almost continuous emission having a color temperature approximating sunlight in the visible wavelengths and also exhibit a complex line spectrum in the 750 to 1000 nm region of the near-infrared spectrum. These spectral lines are responsible for modulations in the coherence function, which may generate artifacts in the images. Short light pulses can be produced by an arc lamp, which may be of interest in FF-OCM for high-speed imaging [43].

Another type of spontaneously emitting light source that is appropriate for FF-OCM is a light-emitting diode (LED). An LED is made of a combination of semiconducting materials that are doped with impurities to allow conversion of electrical energy into light through a reversible process called injection electroluminescence. The wavelength of generated light is determined by the energy bandgap at the p-n junction in the semiconductor. Initial FF-OCM systems used an infrared LED as the light source [31, 37]. In Ref. [31] the emission of the LED was centered at a wavelength of 840 nm with an FWHM of 50 nm. The spectrum and the temporal coherence function of this LED are shown in Fig. 1.5. Several LEDs emitting at distinct wavelengths can be combined to increase the spectral width, that is, to reduce the temporal coherence length [44–46]. In Ref. [46], the emissions of five infrared LEDs were mixed with optimized proportions in a multimode fiber to synthesize a nearly Gaussian-shaped spectrum centered at 840 nm with an FWHM of 100 nm (see Fig. 1.5). Recent advances have been made in the growth and the choice of the semiconductor materials, allowing for brighter LEDs over a broader emission spectrum. Recently, a commercially available broadband LED (FWHM = 265 nm, centered

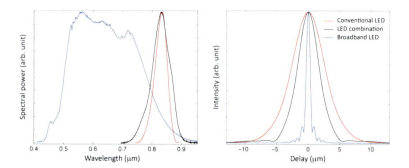

Figure 1.5 Left: Spectral power distribution of a conventional LED, a combination of 5 LEDs, and a broadband LED. Right: Corresponding temporal coherence functions.

at 650 nm) was used in FF-OCM [47] to achieve an ultrahigh axial resolution of 0.7 µm, similar to the resolution achieved with a halogen lamp. High-power broadband LEDs will certainly become a valuable alternative to halogen lamps in FF-OCM. LEDs avoid unnecessary heat generation. They have the advantage of higher energy conversion efficiency and longer lifetime.

Fluorescence is the spontaneous emission of light by a material that has absorbed light at a shorter wavelength. Depending on the nature of the material, light emitted by fluorescence may have a low temporal coherence. Titanium-doped sapphire (Ti:Al$_2$O$_3$) is a popular crystal used for making ultrashort-pulse solid-state or wavelength-tunable lasers. It combines the excellent thermal, physical, and optical properties of the sapphire crystal with the broadest tunable range of any known material, from 660 nm to 1100 nm. The peak of the absorption band being at 490 nm, it can be excited with a variety of laser sources emitting green light. A broadband fluorescence-based light source, based on a titanium-doped sapphire crystal was specifically developed for FF-OCM [48]. A diagram of this light source is depicted in Fig. 1.6. The beam of a frequency-doubled neodymium vanadate (Nd:YVO$_4$) laser emitting at a wavelength of 532 nm was focused in the crystal for longitudinal pumping. The opposite face of the crystal was high reflection coated at the wavelength of the fluorescence emission. The crystal was cooled to 18°C to minimize thermal effects that might reduce

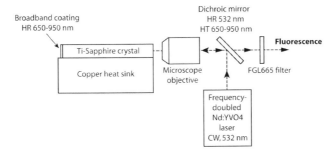

Figure 1.6 Low-coherence light source based on the fluorescence of a Ti:Al$_2$O$_3$ crystal excited by a frequency-doubled Nd:YVO$_4$ laser.

the fluorescence efficiency. A fluorescence power of 75 mW was measured, which corresponds to a conversion efficiency of 2%. The fluorescence power was then limited by the available pump power.

FF-OCM systems are usually limited in image acquisition speed due to the relatively weak radiance of low-spatial-coherence light sources. Attempts have been reported to use broadband spatially coherent light sources such as SLDs, short-pulsed lasers, and supercontinuum laser sources. FF-OCM using spatially coherent light however suffers from a loss of resolution and signal-to-noise ratio (SNR) resulting from coherent multiple scattering. This phenomenon, known as "cross-talk," is a serious limitation to the method, since it prevents shot-noise-limited detection and diffraction-limited imaging in scattering samples [49, 50]. The cross-talk effects in FF-OCM can be minimized by reducing the spatial coherence of light. A method consists of passing the laser beam through a spinning ground-glass diffuser [51]. This simple approach can, however, result in considerable loss of light at the diffuser. Another method consists of injecting the laser beam in a multimode optical fiber. The propagation in the fiber tends to make the beam highly multimodal. Mechanical vibrations of the fiber can be applied for mode scrambling [52].

1.2.4 Image Sensor

The image sensor used in FF-OCM should be selected with great care because its performance is crucial for the quality of the images. Different types of array detectors can be used, including silicon cameras based on the charge-coupled device (CCD) or CMOS technologies or InGaAs cameras. Several camera parameters are of particular importance for FF-OCM, such as the spectral sensitivity, dynamic range, frame rate, and number of pixels.

Silicon-based cameras cannot detect photons at wavelengths greater than 1.1 μm because the photons do not have enough energy to create a free-electron charge. Silicon-based cameras are therefore only used for FF-OCM operation in the visible/near-infrared spectral region. To image at longer wavelengths, InGaAs array detectors must be used. Typical InGaAs cameras are sensitive between 900 nm and 1700 nm. Using this kind of detector, FF-OCM imaging around the 1300 nm center wavelength is possible, which has superior penetration in highly scattering samples [42, 53, 54]. The spectral response of InGaAs cameras can be extended to the near-infrared and even to the visible range, down to 400 nm. The extremely broad spectral sensitivity of this kind of camera offers the flexibility of choice for the operation wavelength required for high axial imaging resolution [55]. They, however, suffer a relatively lower definition compared to silicon-based cameras and they are more expensive.

A high dynamic range detector is particularly important in FF-OCM in order to resolve small modulations on a large background signal with low noise. An appropriate camera for FF-OCM should therefore have a large full-well capacity. Silicon-based cameras can work close to the shot-noise limit. Although they generally offer a larger full-well capacity, conventional InGaAs cameras have more electrical noise, including the read-out noise and the dark noise, that may not be negligible compared to the shot noise [42, 53–55].

High frame rates and high definition are other two requirements. A compromise, however, has to be done since these two parameters are roughly inversely proportional to each other. CMOS cameras have the advantage of being faster than CCD cameras. With the advances in CMOS technology, the disadvantage in terms of noise, sensitivity, and dynamic range compared to CCD may disappear, making scientific CMOS cameras a valuable choice for high-performance FF-OCM.

Furthermore, the flexibility of design in CMOS detectors can be exploited to develop arrays with custom pixels for heterodyne detection, canceling out the large reference signal component and directly providing the envelope of the interferometric signal. Very high image acquisition rates were demonstrated with a prototype of such detectors [56, 57]. Presently, this can be achieved with a commercially available custom camera [58].

1.3 Image Acquisition

1.3.1 *Axial Scan*

FF-OCM acquires en face-oriented tomographic images. To vary the depth of the imaged section within the sample, it is necessary to modify the position of the coherence plane (plane defined by an optical path length difference of zero in the interferometer) relative to the sample. Usually, defocus then occurs [53, 59]. An illustration of this phenomenon is shown in Fig. 1.7. In this example, the sample is elevated to image deeper. If the refractive index n of the sample differs from the refractive index n_{im} of the microscope objective immersion medium, the coherence plane and the focal plane separate from each other. This phenomenon yields degradation of the contrast of the tomographic image and blur.

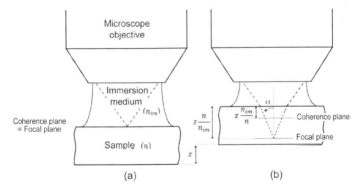

Figure 1.7 (a) The interference microscope is adjusted so that the coherence plane and the focal plane coincide when imaging the surface of the sample. (b) When the sample is elevated a distance z, the coherence and focal planes separate from each other, except if the refractive index of the sample (n) is equal to the refractive index of the immersion medium (n_{im}).

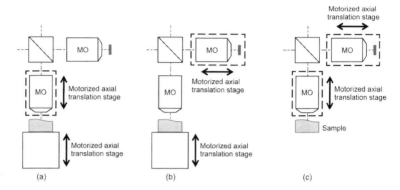

Figure 1.8 Examples of mechanical adjustments applied for dynamic focusing in FF-OCM based on a Linnik interferometer.

Usually, two adjustments are required in order to both vary the depth of the imaged section in the sample and avoid mismatch between the coherence plane and the focal plane. With the Linnik-type interferometer, these two adjustments can be achieved easily. Possible configurations where the sample is displaced and the focus or the length of the reference arm is modified are shown in Fig. 1.8a and Fig. 1.8b. If the sample is displaced axially by a distance z, the microscope objective has then to be translated an approximate distance

$$z_{\text{obj}} = z \frac{\left(n_{\text{im}}^2 - n^2\right)}{\left(n_{\text{im}}^2 - n_{\text{im}} - n^2\right)}, \quad (1.1)$$

or the length of the reference arm has to be modified approximately by

$$z_{\text{ref}} = z \frac{\left(n^2 - n_{\text{im}}^2\right)}{n_{\text{im}}}. \quad (1.2)$$

A configuration where the sample is immobile is shown in Fig. 1.8c, where both the length of the reference arm and the focus on the sample are adjusted. To change the depth of the tomographic image in the sample of a quantity z, the microscope objective is displaced axially of a distance

$$z_{\text{obj}} = z \frac{n_{\text{im}}}{n}. \quad (1.3)$$

To ensure that the focus and coherence plane match, the length of the reference arm has to be modified by a quantity

$$z_{\text{ref}} = z \frac{\left(n^2 + n_{\text{im}} - n_{\text{im}}^2\right)}{n}. \quad (1.4)$$

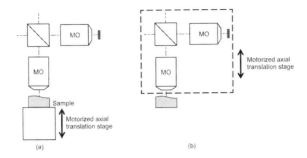

Figure 1.9 Possible experimental configurations for depth scanning the sample, while maintaining the focus when immersion microscope objectives are used.

A way to avoid the two adjustments is to use immersion microscope objectives whose immersion medium has a refractive index similar to the one of the sample ($n_{im} \approx n$). Only one adjustment is then required. This can be accomplished for example by moving the sample (Fig. 1.9a) or by moving the whole interferometer (Fig. 1.9b). In addition to simplifying the experimental setup, a significant advantage of using immersion microscope objectives with an appropriate immersion medium is to minimize optical dispersion mismatch in the interferometer arms. Optical dispersion mismatch is a real issue in broadband FF-OCM that severely degrades the axial resolution and the signal contrast if it is not carefully addressed. By using an immersion medium, changes of the optical dispersion in the sample arm can be minimized as the imaging depth is varied. With dry objectives, optical dispersion can be balanced between the interferometer arms by inserting a variable-thickness medium of appropriate dispersion in one of the arms. This method was implemented in Ref. [55] by rotating a fused silica glass plate placed in the reference arm, using a motorized rotation stage.

1.3.2 *Phase-Shifting Interferometry*

Tomographic imaging with FF-OCM, as with all OCT systems, relies on the extraction of the amplitude of the interferometric signal generated by the backscattering from the sample of broadband light.

The method for extraction of the interferometric signal amplitude, that is, the interference fringe envelope, however, differs depending on the type of OCT technique.

In time-domain OCT, the optical length of the reference arm is scanned to probe the depth in the sample. This scanning generates a modulation of the interferometric signal at a Doppler beat frequency related to the speed of the scan. The signal from the photodetector is electronically filtered at this frequency to get the amplitude of the interferometric signal.

In frequency-domain OCT, the power spectral density of light at the interferometer output is acquired, by encoding the optical frequency in time either with a spectrally scanning source or with a dispersive detector like a grating and a linear detector array. The depth information is then calculated by a Fourier transform from the acquired spectra.

In conventional FF-OCM, the amplitude of the interferometric signal is obtained by arithmetic combination of several phase-shifted interferometric images acquired by a camera. This method is based on phase-shifting interferometry, a concept that has been performed with many types of interferometric systems, using a variety of algorithms to get a phase map from several intensity fringe patterns acquired with an area camera. Phase-shifting algorithms have also been developed for digital fringe envelope detection in low-coherence (white-light) interference microscopy. Combined with a procedure for determining the position of the fringe envelope peak, this method enables sample surface topography measurements with theoretically unlimited height range [60–62]. The extraction of the fringe envelope performed in white-light interference microscopy is similar to the method applied in FF-OCM for tomographic imaging.

For all phase-shifting algorithms, a discrete or continuous temporal phase shift is introduced in the interferometer and several interferometric images are acquired. The most common method used in FF-OCM to generate the required phase shift consists of displacing a reference reflector using a piezoelectric transducer (PZT) [32, 33]. Other methods have been demonstrated to achieve achromatic phase shifting using a rotating polarizer [63, 64], a rotating half-wave plate [65], or a nematic liquid-crystal phase shifter [66]. Alternatively, sinusoidal phase modulation has been achieved using a photoelastic phase modulator in association with stroboscobic illumination [31, 37].

In the so-called phase-stepping technique, the phase is stepped by a known amount between each interferometric image acquisition. A large number of phase-stepping algorithms have been proposed for white-light interference microscopy. Several phase-stepping algorithms have been used in FF-OCM [33, 53, 65–67]. This method may, however, be limited in operation speed when the phase shift is induced by a mechanical displacement of an element in the interfermeter.

In the so-called integrating-bucket technique, the interferometric images are acquired while the phase is being shifted continuously. The bandwidth limitation on stepped phase shift techniques is then significantly reduced, enabling higher operation speed at the price of a certain fringe washout. In this method, the phase is usually shifted linearly in a sawtoothlike manner, and several integrated intensity values (or buckets) are recorded by the camera. Phase-shifting interferometry that uses sinusoidal phase modulation is less usual. An algorithm with sinusoidal phase modulation and four integrating buckets was initially proposed for phase measurements [68, 69]. This algorithm was revisited later for fringe envelope detection and applied to FF-OCM [31, 32, 34, 42]. A version with only two integrating buckets was proposed in Refs. [35, 36]. The main interest of sinusoidally modulated phase-shifting interferometry is the high operation speed that can be reached even with mechanical modulation of the phase using a PZT. A drawback of this method is the relative difficulty to adjust the modulation parameters, that is, the amplitude of the PZT oscillation and the synchronization phase between the PZT oscillation and the CCD triggering signal. The method of sinusoidal phase modulation and four integrating buckets used in many FF-OCM systems is presented in detail in the next section.

As will be discussed later in this chapter, the drawback of conventional phase-shifting methods in FF-OCM is the sensitivity to motion of the sample, which constitutes a serious handicap for in vivo imaging. One of the problems arises from the fact that the tomographic image is calculated from several interferometric images that are acquired sequentially in time. Because of sample motion, the intensity of light received by some pixels of the camera may change between successive frames, thus generating an intrusive signal in the calculated tomographic image. To solve this problem, several types of instantaneous phase-shifting methods have been implemented in FF-OCM [43, 48, 70–72]. The detection sensitivity of the reported systems was, however, too low for deep imaging. Work on single-shot FF-OCM is presented and discussed in Chapter 7 of this book.

1.3.3 Sinusoidal Phase Shifting and Four Integrating Buckets

We describe in this paragraph the sinusoidal phase-shifting algorithm implemented in many FF-OCM systems [31, 32, 34, 42]. As will be explained later in this chapter, the amplitude of the interferometric signal as a function of the axial (z) position of a reflector (or backscattering structure) within the sample is modulated by an envelope determined by both the temporal coherence of illumination and the depth of field of the microscope objective(s). To derive simple analytical equations, the problem is treated here by considering a square-shaped envelope. In that case, light returning from the sample and collected by the microscope objective can be decomposed into two parts, depending on whether it interfere or not. Light that interferes originates from a coherence slice within the sample, the thickness of this slice being the width of the envelope. The reflectivity of the backscattering structures located in the coherence slice at a given depth z_0 is represented by two-dimensional distribution $R(x, y, z_0)$. The light originating from the coherence slice interferes with the light reflected by the plane reference surface. The rest of the light backscattered from the sample and collected by the microscope objective that does not interfere is represented by an equivalent reflectivity coefficient denoted as R_{inc}. This quantity, made up from all reflected and backscattered components throughout the sample except those in the coherence slice, can be considered as constant. The intensity of light incident onto the image sensor placed at the output of a Linnik-type FF-OCM system can thus be written as

$$I = \frac{I_0}{4}\left(R + R_{\text{inc}} + R_{\text{ref}} + 2\sqrt{RR_{\text{ref}}}\cos\phi\right), \quad (1.5)$$

where I_0 is the intensity of light incident on the interferometer beam splitter and ϕ the optical phase. Due to the narrow width of the coherence slice, the amount of light returning from the sample that interfere and that is detected is small compared to all the light that is incident on the image sensor. This implies that $R \ll R_{\text{inc}}$ and $R \ll R_{\text{ref}}$. With this assumption, the signal can be rewritten as

$$I = \bar{I}\left(1 + V\cos\phi\right), \quad (1.6)$$

where

$$\bar{I} = I_0\left(R_{\text{inc}} + R_{\text{ref}}\right)/4 \quad (1.7)$$

is the bias intensity, and

$$V = \frac{2\sqrt{RR_{\text{ref}}}}{(R_{\text{inc}} + R_{\text{ref}})} \quad (1.8)$$

is the fringe visibility.

By introducing a sinusoidal modulation of the optical phase in the interferometer, of amplitude ψ and angular frequency ω, the intensity becomes

$$I(t) = \bar{I}\left\{1 + V\cos\left[\phi + \psi\sin(\omega t + \theta)\right]\right\}, \quad (1.9)$$

the parameter θ being determined by the time origin. The time-varying light intensity is integrated successively over the four quarters of the modulation period $T = 2\pi/\omega$ by the image sensor. The time integration of $I(t)$ is performed in parallel by all the pixels of the camera (frame-transfer or full-frame camera). The charge storage period of the camera is set to be one-quarter of the period T of the sinusoidal phase modulation. Four images (four frames of interferogram) are thus recorded. Each of them is accumulated N times in order to increase the SNR. The mean quantum efficiency of the detector being η, the four accumulated frames are

$$E_p = \eta N \int_{(p-1)T/4}^{pT/4} I(t)\, dt, \quad p = 1, 2, 3, 4. \quad (1.10)$$

To calculate the integral in Eq. 1.10, a Fourier decomposition of $I(t)$ is written using Bessel functions of the first kind, according to

$$I(t)/\bar{I} = 1 + V\cos\phi\left\{J_0(\psi) + 2\sum_{k=1}^{\infty} J_{2k}(\psi)\cos[2k(\omega t + \theta)]\right\}$$

$$- 2V\sin\phi\sum_{k=0}^{\infty} J_{2k+1}(\psi)\sin[(2k+1)(\omega t + \theta)]. \quad (1.11)$$

The following relations can then be expressed:

$$E_1 - E_2 = \frac{2}{\pi}\eta NT\bar{I}V(\Gamma_1 \cos\phi - \Gamma_2 \sin\phi) \qquad (1.12a)$$

and

$$E_3 - E_4 = \frac{2}{\pi}\eta NT\bar{I}V(\Gamma_1 \cos\phi + \Gamma_2 \sin\phi), \qquad (1.12b)$$

where

$$\Gamma_1 = \sum_{k=0}^{\infty} J_{4k+2}(\psi)[\sin 2(2k+1)\theta]/(2k+1) \qquad (1.13a)$$

and

$$\Gamma_2 = \sum_{k=0}^{\infty} J_{2k+1}(\psi)[\sin(2k+1)\theta](-1)^k/(2k+1). \qquad (1.13b)$$

If $\Gamma_1 = \Gamma_2 = \Gamma$, the visibility V can be calculated according to the following frame combination:

$$(E_1 - E_2)^2 + (E_3 - E_4)^2 = \left(\frac{2\sqrt{2}\Gamma}{\pi}\right)^2 (\eta NT\bar{I})^2 V^2. \qquad (1.14)$$

The illumination flux is adjusted so that the camera pixel wells are close to saturation. Since $V \ll 1$, one can write

$$\xi_{sat} \approx \eta \int_0^{T/4} \bar{I}dt = \eta T\bar{I}/4. \qquad (1.15)$$

Considering the above equation and the expression of the fringe visibility, Eq. 1.14 can be rewritten as

$$(E_1 - E_2)^2 + (E_3 - E_4)^2 = \left(\frac{16\sqrt{2}\Gamma}{\pi}\right)^2 (N\xi_{sat})^2 \frac{R_{ref}}{(R_{inc}+R_{ref})^2}R. \qquad (1.16)$$

The image combination expressed in the above equation is proportional to $R(x,y)$, that is, to the reflectivity of the structures within the sample that are located in the coherence slice. This formula therefore provides an image of these structures, that is, an en face tomographic image at a given depth.

Equation 1.16 is valid, provided that the modulation parameters (ψ, θ) are adjusted so that $\Gamma_1 = \Gamma_2$. Numerical simulations show that an infinity of couples (ψ, θ) can be chosen so that $\Gamma_1 = \Gamma_2$ (see Fig. 1.10a). It is interesting to select the couple (ψ, θ) that maximizes

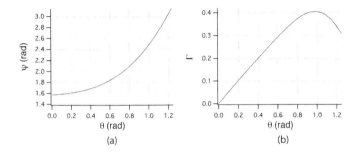

Figure 1.10 Parameters of the sinusoidal phase modulation. (a) Plot of ψ versus θ so that $\Gamma_1 = \Gamma_2$. (b) Plot of Γ versus θ so that $\Gamma_1 = \Gamma_2$.

the parameter Γ so that the image combination expressed in Eq. 1.16 reaches its maximum. The variation of Γ as a function of θ is plotted in Fig. 1.10b. The maximum of Γ, noted Γ_{max}, is reached when $\psi \approx 2.45$ and $\theta \approx 0.98$. Then, $\Gamma_1 = \Gamma_2 = \Gamma_{max} \approx 0.41$.

In the general case of a fringe envelope with arbitrary shape, the previous calculations are in principle no longer valid. Numerical simulations can be carried out to study the efficiency of the sinusoidal phase-shifting algorithm presented previously for arbitrary fringe envelope detection. We consider in the simulations the practical case of a Gaussian-shaped source spectrum as detected typically with a halogen lamp as a light source, using microscope objectives with a moderate numerical aperture of 0.3. The true fringe envelope, calculated using Eq. 1.23 introduced later in this chapter, is then close to a Gaussian function. The extracted envelope (see Fig. 1.11) deviates from the true response by the presence of a modulation. Other algorithms such as the five-frame Larkin algorithm [73] happen to be more accurate (see Fig. 1.11). The presence of modulations in the calculated fringe envelope is, however, not a real issue in FF-OCM. The main purpose of phase-shifting interferometry implemented in FF-OCM is to extract the weak interferometric component of the signal detected by the camera. This is accomplished by calculating differences of phase-shifted images, regardless of the algorithm. The interest of a sinusoidal phase modulation is the possibility to induce the phase shift by a mechanical oscillation at high frequency. This method of sinusoidal phase modulation with four integrating buckets has been

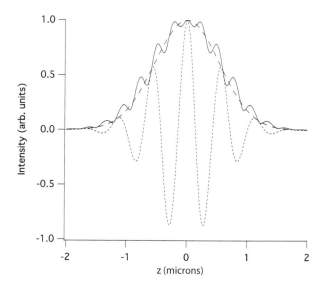

Figure 1.11 Fringe envelope extraction using sinusoidal phase modulation with four integrating buckets. The dotted line shows the interference fringes. The dashed line represents the actual envelope as can be extracted with the Larkin algorithm.

implemented in many FF-OCM systems for the production of en face tomographic images in real time.

1.4 Performance

1.4.1 *Detection Sensitivity*

The smallest signal that can be detected in FF-OCM, defined as detection sensitivity, is a key parameter that affects the imaging contrast and the penetration depth. Several possible sources of noise need to be taken into account to determine the detection sensitivity. In addition to the shot noise induced by the fundamental photon noise, other kinds of noise originating from the camera may not be negligible, such as the read-out noise and the dark noise [42]. In contrast to conventional InGaAs cameras, scientific silicon-based cameras are almost shot-noise limited. When their pixel wells are close to saturation, the electrical noise is small compared to the shot noise resulting from the photon noise. Ordinarily, the photon noise is Poisson distributed, implying that the variance of the number of detected photons is equal to the number of photons itself. The detector shot noise can be modeled as an additive component, v, such that the number of photoelectrons delivered by each pixel of the camera is $E + v$. The actual signal is then (see Eq. 1.16)

$$\text{Signal} = [(E_1 + v_1) - (E_2 + v_2)]^2 + [(E_3 + v_3) - (E_4 + v_4)]^2. \quad (1.17)$$

The additive noise $v_{i=1,2,3,4}$ has the following properties:

$$\langle v_i \rangle = 0, \ \langle v_i^2 \rangle = \sigma^2, \ \langle v_i v_j \rangle_{i \neq j} = 0, \quad (1.18)$$

where the angle brackets in Eq. 1.18 denote a time average. The illumination flux is adjusted such that the camera pixel wells are close to their saturation, ξ_{sat}. The variance of the shot noise is then $\sigma^2 \approx N\xi_{\text{sat}}$. The background noise in the tomographic image can be regarded as the average value of the signal when there is no interference. By averaging Eq. 1.17 when $E_1 = E_2 = E_3 = E_4$ (no interference), and considering the properties of the additive noise v, an expression of the background noise can be written as

$$\text{Background noise} = 4N\xi_{\text{sat}}. \quad (1.19)$$

Combining the previous equation with the expression of the signal given in Eq. 1.16, the shot-noise-limited signal-to-background-noise ratio is

$$\text{SNR}_{\text{shot}} = \left(\frac{8\sqrt{2}\Gamma_{\text{max}}}{\pi}\right)^2 N\xi_{\text{sat}} \frac{R_{\text{ref}}}{(R_{\text{inc}} + R_{\text{ref}})^2} R. \quad (1.20)$$

If the electric noise, including the read-out noise and the dark noise, is not negligible compared to the shot noise, the expression of the signal-to-background-noise ratio should be modified. The noise-equivalent electrons (NEEs) representing total electrical noise being χ, the signal-to-background-noise ratio can be written, assuming the pixel wells to be full upon maximum illumination and neglecting the relative intensity noise, as

$$\text{SNR} = \text{SNR}_{\text{shot}} \left(\frac{\xi_{\text{sat}}}{\xi_{\text{sat}} + \chi^2}\right). \quad (1.21)$$

The minimum detectable reflectivity, reached at the detection threshold, that is, when SNR = 1, is then

$$R_{\min} = K \left(\frac{\xi_{sat} + \chi^2}{N\xi_{sat}^2} \right) \frac{(R_{inc} + R_{ref})^2}{R_{ref}}, \text{ with } K = \left(\frac{\pi}{8\sqrt{2}\Gamma_{\max}} \right)^2 \approx 0.46. \quad (1.22)$$

The detection sensitivity is defined as $10 \times \log(R_{\min})$.

According to Eq. 1.22, the sensitivity can be improved by increasing the full-well-capacity ξ_{sat}. However, increasing the full-well capacity, while maintaining the image acquisition time, requires the availability of a more powerful light source, which represents a practical limitation. Moreover, the maximal permissible exposure level of biological samples is limited. Another way to increase the detection sensitivity is to accumulate a larger number N of interferometric images. Since accumulating images increases the acquisition time, this way of improving the detection sensitivity is however limited if the sample is likely to move during the acquisition. In vivo imaging, in particular, is incompatible with the accumulation of a large number of images. In order to minimize parameter R_{inc}, the reflection on the sample surface can be minimized by index matching, which is achieved by using water-immersion objectives. The reference mirror reflectivity, R_{ref}, also has an influence on the detection sensitivity. By calculating the derivative of Eq. 1.22 with respect to R_{ref}, one can easily establish that the optimal value of R_{ref} is reached when $R_{ref} = R_{inc}$, which corresponds to a value of a few percent. In practice, the detection sensitivity of FF-OCM ranges between -75 dB and -90 dB [34, 35, 53].

Scanning OCM combines two physical mechanisms to achieve optical sectioning, confocal gating and coherence gating. Thanks to the confocal gate, light coming from out-of-focus regions in the sample is (partially) eliminated before arriving at the detector. In FF-OCM, there is no confocal gate to eliminate light that does not contribute to interference. This unwanted light is detected by the camera, which leads to a reduction of the useful dynamic of the detection. Full-field illumination thus leads to a significant reduction of the detection sensitivity due to the absence of confocal gate [74]. Furthermore, multiple light scattering may generate interferometric components that are detected by the camera. The collected fraction of multiply scattered, relative to singly scattered, light increases with depth, thus reducing image contrast [75] and degrading the axial and lateral resolutions [76–78]. This phenomenon is reduced in scanning OCM due to the presence of the confocal gate in addition to the coherence gate.

1.4.2 Axial Resolution

The interferometric signal measured by a two-beam interference microscope (Michelson or Linnik type) as a function of sample axial position z (considered here as a single reflective surface) is proportional to [31]

$$I(z) = \int_0^\infty S(k) \int_0^{\alpha_{max}} \cos(2kz\cos\alpha) \cos\alpha \sin\alpha \, d\alpha \, dk. \quad (1.23)$$

In the above equation, $S(k)$ represents the detected power spectral density of the light source (k being the wavenumber) and α_{max} denotes the maximum incidence angle of illumination upon the sample. The angle α_{max} is related to the numerical aperture, NA of the microscope objective(s) by $\text{NA} = n\sin\alpha_{max}$, n being the refractive index of the sample.

The axial resolution in FF-OCM can be defined as the width of the axial response $I(z)$. To provide a formula for the axial resolution, an analytical expression of $I(z)$ has to be established. The integral

$$I_{NA}(z) = \int_0^{\alpha_{max}} \cos(2kz\cos\alpha)\cos\alpha\sin\alpha \, d\alpha \quad (1.24)$$

can be calculated analytically [79]. The complicated analytical expression can be very accurately approximated by

$$I_{NA}(z) = \text{sinc}[kz(1-\cos\alpha_{max})]\cos[kz(1+\cos\alpha_{max})]. \quad (1.25)$$

The function $I_{NA}(z)$ can be regarded as an oscillating function of frequency $k(1+\cos\alpha_{max})/2\pi$, modulated by an envelope varying like the cardinal sine function "sinc". Since the envelope oscillates slower with k than the carrier, the function $I(z)$ in Eq. 1.23 can be approximately rewritten as

$$I(z) \approx \text{sinc}[k_0 z(1-\cos\alpha_{max})] \int_0^\infty S(k)\cos[kz(1+\cos\alpha_{max})] \, dk, \quad (1.26)$$

where k_0 is the center wavenumber. The integral in Eq. 1.26 involves the real part of the Fourier transform of $S(k)$. Assuming a Gaussian-shaped function $S(k)$ of FWHM Δk, an expression of the axial response is then

$$I(z) = \text{sinc}\left[k_0 z (1 - \cos\alpha_{max})\right] \exp\left[-\frac{(1+\cos\alpha_{max})^2 \Delta k^2}{16\ln 2} z^2\right]$$
$$\times \cos\left[k_0 (1+\cos\alpha_{max}) z\right]. \quad (1.27)$$

For moderate values of NA, the dependence of the Gaussian function with α_{max} can be neglected. For high values of NA, the "sinc" function becomes narrower that the Gaussian function, which implies that the dependence of the Gaussian function with α_{max} can be ignored without significant modification of the global envelope of $I(z)$. Finally, a good approximation of $I(z)$ is

$$I(z) = V_S(z) \times V_{NA}(z) \cos\left[k_0 (1+\cos\alpha_{max}) z\right], \quad (1.28)$$

with

$$V_S(z) = \exp\left(-\frac{\Delta k^2}{4\ln 2} z^2\right), \quad (1.29)$$

and

$$V_{NA}(z) = \text{sinc}\left[k_0 z (1 - \cos\alpha_{max})\right]. \quad (1.30)$$

The axial response of the interference microscope with arbitrary temporal coherence length and numerical aperture can be interpreted as a carrier of frequency $k(1+\cos\alpha_{max})/2\pi$ modulated by an envelope that is the product of two independent functions. One of them, $V_{NA}(z)$, is determined by the numerical aperture and the other one, $V_S(z)$, is determined by the spectral width of light.

The FWHM of $V_S(z)$ is

$$\Delta z_S = \frac{4\ln 2}{\Delta k} = \frac{l_c}{2n} \approx \frac{2\ln 2}{n\pi}\left(\frac{\lambda_0^2}{\Delta\lambda}\right), \quad (1.31)$$

where $\Delta\lambda$ represents the FWHM of the spectral power distribution expressed as a function of wavelength and λ_0 is the center wavelength. Quantity l_c denotes the temporal coherence length.

The FWHM of $V_{NA}(z)$ is

$$\Delta z_{NA} \approx \frac{\pi}{k_0 (1 - \cos\alpha_{max})}, \quad (1.32)$$

which can be expressed as a function of the numerical aperture at third order in the incidence angle as

$$\Delta z_{NA} = \frac{n\lambda_0}{NA^2}. \qquad (1.33)$$

Equation 1.23 identifies to the usual expression of the depth of field of a microscope objective.

The axial resolution Δz of FF-OCM can be defined as the width of the global envelope $V_S(z) \times V_{NA}(z)$. Assuming the central lobe of $V_{NA}(z)$ to be close to a Gaussian function, a general formula for the axial resolution of FF-OCM can be expressed as

$$\Delta z = \left(\frac{1}{\Delta z_S{}^2} + \frac{1}{\Delta z_{NA}{}^2} \right)^{-\frac{1}{2}}. \qquad (1.34)$$

Results of numerical simulations based on Eqs. 1.31–1.34 are shown in Fig. 1.12. In general, for a given light source, the axial resolution of FF-OCM is improved by increasing the numerical aperture of the objective(s). However, when using a broadband

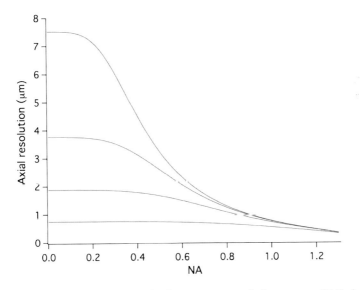

Figure 1.12 Axial resolution (Δz) versus numerical aperture (NA) for different values of the temporal coherence length ($l_c = 20$ μm, 10 μm, 5 μm, 2 μm from top to bottom line plots). The center wavelength is $\lambda_0 = 700$ nm. The medium is water.

28 | Introduction to Full-Field Optical Coherence Microscopy

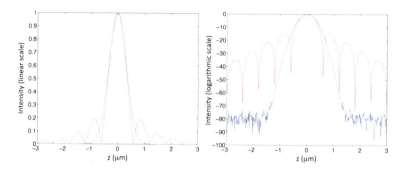

Figure 1.13 Envelope of the axial response of FF-OCM for (NA = 1.1, l_c = 10 µm) in red, and (NA = 0.3, l_c = 2 µm) in blue, at λ_0 = 700 nm.

light source like a halogen lamp, the numerical aperture has almost no effect on the axial resolution that is then governed by the temporal coherence. With narrow-band light, in contrast, the numerical aperture may then have a significant impact on the axial resolution. For example, with NA = 1.1 using monochromatic light at λ_0 = 700 nm, the theoretical axial resolution in water is 0.6 µm. The same resolution can be obtained with low NA values by increasing the spectral width of illumination to $\Delta\lambda$ = 270 nm. However, as can be seen in Fig. 1.13, the shape of the axial response is different depending on whether it is determined by the numerical aperture or by the temporal coherence. The function $V_{NA}(z)$ decreases from $z = 0$ with oscillations, whereas the function $V_S(z)$ decreases monotically in the case of a Gaussian-shaped spectral power distribution. When represented in logarithmic scale in presence of additive noise, one can see that the function $V_S(z)$ reaches the background noise floor more rapidly than $V_{NA}(z)$. The coherence gate (V_S) rejects scattered light with efficiency orders of magnitude better than the spatial gate (V_{NA}) imposed by the depth of field of the microscope objective(s). The combined effects of the two gates are generally stronger than either gate individually, achieving greater image contrast and image penetration in scattering tissue [24]. The efficiency of the coherence gate degrades with depth in the sample due to dispersion mismatch in the interferometer arms. In practice, dispersion mismatch is unavoidable due to the nonuniform distribution of the refractive index within the sample.

This phenomenon leads to a degradation of the axial resolution as the imaging depth in the sample increases [34]. For example, at a depth of 250 µm in human dermis, the axial resolution was estimated to be degraded by a factor of ∼1.4 using broadband light ($\Delta\lambda = 270$ nm at 750 nm center wavelength). We may then think that it would be preferable to use narrow band light to reduce/eliminate the effect of dispersion and employ high NA objective(s) to maintain high axial resolution [80, 81]. This is, however, at the cost of a reduced lateral field of view (since high NA objectives generally have a small field of view) and reduced efficiency of scattered light rejection compared to the coherence gate, as discussed previously. Moreover, the $V_{NA}(z)$ response alone, requiring high values of NA to be efficient, is affected by optical aberrations when imaging into tissue, which limits the performance for deep imaging.

The effect of both spatial coherence and temporal coherence on the spatial resolution of FF-OCM is discussed in detail in Chapters 2 and 3 of this book. A nonconventional FF-OCM system using spatially incoherent monochromatic light is reported in Chapter 10.

1.4.3 Lateral Resolution

The lateral (transverse) resolution of an optical imaging system can be defined as the width of the transverse point-spread function (PSF). In an interference microscope, as involved in FF-OCM, the amplitude PSF identifies to the intensity PSF of a reflected light microscope [31, 62]. Assuming an aberration-free microscope, the PSF is determined by diffraction. The amplitude PSF of FF-OCM is then described by the Airy function

$$I(r) = \left[\frac{2J_1(2\pi r \text{NA}/\lambda_0)}{2\pi r \text{NA}/\lambda_0}\right]^2, \quad (1.35)$$

where λ_0 is the center wavelength and NA the numerical aperture of the microscope objective(s). The FWHM of the PSF yields the lateral resolution

$$\Delta r \approx \frac{\lambda_0}{2\text{NA}}. \quad (1.36)$$

The lateral resolution can be adjusted by the choice of the microscope objective(s) (value of NA). In practice, it is interesting to have an isotropic resolution, that is, the same resolution in the lateral and axial directions, for three-dimensional imaging. This is an interesting feature of FF-OCM, in contrast to confocal microscopy where the axial resolution is always lower than the lateral resolution. In practice, FF-OCM can offer an isotropic spatial resolution on the order of 1 µm by using relatively modest values of NA (0.3–0.5), thus avoiding optical aberrations induced by the propagation of light through the turbid biological sample being imaged.

1.4.4 Sample Motion Artifacts

In interferometry, path length modifications of a fraction of the wavelength generate significant phase changes that may blur the measured interference signal if the detector is not fast enough. In FF-OCM, sample motions along the axial direction must not exceed a quarter of the center wavelength typically during the image acquisition time to avoid interferometric signal washout, which typically represents a maximal permissible axial speed of the sample of 1–10 µm/s [48]. The relatively long acquisition time of the interferometric signal in FF-OCM is currently a limitation for efficient real-time in vivo applications. For example, the speed of blood flow is a few hundreds of µm/s. Physiological motions such as the cardiac motion may even reach 100 mm/s. For these kinds of application, a shorter acquisition time is required. High-speed FF-OCM systems have been proposed, making possible in vivo imaging [41, 43, 71, 72, 82, 83]. However, the detection sensitivity of these systems was quite low. A real progress would be to use a camera with both high dynamic range and high frame rate, provided that the radiance of the light source could be sufficiently increased to fill the pixel wells and that the sample could tolerate the light exposure. Although the time to produce a single en face tomographic image is similar in conventional FF-OCM (with image accumulation) and scanning OCM, this latest technique is less sensitive to axial motion since the acquisition of the interferometric signal for each pixel of the tomographic image is considerably faster. For this aspect, parallel acquisition using a camera, as it is done in conventional FF-OCM, constitutes a drawback compared to point-by-point acquisition using a high-speed high-dynamic single detector.

Transverse motion of the sample has other consequences in FF-OCM. The tomographic signal in FF-OCM is generated by the variation of the interferometric signal amplitude induced by the phase shift introduced in the interferometer between interferometric images acquired sequentially in time. However, the intensity of light upon each pixel of the camera may also vary if the backscattering structures of the sample move transversally. This variation of intensity generates a spurious signal, which superimposes on the tomographic signal. Axial motion of the sample has much less influence on this spurious signal than transverse motion, since it only causes defocusing, which does not change significantly the intensity of light received by each pixel. One can consider that the motion of the scatterers within the sample become visible when the resulting intensity variation is greater than the minimal detectable reflectivity R_{min} given by Eq. 1.22. In the experimental setup reported in Ref. [48], transverse motion of a particle with equivalent reflectivity of 10^{-5} had no effect if it moved slower than $\sim$100 μm/s, indicating that the effect of transverse motion was less severe than the effect of axial motion (speed less that 1–10 μm/s). Artifacts resulting from transverse motion can be completely removed if the interferometric images used to calculate the tomographic image are acquired simultaneously and not sequentially in time. For that purpose, instantaneous phase-shifting interferometry was successfully applied in FF-OCM, at the price of an increase in system complexity, calibration and cost [43, 48, 70, 72, 82]. Another approach has been investigated using a single interferometric image to produce a tomographic image by data processing based on the Riesz transform [83]. It should be noted that in scanning OCM, transverse motion does not generate a spurious signal as in FF-OCM but may, however, lead to a distortion of the image. This artifact can be reduced by image processing using motion correction algorithms.

1.5 Clinical Applications

Biopsy and histology are part of the standard method for disease assessment at the tissue and cellular levels. A biopsy is an invasive medical procedure to remove a small amount of tissue. Histology involves long and delicate processing of the excised tissue for sectioning, staining, and diagnosis. The usual method of tissue processing consists of several steps: fixation that preserves the tissue; processing that dehydrates, clears, and infiltrates the tissue with paraffin, embedding that allows orientation of the specimen in a block; and sectioning using a microtome to produce very thin sections. An alternative technique for tissue processing consists of quickly freezing the tissue to preserve it and provide sufficient hardness so that it can be sectioned immediately. This faster technique is often used during surgery to locate a tumor margin and ensure that it has all been removed. This method is, however, also time consuming, especially for slice-by-slice observations.

An optical technique that would provide tomographic images of spatial resolution and contrast similar to that of histology images, without the need for tissue processing and preparation, would be of great interest. The capability of FF-OCM for noninvasive imaging of tissues has been extensively reported and discussed in the literature. A review of the main applications and the potential of the technology in the biomedical field is provided in this section.

1.5.1 *Ophthalmic Tissue Imaging*

FF-OCM has been used for the characterization of various ophthalmic tissues from the anterior and posterior segments of animal and human eyes [41, 84–89].

In Ref. [84], cellular-level resolution imaging of isolated tissue samples from rat, mouse, and pig eyes was reported. En face (xy) tomographic images revealed corneal epithelial and stromal cells, lens fibers, nerve fibers, major vessels, and retinal pigment epithelial cells. In vertical (xz) reconstructed sections, cellular layers within the cornea and retina and arterioles and venules were clearly defined. Transscleral retinal imaging was also achieved in albino animals.

The possibility of in vivo FF-OCM imaging in ophthalmology was demonstrated for the first time in Ref. [85] using a xenon arc lamp source and a fast CMOS camera. Each en face tomographic image was acquired in a 4 ms period at a repetition rate of 250 Hz. Images of the anterior segment of the eye of an anesthetized rat revealed some cellular features in the corneal layers.

In Ref. [87], FF-OCM was used for ex vivo human cornea imaging. Subcellular imaging of the corneal structure and pathologies on the entire thickness of the tissue was demonstrated, even in edematous corneas, as well as interface quality and changes in the collagen structure due to laser incisions.

Numerous examples of FF-OCM imaging of ophthalmic tissues are shown in Chapter 19 of this book, entirely devoted to the application of FF-OCM in ophthalmology.

1.5.2 Skin Tissue Imaging

The interest of FF-OCM in dermatology and cosmetology has been investigated more recently [90–93].

The potential of the technique for the understanding and monitoring of skin hydration and pigmentation, as well as skin inflammation was reported in Refs. [90, 92]. These studies have shown that FF-OCM images can provide various parameters of the skin epidermis and dermis, including cell size, organization and density, layer thickness, melanin caps, and melanocytes.

A pilot study was carried out to evaluate the ability of FF-OCM to serve as a tool for dermatopathologists to identify or exclude malignancy in Mohs surgery specimens [91]. In this study, a dermatopathologist blinded to the diagnosis examined eighteen FF-OCM images. The nine images interpreted as negative for malignancy were in agreement with the hematoxylin and eosin (H&E)-stained frozen sections. Six of the remaining FF-OCM images were correctly interpreted as positive for malignancy, and three were deferred because malignancy could not be confirmed or excluded.

An evaluation of the clinical value of FF-OCM in the management of patients with head and neck cancers was reported in Ref. [93]. Freshly excised samples from patients, of mouth, tongue, epiglottis, and larynx tissues, both healthy and cancerous were imaged with FF-OCM and compared with histology. Common

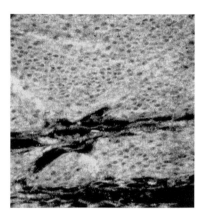

Figure 1.14 En face FF-OCM image of human skin, in vitro, at depth of 50 μm. Field of view: 500 μm × 500 μm.

features were identified and characteristics of each tissue type were matched in order to form an image atlas for pathologist training. Indicators of tumors such as heterogeneities in cell distribution, surrounding stroma, and anomalous keratinization were successfully identified.

An example of a typical FF-OCM image of freshly excised human skin is shown in Fig. 1.14. This en face tomographic image was acquired at a depth of 50 μm below the surface with an isotropic spatial resolution of 0.8 μm. The shape and distribution of cell nuclei in the stratum spinosum layer are clearly revealed.

1.5.3 Breast Tissue Imaging

The potential of FF-OCM as a tool for per-operative diagnoses of breast tumors has been reported. In the study published in Ref. [94], two pathologists were able to distinguish between FF-OCM images of normal and benign breast tissues excised during the operation, with a sensitivity of 94% and 90% and a specificity of 75% and 79%, respectively.

Figures 1.15, 1.16, and 1.17 show FF-OCM images of human breast tissues from biopsies. All these tomographic images are in the en face orientation. Figure 1.15 shows an FF-OCM image

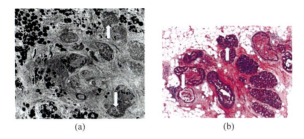

Figure 1.15 FF-OCM image (a) and corresponding histology (b) of freshly excised human breast tissue. The width of the images is 400 μm. Courtesy of LLTech.

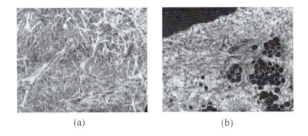

Figure 1.16 FF-OCM images of human breast tissue showing healthy fibrous tissue with normal adipocytes (a) and diseased tissue with nodular invasive carcinoma (b). Courtesy of LLTech.

and the corresponding stained histological section. The downward arrow indicates the presence of a lobule and the upward arrow the presence of a milk duct with necrosis. Figure 1.16 compares normal and abnormal tissues. Figure 1.16a shows the fibrous structure of healthy tissue with normal adipocytes. Figure 1.16b shows nodular invasive carcinoma. Carcinomatous cells intermingled with trabecula of the stroma reaction are visible. These images demonstrate the ability of FF-OCM to reveal architectural changes in tissue structures associated with cancer development. Figure 1.17 shows several features of human breast tissue such as duct with calcification (a), blood vessel (b), and fat cells (c). Breast tissue imaging using FF-OCM is discussed in more detail in Chapter 18 of this book.

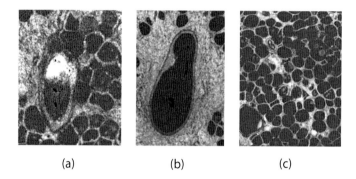

(a)   (b)   (c)

Figure 1.17 Several biological features of human breast tissue can be visualized using FF-OCM. Duct with calcification (a), blood vessel (b), and fat cells (c). Courtesy of LLTech.

1.5.4 Other Tissue Imaging

Various other human tissues have been imaged with FF-OCM [95]. Imaging of tumorous and nontumorous human brain tissues has been reported in Ref. [96] and is presented in detail in Chapter 18 of this book. Prostate biopsies have been imaged recently with FF-OCM [97]. Other tissues have been imaged such as human thyroid [36] and human esophagus [34], as an illustration of the FF-OCM capabilities in terms of spatial resolution and imaging penetration depth.

In Ref. [98], ex vivo imaging with FF-OCM of testicular tissues from normal and busulfan-treated rats was reported. Normal adult rats exhibited tubules with uniform size and shape, whereas the busulfan-treated animals showed marked heterogeneity in tubular size and shape and only 10% contained sperm within the lumen. This study has shown the potential of FF-OCM to facilitate real-time visualization of spermatogenesis in humans and aid in microdissection testicular sperm extraction for men suffering from infertility.

A study of the clinical interest of FF-OCM in identifying and differentiating lung tumors from nonneoplastic lung tissue has been carried out [99]. Structures of healthy lung tissue such as alveoli, bronchi, pleura, and blood vessels were identified. Using FF-OCM, pathologists were able to identify all tumor specimens and in most cases the histological subtype of tumor. Benign diagnosis was established with high confidence in about half the tumor-free specimens. Further analysis revealed two major confounding features, extensive lung collapse and presence of smoker macrophages. However, the smoker macrophages could often be identified as distinct from tumor cells on the basis of their relative location in the alveoli, size, and presence of anthracosis. These results demonstrated the potential of FF-OCM as an adjunct to intrasurgical frozen section analysis for margin assessment, especially in limited lung resections.

1.6 Applications in Biology

1.6.1 *Developmental Biology*

The high-resolution and noninvasive imaging capability of FF-OCM makes this technology particularly suitable for applications in the field of embryology and developmental biology [34, 36, 100–103]. Chapter 20 of this book reports on the use of FF-OCM for studies in mammalian developmental biology.

Early patterning and polarity of mammalian embryos are fundamental and important in developmental biology. In Ref. [100], FF-OCM was used to study how the establishment of the anterior-posterior polarity relates to the morphology of the postimplantation embryo. It was shown that the change in the orientation of the anterior-posterior polarity axis in the gastrulating mouse embryo is only apparent and results from a dramatic remodeling of the whole epiblast, in which cell migrations take no part. These results revealed a level of regulation and plasticity so far unsuspected in the mouse gastrula.

The potential of FF-OCM in providing new insights into and potential breakthroughs to the controversial issues of early patterning and polarity in mammalian developmental biology was reported in Refs. [101, 102]. FF-OCM was used to study the dynamics of developmental processes in mouse preimplantation lives. Images were acquired at different times to investigate the three-dimensional spatial relationship between the second polar body (2PB) and the first cleavage plane. The study showed that

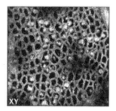 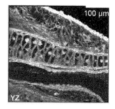 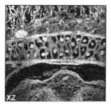

Figure 1.18 FF-OCM imaging of *Xenopus laevis*.

only 25% of the predicted first cleavage planes pass through the 2PB, whereas only 27% of the real cleavage planes pass through the 2PB.

Xenopus is an invaluable tool to study vertebrate embryology and development, basic cell and molecular biology, genomics, neurobiology, and toxicology and to model human diseases. FF-OCM images of a *Xenopus laevis* (African frog) tadpole, in the region of the head, are shown in Fig. 1.18. Three sections in orthogonal orientations are presented. Mesenchymal stem cells are revealed with their membrane and nucleus morphology. Highly contrasted tissues appear, such as the epidermis and one of the olfactory tubes.

1.6.2 Cell Biology

The feasibility of identifying cancer cells by measuring the refractive index distribution across a single live cell using FF-OCM was reported in Ref. [104]. Refractive index maps of several living cell lines of normal and cancer cells were constructed and quantitatively analyzed. The experiments showed that cancer cells had a higher refractive index than normal ones.

In Ref. [105], FF-OCM was combined with an optical tweezer for simultaneously holding a moving specimen and imaging it at high resolution using a high-numerical-aperture microscope objective. The optical trapping was achieved by focusing a 1064-nm Q-switched laser beam, whereas a broadband Ti:sapphire femtosecond laser was used as the light source for imaging. Feasibility of the combined system was demonstrated by imaging micron-sized polystyrene beads and a living suspension cell in medium.

1.6.3 *Plant Studies*

FF-OCM has been used for plant imaging at the cellular level [34, 35, 106]. In Ref. [106], FF-OCM was used as a noninvasive imaging technique to follow early events in the establishment of the hypersensitive response in tobacco (*Nicotiana sp.*) induced by the inoculation of the bacterial elicitor harpin protein. A decrease in chloroplast backscattered signal as early as 30 min after harpin infiltration was observed. A simple physical model, which accounted for the structure and distribution of thylakoid membranes, suggested that this loss of scattering could be associated with a modification in the refractive index of the thylakoid membranes. These observations were correlated with a decrease in photosynthesis, emphasizing changes in chloroplast structure as one of the earliest hallmarks of plant hypersensitive cell death.

1.7 Material Characterization

Application of FF-OCM to material characterization and evaluation is treated in Chapter 21 of this book. We present in this section a few other examples of application of the technique in this field.

Monitoring cell profiles in three-dimensional porous scaffolds presents a major challenge in tissue engineering. In Refs. [107, 108], FF-OCM was investigated as an imaging modality to monitor noninvasively structures and cells in engineered tissues. FF-OCM was employed to visualize macrostructural morphology and to delineate the morphology of cells and constructs in a developing in vitro–engineered bone tissue. The results have shown valuable potential for the use of the FF-OCM technology in noninvasive monitoring of cellular activities in three-dimensional developing engineered tissues.

Chemical mechanical polishing (CMP) is a key process for planarization of silicon and AlTiC wafers for semiconductors and magnetic heads. Removal rate of wafer material is directly dependent on the surface roughness of a CMP pad. The evaluation of the structure of the pad surface, under an in situ CMP process, is, however, limited due to the existence of polishing fluids on the pad surface. In Ref. [109], FF-OCM was used to investigate the surface of wet pads. The wet pad surface could be quantitatively characterized in terms of the polishing pad lifetime and also be three-dimensionally visualized. Reasonable polishing span could be evaluated from the surface roughness measurement and the groove depth measurement made by FF-OCM.

In Ref. [110], the use of swept-source FF-OCM was reported for simultaneous tomography and topography of microelectromechanical systems (MEMS) based on silicon integrated circuits. In Ref. [111], the technology was used for tomographic and volumetric reconstruction of composite materials.

Chapter 11 of this book reports on the development of a variant of FF-OCM for measuring surface roughness in real time. The technology is based on continuous scanning white-light interference microscopy used in combination with a high-speed CMOS camera and FPGA-cabled logic processing. On-line results have been demonstrated of the real-time measurement of a CMOS chip and a rough GaN surface after chemical etching being translated laterally at speed of several tens of micrometers per seconds. The capacity of the system has also been demonstrated with the measurement in the change in shape of a drop of liquid drying over a period of more than 8 min. Such a real-time, online measurement system with flexibility in the choice of image size, frame rate, and scan depth opens up new applications in surface metrology for the quantitative characterization of a large variety of time-varying surfaces, such as with transients in MEMS, soft materials, surface chemical reactions, and layer growth.

1.8 Conclusion

FF-OCM was introduced in the early 2000s as a new optical imaging modality for acquiring en face tomographic images of semitransparent samples with micrometer-scale spatial resolution.

FF-OCM can be regarded as an alternative to conventional OCT (or OCM), with full-field illumination of the sample and parallel detection of the interferometric signal using an area camera, thus avoiding any beam scanning. A broadband light source preferentially of low spatial coherence is used such as a halogen lamp or a light-emitting diode (LED) instead of a sophisticated and expensive spatially coherent light source. In addition to its simplicity and robustness, FF-OCM is of particular interest for the imaging spatial resolution it offers. Owing to the broad spectrum of thermal light and the possibility of using relatively high numerical aperture microscope objectives, submicrometer spatial resolution imaging can be achieved, thus outperforming conventional OCT.

Identical to an optical microscope in terms of offered spatial resolution, FF-OCM can produce images of biological tissue that are similar to histology images, with the great advantage of avoiding the long and delicate sample preparation required in histology. FF-OCM has been applied successfully to the imaging of various excised biological tissues with a subcellular-level resolution. Several studies of the clinical value of FF-OCM in identifying and differentiating tumors have been reported. Imaging of ex vivo animal embryos has been performed with FF-OCM for research in developmental biology. The technology has also been applied to the nondestructive characterization and evaluation of various semitransparent materials.

FF-OCM is highly effective in high-resolution imaging, provided that the sample is not moving. However, the technique is usually ill-suited to dynamic imaging. Compared to OCT, FF-OCM suffers from a disadvantageous sensitivity to motion of the sample that could generate artifacts in the images. Despites efforts to solve this problem, high-quality FF-OCM imaging is presently limited to stationary samples. One of the main challenges in the current development of FF-OCM is to make this technique suitable to dynamic imaging. With the constant advances in the technology of image sensors and light sources, one can reasonably think that the current limitation of FF-OCM will be overcome, while maintaining the image quality in terms of resolution and contrast. This progress would considerably enlarge the application domains of FF-OCM, especially in biology and medicine. FF-OCM would then constitute a powerful tool for high-resolution real-time in vivo imaging without a contrast agent, making possible noninvasive in situ histopathology without the need for excision and histological processing of tissues. FF-OCM would find use in a wide range of clinical situations ranging from the visualization of tissue pathology, where excisional biopsies are hazardous or impossible and help reduce significantly the number of unnecessary biopsies, to guiding of surgical procedures.

Acknowledgments

This work would not have been possible without the invaluable contribution of postdoctoral associates and PhD students, including Delphine Sacchet, Houssine Makhlouf, Antoine Federici, Anton Grebenyuk, and Jonas Ogien.

References

1. König, K. (2000). Multiphoton microscopy in life sciences. *J. Microsc.*, **200**, pp. 83–104.
2. Helmchen, F., Denk, W. (2005). Deep tissue two-photon microscopy. *Nat. Methods*, **2**, pp. 932–940.
3. Huang, D., Swanson, E. A., Lin, C. P., Schuman, J. S., Stinson, W. G., Chang, W., Hee, M. R., Flotte, T., Gregory, K., Puliafito, C. A., Fujimoto, J. G. (1991). Optical coherence tomography. *Science*, **254**, pp. 1178–1181.
4. Fujimoto, J. G., Brezinski, M. E., Tearney, G. J., Boppart, S. A., Bouma, B. E., Hee, M. R., Southern, J. F., Swanson, E. A. (1995). Optical biopsy and imaging using optical coherence tomography. *Nat. Med.*, **1**, pp. 970–972.
5. Fercher, A. F. (1996). Optical coherence tomography. *J. Biomed. Opt.*, **1**, pp. 157–173.
6. Tearney, G. J., Bouma, B. E., Boppart, S. A., Golubovic, B., Swanson, E. A., Fujimoto, J. G. (1996). Rapid acquisition of in-vivo biological images by use of optical coherence tomography. *Opt. Lett.*, **21**, pp. 1408–1410.
7. Swanson, E. A., Izatt, J. A., Hee, M. R., Huang, D., Lin, C. P., Schuman, J. S., Puliafito, C. A., Fujimoto, J. G. (1993). In-vivo retinal imaging by optical coherence tomography. *Opt. Lett.*, **18**, pp. 1864–1866.
8. Wojtkowski, M., Leitgeb, R., Kowalczyk, A., Bajraszewski, T., Fercher, A. F. (2002). In-vivo human retinal imaging by Fourier domain optical coherence tomography. *J. Biomed. Opt.*, **7**, pp. 457–463.

9. Hitzenberger, C. K., Trost, P., Lo, P. W., Zhou, Q. (2003). Three-dimensional imaging of the human retina by high-speed optical coherence tomography. *Opt. Express*, **11**, pp. 2753–2761.
10. Nassif, N., Cense, B., Park, B. H., Yun, S. H., Chen, T. C., Bouma, B. E., Tearney, G. J., de Boer, J. F. (2004). In-vivo human retinal imaging by ultrahigh-speed spectral domain optical coherence tomography. *Opt. Lett.*, **29**, pp. 480–482.
11. Izatt, J. A., Hee, M. R., Swanson, E. A., Lin, C. P., Huang, D., Schuman, J. S., Puliafito, C. A., Fujimoto, J. G. (1994). Micrometer-scale resolution imaging of the anterior eye in vivo with optical coherence tomography. *Arch. Ophthalmol.*, **112**, pp. 1584–1589.
12. Trefford, S., Desmond, F. (2008). Optical coherence tomography of the anterior segment. *Ocul. Surf.*, **6**, pp. 117–127.
13. Fujimoto, J. G. (2003). Optical coherence tomography for ultrahigh resolution in vivo imaging. *Nat. Biotechnol.*, **21**, pp. 1361–1367.
14. Zysk, A. M., Nguyen, F. T., Oldenburg, A. L., Marks, D. L., Boppart, S. A. (2007). Optical coherence tomography: a review of clinical development from bench to bedside. *J. Biomed. Opt.*, **12**, 051403.
15. Drexler, W., Morgner, U., Kärtner, F. X., Pitris, C., Boppart, S. A., Li, X. D., Ippen, E. P., Fujimoto, J. G. (1999). In-vivo ultrahigh-resolution optical coherence tomography. *Opt. Lett.*, **24**, pp. 1221–1223.
16. Povazay, B., Bizheva, K., Unterhuber, A., Hermann, B., Sattmann, H., Fercher, A. F., Drexler, W., Apolonski, A., Wadsworth, W. J., Knight, J. C., Russell, P. St. J., Vetterlein, M., Scherzer, E. (2002). Submicrometer axial resolution optical coherence tomography. *Opt. Lett.*, **27**, pp. 1800–1802.
17. Wang, Y., Zhao, Y., Nelson, J. S., Chen, Z., Windeler, R. S. (2003). Ultrahigh-resolution optical coherence tomography by broadband continuum generation from a photonic crystal fiber. *Opt. Lett.*, **28**, pp. 182–184.
18. Lexer, F., Hitzenberger, C. K., Drexler, W., Molebny, S., Sattmann, H., Sticker, M., Fercher, A. F. (1999). Dynamic coherent focus OCT with depth-independent transversal resolution. *J. Mod. Opt.*, **46**, pp. 541–553.
19. Ding, Z., Ren, H., Zhao, Y., Nelson, J. S., Chen, Z. (2002). High-resolution optical coherence tomography over a large depth range with an axicon lens. *Opt. Lett.*, **27**, pp. 243–245.
20. Leitgeb, R. A., Villiger, M., Bachmann, A. H., Steinmann, L., Lasser, T. (2006). Extended focus depth for Fourier domain optical coherence microscopy. *Opt. Lett.*, **31**, pp. 2450–2452.

21. Ralson, T. S., Marks, D. L., Kamalabadi, F., Boppart, S. A. (2005). Deconvolution methods for mitigation of transverse blurring in optical coherence tomography. *IEEE Trans. Image Proc.*, **14**, pp. 1254–1264.
22. Hamilton, D. K., Sheppard, C. J. R. (1982). A confocal interference microscope. *Opt. Acta*, **29**, pp. 1573–1577.
23. Gu, M., Sheppard, C. J. R. (1993). Fiber-optical confocal scanning interference microscopy. *Opt. Commun.*, **100**, pp. 79–86.
24. Izatt, J. A., Hee, M. R., Owen, G. M., Swanson, E. A., Fujimoto, J. G. (1994). Optical coherencemicroscopy in scattering media. *Opt. Lett.*, **19**, pp. 590–592.
25. Podoleanu, A. G., Dobre, G. M., Jackson, D. A. (1998). En-face coherence imaging using galvanometer scanner modulation. *Opt. Lett.*, **23**, pp. 147–149.
26. Kempe, M., Rudolph, J. (1996). Analysis of heterodyne and confocal microscopy. *J. Mod. Opt.*, **43**, pp. 2189–2204.
27. Aguirre, A. D., Hsiung, P., Ko, T. H., Hartl, I., Fujimoto, J. G. (2003). High-resolution optical coherence microscopy for high-speed, in vivo cellular imaging. *Opt. Lett.*, **28**, pp. 2064–2066.
28. Clark, A. L., Gillenwater, A., Alizadeh-Naderi, R., El-Naggar, A. K., Richards-Kortum, R. (2004). Detection and diagnosis of oral neoplasia with an optical coherence microscope. *J. Biomed. Opt.*, **9**, pp. 1271–1280.
29. Izatt, J. A., Kulkarni, M. D., Wang, H. W., Kobayashi, K., Sivak, M. V. (1996). Optical coherence tomography and microscopy in gastrointestinal tissues. *IEEE J. Sel. Top. Quantum Electron.*, **2**, pp. 1017–1028.
30. Aguirre, A.D., Chen, Y., Bryan, B., Mashimo, H., Huang, Q., Connolly, J. L., Fujimoto, J. G. (2010). Cellular resolution *ex vivo* imaging of gastrointestinal tissues with optical coherence microscopy. *J. Biomed. Opt.*, **15**, 016025.
31. Dubois, A., Vabre, L., Boccara, A. C., Beaurepaire, E. (2002). High-resolution full-field optical coherence tomography with a Linnik microscope. *Appl. Opt.*, **41**, pp. 805–812.
32. Vabre, L., Dubois, A., Boccara, A. C. (2002). Thermal-light full-field optical coherence tomography. *Opt. Lett.*, **27**, pp. 530–532.
33. Laude, B., De Martino, A., Drévillon, B., Benattar, L., Schwartz, L. (2002). Full-field optical coherence tomography with thermal light. *Appl. Opt.*, **41**, pp. 6637–6645.

34. Dubois, A., Grieve, K., Moneron, G., Lecaque, R., Vabre, L., Boccara, A. C. (2004). Ultrahigh-resolution full-field optical coherence tomography. *Appl. Opt.*, **43**, pp. 2874–2882.
35. Dubois, A., Moneron, G., Grieve, K., Boccara, A. C. (2004). Three-dimensional cellular-level imaging using full-field optical coherence tomography. *Phys. Med. Biol.*, **49**, pp. 1227–1234.
36. Oh, W. Y., Bouma, B. E., Iftimia, N., Yelin, R., Tearney, G. J. (2006). Spectrally-modulated full-field optical coherence microscopy for ultrahigh-resolution endoscopic imaging. *Opt. Express*, **14**, pp. 8675–8684.
37. Beaurepaire, E., Boccara, A. C., Lebec, M., Blanchot, L., Saint-Jalmes, H. (1998). Full-field optical coherence microscopy. *Opt. Lett.*, **23**, pp. 244–246.
38. Sheng-Hua, L., Chia-Jung, C., Ching-Fen, K. (2013). Full-field optical coherence tomography using immersion Mirau interference microscope. *Appl. Opt.*, **52**, pp. 4400–4403.
39. Chien-Chung, T., Chia-Kai, C., Kuang-Yu, H., Tuan-Shu, H., Ming-Yi, L., Jeng-Wei, T., Sheng-Lung, H. (2014). Full-depth epidermis tomography using a Mirau-based full-field optical coherence tomography. *Biomed. Opt. Express*, **5**, pp. 3001–3010.
40. Karamata, B., Lambelet, P., Laubscher, M., Salathé, R. P., Lasser, T. (2004). Spatially incoherent illumination as a mechanism for crosstalk suppression in wide-field optical coherence tomography. *Opt. Lett.*, **29**, pp. 736–738.
41. Grieve, K., Dubois, A., Simonutti, M., Paques, M., Sahel, J., Gargasson, J. F. L., Boccara, C. (2005). In vivo anterior segment imaging in the rat eye with high speed white light full-field optical coherence tomography. *Opt. Express*, **13**, pp. 6286–6295.
42. Oh, W. Y., Bouma, B. E., Iftimia, N., Yun, S. H., Yelin, R., Tearney, G. J. (2006). Ultrahigh-resolution full-field optical coherence microscopy using InGaAs camera. *Opt. Express*, **14**, pp. 726–735.
43. Moneron, G., Boccara, A. C., Dubois, A. (2005). Stroboscopic ultrahigh-resolution full-field optical coherence tomography. *Opt. Lett.*, **30**, pp. 1351–1353.
44. Schmitt, J. M., Lee, S. L., Yung, K. M. (1997). An optical coherence microscope with enhanced resolving power in thick tissue. *Opt. Commun.*, **142**, pp. 203–207.
45. Zhang, Y., Sato, M., Tanno, N. (2001). Resolution improvement in optical coherence tomography by optimal synthesis of light-emitting diodes. *Opt. Lett.*, **26**, pp. 205–207.

46. Sacchet, D., Moreau, J., Georges, P., Dubois, A. (2009). Multi-band ultrahigh resolution full-field optical coherence tomography. *Proc. SPIE* 7372, Optical Coherence Tomography and Coherence Techniques IV, 73721F.

47. Ogien, J., Dubois, A. (2016). High-resolution full-field optical coherence microscopy using a broadband light-emitting diode. *Opt. Express*, **24**, pp. 9922–9931.

48. Sacchet, D., Brzezinski, M., Moreau, J., Georges, P., Dubois, A. (2010). Motion artifact suppression in full-field optical coherence tomography. *Appl. Opt.*, **49**, pp. 1480–1488.

49. Karamata, B., Lambelet, P., Laubscher, M., Leutenegger, M., Bourquin, Lasser, T. (2005). Multiple scattering in optical coherence tomography. I. Investigation and modeling. *J. Opt. Soc. Am. A,* **22**, pp. 1369–1379.

50. Karamata, B., Lambelet, P., Leutenegger, M., Laubscher, M., Bourquin, S., Lasser, T. (2005). Multiple scattering in optical coherence tomography. II. Experimental and theoretical investigation of cross talk in wide-field optical coherence tomography. *J. Opt. Soc. Am. A*, **22**, pp. 1380–1388.

51. Grebenyuk, A., Federici, A., Ryabukho, V., Dubois, A. (2014). Numerically focused full-field swept-source optical coherence microscopy with low spatial coherence illumination. *Appl. Opt.*, **53**, pp. 1697–1708.

52. Považay, B., Unterhuber, A., Hermann, B., Sattmann, H., Arthaber, H., Drexler, W. (2006). Full-field time-encoded frequency-domain optical coherence tomography. *Opt. Express*, **14**, pp. 7661–7669.

53. Dubois, A., Moneron, G., Boccara, C. (2006). Thermal-light full-field optical coherence tomography in the 1.2 μm wavelength region. *Opt. Commun.*, **266**, pp. 738–743.

54. Sacchet, D., Moreau, J., Georges, P., Dubois, A. (2008). Simultaneous dual-band ultra-high resolution full-field optical coherence tomography. *Opt. Express*, **16**, pp. 19434–19446.

55. Federici, A., Dubois, A., (2014). Three-band, 1.9-μm axial resolution full-field optical coherence microscopy over a 530–1700 nm wavelength range using a single camera. *Opt. Lett.* **39**, pp. 1374–1377.

56. Bourquin, S., Seitz, P., Salathé, R. P. (2001). Optical coherence topography based on a two-dimensional smart detector array. *Opt. Lett.*, **26**, pp. 512–514.

57. Laubscher, M., Ducros, M., Karamata, B., Lasser, T., Salathe, R. (2002). Video-rate three-dimensional optical coherence tomography. *Opt. Express*, **10**, pp. 429–435.

58. Lambelet P. (2011). Parallel optical coherence tomography (pOCT) for industrial 3D inspection. *Proc. SPIE 8082*, Optical Measurement Systems for Industrial Inspection VII, 80820X.
59. Labiau, S., David, G., Gigan, S., Boccara, A. C. (2009). Defocus test and defocus correction in full-field optical coherence tomography. *Opt. Lett.*, **34**, pp. 1576–1578.
60. Caber P. J. (1993). Interferometric profiler for rough surfaces. *Appl. Opt.*, **32**, pp. 3438–3441.
61. Davidson, M., Kaufman, K., Mazor, I., Cohen, F. (1987) Integrated circuit metrology, inspection, and process control. *Proc SPIE*, **775**, 233.
62. Kino, G. S., Chim, S. C. (1990). Mirau correlation microscope. *Appl. Opt.*, **29**, pp. 3775–3783.
63. Roy, M., Svahn, P., Cherel, L., Sheppard, C. J. R. (2002). Geometric phase-shifting for low-coherence interference microscopy. *Opt. Lasers Eng.*, **37**, pp. 631–641.
64. Watanabe, Y., Hayasaka, Y., Sato, M., Tanno, N. (2005). Full-field optical coherence tomography by achromatic phase-shifting with a rotating polarizer. *Appl. Opt.*, **44**, pp. 1387–1392.
65. Ya-Liang, Y., Zhi-Hua, D., Kai, W., Ling, W., Lan, W. (2010). Full-field optical coherence tomography by achromatic phase-shifting with a rotating half-wave plate. *J. Opt.*, **12**, 035301.
66. Sheng-Hua, L., Chien-Yell, W., Cho-Yen, H., Kuan-Yu, C., Hui-Yu, C. (2012). Full-field optical coherence tomography using nematic liquid-crystal phase shifter. *Appl. Opt.*, **51**, pp. 1361–1366.
67. Shoude, C., Xianyiang, C., Costel, F. (2007). An efficient algorithm used for full-field optical coherence tomography. *Opt. Lasers Eng.*, **45**, pp. 1170–1176.
68. Sasaki, O., Okazaki, H., Sakai, M. (1987). Sinusoidal phase modulating interferometer using the integrating-bucket method. *Appl. Opt.*, **26**, pp. 1089–1093.
69. Dubois, A. (2001). Phase-map measurements by interferometry with sinusoidal phase modulation and four integrating-buckets. *J. Opt. Soc. Am. A*, **18**, pp. 1972–1979.
70. Dunsby, C., Gu, Y. French, P. (2003). Single-shot phase-stepped wide-field coherence-gated imaging. *Opt. Express*, **11**, pp. 105–115.
71. Hrebesh, M. S., Dabu, R. Sato, M. (2009). In vivo imaging of dynamic biological specimen by real-time single-shot full-field optical coherence tomography. *Opt. Commun.*, **282**, pp. 674–683.

72. Hrebesh, M. S. (2012). Full-field and single-shot full-field optical coherence tomography: a novel technique for biomedical imaging applications. *Adv. Opt. Technol.*, **2012**, Article ID 435408, 26 pages.
73. Larkin, K. G. (1996). Efficient nonlinear algorithm for envelope detection in white light interferometry*J. Opt. Soc. Am. A*, **13**, pp. 832–843.
74. Chen, Y., Huang, S. W., Aguirre, A. D., Fujimoto, J. G. (2007). High-resolution line-scanning optical coherence microscopy. *Opt. Lett.*, **32**, pp. 1971–1973.
75. Pan, Y. T., Birngruber, R., Engelhardt, R. (1997). Contrast limits of coherence-gated imaging in scattering media. *Appl. Opt.*, **36**, pp. 2979–2983.
76. Yadlowsky, M. J., Schmitt, J. M., Bonner, R. F. (1995). Multiple-scattering in optical coherence microscopy. *Appl. Opt.*, **34**, pp. 5699–5707.
77. Schmitt, J. M., Knüttel, A. (1997). Model of optical coherence tomography of heterogeneous tissue. *J. Opt. Soc. Am. A*, **14**, pp. 1231–1242.
78. Lu, Q., Gan, X. S., Gu, M., Luo, Q. M. (2004). Monte Carlo modeling of optical coherence tomography imaging through turbid media. *Appl. Opt.*, **43**, pp. 1628–1637.
79. Dubois, A., Selb, J., Vabre, L., Boccara, A. C. (2000). Phase measurements with wide-aperture interferometers. *Appl. Opt.*, **39**, pp. 2326–2331.
80. Safrani, A., Abdulhalim, I. (2012). Ultrahigh-resolution full-field optical coherence tomography using spatial coherence gating and quasi-monochromatic illumination. *Opt. Lett.*, **37**, pp. 458–460.
81. Srivastava, V., Nandy, S., Mehta, D. S. (2013). High-resolution full-field optical coherence tomography using a spatially incoherent monochromatic light source. *Appl. Phys. Lett.*, **103**, pp. 103702–103706.
82. Watanabe, Y., Sato, M. (2008). Three-dimensional wide-field optical coherence tomography using an ultrahigh-speed CMOS camera. *Opt. Commun.*, **281**, pp. 1889–1895.
83. Schausberger, S. E., Heise, B., Bernstein, S., Stifter, D. (2012). Full-field optical coherence microscopy with Riesz transform-based demodulation for dynamic imaging. *Opt. Lett.*, **37**, pp. 4937–4939.
84. Grieve, K., Paques, M., Dubois, A., Sahel, J., Boccara, C. Le Gargasson, J.-F. (2004). Ocular tissue imaging using ultrahigh-resolution full-field optical coherence tomography. *Invest. Ophthalmol. Vis. Sci.*, **45**, pp. 4126–4131.

85. Grieve, K., Moneron, G., Dubois, A., Le Gargasson, J.-F., Boccara, C. (2005). Ultrahigh resolution ex-vivo ocular imaging using ultrashort acquisition time en face optical coherence tomography. *J. Opt. A*, **7**, pp. 368–373.
86. Akiba, M., Maeda, N., Yumikake, K., Soma, T., Nishida, K., Tano, Y., Chan, K. P. (2007). Ultrahigh-resolution imaging of human donor cornea using full-field optical coherence tomography. *J. Biomed. Opt.*, **12**, 041202.
87. Latour, G., Georges, G., Siozade-Lamoine, L., Deumie, C., Conrath, J., Hoffart, L. (2010). Human graft cornea and laser incisions imaging with micrometer scale resolution full-field optical coherence tomography. *J. Biomed. Opt.*, **15**, 056006.
88. Auksorius, E., Bromberg, Y., Motiejunaite, R., Pieretti, A., Liu, L., Coron, E., Aranda, J., Goldstein, A. M., Bouma, B. E., Kazlauskas, A., Tearney, G. J. (2012). Dual-modality fluorescence and full-field optical coherence microscopy for biomedical imaging applications. *Biomed. Opt. Express*, **3**, pp. 661–666.
89. Ghouali, W., Grieve, K., Bellefqi, S., Sandali, O., Laroche, L., Harms, F., Paques, M., Borderie, V. (2015). Human donor cornea imaging with full-field optical coherence tomography. *Curr. Eye Res.*, (unpublished).
90. Dalimier, E., Salomon, D. (2012). Full-field optical coherence tomography: a new technology for 3D high-resolution skin imaging. *Dermatology*, **224**, pp. 84–92.
91. Durkin, J. R., Fine, J. L., Sam, H., Pugliano-Mauro, M., Ho, J. (2013). Imaging of Mohs micrographic surgery sections using full-field optical coherence tomography: a pilot study. *Dermatol. Surg.*, **40**, pp. 266–274.
92. Dalimier, E., Bruhat, A., Grieve, K., Harms, F., Martins, F., Boccara, A. C. (2014). High resolution in-vivo imaging of skin with full field optical coherence tomography. *Proc. SPIE*, **8926**, 8926P.
93. De Leeuw, F., Latrive, A., Casiraghi, O., Ferchiou, M., Harms, F, Boccara, C., Laplace-Builhé, C. (2014). Optical biopsy on head and neck tissue using full-field OCT: a pilot study. *Proc. SPIE*, **8926**, 892626.
94. Assayag, A, Antoine, M., Sigal-Zafrani, B., Riben, M., Harms, F., Burcheri, A., Grieve, K., Dalimier, E., Le Conte de Poly, B., Boccara, C. (2014). Large field, high resolution full-field optical coherence tomography: a pre-clinical study of human breast tissue and cancer assessment. *Technol. Cancer Res. Treat.*, **13**, pp. 455–468.
95. Jain, M., Shukla, N., Manzoor, M., Nadolny, S., Mukherjee, S. (2011). Modified full-field optical coherence tomography: a novel tool for rapid histology of tissues. *J. Pathol. Inform.*, **2**, 28.

96. Assayag, O., Grieve, K., Devaux, B., Harms, F., Pallud, J., Chretien, F., Boccara, C., Varlet, P. (2013). Imaging of non tumorous and tumorous human brain tissue with full-field optical coherence tomography. *Neuroimage Clin.*, **2**, pp. 549–557.
97. Beuvon, F., Dalimier, E., Cornud, F., Barry Delongchamps, N. (2014). Full-field optical coherence tomography of prostate biopsies: a step towards pre-histological diagnosis. *Prog. Urol.*, **24**, pp. 22–30.
98. Ramasamy, R., Sterling, J., Manzoor, M., Salamoon, B., Jain, M., Fisher, E., Li, P. S., Schlegel, P. N., Mukherjee, S. (2012). Full field optical coherence tomography can identify spermatogenesis in a rodent sertoli-cell only model. *J. Pathol. Inform.*, **3**, 4.
99. Jain, M., Narula, N., Salamon, B., Shevchuk, M. M., Aggarwal, A., Altorki, N., Stiles, B., Boccara, C., Mukherjee, S. (2013). Full-field optical coherence tomography for the analysis of fresh unstained human lobectomy specimens. *J. Pathol. Inform.*, **4**, 26.
100. Perea-Gomez, A., Camus, A., Moreaua, A., Grieve, K., Moneron, G., Dubois, A., Cibertb, C., Collignona, J. (2004). Initiation of gastrulation in the mouse embryo is preceded by an apparent shift in the orientation of the anterior-posterior axis. *Curr. Biol.*, **14**, pp. 197–207.
101. Zheng, J. G., Lu, D. Y., Chen, T. Y., Wang, C. M., Tian, N., Zhao, F. Y., Huo, T. C., Zhang, N., Chen, D. Y., Ma, W. Y., Sun, J. L., Xue, P. (2012). Label-free subcellular 3D live imaging of preimplantation mouse embryos with full-field optical coherence tomography. *J. Biomed. Opt.*, **17**, 070503.
102. Zheng, J. G., Huo, T. C., Chen, T. Y., Wang, C. M., Zhang, N., Tian, N., Zhao, F. Y., Lu, D. Y., Chen, D. Y., Ma, W. Y., Sun, J. L., Xue, P. (2013). Understanding three-dimensional spatial relationship between the mouse second polar body and first cleavage plane with full-field optical coherence tomography. *J. Biomed. Opt.*, **18**, 010503.
103. Zheng, J. G., Huo, T. C., Tian, N., Chen, T. Y., Wang, C. M., Zhang, N., Zhao, F. Y., Lu, D. Y., Chen, D. Y., Ma, W. Y., Sun, J. L., Xue, P. (2013). Noninvasive three-dimensional live imaging methodology for the spindles at meiosis and mitosis. *J. Biomed. Opt.*, **18**, 050505.
104. Choi, W. J., Jeon, D. I., Ahn, S. G., Yoon, J. H., Kim, S., Lee, B. H. (2010). Full-field optical coherence microscopy for identifying live cancer cells by quantitative measurement of refractive index distribution. *Opt. Express*, **18**, pp. 23285–23295.
105. Choi, W. J., Park, K. S., Eom, T. J., Oh, M. K., Lee, B. H. (2012). Tomographic imaging of a suspending single live cell using optical tweezer-combined full-field optical coherence tomography. *Opt. Lett.*, **37**, pp. 2784–2786.

106. Boccara, M., Schwartz, W., Guiot, E., Vidal, G., De Paepe, R., Dubois, A., Boccara, A. C. (2007). Early chloroplastic alterations analysed by optical coherence tomography during a hairpin-induced hypersensitive response. *Plant J.*, **50**, pp. 338–346.
107. Yang, Y., Wang, R. K., Guyot, E., El Haj, A., Dubois, A. (2005). Application of optical coherence tomography for tissue engineering. *Proc. SPIE*, **5690**, Coherence Domain Optical Methods and Optical Coherence Tomography in Biomedicine IX.
108. Yang, Y., Dubois, A., Qin, X. P., Li, J., El Haj, A., Wang, R. K. (2006). Investigation of optical coherence tomography as an imaging modality in tissue engineering. *Phys. Med. Biol.*, **51**, pp. 1649–1659.
109. Choi, W. J., Jung, S. P., Shin, J. G., Yang, D., Lee, B. H. (2011). Characterization of wet pad surface in chemical mechanical polishing (CMP) process with full-field optical coherence tomography (FF-OCT). *Opt. Express*, **19**, pp. 13343–13350.
110. Tulsi, A., Chandra, S., Dalip Singh, M. (2009). Simultaneous tomography and topography of silicon integrated circuits using full-field swept-source optical coherence tomography. *J. Opt. A: Pure Appl. Opt.*, **11**, 045501.
111. Srivastava, V., Tulsi, A., Madhu, S., Dalip Singh, M. (2012). Tomographic and volumetric reconstruction of composite materials using full-field swept-source optical coherence tomography. *Meas. Sci. Technol.*, **23**, 055203.

PART I
THEORETICAL ASPECTS

Chapter 2

Theory of Imaging and Coherence Effects in Full-Field Optical Coherence Microscopy

Anton A. Grebenyuk[a] and Vladimir P. Ryabukho[a,b]

[a] *Department of Optics and Biophotonics, Saratov State University, Astrakhanskaya ul. 83, Saratov 410012, Russia*
[b] *Institute of Precision Mechanics and Control, Russian Academy of Sciences, Rabochaya ul. 24, Saratov 410028, Russia*
grebenyukaa@yandex.ru

This chapter considers the theory of image formation in full-field optical coherence microscopy (FF-OCM). The equations describing image formation, including main experimental parameters of the FF-OCM setup, are derived. The principle of coherence-gated imaging is further analyzed and the terminology of coherence effects in FF-OCM is discussed. The coherence effects related to longitudinal and transverse resolution are analyzed. Finally, the influence of a sample refractive index on the detection of a coherence signal from the depth of a sample is considered.

Handbook of Full-Field Optical Coherence Microscopy: Technology and Applications
Edited by Arnaud Dubois
Copyright © 2016 Pan Stanford Publishing Pte. Ltd.
ISBN 978-981-4669-16-0 (Hardcover), 978-981-4669-17-7 (eBook)
www.panstanford.com

2.1 Introduction

Optical coherence tomography (OCT) is a 3D imaging technique [1–3] based on the principal of longitudinal sectioning by low-coherence interferometry [4]. OCT imaging has found a lot of applications in biomedical research and clinical diagnostics [2, 3].

Conventional OCT utilizes objectives with relatively small numerical apertures (NAs) and narrow light beams to preserve a sufficient depth of field without displacement of the optical focus position in the sample [5]. In such systems the main physical quantity governing the longitudinal resolution is the temporal spectrum of the optical field. The angular spectrum of the optical field is considered quite narrow and its effect is often neglected in conventional OCT analysis.

Although high longitudinal resolution can be achieved with this approach [5], it has a major drawback—that the low NA objectives don't provide high transverse resolution. The version of OCT utilizing objectives with high NAs is often called optical coherence microscopy (OCM). It has been realized in both confocal [6] and full-field [7–9] modalities.

With moderate and high NAs, OCM gains wide angular spectrum of the optical field and achieves high transverse resolution. However, the wide angular spectrum has also a number of other important effects on the coherence signal [10–13]. Therefore the simplified theoretical analysis from conventional OCT, taking into account only the temporal spectrum, is not applicable to imaging with high NAs. More strict theoretical analysis is necessary for understanding of the coherence effects, selection of necessary setup parameters, adjustment and working with OCM.

The full-field modality of OCM, called full-field optical coherence microscopy (FF-OCM) or full-field optical coherence tomography (FF-OCT), has attracted a lot of attention due to the possibility of parallel en face data acquisition and its potentially faster operation speed [7–9, 14–18]. It is based on the principle of full-field interference microscopy, that is, a combination of the possibilities of full-field optical microscopy and interferometry.

A number of theories have been developed for analyzing the properties of the coherence signal in full-field interference microscopy and particularly in FF-OCM, e.g. [8, 10, 13, 19–21]. They allowed studying various effects, including the influence of the temporal and angular spectra on the longitudinal resolution of the coherence signal [8, 10, 13, 19, 20] and the influence of the sample refractive index on the coherence signal from structures inside the sample [13, 21].

To provide a general theoretical analysis, applicable to various types of optical schemes and imaged samples, we have proposed recently an approach [22, 23] based on careful tracing of the optical fields' propagation in the sample and reference arms and analysis of their interference in the registration plane. This approach allows for analysis on a unified basis of various coherence effects, provides the possibility to distinguish between the illumination and imaging aperture effects on the coherence signal [23] and the possibility of numerical focusing in FF-OCM even with partial spatial coherence of illumination [24].

On the basis of this approach, this chapter describes the principles of image formation in FF-OCM and provides equations for investigation of various FF-OCM modalities (Section 2.2). Using the presented equations, Section 2.3 considers the physical origins and the properties of FF-OCM imaging.

In Section 2.3.1 the influence of the temporal and angular spectra on the longitudinal resolution is considered and the terminology suitable for clear interpretation of the FF-OCM coherence effects is discussed. In Section 2.3.2 the effects of the illumination and imaging apertures on the impulse response of the coherence signal are analyzed. Finally, Section 2.3.3 discusses the influence of sample refractive index on the localization and amplitude of the FF-OCM coherence signal from the depth of a sample.

2.2 Signal Formation in FF-OCM

The majority of FF-OCM systems are based on the principle of the Linnik interference microscope [8, 9, 14–18], which is based on the combination of the Michelson interferometer with two optical microscopes, introduced in the interferometer arms. Although many conclusions discussed in this chapter are applicable to other kinds of interference microscopes, for definiteness we shall discuss them in the Linnik-type FF-OCM, presented in Fig. 2.1.

2.2.1 Principle of the Analysis

The signal of a matrix photodetector consisting of an array of individual detectors with linear response can be considered as an array of sampled intensity values of the optical field in the registration plane. In this chapter we shall assume that the optical fields can be considered stationary and ergodic.

The intensity distribution, resulting from the interference of the fields, formed by the sample and reference arms, can be written in the following form:

$$I(x,y) = I_S(x,y) + I_R(x,y) + \Gamma(x,y) + \Gamma^*(x,y), \quad (2.1a)$$

where $I_S(x,y)$ and $I_R(x,y)$ are the intensities of the optical fields formed by the sample and reference arms, and $\Gamma(x,y)$ is their mutual coherence function;

$$I(x,y) = \frac{1}{2\pi} \int_0^\infty I(\omega;x,y) d\omega, \quad (2.1b)$$

$$\Gamma(x,y) = \frac{1}{2\pi} \int_0^\infty \Gamma(\omega;x,y) d\omega, \quad (2.1c)$$

where [23]

$$I(\omega;x,y) = I_S(\omega;x,y) + I_R(\omega;x,y) + \Gamma(\omega;x,y) + \Gamma^*(\omega;x,y), \quad (2.2)$$

$$\langle V_S(\omega;x,y) V_S^*(\omega';x,y) \rangle = 2\pi\delta(\omega-\omega') I_S(\omega;x,y), \quad (2.3a)$$

$$\langle V_R(\omega;x,y) V_R^*(\omega';x,y) \rangle = 2\pi\delta(\omega-\omega') I_R(\omega;x,y), \quad (2.3b)$$

$$\langle V_S(\omega;x,y) V_R^*(\omega';x,y) \rangle = 2\pi\delta(\omega-\omega') \Gamma(\omega;x,y), \quad (2.3c)$$

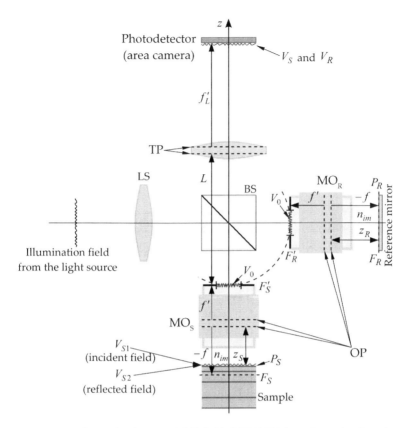

Figure 2.1 General scheme of full-field OCT/OCM based on the Linnik interferometer. Illumination field (usually of low spatial coherence) comes from a light source and enters the lens system, LS. This lens system contains field and illumination aperture diaphragms for shaping the illumination field and can be realized in different ways, for example, for critical or Köhler illumination. MO_S, MO_R, microscope objectives in the sample and reference arms; OP, principal planes of the objectives; F_S and F_R, front focal planes; F'_S and F'_R, back focal planes, where the objectives' apertures are located; f and f', front and back focal distances of the objectives; f'_L, back focal distance of the tube lens; TP, principal planes of the tube lens; L, distance from the objectives' back focal planes to the tube lens front principal plane; BS, beam splitter. Other notations are defined in the text.

ω is the circular temporal frequency, $V_S(\omega; x, y)$ and $V_R(\omega; x, y)$ are proportional to the complex amplitude distributions of the monochromatic components of the optical fields formed, respectively, by the sample and reference arms in the registration plane, and angular brackets denote statistical or time averaging. $I_S(\omega; x, y)$ and $I_R(\omega; x, y)$ are the spectral densities of the sample and reference fields, and $\Gamma(\omega; x, y)$ is their mutual spectral density. Equations 2.1–2.3 are based on the Wiener–Khintchin theorem derivation of Ref. [25], which provides rather strict yet simple analysis.

The functions $V_S(\omega; x, y)$ and $V_R(\omega; x, y)$ are affected by many factors related to the field propagation through the interference microscope. The properties of these factors can be known (e.g., optical components) or unknown (e.g., sample properties) deterministic (the setup parameters) or stochastic (random instantaneous structure of the illumination field).

To clarify this issue, we need to sort out the deterministic and stochastic properties, the known and unknown parameters. The first thing to be done here is to proceed from the mutual spectral density $\Gamma(\omega; x, y)$ of two unknown functions $V_S(\omega; x, y)$ and $V_R(\omega; x, y)$ to some quantity, which can be considered as a known parameter, characteristic of the setup. It can be done by application of the optical transmission function approach, described in Refs. [26, 27]. This means that we need to select some initial surface in the interferometer, with known properties of the optical field over it, and trace the field changes, as the field propagates through the two arms till the registration plane.

In fact, the initial surface is not necessarily to be chosen the same for both arms. Physically different initial planes can be chosen for each arm, provided that the optical field distribution in these planes is identical. As has been shown in a number of works [23, 24, 28], for taking into account the properties of illumination and imaging systems, but in the same time avoiding unnecessary complication of equations, it is convenient to select the aperture planes (back focal planes) of the objectives during illumination (before reflection from the sample and reference mirror) as the initial surfaces (Fig. 2.1).

Considering the illumination process, if we denote by V_0 the illumination field immediately after passing the objective aperture, then its propagation to the objective's front focal plane can be described by means of scalar diffraction analysis (the Fresnel–Kirchhoff integral and the angular spectrum approach in parabolic approximation, and considering lenses as phase modulators) [25, 29, 30], as follows [22]:

$$\tilde{V}_1(\omega;k_x,k_y) \approx -i\lambda\, f' \exp[ik\, f'(1+n_{\mathrm{im}}^2)]\, V_0\left(\omega;-f'\frac{k_x}{k},-f'\frac{k_y}{k}\right), \tag{2.4}$$

where $\lambda = 2\pi c/\omega$ is the wavelength, c is the light speed in vacuum, $k = \omega/c$ is the wavenumber, k_x and k_y are the transversal circular spatial frequencies, n_{im} is the refractive index of immersion, and the tilde sign denotes the transversal spatial spectrum of the respective function, defined as

$$\tilde{V}_1(\omega;k_x,k_y) = \iint V_1(\omega;x_S,y_S)\exp[-i(x_S k_x + y_S k_y)]dx_S dy_S \tag{2.5}$$

(the integration limits are infinite, unless explicitly noted otherwise), $V_1(\omega;x_S,y_S)$ is the complex amplitude distribution of the illumination optical field in the objective's front focal plane.

Considering the imaging process, if we denote by V_f the effective optical field, reflected from the sample, in the objective's front focal plane (the meaning of "effective" will be described below), then the field, incident on the registration plane, can be described as [22]

$$V(\omega;x,y) \approx \mu_i(\omega;x,y) \iint \tilde{V}_f\left(\omega;k\frac{x_3}{f'},k\frac{y_3}{f'}\right) A(\omega;x_3,y_3)$$

$$\times \exp\left[-i\frac{k}{f'_L}(x_3 x + y_3 y)\right] dx_3 dy_3, \tag{2.6a}$$

where $\mu_i(\omega;x,y)$ is a complex coefficient

$$\mu_i(\omega;x,y) = -\frac{\exp\{ik[f'(1+n_{\mathrm{im}}^2)+L+f'_L]\}}{\lambda^2 f' f'_L}$$

$$\times \exp\left[\frac{i\pi}{\lambda f'_L}\left(1-\frac{L}{f'_L}\right)(x^2+y^2)\right]. \tag{2.6b}$$

It can be noted, that in the case of a cube beam splitter (Fig. 2.1), a glass cube is present between the objective and the tube lens in the optical unfolding of either interferometer arm, while Eqs. 2.6 have been derived for the case of free space between the objective and the tube lens. However, it can be shown, that the presence of this glass cube leads to an additional phase modulation of $V(\omega;x,y)$, which is similar for the two arms and nevertheless vanishes in resultant equations for the coherence function and intensities.

Equation 2.4 describes the optical field propagation from the objective's back focal plane to its front focal plane and Eq. 2.6 describes the propagation from the objective's front focal plane to the registration plane. However, the sample of interest in FF-OCM is three-dimensional and may likely be located before this focal plane.

To solve this problem we can do the following: select some plane P_S between the objective and the sample (similarly P_R between the objective and the mirror in the reference arm) and derive such illuminating and reflected fields in this plane, which would correspond to the fields V_l and V_f in the focal plane F_S (see Fig. 2.1).

Being located in the P_S plane, the field V_{S1} defined by

$$\tilde{V}_{S1}(\omega; k_x, k_y) = \tilde{V}_l(\omega; k_x, k_y)\Phi_S(\omega; k_x, k_y), \quad (2.7a)$$

$$\tilde{V}_{Sf}(\omega; k_x, k_y) = \tilde{V}_{S2}(\omega; k_x, k_y)\Phi_S(\omega; k_x, k_y), \quad (2.7b)$$

corresponds to the field V_l in the objective's front focal plane in the absence of a sample (z_S is the distance from the objective's front principal plane to the P_S plane).

Similarly, instead of the field V_{S2}, actually reflected/back-scattered from the sample, we can use

$$\Phi_S(\omega; k_x, k_y) = \exp[i(z_S - |f|)(k^2 n_{im}^2 - k_x^2 - k_y^2)^{1/2}], \quad (2.8)$$

where V_{Sf} corresponds to V_f in Eq. 2.6a, when applied to the sample arm. V_{Sf} is such an optical field, which being located in the objective's front focal plane in the absence of a sample, would produce the same field V_{S2} in the P_S plane, as if the sample was really present.

If the sample is located closer to the objective, than the focal plane F_S, then the field V_{Sf} is a mathematical abstraction, rather than an existing optical field. Therefore we call it an "effective" optical field in the objective's front focal plane.

Such separate treatment of the illumination system, sample, and imaging system may seem complicated at first, but in fact appears to be very convenient, when analyzing various kinds of samples is necessary. Another advantage of this approach is the following: Equations 2.4 and 2.6 for the illuminating and imaging optical fields have been derived, based on the principles of scalar diffraction analysis [25, 29, 30] in parabolic approximation. Applied to these equations, we can assume that the optical system is corrected for aberrations in such a way that the shape of the equations hold when we proceed to higher NAs (some aberrations can still be taken into account with the aperture function $A(\omega; x_3, y_3)$).

An assumption of this kind cannot be applied to the analysis of optical field propagation to out-of-focus regions or through the sample, which has to be done more strictly, like Eqs. 2.7 and 2.8. Here appears the benefit of using the "effective" field and separate analysis of the field propagation through the sample and optical system: the analysis of the whole propagation process at once in parabolic approximation would lead to losing the applicability to high NA analysis. The separate consideration preserves the possibility to analyze relatively high NA systems, with the ultimate limitation being the scalar approximation.

2.2.2 Optical Transmission Functions of the Sample and Reference Arms

To complete the analytic description of the optical transmission function of an interference microscope arm, we need to analyze what happens with the optical field in the object space using the principles described above and Eqs. 2.7 and 2.8.

Let's consider this analysis by the example of the reference arm, where the illumination field is reflected by a simple plain mirror. It is convenient then to select the P_R plane on the mirror surface so that

$$\tilde{V}_{R2}(\omega, k_x, k_y) = r_R(\omega; k_x, k_y)\tilde{V}_{R1}(\omega; k_x, k_y), \quad (2.9)$$

where the functions V_{R1} and V_{R2} are determined with respect to the P_R plane in a similar manner as V_{S1} and V_{S2} with respect to the P_S plane; r_R is the amplitude reflection coefficient of the mirror in reference arm.

Substituting Eq. 2.4 into the equivalent of Eq. 2.7a for the reference arm, using Eqs. 2.9 and 2.7b, and finally Eq. 2.6, the following expression can be obtained for the complex amplitude of the optical field, formed by the reference arm in the registration plane:

$$V_R(\omega;x,y) \approx \mu(\omega;x,y) \iint dx_0 dy_0 V_0(\omega;x_0,y_0) \Phi_R^2\left(\omega;-k\frac{x_0}{f'},-k\frac{y_0}{f'}\right)$$
$$\times r_R\left(\omega;-k\frac{x_0}{f'},-k\frac{y_0}{f'}\right) A(\omega;-x_0,-y_0) \exp[ik(x_0 x + y_0 y)/f_L'],$$
(2.10)

where

$$\mu(\omega;x,y) = \frac{i}{\lambda f_L'} \exp\{ik[2f'(1+n_{im}^2) + L + f_L']\}$$
$$\times \exp\left[\frac{i\pi}{\lambda f_L'}\left(1 - \frac{L}{f_L'}\right)(x^2+y^2)\right] \quad (2.11)$$

and Φ_R is defined similarly to Eq. 2.8 as

$$\Phi_R(\omega;k_x,k_y) = \exp[i(z_R - |f|)(k^2 n_{im}^2 - k_x^2 - k_y^2)^{1/2}], \quad (2.12)$$

z_R is the distance from the front principal plane of the objective in reference arm to the P_R plane on the mirror surface.

Equation 2.10 describes the change of the complex amplitude of the field V_0 as it propagates from the reference objective's aperture plane to the mirror, reflects, and propagates back to the registration plane. Apart from the $V_0(\omega;x_0,y_0)$ function and the integral over (x_0,y_0), Eq. 2.10 represents the optical transmission function of the reference arm from the reference objective's aperture plane to the registration plane [23]. Meaning this, we shall call the expressions of this kind as expressions for transmission functions, though it should be borne in mind that the transmission function itself is only a part of such expression.

In a similar way we can analyze the transmission function for the sample arm. For clarity some model sample should be selected, which would allow for analysis of the coherence effects of interest but be simple enough to provide analytic expressions and clear physical interpretation of these effects (for more complicated samples numerical simulation using the above equations can be applied, however, at the cost of losing the clarity of physical interpretation). For this purpose two types of model samples are usually analyzed in the literature:

(1) Flat mirror [10, 13] or the more general case of a layer [13] or a multilayered structure with plane parallel uniform interfaces,

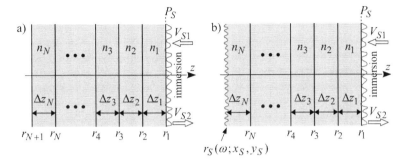

Figure 2.2 Schemes of the multilayered model samples with N layers of thicknesses $\{\Delta z_j\}$ and refractive indices $\{n_j\}$ for analysis: (a) all interfaces are plain (have uniform reflection and transmission coefficients); (b) $(N+1)$th interface has a transversal variation (structure) $r_S(\omega; x_S, y_S)$ of the amplitude reflection coefficient distribution.

perpendicular to the optical axis [21, 31] (Fig. 2.2a). This type of model sample is usually utilized for analysis of the longitudinal resolution of OCT/OCM and the effect of a sample refractive index.

(2) Scattering medium with uniform refractive index with the effect of scattering by outer structures neglected in the illumination field passing to the inner structures [32, 33]. This model sample is usually utilized for analysis of the defocus effects and development of numerical focusing techniques.

These model samples allow analyzing the coherence effects from different points of view and thence both deserve consideration. For versatile analysis of the coherence imaging effects, we shall use both types of model samples in investigation of the signal formation. However, in the second case instead of a medium with uniform refractive index, a multilayered structure will be considered for more generality, with the outer N interfaces being uniform and the $(N+1)$th having transversal variation (structure) of the amplitude reflection coefficient $r_S(\omega; x_S, y_S)$ (Fig. 2.2b) (it is somewhat similar to the case if we would consider all interfaces structured, but neglect the disturbing effect of the outer ones on the signal from the inner ones).

Then, using the angular spectrum approach and applying Eqs. 2.7 and 2.8, we can obtain for the case of the sample with uniform interfaces (Fig. 2.2a)

$$\tilde{V}_{S2}(\omega; k_x, k_y) = \tilde{V}_{S1}(\omega; k_x, k_y) \sum_{m=1}^{N+1} \Re_m(\omega; k_x, k_y)$$

$$\times \exp\left[2i \sum_{j=1}^{m-1} \Delta z_j \sqrt{k^2 n_j^2 - k_x^2 - k_y^2}\right], \quad (2.13a)$$

$$\tilde{V}_{S2}(\omega; k_x, k_y) = (2\pi)^{-2} \prod_{j=1}^{N} t_{j,j-1}(\omega; k_x, k_y)$$

$$\times \exp[i \Delta z_j (k^2 n_j^2 - k_x^2 - k_y^2)^{1/2}]$$

$$\times \iint dk'_x dk'_y \tilde{V}_{S1}(\omega; k'_x, k'_y) \prod_{j=1}^{N} t_{j-1,j}(\omega; k'_x, k'_y)$$

$$\times \exp[i \Delta z_j (k^2 n_j^2 - k'^2_x - k'^2_y)^{1/2}]$$

$$\times \iint r_S(\omega; x_S, y_S)$$

$$\times \exp\{-i[x_S(k_x - k'_x) + y_S(k_y - k'_y)]\} dx_S dy_S,$$

(2.13b)

where r_m is the amplitude reflection coefficient from the mth interface; $t_{j-1,j}$ and $t_{j,j-1}$ are the amplitude transmission coefficients from the $j-1$th layer to the jth and backward. The field V_{S2} here takes into account the signals due to the reflection from all the interfaces, although the re-reflections are neglected.

For the sample with the $(N+1)$th interface having a transversal structure (Fig. 2.2b)

$$\Re_m(\omega; k_x, k_y) = r_m(\omega; k_x, k_y) \prod_{j=1}^{m-1} t_{j-1,j}(\omega; k_x, k_y) t_{j,j-1}(\omega; k_x, k_y).$$

(2.14)

The field V_{S2} in Eq. 2.13b corresponds only to the part of the total field reflected from the sample, corresponding to the reflection from the structured $(N+1)$th interface. This simplification will help us in preserving the clarity of interpretation; however, if necessary, the signals due to the reflection from the other interfaces can as well be included in Eq. 2.13b similarly to Eq. 2.13a.

Joining Eqs. 2.4 and 2.6–2.8 with expression for the V_{S2} field in the case of a sample with uniform interfaces (Eq. 2.13a), we can obtain for complex amplitude of the sample field in the registration plane

$$V_S(\omega;x,y) \approx \mu(\omega;x,y) \sum_{m=1}^{N+1} \iint V_0(\omega;x_0,y_0) A(\omega;-x_0,-y_0)$$

$$\times \Re_m \left(\omega; -k\frac{x_0}{f'}, -k\frac{y_0}{f'}\right)$$

$$\times \exp\left[2ik \sum_{j=0}^{m-1} \Delta z_j \left(n_j^2 - \frac{x_0^2 + y_0^2}{f'^2}\right)^{1/2}\right]$$

$$\times \exp\left[i\frac{k}{f'_L}(x_0 x + y_0 y)\right] dx_0 dy_0, \qquad (2.15a)$$

where for shortness the following notations have been used: $z_S - |f| = \Delta z_0$, $n_{im} = n_0$.

Similarly, using the expression for the V_{S2} field in the case of a sample having a transversal structure (Eq. 2.13b), the following expression can be obtained for the complex amplitude of the sample field in the registration plane [24]

$$V'_S(\omega;x,y) \approx \frac{\mu(\omega;x,y)}{(\lambda f')^2} \iint dx_S dy_S r_S(\omega;x_S,y_S) \iint A(\omega;x_3,y_3)$$

$$\times T_2\left(\omega; k\frac{x_3}{f'}, k\frac{y_3}{f'}\right) \exp\left\{-ik\left[x_3\left(\frac{x}{f'_L} + \frac{x_S}{f'}\right)\right.\right.$$

$$\left.\left. + y_3\left(\frac{y}{f'_L} + \frac{y_S}{f'}\right)\right]\right\} dx_3 dy_3$$

$$\times \iint V_0(\omega;x_0,y_0) T_1\left(\omega; -k\frac{x_0}{f'}, -k\frac{y_0}{f'}\right)$$

$$\times \exp[-ik(x_0 x_S + y_0 y_S)/f'] dx_0 dy_0, \qquad (2.15b)$$

where

$$T_1(\omega;k_x,k_y) = \prod_{j=1}^{N} t_{j-1,j}(\omega;k_x,k_y) \prod_{j=0}^{N} \exp[i\Delta z_j (k^2 n_j^2 - k_x^2 - k_y^2)^{1/2}],$$
$$(2.16a)$$

$$T_2(\omega;k_x,k_y) = \prod_{j=1}^{N} t_{j,j-1}(\omega;k_x,k_y) \prod_{j=0}^{N} \exp[i\Delta z_j (k^2 n_j^2 - k_x^2 - k_y^2)^{1/2}].$$
$$(2.16b)$$

Having derived the optical transmission functions for the reference arm (Eq. 2.10) and for the sample arm (in two different cases) (Eq. 2.15), we know how the optical field is being changed when propagating from the initial plane to the registration plane through the two arms. To complete the signal analysis we need to proceed from the stochastic distributions $V_0(\omega; x_0, y_0)$ to some deterministic function—the autocorrelation function $<V_0(\omega; x_0, y_0) V_0^*(\omega'; x_0', y_0')>$.

2.2.3 Resultant Signal of FF-OCM

When an optical field is incident on a lens, the transversal spatial coherence length of this field in the focal plane before the lens is inversely related to the illuminated area size in the focal plane after the lens [34]. In other words, the more spatially incoherent the illuminating field is before the lens, the larger the area that is illuminated after the lens. Therefore the autocorrelation function $<V_0(\omega; x_0, y_0) V_0^*(\omega'; x_0', y_0')>$ (cross-spectral density of the illuminating field V_0) determines the size of the illuminated part of the field of view. Thence we can use the following approximate expression:

$$\langle V_0(\omega; x_0, y_0) V_0^*(\omega'; x_0', y_0') \rangle \approx 2\pi \delta(\omega - \omega')$$
$$\times I_0(\omega; x_0, y_0) \delta(x_0 - x_0') \delta(y_0 - y_0'), \quad (2.17)$$

if the whole part of the field of view, captured by the matrix photodetector, is uniformly illuminated. Equation 2.17 represents an assumption of total transversal spatial incoherence of the illuminating optical field in the objectives' aperture planes. It is usual to be done in analysis of full-field low-coherence interference microscopes (though not always explicitly noted) [8, 10, 13, 20, 21, 23, 24, 31].

Substituting Eqs. 2.10 and 2.15a into Eq. 2.3c and applying Eq. 2.17, the following expression can be obtained for the mutual spectral density $\Gamma(\omega; x, y)$ in the case of imaging a layered sample with uniform interfaces (in Fig. 2.2a):

$$\Gamma(\omega;x,y) = |\mu(\omega;x,y)|^2 \sum_{m=1}^{N+1} \iint dx_0 dy_0 I_{0m}(\omega;x_0,y_0)$$

$$\times \exp\left\{2ik\left[(z_S - z_R)\left(n_{im}^2 - \frac{x_0^2 + y_0^2}{f'^2}\right)^{1/2}\right.\right.$$

$$\left.\left. + \sum_{j=1}^{m-1} \Delta z_j \left(n_j^2 - \frac{x_0^2 + y_0^2}{f'^2}\right)^{1/2}\right]\right\}, \quad (2.18a)$$

$$\Gamma'(\omega;x,y) \approx \mu_0(\omega) \iint dx_S dy_S r_S(\omega;x_S, y_S)$$

$$\times \iint A_2(\omega;x_3,y_3) \exp\left\{-ik\left[x_3\left(\frac{x}{f'_L} + \frac{x_S}{f'}\right)\right.\right.$$

$$\left.\left. + y_3\left(\frac{y}{f'_L} + \frac{y_S}{f'}\right)\right]\right\} dx_3 dy_3$$

$$\times \iint A_1(\omega;x_0,y_0) \exp\left\{-ik\left[x_0\left(\frac{x}{f'_L} + \frac{x_S}{f'}\right)\right.\right.$$

$$\left.\left. + y_0\left(\frac{y}{f'_L} + \frac{y_S}{f'}\right)\right]\right\} dx_0 dy_0. \quad (2.18b)$$

It can be noticed that the $|\mu(\omega;x,y)|^2$ here is constant over (x,y) and thence $\Gamma(\omega;x,y) = \Gamma(\omega)$; similarly $|\mu(\omega;x,y)|^2$ will also be henceforth denoted as $|\mu(\omega)|^2$.

Similarly using Eqs. 2.10, 2.15b, 2.3c, and 2.17 the following expression can be obtained for the mutual spectral density $\Gamma'(\omega;x,y)$ in the case of imaging the sample with transversal structure (in Fig. 2.2b) [24]

$$I_{0m}(\omega;x_0,y_0) = I_0(\omega;x_0,y_0)|A(\omega;-x_0,-y_0)|^2$$

$$\times \Re_m\left(\omega;-k\frac{x_0}{f'},-k\frac{y_0}{f'}\right) r_R^*\left(\omega;-k\frac{x_0}{f'},-k\frac{y_0}{f'}\right), \quad (2.19)$$

where

$$\mu_0(\omega) = \frac{|\mu(\omega;x,y)|^2}{(\lambda f')^2} = \lambda^{-4} f'^{-2} f'^{-2}_L, \quad (2.20)$$

$$A_1(\omega;x_0,y_0) = I_0(\omega;x_0,y_0) A^*(\omega;-x_0,-y_0) r_R^*\left(\omega;-k\frac{x_0}{f'},-k\frac{y_0}{f'}\right)$$

$$\times T_1\left(\omega;-k\frac{x_0}{f'},-k\frac{y_0}{f'}\right)\left[\Phi_R^*\left(\omega;-k\frac{x_0}{f'},-k\frac{y_0}{f'}\right)\right]^2, \quad (2.21a)$$

$$A_2(\omega;x_3,y_3) = A(\omega;x_3,y_3)\mathrm{T}_2\left(\omega;k\frac{x_3}{f'},k\frac{y_3}{f'}\right). \quad (2.21\mathrm{b})$$

Finally the mutual coherence function $\Gamma(x,y)$ can be calculated from the mutual spectral density $\Gamma(\omega;x,y)$ determined by Eq. 2.18a or 2.18b by integration over the temporal spectrum (Eq. 2.1c).

The derived expressions for the transmission functions (Eq. 2.10 and Eq. 2.15) together with Eqs. 2.1–2.3 and 2.17 allow for analysis in a similar manner of the spectral densities $I_S(\omega;x,y)$ and $I_R(\omega;x,y)$, intensities formed by the sample and reference arms $I_S(x,y)$ and $I_R(x,y)$, and finally the total intensity $I(x,y)$. However, for FF-OCM imaging the main quantity of interest is the mutual coherence function $\Gamma(x,y)$ and so it will be analyzed henceforth.

To facilitate the analysis of coherence effects in FF-OCM imaging in the next section, a useful representation of the expressions for $\Gamma'(\omega;x,y)$ will be presented here. For understanding of the resolution effects in FF-OCM it's helpful to consider the transversal spatial spectrum of the mutual spectral density in the case of sample with transversal structure [24]:

$$\tilde{\Gamma}'(\omega;k_x,k_y) \approx \mu_0(\omega)M^2\left[\prod_{j=1}^{N}t_{j-1,j}(\omega)t_{j,j-1}(\omega)\right]$$

$$\times \tilde{r}_S(\omega;-Mk_x,-Mk_y)\,\Xi(\omega;k_x,k_y), \quad (2.22\mathrm{a})$$

where

$$\Xi(\omega;k_x,k_y) = \iint dx_S dy_S \exp[-iM(k_x x_S + k_y y_S)]$$

$$\times \iint A(\omega;x_3,y_3)\exp\left[ik\sum_{j=0}^{N}\Delta z_j\sqrt{n_j^2 - \frac{x_3^2+y_3^2}{f'^2}}\right]$$

$$\times \exp[-i\frac{k}{f'}(x_3 x_S + y_3 y_S)]dx_3 dy_3$$

$$\times \iint A_i(\omega;x_0,y_0)\exp\left\{ik\left[(z_S - 2z_R + |f|)\sqrt{n_0^2 - \frac{x_0^2+y_0^2}{f'^2}}\right.\right.$$

$$\left.\left.+\sum_{j=1}^{N}\Delta z_j\sqrt{n_j^2 - \frac{x_0^2+y_0^2}{f'^2}}\right]\right\}$$

$$\times \exp[-i\frac{k}{f'}(x_0 x_S + y_0 y_S)]dx_0 dy_0, \quad (2.22\mathrm{b})$$

$$A_i(\omega; x_0, y_0) = I_0(\omega; x_0, y_0) A^*(\omega; -x_0, -y_0) r_R^* \left(\omega; -k\frac{x_0}{f'}, -k\frac{y_0}{f'}\right),$$
(2.22c)

the angular dependence of the transmission coefficients of the outer interfaces $t_{j-1,j}$ and $t_{j,j-1}$ has been neglected in Eq. 2.22.

The function $\Xi(\omega; k_x, k_y)$ plays a very important role in the image formation process—it acts as a varying coefficient for different temporal and spatial frequencies.

It can be noticed that in Eq. 2.22b the function $A_i(\omega; x_0, y_0)$ has a very similar role to the aperture function $A(\omega; x_3, y_3)$; however, A_i is determined by the properties of the illumination, rather than the imaging system. Therefore it is convenient to call it an *illumination aperture function* [23, 24]. The importance of this illumination aperture and its role in FF-OCM imaging will be discussed in the next section.

FF-OCM can be realized in time domain [7–9, 14–16, 18] as well as in Fourier domain (with frequency swept sources) [17, 24, 32, 33, 35, 36]. The approach to theoretical analysis described in this section allows for analysis of various effects related to image formation in FF-OCM and interplay of different components of the interference signal in both time domain and Fourier domain FF-OCM.

2.3 Coherence Effects in FF-OCM

Section 2.2 presented the equations allowing for analysis of the total FF-OCM signal and particularly of the mutual coherence function of the sample and reference arms' fields. This mutual coherence function has the property of longitudinal sectioning of the signal from different sample depths, as will be discussed below. It is therefore the most important part of the FF-OCM signal and is extracted from the latter to provide longitudinally sectioned (coherence-gated) 3D imaging.

Time domain FF-OCM provides the total coherence function $\Gamma(x, y)$. Fourier domain FF-OCM gives access to the mutual spectral density $\Gamma(\omega; x, y)$ at many temporal frequencies ω, which is then processed to produce the total coherence function with modified properties. Therefore it is more informative to consider the mutual spectral density in the first place. These results are readily applicable to Fourier domain and quasi-monochromatic time domain FF-OCM and if necessary can be easily integrated over ω to provide the total $\Gamma(x, y)$ for analysis of broadband time domain FF-OCM.

The coherence function $\Gamma(x, y)$ or the mutual spectral density $\Gamma(\omega; x, y)$ can be reconstructed from the interference signal in a number of ways, including the phase-shifting approach, the off-axis approach, or Fourier filtration in temporal spectrum domain. For generality, we shall consider henceforth the properties of the mutual spectral density (or the mutual coherence function) itself, without concern of the particular way it has been reconstructed from the raw interferometric data.

2.3.1 Principle of Coherence Gating

Using the results of Section 2.2, we can consider the principles of longitudinal sectioning of signals from different sample depths, provided by the mutual coherence function of the sample and reference arms' fields.

2.3.1.1 Sample with uniform interfaces

Let's start with the case of a layered sample with uniform interfaces (Fig. 2.2a), described by Eq. 2.18a. For the sake of interpretation, it is convenient to proceed in Eq. 2.18a to parabolic approximation, which yields

$$\Gamma(\omega) \approx |\mu(\omega)|^2 \sum_{m=1}^{N+1} \exp\{i\omega \Delta t_m\}$$

$$\times \iint I_{0m}(\omega; x_0, y_0) \exp\left\{-ik\left[\frac{z_S - z_R}{n_{\text{im}}} + \sum_{j=1}^{m-1} \frac{\Delta z_j}{n_j}\right] \frac{x_0^2 + y_0^2}{f'^2}\right\} dx_0 dy_0,$$

(2.23a)

where Δt_m denotes the following quantity:

$$\Delta t_m = \frac{2}{c}\left[n_{\text{im}}(z_S - z_R) + \sum_{j=1}^{m-1} n_j \Delta z_j\right].$$
(2.23b)

Increase of the value $n_{\text{im}}(z_S - z_R) + \sum_{j=1}^{m-1} n_j \Delta z_j$ leads to stronger linear phase modulation $\exp\{i\omega\Delta t_m\}$ of $\Gamma(\omega)$. The total coherence function Γ is an integral of $\Gamma(\omega)$ over the temporal frequencies ω and is the smaller, the stronger is the phase modulation $\exp\{i\omega\Delta t_m\}$. Minimal (absent) phase modulation $\exp\{i\omega\Delta t_m\}$ provides the biggest $|\Gamma|$, corresponding to $\Delta t_m = 0$. This means the presence of longitudinal sectioning in the coherence signal. When performing a longitudinal A-scan (scanning $z_S - z_R$), the positions of the individual signals, governed by the phase modulation $\exp\{i\omega\Delta t_m\}$, from each mth interface are determined by the following equations:

$$z_S - z_R = -\sum_{j=1}^{m-1} \Delta z_j \frac{n_j}{n_{\text{im}}}. \tag{2.24a}$$

The strength of this sectioning effect is determined by the width of the temporal spectrum of the field: the wider is the temporal spectrum, the stronger sectioning effect is produced by the linear phase modulation $\exp\{i\omega\Delta t_m\}$. Besides, Δt_m allows for a simple physical interpretation as the temporal delay between the reference wave field and the wave field reflected from the mth interface in the sample arm. Therefore in conventional OCT with low NAs this sectioning effect is often termed as related to the time delay between the interfering fields [2]. On the other hand, this Δt_m arises just from the difference of the sample and reference arms' transmission functions, without introducing in the derivations any temporal delays between the sample and reference arms' fields. This difference becomes significant, when higher NAs are utilized and another kind of sectioning comes into effect.

This other kind of sectioning is related to the other phase modulation and integral in $\Gamma(\omega)$. This is the phase modulation over (x_0, y_0), affecting $I_{0m}(\omega; x_0, y_0)$. Increase of this phase modulation leads to attenuation of the integral over (x_0, y_0) and thence of $\Gamma(\omega)$, and this attenuation is the stronger, the larger is the illuminated part of the aperture $I_0(\omega; x_0, y_0)$. The position of the individual signal from each mth interface in the A-scan is determined by the minimal value of this phase modulation when $(z_S - z_R)/n_{\text{im}} + \sum_{j=1}^{m-1} \Delta z_j/n_j = 0$, that is,

$$z_S - z_R = -\sum_{j=1}^{m-1} \Delta z_j \frac{n_{\text{im}}}{n_j}. \tag{2.24b}$$

This sectioning effect is related to the angular spectrum of the optical field and does not provide a simple interpretation as related to a time delay. The Eq. 2.24b is similar to the one obtained previously with a different theoretical approach in [21].

Figure 2.3 illustrates the sectioning effects related to the temporal and angular spectra and their interplay in the resultant coherence signal. The graphs in Fig. 2.3 have been obtained using Eqs. 2.18a and 2.1c, where the sample was considered as

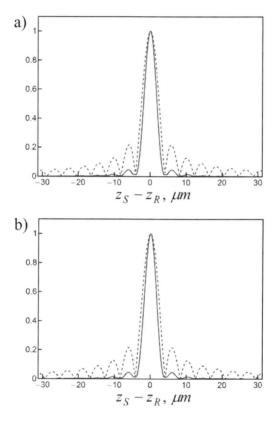

Figure 2.3 Absolute value of the coherence function Γ (normalized to unity maximum) when scanning $z_S - z_R$ in the case of a sample consisting of a single reflecting surface. The dashed line corresponds to $\Delta\omega = 2.3 \times 10^{14}$ rad/s ($\Delta\lambda \approx 60$ nm), NA $= 0.05$ in (a) and $\Delta\omega = 3.83 \times 10^{12}$ rad/s ($\Delta\lambda \approx 1$ nm), NA $= 0.4$ in (b). The solid line in both images corresponds to $\Delta\omega = 2.3 \times 10^{14}$ rad/s, NA $= 0.4$.

consisting of a single reflecting surface, the temporal spectrum of the illumination is rectangular, centered at $\omega_0 = 2.69 \times 10^{15}$ rad/s (corresponding to $\lambda_0 \approx 700$ nm), the objective aperture is uniformly filled with illumination. (The graphs presented in Figs. 2.3–2.8 have been computed using home-made programs in Wolfram Mathematica software.)

It can be seen from comparison of Figs. 2.3a and 2.3b that a rather similar sectioning effect can be achieved when using illumination with wide temporal spectrum and small angular aperture (Fig. 2.3a) or narrow temporal spectrum and large angular aperture (Fig. 2.3b). The sectioning effect becomes even stronger when illumination with both wide temporal spectrum and large angular aperture is used.

Theoretical and experimental studies of these coherence effects related to the temporal and angular spectra for different kinds of interferometer arrangements and experimental conditions can be found in e.g. [8, 10–13, 18–21, 23, 31, 37–41].

It can be noted that the sectioning effect, related to the angular spectrum, is mostly determined by $I_0(\omega; x_0, y_0)$—the illuminated part of the objective aperture, almost regardless of the actual aperture size. This property has led to the use of the concept of an effective NA in some works [18, 40].

2.3.1.2 Sample with transversal structure

Now let's consider the more general case of a layered sample having an interface with a transversal structure (Fig. 2.2b) using Eq. 2.22 for the transversal spatial spectrum of the mutual spectral density. The function $\Xi(\omega; k_x, k_y)$ can be represented in the following form [24]:

$$\Xi(\omega; k_x, k_y) = \Xi_t(\omega)\Xi_a(\omega; k_x, k_y), \qquad (2.25a)$$

where

$$\Xi_t(\omega) = \exp\{2ik\,[n_{\text{im}}(z_S - z_R) + \sum_{j=1}^{N} n_j \Delta z_j]\}, \qquad (2.25b)$$

$$\Xi_a(\omega; k_x, k_y) = \Xi(\omega; k_x, k_y)\Xi_t^*(\omega). \qquad (2.25c)$$

For the clarity of interpretation, let's consider the parabolic approximation in $\Xi_a(\omega; k_x, k_y)$:

$$\Xi_a(\omega; k_x, k_y) \approx \iint dx_S dy_S \exp[-iM(k_x x_S + k_y y_S)]$$

$$\times \iint A(\omega; x_3, y_3) \exp\left[-i\frac{k}{2}\sum_{j=0}^{N}\frac{\Delta z_j}{n_j}\frac{x_3^2 + y_3^2}{f'^2}\right]$$

$$\exp[-i\frac{k}{f'}(x_3 x_S + y_3 y_S)]dx_3 dy_3$$

$$\times \iint A_i(\omega; x_0, y_0) \exp\left\{-i\frac{k}{2}\left[\frac{z_S - 2z_R + |f|}{n_0}\right.\right.$$

$$\left.\left. + \sum_{j=1}^{N}\frac{\Delta z_j}{n_j}\right]\frac{x_0^2 + y_0^2}{f'^2}\right\}$$

$$\times \exp[-i\frac{k}{f'}(x_0 x_S + y_0 y_S)]dx_0 dy_0. \qquad (2.26)$$

The function $\Xi_t(\omega)$, producing linear phase modulation over the temporal frequencies ω, leads to the presence of a longitudinal sectioning effect related to the temporal spectrum, similarly to the previously considered case of the sample with uniform interfaces.

As for the sectioning effect, related to the angular spectrum, the situation becomes sufficiently different. In the case of a sample having a transversal structure, this sectioning effect is determined by $\Xi_a(\omega; k_x, k_y)$, which comprises two integrals: over the imaging aperture function $A(\omega; x_3, y_3)$ and over the illumination aperture function $A_i(\omega; x_0, y_0)$. These aperture functions are phase-modulated: the stronger are the phase modulations, the weaker is the respective integral and hence $\Xi(\omega; k_x, k_y)$. However, unlike the case of the sample with uniform interfaces, the properties of these phase modulations are dependent on the focus position, comprising $\Delta z_0 = z_S - |f|$ for the integral over $A(\omega; x_3, y_3)$, and $z_S - 2z_R + |f|$ for the integral over $A_i(\omega; x_0, y_0)$. Moreover, these phase modulations are not identical: one independent and the other dependent on the reference mirror position z_R.

For specific applications, different kinds of interplay between these phase modulations can be useful. However, in general two important conclusions can be made:

(1) When performing an A-scan in FF-OCT/FF-OCM with objectives, which NA size is not negligible, it is not indifferent whether to make the scanning by shifting the sample or the reference mirror. For correct performance, the sample should be shifted in the process of an A-scan.
(2) If the size of the numerical aperture of illumination (NAi) is not negligible, the surface of the reference mirror should be placed in the focus of the respective objective, so that $z_R = |f|$ and the phase modulations over $A(\omega; x_3, y_3)$ and $A_i(\omega; x_0, y_0)$ are identical.

Thus, in the more general case of imaging a sample with transversal structure, both imaging and illumination aperture functions have important effects on the coherence signal and its properties can't be described with a single effective aperture. The influence of the illumination and imaging aperture functions on the shape of the coherence signal will be discussed in the next section.

Although $\Xi_a(\omega; k_x, k_y)$ is dependent on ω, its longitudinal sectioning properties are determined mostly by the transversal spatial frequencies (k_x, k_y). Therefore its sectioning effect can be termed as the longitudinal coherence gating related to transversal spatial spectrum. Since for a given ω, the values of (k_x, k_y) determine a certain angle, this sectioning effect may be also termed as longitudinal coherence gating related to the angular spectrum.

To distinguish the coherence effects related to the temporal and angular spectra, different terminological approaches are used in the literature [8, 10–13, 20, 21, 23, 40]. The coherence gating effect related to the temporal spectrum is often termed as effect of temporal coherence, while the coherence gating effect related to the angular spectrum is often termed as effect of spatial coherence [12, 13, 20, 21, 40]. This terminology has several advantages, being simple and compatible with the conventional (low-NA) OCT theory.

On the other hand both the effects related to the temporal and angular spectra are produced by the differences in the spatial configuration of the two interferometer arms (not temporal delays) and affect the same function—mutual coherence function of the sample and reference arms' fields. Therefore to avoid opposing these effects as related to temporal and spatial coherence, another terminological approach can be used: term them as "longitudinal coherence gating due to the temporal spectrum" and "longitudinal coherence gating due to the angular spectrum" or, in a simplified way, "temporal spectrum gating" and "angular spectrum gating."

Both these terminological approaches have certain advantages and require further investigation. For the avoidance of doubt in this chapter we shall use the latter approach.

2.3.2 Impulse Response and Illumination Aperture Function

As has been shown in Section 2.3.1, the properties of the coherence signal related to the angular spectrum are dependent on whether we consider the signal from a uniform reflecting interface or an interface having a transversal structure. To investigate further the effects related to the angular spectrum, in this section we shall consider the mutual spectral densities. In the case of quasi-monochromatic illumination $\Delta\omega \to 0$ a mutual spectral density can be considered proportional to the respective mutual coherence function.

The mutual spectral density $\Gamma'(\omega; x, y)$ in the case of the sample being a point scatterer ($N = 0$, $r_S(\omega; x_S, y_S) = r_S(\omega)\delta(x_S)\delta(y_S)$) can be termed as the impulse response (point-spread function, PSF) for quasi-monochromatic illumination, and after scaling to the object space coordinates can be denoted as $\Gamma'_{PSF}(\omega; x/M, y/M) = \Gamma'_{PSF}(\omega; x_S, y_S)$.

Figure 2.4 illustrates the difference between the mutual spectral density signal from a single uniform reflecting surface (mirror), calculated according to Eq. 2.18a, and the signal from a point scatterer in the center of the scatterer's image $\Gamma'_{PSF}(\omega; 0, 0)$, which corresponds to the longitudinal change of the impulse response for quasi-monochromatic illumination (it is calculated according to Eq. 2.18b). As discussed in the previous section, the reference mirror is located in the focus of the respective objective for correct performance, so that $z_S - z_R = z_S - |f| = \Delta z_0$.

For the signal from a uniform reflecting surface (dash line) the angular spectrum gating effect is determined by the illumination aperture size: when NAi is close to zero (Fig. 2.4a), there is no

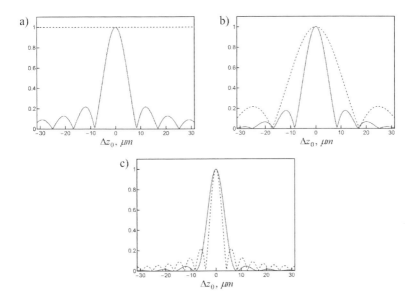

Figure 2.4 Absolute value of the normalized (to unity maximum) mutual spectral density (corresponding to coherence function in the case of quasi-monochromatic illumination) when scanning a uniform reflecting surface (mirror) (dash line) and a point scatterer (continuous line). $\omega = 2.69 \times 10^{15}$ rad/s, NA = 0.4; (a) NAi = 0.005, (b) NAi = 0.2, and (c) NAi = 0.4.

sectioning effect present. However, when considering reflection from a point scatterer, the signal is dependent on the variation of Δz_0 even with extremely small NAi.

This effect can be explained in the following way: NAi close to zero corresponds to almost spatially coherent (plane wave) illumination, which does not provide angular spectrum gating. But the presence of some transversal structure in the sample leads to scattering of the illuminating field, so that it fills to some extent the imaging aperture, even though the illumination aperture was very small. This means broadening of the angular spectrum of the reflected wavefield, which is the stronger, the smaller is the size of the scattering structures. Therefore the sharpening of the mutual spectral density signal with increase of NAi in the case of imaging a point scatterer is not as dramatic as in the case of imaging a plain reflecting surface.

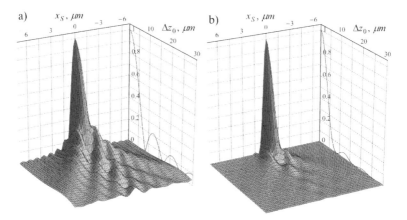

Figure 2.5 3D plots present the evolution of the absolute value of the normalized $\Gamma'_{PSF}(\omega; x_S, 0)$ with defocus Δz_0; 2D plots on the right in each image present the evolution of its central part $|\Gamma'_{PSF}(\omega; 0, 0)|$. (a) NAi = 0.005; (b) NAi = 0.4. For both images NA = 0.4, $\omega = 2.69 \times 10^{15}$ rad/s.

On the other hand it can be noted, that the decrease of the signal from a point scatterer with an increase of $|\Delta z_0|$ in case of low NAi (continuous line in Fig. 2.4a) does not mean that the information about the defocus regions is completely lost in the signal, since these graphs present only the signal in the center of the scatterer's image $\Gamma'_{PSF}(\omega; 0, 0)$.

For more detailed analysis Fig. 2.5 presents 3D graphs illustrating the evolution of the mutual spectral density signal from a point scatterer (impulse response for quasi-monochromatic illumination $\Gamma'_{PSF}(\omega; x_S, y_S)$) with the defocus Δz_0 (the 2D plots in Figs. 2.5a and 2.5b are similar to the continuous line plots in Figs. 2.4a and 2.4c respectively).

It can be seen from Fig. 2.5a that in the case of narrow NAi, the quasi-monochromatic impulse response is rather spread over the transverse coordinates (blurred) with defocus than lost completely. Application of appropriate numerical focusing algorithms allows for recovery of the sharpness and even amplitude of the signal [24, 42]. With increase of NAi, the signal gets not only blurred with defocus but loses the energy (Fig. 2.5b). Even if appropriate numerical correction is applied, the signal amplitude cannot be completely restored [24].

The size and shape of the illumination aperture have also an important effect on the transversal shape of impulse response [24]. Figure 2.6a presents the absolute value of the focused $\Gamma'_{PSF}(\omega; x_S, 0)$ (with $\Delta z_0 = 0$) for a fixed imaging aperture and different sizes of the illumination aperture, while Fig. 2.6b presents the respective $|\Xi_a(\omega; k_x, 0)|$ distributions (for the considered circular imaging and

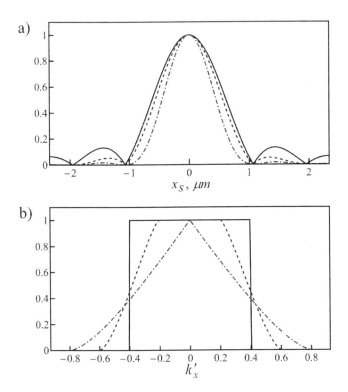

Figure 2.6 (a) Absolute value of the normalized $\Gamma'_{PSF}(\omega; x_S, 0)$ versus x_S; (b) absolute value of the normalized $\Xi_a(\omega; k_x, 0)$ function (which determines transmittance of different transversal spatial frequencies) versus the normalized transversal spatial frequency $k'_x = M \times k_x/k$. NA = 0.4, $\omega = 2.69 \times 10^{15}$ rad/s. The continuous line corresponds to NAi = 0.005, the dashed line corresponds to NAi = 0.2, and the point-dash line corresponds to NAi = 0.4.

illumination apertures both $\Gamma'_{PSF}(\omega;x_S,y_S)$ and $\Xi_a(\omega;k_x,k_y)$ have circular symmetry).

For the considered circular imaging and illumination apertures, the region of nonzero values of $\Xi_a(\omega;k_x,k_y)$ is determined in the normalized spatial frequencies by the relation $(k_x'^2 + k_y'^2)^{1/2} \leq NA + NAi$. For very narrow illumination aperture $NAi = 0.005$ and hence quasi spatially coherent illumination over the sample, $|\Xi_a(\omega;k_x,0)|$ is different from zero for k_x' within approximately $-NA$ and NA (the continuous line in Fig. 2.6b). Increase of the illumination aperture leads to broadening and smoothing of the $\Xi_a(\omega;k_x,k_y)$ function. With increase of NAi the broadening of $\Xi_a(\omega;k_x,k_y)$ leads to sharpening of the impulse response, while the smoothing of $\Xi_a(\omega;k_x,k_y)$ leads to suppression of the side lobes of the impulse response (Fig. 2.6a). Both these effects are important for enhancement of the FF-OCM imaging quality.

2.3.3 Influence of the Sample Refractive Index

As can be seen from Eq. 2.24, the refractive indices of the sample layers and immersion medium have different effects on the position of the coherence signal in the longitudinal A-scan, when the coherence signal is determined by the temporal or the angular spectrum (the refractive indices of the sample layers appear in the numerator and denominator respectively).

In this section we shall consider this question in more detail using the model of a layered sample with uniform interfaces (Fig. 2.2a) with Eqs. 2.18a and 2.1c. To make the general conclusions applicable also to the case of a sample with transversal structure, we shall assume that $z_R = |f|$ and hence $z_S - z_R = z_S - |f| = \Delta z_0$.

Figure 2.7 presents the coherence function, corresponding to an A-scan of a sample with 3 layers with layers thicknesses $\Delta z_1 = 30$ μm, $\Delta z_2 = 20$ μm, and $\Delta z_3 = 40$ μm and refractive indices $n_1 = 1.3$, $n_2 = 1.5$, and $n_3 = 1.3$. The refractive index of immersion $n_{im} = n_0 = 1$, refractive index of the substrate $n_{sub} = 1$. The actual positions of the interfaces in the sample are denoted by vertical dash lines. The angular dependence of the reflection and transmission coefficients of different interfaces has been neglected in the computations.

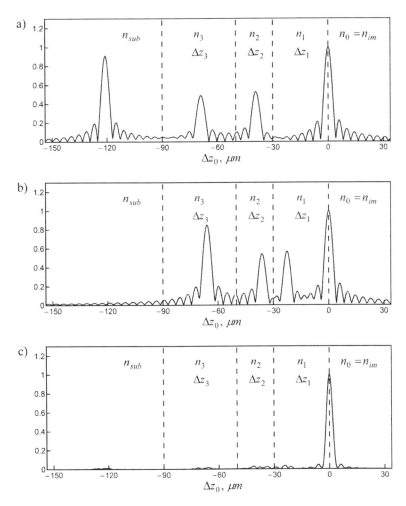

Figure 2.7 Absolute value of the normalized coherence function in an A-scan of a sample with three parallel layers: (a) $\Delta\omega = 2.3 \times 10^{14}$ rad/s, NAi = 0.05; (b) $\Delta\omega = 3.03 \times 10^{17}$ rad/s, NAi = 0.4; and (c) $\Delta\omega = 2.3 \times 10^{14}$ rad/s, NAi = 0.4. For all images NA = 0.4 and $\omega_0 = 2.69 \times 10^{15}$ rad/s.

When the temporal spectrum is broad and the angular spectrum is narrow (Fig. 2.7a), so that the temporal spectrum gating effect is predominant, the signals from the individual interfaces are located further from the sample surface in the A-scan in accordance with Eq. 2.24a. In the opposite case of predominant influence of the angular spectrum (Fig. 2.7b), the signals from the individual interfaces are located closer to the sample surface in accordance with Eq. 2.24b. When both wide temporal and angular spectra are utilized, this mismatch of the temporal and angular spectrum gates positions leads to attenuation and loss of the signal from the inner sample structures (Fig. 2.7c).

These effects of coherence signal localization and fading have been studied theoretically and experimentally in a number of works [13, 15, 18, 21, 23, 24, 31, 37, 42, 43]. Several solutions for overcoming the negative fading effect have been demonstrated, including using an immersion medium matching the sample refractive index [14], dynamic adjustment of the focus position [15], adjustment of the length of the reference arm [43], or using quasi-monochromatic illumination with only angular spectrum gating [18]. This effect can also be overcome by appropriate numerical processing in Fourier domain (swept-source) FF-OCM [24, 42].

Figure 2.8 illustrates one of the possible solutions of this problem by using an immersion medium with a refractive index matching the refractive index of the sample. This solution can be helpful if the refractive index does not change significantly over the imaged depth in the sample.

The model sample has the same layers thickness as in Fig. 2.7; the refractive indices are $n_1 = 1.3$, $n_2 = 1.35$, $n_3 = 1.33$, and $n_\text{sub} = 1.35$. With $n_\text{im} = 1$, the mismatch of the positions of the temporal and angular spectrum gates leads to loss of the signal from the inner sample structures similarly to Fig. 2.7c. However, when the immersion refractive index approximately matches that of the sample layers, the signal from the depth of the sample can be recovered, at least to some extent (the better is the refractive index matching, the better is the signal recovery).

Also the positions of the signals due to the resultant coherence gate in the case of matched refractive indices of the sample and immersion medium correspond to the actual positions of the sample interfaces (though, again, the degree of the positions matching is subject to the degree of matching the refractive indices).

Finally, it can be noted that Eq. 2.24b (where we should in general set $z_R = |f|$) is approximate, as it has been obtained in the parabolic approximation. If the sample and the immersion refractive indices

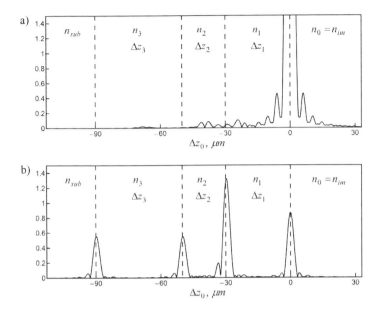

Figure 2.8 Absolute value of the coherence function in an A-scan of a sample with three parallel layers; $\omega_0 = 2.69 \times 10^{15}$ rad/s, $\Delta\omega = 2.3 \times 10^{14}$ rad/s, NA = 0.4, NAi = 0.4. (a) $n_{im} = 1$; (b) $n_{im} = 1.33$.

are not matched, for higher NAs the signal from the depth of the sample, related to the angular spectrum gate, can be aberrated and its position deviating from that determined by Eq. 2.24b.

2.4 Conclusions

This chapter considered the theory of image formation in a full-field optical coherence microscope based on the Linnik interferometer. The derived equations allow for analysis of the total intensity signal $I(x, y)$ as well as its components, of which the mutual coherence function of the sample and reference fields $\Gamma(x, y)$ is of primary interest. Two general model samples have been considered, both consisting of parallel layers with different refractive indices. One of the considered samples has only uniform interfaces and the other represents a more general case when one of the interfaces has a transversal variation (structure) of the reflection coefficient.

Since the signal of interest (the coherence function) is determined by the differences in the spatial configuration of the sample and the reference arms, rather than any temporal delays, it is convenient to term the coherence gating effects related to the temporal and angular spectra as "longitudinal coherence gating due to the temporal spectrum" and "longitudinal coherence gating due to the angular spectrum" or, in a simplified way, "temporal spectrum gating" and "angular spectrum gating."

In the coherence signal of FF-OCM an important role belongs not only to the aperture function of the microscope objectives (imaging aperture function) but also to the illumination aperture function, which is determined mostly by the intensity distribution of the illumination field over the objective aperture (Eq. 2.22c). This illumination aperture function affects both the longitudinal and the transversal shape of the impulse response. In particular, the higher the numerical aperture of illumination, the higher transversal spatial frequencies take part in formation of the coherence signal, which leads to improvement of the transverse resolution by sharpening the impulse response and decreasing its side lobes.

An important fact should be noticed hereby: the impulse response of the coherence function $\Gamma(x, y)$ is not similar to the impulse response of the intensity of the sample field $I_S(x, y)$. Therefore, the transverse resolution provided by the interference term—the coherence function—may in general be different from the resolution provided by the intensity signal in the same interference microscope.

Some notes can be made regarding the range of applicability of the utilized theoretical model. Even if the microscope optics is corrected for aberrations well enough, so that the model is applicable to relatively high NA imaging, it should be borne in mind that the model is based on the scalar approximation, which ultimately limits the applicability to high NA imaging.

In this chapter the interference microscope signal has been considered as simply proportional to the resultant intensity $I(x, y)$. However, strictly speaking, the sampling process is affected by the finite size of the photosensitive area of the individual pixels of the matrix photodetector. If this effect is significant in particular experimental conditions, it can be taken into account by integrating the resultant intensity $I(x, y)$ determined by Eq. 2.1a or 2.1b over each pixel's photosensitive area. Also, if necessary, the spectral sensitivity of the photodetector and transmittance of the optical elements can be taken into account by inclusion of respective ω-dependent factor in the integrand of Eqs. 2.1b and 2.1c.

An important area of FF-OCM analysis which has almost not been discussed in this chapter is the numerical processing of the FF-OCM signal for numerically focused 3D imaging. Different theoretical analyses and approaches to numerically focused imaging in FF-OCM can be found e.g. in [24, 32, 33, 35, 36, 42, 44, 45].

Finally, it should be noted that the range of applications of the described theoretical approach and the presented equations is not limited to the considered examples and they can be applied, for example, to Fourier domain FF-OCM imaging (using a frequency-swept source) and investigation of the possibilities of numerical correction of the FF-OCM signal (e.g., for the defocus effects), even with partial spatial coherence of illumination, as has been shown recently in Ref. [24].

Acknowledgments

We thank Dmitry Lyakin (Institute of Precision Mechanics and Control of the Russian Academy of Sciences, Saratov) and Vladislav Lychagov (Department of Optics and Biophotonics, Saratov State University, Saratov) for helpful discussions and fruitful collaboration on optical coherence effects.

We are also grateful to Arnaud Dubois and Antoine Federici (Laboratoire Charles Fabry, Institut d'Optique Graduate School, Paris) for fruitful discussions and collaboration on optical coherence microscopy.

References

1. Huang, D., Swanson, E. A., Lin, C. P., Schuman, J. S., Stinson, W. G., Chang, W., Hee, M. R., Flotte, T., Gregory, K., Puliafito, C. A., and Fujimoto, J. G. (1991). Optical coherence tomography. *Science*, **254**, pp. 1178–1181.

2. Drexler, W., and Fujimoto, J. G., eds. (2008). *Optical Coherence Tomography* (Springer, Berlin, Heidelberg, New York).
3. Tuchin, V. V., ed. (2013). *Handbook of Coherent-Domain Optical Methods*, 2nd Ed. (Springer, New York, Heidelberg, Dordrecht, London).
4. Ivanov, A. P., Chaikovskii, A. P., Kumeisha, A. A., and Shcherbakov, V. N. (1978). Interferometric study of the spatial structure of a light-scattering medium. *J. Appl. Spectrosc.*, **28**, pp. 359–364.
5. Fujimoto, J., and Drexler, W. (2008). *Optical Coherence Tomography*, eds. Drexler, W., and Fujimoto, J. G., Chapter 1: Introduction to optical coherence tomography (Springer, Berlin, Heidelberg, New York), pp. 1–45.
6. Aguirre, A. D., and Fujimoto, J. G. (2008). *Optical Coherence Tomography*, eds. Drexler, W., and Fujimoto, J. G., Chapter 17: Optical coherence microscopy (Springer, Berlin, Heidelberg, New York), pp. 505–542.
7. Beaurepaire, E., Boccara, A. C., Lebec, M., Blanchot, L., and Saint-Jalmes, H. (1998). Full-field optical coherence microscopy. *Opt. Lett.*, **23**, pp. 244–246.
8. Dubois, A., Vabre, L., Boccara, A. C., and Beaurepaire, E. (2002). High-resolution full-field optical coherence tomography with a Linnik microscope. *Appl. Opt.*, **41**, pp. 805–812.
9. Vabre, L., Dubois, A., and Boccara, A. C. (2002). Thermal-light full-field optical coherence tomography. *Opt. Lett.*, **27**, pp. 530–532.
10. Kino, G. S., and Chim, S. S. C. (1990). Mirau correlation microscope. *Appl. Opt.*, **29**, pp. 3775–3783.
11. Rosen, J., and Yariv, A. (1995). Longitudinal partial coherence of optical radiation. *Opt. Commun.*, **117**, pp. 8–12.
12. Ryabukho, V., Lyakin, D., and Lobachev, M. (2004). Influence of longitudinal spatial coherence on the signal of a scanning interferometer. *Opt. Lett.*, **29**, pp. 667–669.
13. De Groot, P., and de Lega, X. C. (2004). Signal modeling for low-coherence height-scanning interference microscopy. *Appl. Opt.*, **43**, pp. 4821–4830.
14. Dubois, A., Grieve, K., Moneron, G., Lecaque, R., Vabre, L., and Boccara, C. (2004). Ultrahigh-resolution full-field optical coherence tomography. *Appl. Opt.*, **43**, pp. 2874–2883.
15. Dubois, A., Moneron, G., and Boccara, C. (2006) Thermal-light full-field optical coherence tomography in the 1.2 μm wavelength region. *Opt. Commun.*, **266**, pp. 738–743.

16. Makhlouf, H., Perronet, K., Dupuis, G., Lévêque-Fort, S., and Dubois, A. (2012). Simultaneous optically sectioned fluorescence and optical coherence microscopy with full-field illumination. *Opt. Lett.*, **37**, pp. 1613–1615.
17. Povazay, B., Unterhuber, A., Hermann, B., Sattmann, H., Arthaber, H., and Drexler, W. (2006). Full-field time-encoded frequency-domain optical coherence tomography. *Opt. Express*, **14**, pp. 7661–7669.
18. Safrani, A., and Abdulhalim, I. (2012). Ultrahigh-resolution full-field optical coherence tomography using spatial coherence gating and quasi-monochromatic illumination. *Opt. Lett.*, **37**, pp. 458–460.
19. Abdulhalim, I. (2001). Theory for double beam interference microscopes with coherence effects and verification using the Linnik microscope. *J. Mod. Opt.*, **48**, pp. 279–302.
20. Zeylikovich, I. (2008). Short coherence length produced by a spatial incoherent source applied for the linnik-type interferometer. *Appl. Opt.*, **47**, pp. 2171–2177.
21. Safrani, A., and Abdulhalim, I. (2011). Spatial coherence effect on layer thickness determination in narrowband full-field optical coherence tomography. *Appl. Opt.*, **50**, pp. 3021–3027.
22. Grebenyuk, A. A., and Ryabukho, V. P. (2012). Theoretical model of volumetric objects imaging in a microscope. *Proc. SPIE*, **8430**, 84301B.
23. Grebenyuk, A. A., and Ryabukho, V. P. (2012). Coherence effects of thick objects imaging in interference microscopy. *Proc. SPIE*, **8427**, 84271M.
24. Grebenyuk, A., Federici, A., Ryabukho, V., and Dubois, A. (2014). Numerically focused full-field swept-source optical coherence microscopy with low spatial coherence illumination. *Appl. Opt.*, **53**, pp. 1697–1708.
25. Mandel, L., and Wolf, E. (1995). *Optical Coherence and Quantum Optics* (Cambridge University Press).
26. Wolf, E. (1954). A macroscopic theory of interference and diffraction of light from finite sources I. Fields with a narrow spectral range. *Proc. Roy. Soc. A*, **225**, pp. 96–111.
27. Wolf, E. (1955). A macroscopic theory of interference and diffraction of light from finite sources II. Fields with a spectral range of arbitrary width. *Proc. Roy. Soc. A*, **230**, pp. 246–265.
28. Grebenyuk, A. A., and Ryabukho, V. P. (2013). Numerical reconstruction of 3D image in Fourier domain confocal optical coherence microscopy. *Proc. Int. Conf. Adv. Laser Technol. ALT'2012*, Bern Open Publishing, pp. 1–5.

29. Born, M., and Wolf, E. (1968). *Principles of Optics*, 4th Ed. (Pergamon Press).
30. Goodman, J. W. (1996). *Introduction to Fourier Optics*, 2nd Ed. (McGraw-Hill).
31. Grebenyuk, A. A., and Ryabukho, V. P. (2012). Theoretical analysis of stratified media imaging in low-coherence interference microscopy. *Proc. SPIE*, **8337**, 833707.
32. Yu, L. F., and Kim, M. K. (2005). Wavelength-scanning digital interference holography for tomographic three-dimensional imaging by use of the angular spectrum method. *Opt. Lett.*, **30**, pp. 2092–2094.
33. Marks, D. L., Ralston, T. S., Boppart, S. A., and Carney, P. S. (2007). Inverse scattering for frequency-scanned full-field optical coherence tomography. *J. Opt. Soc. Am. A*, **24**, pp. 1034–1041.
34. Goodman, J. W. (1985). *Statistical Optics* (John Wiley and Sons, New York, Chichester, Brisbane, Toronto, Singapore).
35. Kim, M. K. (1999). Wavelength-scanning digital interference holography for optical section imaging. *Opt. Lett.*, **24**, pp. 1693–1695.
36. Montfort, F., Colomb, T., Charriere, F., Kuhn, J., Marquet, P., Cuche, E., Herminjard, S., and Depeursinge, C. (2006). Submicrometer optical tomography by multiple-wavelength digital holographic microscopy. *Appl. Opt.*, **45**, pp. 8209–8217.
37. Ryabukho, V., Lyakin, D., and Lobachev, M. (2005). Longitudinal pure spatial coherence of a light field with wide frequency and angular spectra. *Opt. Lett.*, **30**, pp. 224–226.
38. Ryabukho, V. P., Lyakin, D. V., and Lychagov, V. V. (2009). Longitudinal coherence length of an optical field. *Opt. Spectrosc.*, **107**, pp. 282–287.
39. Ryabukho, V. P., Lyakin, D. V., Grebenyuk, A. A., and Klykov, S. S. (2013). Wiener-Khintchin theorem for spatial coherence of optical wave field. *J. Opt.*, **15**, 025405.
40. Abdulhalim, I. (2006). Competence between spatial and temporal coherence in full field optical coherence tomography and interference microscopy. *J. Opt. A: Pure Appl. Opt.*, **8**, pp. 952–958.
41. Abdulhalim, I. (2012). Spatial and temporal coherence effects in interference microscopy and full-field optical coherence tomography. *Ann. Phys. (Berlin)*, **524**, pp. 787–804.
42. Grebenyuk, A. A., and Ryabukho, V. P. (2012). Numerical correction of coherence gate in full-field swept-source interference microscopy. *Opt. Lett.*, **37**, pp. 2529–2531.

43. Labiau, S., David, G., Gigan, S., and Boccara, A. C. (2009). Defocus test and defocus correction in full-field optical coherence tomography. *Opt. Lett.*, **34**, pp. 1576–1578.
44. Marks, D. L., Davis, B. J., Boppart, S. A., and Carney, P. S. (2009). Partially coherent illumination in full-field interferometric synthetic aperture microscopy. *J. Opt. Soc. Am. A*, **26**, pp. 376–386.
45. Kumar, A., Drexler, W., and Leitgeb, R. A. (2013). Subaperture correlation based digital adaptive optics for full-field optical coherence tomography. *Opt. Express*, **21**, pp. 10850–10866.

Chapter 3

Spatiotemporal Coherence Effects in Full-Field Optical Coherence Tomography

Ibrahim Abdulhalim

Department of Electro-optic Engineering and the Ilse Katz Institute for Nanoscale Science and Technology, Ben Gurion University of the Negev, Beer Sheva 84105, Israel
abdulhlm@bgu.ac.il

3.1 Introduction

Interference of light fields is the basis for many optical measurement techniques, some of which use the large coherence length of lasers, but some also benefit from the finite coherence length of wideband sources. With the emergence of sophisticated methodologies that improve the signal-to-noise ratio, such as common path interferometers, superquiet light sources, phase-shifting devices, and wide-dynamic-range detectors, interferometry can provide axial measurements with precision approaching 1 pm or even better. This fact is being utilized in the large interferometers for detecting gravitational waves such the laser interferometer gravitational wave observatory (LIGO). In optical imaging, interference has been utilized successfully in several well-known classical microscopy techniques

---

Handbook of Full-Field Optical Coherence Microscopy: Technology and Applications
Edited by Arnaud Dubois
Copyright © 2016 Pan Stanford Publishing Pte. Ltd.
ISBN 978-981-4669-16-0 (Hardcover), 978-981-4669-17-7 (eBook)
www.panstanford.com

such as the Zernike phase contrast microscope and the Nomarski differential interference contrast microscope. In the context of finite coherence techniques optical coherence microscopy (OCM) has been known for long time in the context of interference microscopy [1–8], coherence radar [9, 10], and white light interferometry [11–16]. Today OCM is a general term for the optical microscopy techniques that are based on relatively short coherence lengths of the light and can be classified into two types, both acting as a 3D imaging tool. The first is low-temporal-coherence (TC) microscopy and macroscopy, also known as optical coherence tomography (OCT), which is being used for medical diagnostics, particularly in ophthalmology and dermatology [17–20]. The second is full-field OCT (FF-OCT), in which imaging is done both in the reference and sample paths using lenses or microscope objectives [21–25]. Several modes were used such as heterodyne detection [26], wide field configuration [27], phase shift technique [28], and a stroboscopic mode [29]. Applications vary from imaging of the eye [30] and other scattering tissue [31–33]. FF-OCT uses low spatial coherence (SC) and TC similar to the well-known coherence probe microscopy (CPM) that has been in use for a long time in optical metrology [1–8]. "CPM" was the term coined for this technique by the optical metrology division of KLA-Tencor (a company based in Santa Clara, CA, USA) and it has many advantages over conventional microscopy or conventional interferometry in its ability to discriminate between different transparent layers in a multilayered stack. A very-low-coherence microscope (Linnik microscope) that uses high numerical aperture (NA) objectives and broadband white light has been in use for many years within the metrology tools [34] of KLA-Tencor [5, 35–37] used for the inspection of the fabrication processes of microelectronic devices, particularly for autofocus purposes, the measurement of the critical dimension (CD), and the stepper misalignment errors between lithographically overlaid layers. Theoretical modeling of the interferogram from such microscopes has been established recently [36] based on scalar imaging theory, taking into account coherence effects following the treatment by Hopkins [38, 39].

Historically, CPM was of interest mainly to the optical metrology community [1–5] for applications such as step height measurements and contrast-enhanced imaging of transparent objects. The whole subject has attracted wide interest in the last two decades, particularly accelerated by the entrance of OCT as a noninvasive powerful technique for biomedical imaging [17]. However, OCT has been considered so far basically as low-temporal-coherence interferometry and only recently some attention was dedicated to the importance of SC effects [32, 40, 41]. The high signal-to-noise ratios obtained by the different OCT techniques allowed imaging of scattering media such as tissue in spite of the very small signals available. This fact has revolutionized the whole subject of OCM because historically the main applications in optical metrology were imaging of nonscattering layers or surface profiling. Recently [42], it was shown that OCT that uses a fiber interferometer is actually an interference confocal microscope and this fact distinguishes it from the conventional CPM or FF-OCT, which images a whole field at once rather than a single point as it is done with confocal microscopes. The distinction between the two techniques is somehow misleading as many researchers now refer to any double-beam OCM as an OCT system. Coherence effects and interplay between SC and TC are subjects of interest [36, 43–45] because they help to understand the physics behind the behavior of low-coherence interference microscopy (LCIM) systems. In recent works [36, 45] the present author concentrated on the distinction between the SC versus TC effects, showing that in fact each plays a different role in the OCT system. It was shown that the coherence region is determined by the TC only when the path length difference is scanned in a region where the reference and sample beams are perfectly longitudinally spatially coherent (collimated). Longitudinal spatial coherence (LSC) takes an effect when the path length scan is performed with a noncollimated beam such as a defocus scan of a lens where the focal depth of the lens determines the main coherence region. For the latter case we can point out to the importance of the SC in determining the fringe size, showing that it is smaller than half the wavelength by a correction factor, which is a function of the LSC length. Knowledge of the fringe size is important for calibration as the scan distance is usually measured in units of the fringe size. It is also of importance in phase shift interferometric techniques, so-called phase OCT now. Deriving analytic solutions also helps in understanding the physics behind.

3.2 Basic Coherence Concepts

To distinguish between TC and LSC we consider first the wave function of a harmonic wave propagating along $\vec{r}$: $\exp(i\omega t - i\vec{k}.\vec{r})$. The frequencies ω, $\vec{k}$ are the temporal and spatial frequencies; each defines the frequency of oscillation of the wave in time or in space along its propagation direction, respectively. If the wave contains only one spatial frequency (collimated beam, for example, as shown in Fig. 3.1a), TC becomes finite when there is a collection of temporal frequencies $\{\omega_j\}_{j=1-N}$ so that at a certain position $r = 0$ the total wave function is given by the superposition $\sum_{j=1}^{N} a_j \exp(i\omega_j t)$ where we assumed that all the waves have the same phase. When $N \to \infty$, this leads to a wave packet localized in time with a finite TC length. Physically one should consider what a detector measures, meaning the ensemble average, which is the autocorrelation function that leads to the similar conclusion of Wiener–Khinchin theorem which states that the TC function is given by the Fourier transform of the power spectrum of the source. Similarly if the wave is quasi-monochromatic ($\Delta\omega \ll \omega_0$) it can be considered approximately *at least formally* as containing only one temporal frequency but a collection of spatial frequencies $\{\vec{k}_j\}_{j=1-N}$, for example, by having a beam with an angular extent (Fig. 3.1b), then the wave function at $t = 0$ (or at certain moment

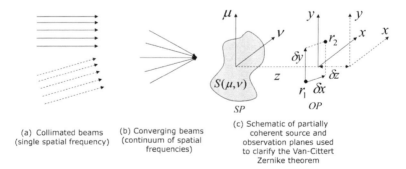

(a) Collimated beams (single spatial frequency) (b) Converging beams (continuum of spatial frequencies) (c) Schematic of partially coherent source and observation planes used to clarify the Van-Cittert Zernike theorem

Figure 3.1 Examples of beams with single spatial frequency (collimated), (b) converging beam containing a continuum of spatial frequencies, and (c) illustration of the Van Cittert–Zernike theorem of the lateral spatial coherence function of a partially coherent source.

of time) is given by the superposition $\sum_{j=1}^{N} b_j \exp(i\vec{k}_j.\vec{r})$. This again leads to a wave packet localized in space with finite LSC length. The LSC can be distinguished from the lateral SC, which results from the fact that the source is extended rather than a point source. The extended source by definition gives rise to many spatial frequencies; however, throughout the chapter the lateral SC of the source is ignored. It will be shown later that although the field is extended, each point in the sample field experiences interference with its conjugate point in the reference field within the Airy disc of the objective lens. This is the reason why when considering a single point the full-field interference microscope behaves similar to a confocal microscope. In fact it is an extension of confocal microscopy to the case of an extended field rather than a single point.

For sources with a continuous spectrum the correlation function becomes the Fourier transform of the frequency spectrum of the source, which leads to the coherence function of the source. For the case of TC this is known as the Wiener–Khinchin theorem:

$$\gamma_{so}(\tau) = \int_{-\infty}^{\infty} S(f) \exp(i 2\pi f \tau) df, \qquad (3.1)$$

where $\gamma_{so}(\tau)$ is the TC function of the source and $S(f)$ is the source spectral distribution function (power spectral density). For the case of lateral SC the equivalent theorem is the Van Cittert–Zernike theorem, which relates the SC function to the spatial frequency spectrum of the source:

$$\gamma_{so}(r_1, r_2) = \int_{-\infty}^{\infty} S(\mu, \nu) \exp\left(i \frac{2\pi}{\lambda z}\right) (\delta x \mu + \delta y \nu) d\mu d\nu, \qquad (3.2)$$

where $\gamma_{so}(r_1, r_2)$ is the SC function of the source and $S(\mu, \nu)$ is the source spatial distribution function, with (μ, ν) being the coordinates on the source plane, z is the distance of the observation plane OP from the source plane SP, and (r_1, r_2) are the two point coordinates on the OP separated in x and y by δx, δy, respectively (Fig. 3.1c). Note that the ratios $(2\pi\mu/\lambda z, 2\pi\nu/\lambda z)$ define the two spatial frequencies in x and y (k_x, k_y) in the OP at distance z from the source. Hence Eq. 3.2 can be converted into an integral over the spatial frequencies or the angular distribution of the source power density. Based on this concept the Michelson stellar interferometer measures the angular size of stars.

The LSC is distinguished from the lateral SC in that only one point is taken in the OP, meaning $\delta x = 0, \delta y = 0$, but the correlation between two points on two planes separated by $z_2 - z_1 = \delta z$ is considered. The integration over all the longitudinal spatial frequencies k_z will then lead to a localized wave packet defining the coherence function in the longitudinal direction. Generalization of the Van Cittert–Zernike theorem to the 3D case was formulated in several works [46–50]; however, the interpretation in terms of the spatial frequencies was considered in our papers [35–37, 45] and the papers by Ryabukho et al. [51, 52]. In FF-OCT both TC and LSC play a role and it is the main focus of this chapter to elucidate the effect of longitudinal coherence on the performance of such systems. Our intention is to distinguish between the two cases and highlight the advantages and disadvantages of SC and TC in the context of FF-OCT.

Based on this discussion we can conclude that the lateral and LSC functions maybe written formally as

$$\gamma_{\text{Lat}}(r_1, r_2) = \int_{-\infty}^{\infty} S(k_x, k_y) \exp(i(k_x \delta x + k_y \delta y)) dk_x dk_y, \quad (3.3)$$

$$\gamma_{\text{Long}}(z_1, z_2) = \int_{-\infty}^{\infty} S(k_z) \exp(i k_z \delta z) dk_z. \quad (3.4)$$

These equations are consistent with the expressions that appear in many textbooks and papers when the correct expression for the spatial frequencies (k_x, k_y, k_z) is introduced. Equation 3.3 is simply the 2D Van Cittert–Zernike theorem, while Eq. 3.4 is the extension to the third dimension. Note that as these represent the Fourier transform of the source spatial distribution, the widths of the coherence functions in the three directions are inversely proportional to the corresponding spatial frequency extents, that is, $(2\pi/\Delta k_x, 2\pi/\Delta k_y, 2\pi/\Delta k_z)$, which represent approximately the SC lengths in the three directions: (l_x, l_y, l_z). Our purpose in this section is not to give an overview of coherence theory but simply to elucidate the difference between TC and SC functions in terms of the temporal and spatial frequency contents that are relevant to FF-OCT.

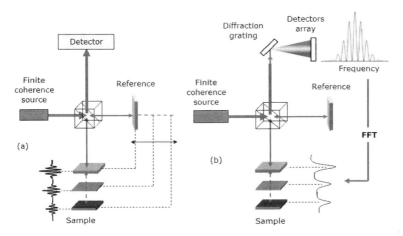

Figure 3.2 Schematic drawings of (a) TD-OCT setup and (b) FD-OCT setup.

3.3 Finite Temporal and Infinite Longitudinal Spatial Coherence Lengths

When optical beams are collimated such as with the standard Michelson interferometer (Fig. 3.2a) one can talk about the existence of a single spatial frequency, but there could be a wide spectral band—hence a spectrum of temporal frequencies. In two-beam interferometry at a single wavelength the interferogram is given by the correlation function between the reference and sample fields:

$$I(\tau) = I_r + I_s + 2\sqrt{I_r I_s} \operatorname{Re}\{\gamma_{rs}(\tau)\}, \quad (3.5)$$

where I_r and I_s are the reference and sample beam intensities, respectively, τ is the time delay between the two beams, and γ_{rs} is the mutual complex correlation function given by

$$\gamma_{rs}(\tau) = \frac{\langle E_r^*(t) E_s(t+\tau)\rangle}{\sqrt{I_r I_s}}, \quad (3.6)$$

which is also known as the TC function, where E_r and E_s are the optical field disturbances of the reference and sample beams, respectively. In this section we assume the reference and sample beams are collimated, that is, contain a single spatial frequency. For monochromatic light, a single temporal frequency ω is involved and Eq. 3.6) leads to sinusoidal infinite fringe pattern varying as $\cos(\omega\tau)$, while for a wideband spectrum this needs to be integrated over the whole spectral range, resulting in a localized fringe pattern within the coherence time τ_c, that is $\Gamma_c = \text{Re}\{\gamma_{rs}(\tau)\}$ is limited to the range of τ_c. The coherence functions have a maximum at $\tau = 0$ and full width at half maximum equals the coherence time τ_c. In time domain OCT (TD-OCT) when scanning the reference mirror a multilayered sample as in Fig. 3.2a exhibits a sequence of such localized fringes at each layer interface. For a Gaussian source at the central frequency ω_c, the normalized spectral distribution is $S(\omega) = (1/\sqrt{2\pi\sigma^2})\exp(-(\omega - \omega_c)^2/2\sigma^2)$, where $\sigma = \delta\omega/2\sqrt{2\ln 2}$ and $\delta\omega$ is the full-width at half-maximum (FWHM). Assuming both the reference and the sample consist of a single ideally reflecting surface then the coherence function is that of the source $\Gamma_c = \text{Re}\{\gamma_{so}(\tau)\}$; however, according to Wiener–Khinchin theorem $\gamma_{so}(\tau) = FT\{S(\omega)\}$; hence for the Gaussian source

$$\gamma_{so}(\tau) = \exp\left\{-\left(\frac{\delta\omega\tau}{4\sqrt{\ln 2}}\right)^2\right\}\exp(-i\omega_c\tau), \qquad (3.7)$$

which has the FWHM of $\tau_c = 8\ln 2/\delta\omega$ or coherence length $l_c = c8\ln 2/\delta\omega \approx 0.88\lambda_c^2/\delta\lambda$. For consistency throughout the chapter we used the definition of the coherence length or time as the FWHM of the correlation function, although some other definitions exist [53, 54]. For a source with a flat spectrum the coherence function is

$$\gamma_{so}(\tau) = \text{sin}\,c\,\{\delta\omega\tau/2\}\exp(-i\omega_c\tau), \qquad (3.8)$$

which gives the coherence length $l_c \approx 1.203\lambda_1\lambda_2/\delta\lambda$. An important difference between the flat and Gaussian spectra are the side lobes of the coherence function for the case of a flat spectrum, which add side lobes to the signal both in TD and in frequency domain OCT (FD-OCT), as can be seen in Fig. 3.2b of the Fourier peaks in FD-OCT. Therefore Gaussian sources are preferable although with flat sources one can get shorter coherence lengths. Another interesting source is the Lorenzian-shaped source $S(\omega) = 2\frac{(\delta\omega/2)}{(\delta\omega/2)^2+(\omega-\omega_0)^2}$, giving the coherence function

$$\gamma_{so}(\tau) = \exp\left\{-\left(\frac{\delta\omega\tau}{2}\right)^2\right\}\exp(-i\omega_c\tau), \qquad (3.9)$$

which gives a coherence length $l_c = \frac{2\ln 2}{\pi}\frac{\lambda_c^2}{\delta\lambda} \approx 0.441\lambda_c^2/\delta\lambda$, twice smaller than that of the Gaussian source.

Assuming the sample is made of a number of scattering layers, N, its response function may then be written as $r_s = \sum_{j=1}^{N} r_j \exp(i\omega\tau_j)$, where r_j is the reflectivity or the scattering amplitude of layer j and τ_j is the time delay of the scattered photons. For multiple interferences not to take place the correlation between photons scattered from different layers j, j' then has to vanish, that is, $\langle r_j^* r_{j'} \rangle = 0$. This requires that the time delay between such photons to be larger than the coherence length, that is, $|\tau_j - \tau_{j'}| > \tau_c$. In this case the coherence function can be written as

$$\Gamma_c = \sqrt{R_r} \text{Re} \left\{ \sum_{j=1}^{N} r_j \gamma_{so}(\tau - \tau_j) \right\}, \quad (3.10)$$

where R_r is the reflectivity of the reference mirror. Hence with the absence of multiple interferences, a source with a short coherence time $\tau_c < |\tau_j - \tau_{j'}|$ reveals a series of interference patterns localized at the location of the scattering interfaces. This is the essence of TD-OCT or white light interferometry (WLI).

In FD-OCT (Fig. 3.2b) the interference signal is measured at each frequency and Fourier-transformed. The FT $I(\tau) = FT\{I(\omega)\}$ in the above example reveals the interferogram

$$I(\tau) = FT\{|S(\omega)|\}(R_r + R_s) + 2\sqrt{R_r} FT$$
$$\times \left\{ \text{Re} \left\{ S(\omega) \sum_{j=1}^{N} r_j \exp(-i\omega(\tau - \tau_j)) \right\} \right\}. \quad (3.11)$$

Where we have used the approximation:

$$\left| \sum_{j=1}^{N} r_j \exp(i\omega\tau_j) \right|^2 \approx R_s.$$

Using the convolution theorem we then get

$$I(\tau) = (R_r + R_s)|\gamma_{so}(\tau)| + 2\sqrt{R_r} \text{Re} \left\{ \gamma_{so}(\tau) \otimes \sum_{j=1}^{N} r_j \delta(\tau - \tau_j) \right\}. \quad (3.12)$$

The first term is just a direct current (DC) term representing the coherence function of the source, while the second term gives Fourier peaks located at positions corresponding to the locations of

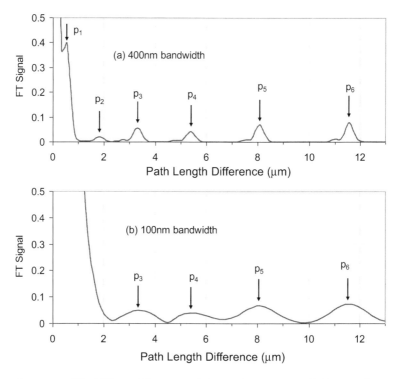

Figure 3.3 FD-OCT signal from a multilayered structure including six thin-film layers, as described in the text. In (a), the short coherence length (bandwidth 400 nm) allows measuring all the six layers, while in (b) the coherence length is four times larger (bandwidth 100 nm), so the first two layers are hidden in the DC peak and the rest have a wide shape. Reproduced from Ref. [55].

the scattering interfaces. The determination of the layers' interfaces requires translation of the time delay τ_j into the spatial position z_j. Because the photon scattered from layer j has to pass through all the layers $q = 1, 2, \ldots, j-1$, the time delay should be written as

$$\tau_j = \frac{2}{c} \sum_{q=1}^{j} n_q z_q, \quad (3.13)$$

where the factor 2 is due to the fact that we work in reflection mode and n_q is the refractive index of layer q. Hence to find the thickness z_q of layer q, one needs to know the refractive indices

and the thicknesses of all the layers prior to its location. Figure 3.3 demonstrates this for the six-layer stack and the layers' indices in the list:
$\{z_q, n_q\} = \{\{0.5, 1.0\}, \{1, 1.332\}, \{1.1, 1.35\}, \{1.6, 1.3\}, \{2, 1.4\}, \{2.5, 1.33\}\}$, where the signal is plotted as a function of the optical path length $p_j = \sum_{q=1}^{j} n_q z_q$. The source used in this calculation is Gaussian with center wavelength 600 nm and two bandwidths, (a) 400 nm and (b) 100 nm, which explains the high axial resolution obtained in (a) and its degradation in (b).

3.4 Finite Longitudinal Spatial and Infinite Temporal Coherence Lengths

When the beam is not collimated such as when a lens is used to obtain full-field imaging in FF-OCT, a spectrum of longitudinal spatial frequencies is involved, which shortens the LSC length such as in the case with the Linnik, Mirau, or Taylor interferometers (Figs. 3.4a, 3.4b, and 3.4c, respectively).

In a previous work [45] we have investigated the interplay between SC and TC and showed that it is possible to rely on SC for obtaining high lateral and axial resolution imaging. Assuming

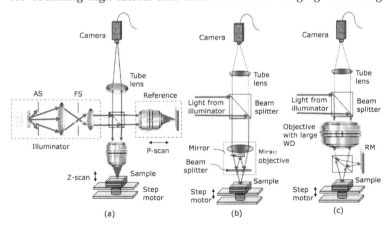

Figure 3.4 Schematic drawings of main FF-OCT schemes: (a) Linnik, (b) Mirau, and (c) Michelson/Taylor. AS, FS, and RM stand for aperture stop, field stop, and reference mirror, respectively.

a monochromatic source is used with a double-beam interference microscope such as the Linnik microscope, the interferogram can be simulated using two approaches: (1) by incorporating the angular dependence of the **k**-vector and integration over all the angles and (2) by calculating the correlation between the reference and sample images. The first approach is approximate because it does not take into account the imaging properties of the lenses. The second approach can handle aberrations; however, it uses the scalar approximation and can be made more accurate if vectorial imaging theory is used. Based on this approach the interferogram may be written as [7, 8]

$$I(\tau) = I_r + I_s + 2\text{Re}\left\{\int_0^{\theta_{max}} \sqrt{I_r I_s} g(\theta)\exp(i\omega\tau\cos\theta)\sin\theta\cos\theta d\theta\right\}. \quad (3.14)$$

Here $\sin\theta_{max} = $ NA is the NA of the objectives, and $g(\theta)$ is the angular distribution of the illumination. Here and through the whole chapter a flat sample is assumed oriented perpendicular to the optic axis. Note that the coherence function based on Eq. 3.14 is in fact an integral over all the longitudinal spatial frequencies $k_z = k_0\cos\theta$, as expected from Eq. 3.4. For small angles, uniform illumination and assuming no angular dependence of the reflectivities of the reference and sample, Eq. 3.14 can be approximated further into

$$I(\tau) \approx 1 + \frac{2\sqrt{R_r R_s}}{R_r + R_s}\sin c(2\omega\tau\sin^2(\theta_{max}/2))\cos(2\omega\tau\cos^2(\theta_{max}/2) + \phi_0). \quad (3.15)$$

The time delay is related to the defocus δz by $\tau = k\delta z/\omega$ and R_r and R_s are the reflectivities of the reference and the sample, respectively. A constant phase ϕ_0 has been introduced to account for any path length differences due to the differences in the objectives, misalignment, or beam splitter phase additions. This phase is also important to introduce when one is using phase shift interferometry (PSI) techniques. Hence the interferogram has an envelope with a width determined by the width of the spatial frequencies spectrum:

$$\delta z_1 \approx \frac{c_1 \lambda}{\sin^2(\theta_{max}/2)}, \quad (3.16)$$

where $c_1 \approx 0.301$, when the FWHM is considered, and $c_1 = 0.5$, when the distance between the first two zeros is considered. Note that δz_1

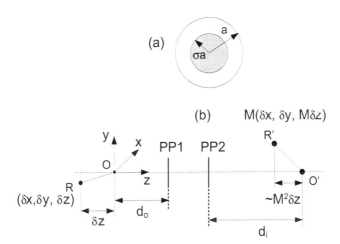

Figure 3.5 (a) The microscope objective aperture showing the illuminated part σa, with a being the radius and σ the spatial coherence factor and (b) equivalent imaging setup replaced by the principal planes PP1 and PP2, showing the main object and image distances (O and O') from the principal planes and a reference object and image points R and R' at certain defocus δz and lateral displacement $\delta r = (\delta x, \delta y)$ (corresponding distances in the image plane are approximately $M^2 \delta z$ and $M \delta r$, respectively).

expresses the focal depth of the system. For low-NA objectives one may write $\sin^2(\theta_{\max}/2) \approx \mathrm{NA}^2/4$ and the envelope width becomes approximately the same as the well-known expression for the focal depth of the objective lens. This shows that even though the light is monochromatic one can obtain high axial resolution based on the LSC. This fact can be used with FF-OCT to obtain both high axial and lateral resolution when high-NA objectives are used.

For the second approach consider the schematic drawing in Fig. 3.5a of the microscope objective aperture showing the illuminated part σa, with a being the radius and σ the SC factor defined as the fraction of the illuminated part of the objective back aperture radius. Figure 3.5b shows the equivalent imaging setup replaced by the principal planes PP1 and PP2, showing the main object and image distances (O and O') from the principal planes and a reference object and image points R and R' at certain defocus δz and lateral displacement δr, which in the image plane correspond to approximately $M^2 \delta z$ and $M \delta z$, respectively, with M being the magnification. Each imaging path is represented by a focal length f, object and image distances d_o, d_i, from the principal planes, $NA = \sin\theta_{max}$, and magnification M. Details of the model are given in Ref. [38], while here we start from the final result of the mutual coherence function for the interferogram of monochromatic light with wavenumber $k = 2\pi/\lambda$:

$$I(\delta r, \delta z, k) = I_r + I_s + \left(4\sqrt{I_r I_s}\Big/(\beta\delta z)\right)$$
$$\times \text{Re}\left[\exp(j(0.5\beta - 2k\cos\theta)\delta z + j\phi_0)\{L_1(-\beta\delta z, \chi\delta r)\right.$$
$$\left. + jL_2(-\beta\delta z, \chi\delta r)\}\right], \quad (3.17)$$

where δz, δr are the defocus and lateral distance between the main and reference points relative to the optic axis in the object space, and $\beta = 4k\sigma^2 \sin^2(\theta_{max}/2)$ and $\chi = kNA\sigma M$, with σ being the SC factor of the illumination. In this expression the following approximation has been used for the deviation from the lens law, $\varepsilon = 1/d_o + 1/d_i - 1/f \approx -4\sin^2(\theta_{max}/2)\delta z/NA^2 f^2$, which for low NA reduces to $-\delta z/f^2$. The functions L_1 and L_2 are Lommel functions of the first and second kind, respectively. Equation 3.17 was derived assuming one single point on a flat surface. For nonmonochromatic light with spectral distribution, the interferogram can be calculated by integrating Eq. 3.17 over the temporal frequency range of the spectrum. Since all the angles θ contribute to the interferogram, the total intensity is obtained by integrating over the solid angle of the aperture cone. This way scattering from the point on the flat sample (into directions outside the illumination cone) and gathered by the full aperture of the lens is taken into account. Using Eq. 3.17 for the case of no lateral mismatch between the two wavefronts, that is, looking at the interference between the two images of points located on the optic axis ($\delta r = 0$), the following expression is obtained [38, 53]:

$$I(\delta z) = 1 + \frac{2\sqrt{R_r R_s}}{R_r + R_s}\cos\left[2\pi(1 + \cos\theta_{max}\right.$$
$$\left. + \sigma^2 \sin^2(\theta_{max}/2))\delta z/\lambda + \phi_0\right]$$
$$\times \text{sinc}\left[4\pi\sin^2(\theta_{max}/2)\delta z/\lambda\right]\text{sinc}\left[2\pi\sigma^2\sin^2(\theta_{max}/2)\delta z/\lambda\right]. \quad (3.18)$$

Equation 3.18 shows that the monochromatic interferogram has an SC envelope composed of two parts, one is due to increase of the curvature of the wavefront as the defocus increases and the other is due to the lateral shift between the two wavefronts as the defocus increases. The monochromatic interferogram width is determined by the width of the SC envelopes. The first sinc function is in fact the same envelope obtained in Eq. 3.15, while the second sinc function is solely an addition of the diffraction effect on imaging. It has the width [38, 53]

$$\delta z_2 \approx \frac{c_2 \lambda}{\sigma^2 \sin^2(\theta_{max}/2)}, \qquad (3.19)$$

where $c_2 \approx 0.603$ when the FWHM is considered and $c_2 = 1$ when the distance between the first two zeros is considered. Hence the width of the first sinc function is smaller by about twice that of the second and therefore the interferogram width is determined mainly by its width δz_1, which is approximately the focal depth of the objective. This is also consistent with the Wolf longitudinal coherence function derived in Ref. [56]. In Wolf's work the difference in the NA of the circular aperture source between the two axial points is considered and appears as a multiplicative factor but it does not affect the width of the LSC function. Similar expressions applied to the Linnik [57] and Mirau [58] microscopes were obtained with slight differences based on the LSC, as appears in the book by Born and Wolf [59]. Hence for perfectly spatially coherent illuminating beam, that is, $\sigma = 0$, the coherence region is infinite, a fact that can be understood as using a point source in the back aperture plane produces a collimated beam in the object plane without dependence on the defocus, apart from producing a phase difference. For the other extreme of maximum spatial incoherence, that is, $\sigma = 1$, the coherence region is the smallest. For a wideband spectrum the interferogram in Eq. 3.16 needs to be multiplied by the source spectral distribution and integrated over the whole spectrum. The interplay between the TC and the SC then takes place, as it is also known in holography [60–62].

3.5 Lateral Mismatch as a Source of Lateral Decoherence

Decoherence can occur when there is a lateral mismatch between the two wavefronts that can originate from misalignment between the two aperture planes of the microscope objectives, for example, in the Linnik interferometer. To consider the effect of this mismatch we consider interference between the two wavefronts originating from two points separated by lateral distance δr in the object plane.

To discuss the coherence effects we start from Eq. 3.17 with $\delta z = 0$, that is, the case of zero defocus, meaning both the object and the reference mirror are in focus. In this case Eq. 3.17 reduces to the following:

$$I(\delta r, 0, k) = I_r + I_s + 2\sqrt{I_r I_s}\left[\frac{2J_1(\chi \delta r)}{\chi \delta r}\right]\cos\phi_0. \quad (3.20)$$

This equation shows that when both the object and reference beams come from the same object plane ($\delta z = 0$), and the interferogram is similar to that from a Michelson interferometer but with smaller fringe visibility due to the SC envelope:

$$V_S = \left|\frac{2J_1(\chi \delta r)}{\chi \delta r}\right|. \quad (3.21)$$

The physical meaning of this behavior is that for two points located at different locations (separated by lateral distance δr) in the object plane with respect to the optic axis, and being imaged by the two objectives (the main and reference), their mutual coherence decreases as δr increases. In a different manner, the two points give two beams with different spatial frequencies in the Fourier plane. The full width of this lateral coherence envelope defines the lateral coherence region, which is given by:

$$\delta r_c = \frac{c_s \lambda}{\sigma \text{NA}}, \quad (3.22)$$

where $c_s \approx 0.704$ for the FWHM case and $c_s \approx 1.22$ when the distance between the first two zeros is considered. The corresponding longitudinal coherence region is then determined approximately by $\delta z_c = c_s \lambda / \sigma^2 \text{NA}^2$, which is basically the focal depth of the microscope objective when the SC effects are taken into account. The determination of δz_c assumes small angles and therefore we believe the expression for δz_1 is a more accurate measure of the LSC. Hence in double beam interference microscopes each point in the image is originating from interference between the two conjugate point images one on the object and another on the reference mirror both at the same distance from the optic axis within a circle of radius δr_c. Points on the reference mirror further away from this distance will contribute only a background to the interferometric image of the on-axis point object.

3.6 Finite Longitudinal and Finite Temporal Coherence Lengths

When both the LSC and the TC lengths are finite the properties of the interferogram are determined by both. The combined coherence length may be estimated as $1/l_c = 1/l_{tc} + 1/l_{sc}$ because when $l_{tc} \rightarrow \infty$ it will be determined by l_{sc} and when $l_{sc} \rightarrow \infty$ it will be determined by l_{tc}. The question to be answered now is, What will be the effect of the TC and SC lengths on the coherence region when both are finite? To answer this question Eq. 3.17 needs to be integrated over the spectral range and the solid angles aperture cone. However, we shall try to answer the question based on the above analytic analysis. It is clear that for each SC length, there is a limit on the TC, above which it stops to affect the interferogram width. This limit is determined from the requirement $\delta z_c < l_c$, which gives the following lower limit on the effective NA, or the minimum value of σNA required for the TC not to influence the interferogram width. For a flat spectrum the condition becomes

$$\sigma \text{NA} > \sqrt{\frac{c_1(\Delta\lambda/\bar{\lambda})}{c_2(1 - 0.25(\Delta\lambda/\bar{\lambda})^2)}}. \quad (3.23)$$

This inequality shows that the limit required depends on the ratio $\Delta\lambda/\bar{\lambda}$, which is a measure of the TC. Figure 3.6 shows the minimum effective NA (i.e., σNA) versus the parameter $\Delta\lambda/\bar{\lambda}$. For a standard fiber-based OCT system where light-emitting diodes (LEDs) are used with $\bar{\lambda} \approx 800$ nm and $\Delta\lambda/\bar{\lambda} \approx 0.02$–$0.05$, the limit on the effective NA is ~ 0.12–0.18, which is not far from the NA of standard single-mode fibers (see the inset). Therefore decreasing $\Delta\lambda/\bar{\lambda}$ in this case will not affect much the interferogram width. This statement is correct even if we consider the OCT fiber-based system as a confocal microscope [42], since the axial resolution of the confocal microscope is determined by the focal depth, which is basically a measure of the SC length.

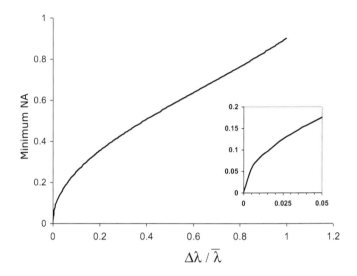

Figure 3.6 The minimum NA value required for the temporal coherence not to affect the interferogram width versus the relative spectral bandwidth $\Delta\lambda/\bar{\lambda}$. The width is taken as the distance between the first two zeros, that is, $c_1 = 1.22$, $c_2 = 2$. Reproduced from Ref. [37].

The present theory does not apply fully to the fiber confocal microscope case because then the fiber modal field distribution needs to be taken into account [4]; however, the physics explaining the interferogram width and the interplay between the TC and the SC is basically the same. Hence for each TC length there is a minimum NA value above which the interferogram starts to narrow due to the SC. Saying it differently, for each NA value there is a limit on the TC length above which it stops affecting the interferogram. If the TC decreases, the interferogram will start to narrow when the TC length becomes shorter than the SC length. Similarly increasing the NA to decrease the SC length, the interferogram starts to narrow when the SC length becomes shorter than the TC length.

To further confirm these analytics-based predictions we have integrated Eq. 3.18 numerically. Figure 3.7 shows two different cases where in (a) monochromatic light is used but a high NA showing that the width is less than 0.5 μm, even though the TC is perfect, while in (b) the same width is obtained, even though white light is

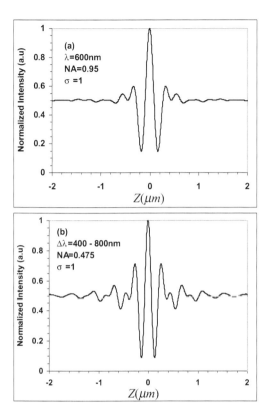

Figure 3.7 Z-scan interferogram with the illumination filling the whole back aperture of the objective ($\sigma = 1$) (a) using a high NA of 0.95 and monochromatic light and (b) wide-band illumination 400–800 nm but the NA is lower by a factor of 2. This figure demonstrates the interplay between the SC and the TC when a Z-scan is performed. Reproduced from Ref. [37].

used with 400 nm spectral width but with an NA value smaller by a factor of 2. Hence in (a) the interferogram width is determined by the SC, while in (b) it is determined by the TC. In Fig. 3.8 an experimental verification to this behavior is presented using the Mirau interference microscope. In order to demonstrate the effect of the TC one may then perform path length scan (P-scan) instead of the defocus scan (Z-scan). In this case only the axial phase varies instead of the lateral variation due to wavefront distortion, as can be seen in Eq. 3.18 for the monochromatic case when $\phi_0 = 4\pi \delta p/\lambda$

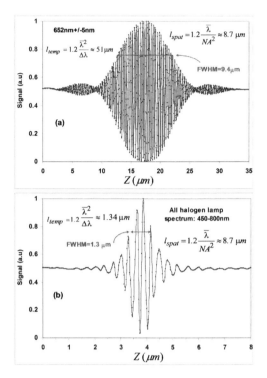

Figure 3.8 Experimental interferograms demonstrating the competition between temporal coherence and spatial coherence. In (a) the interferogram width is determined by the longitudinal spatial coherence, while in (b) it is determined by the temporal coherence length.

is varied by varying δp rather than δz. As it can be seen in the Linnik case this can be done by varying the path length between the beam splitter cube and the objectives either by scanning of the beam splitter cube or any other phase modulation means. Figure 3.9 demonstrates this behavior in which the P-scan interferogram width is determined only by the TC length. This mode of operation is in fact similar to the OCT mode where collimated beams are used such as with Michelson interferometer. When a defocus is introduced the contrast of the P-scan interferogram is reduced but its width remains unaltered and for a monochromatic light a pure sinusoidal interferogram is obtained. This behavior was in fact observed experimentally [36].

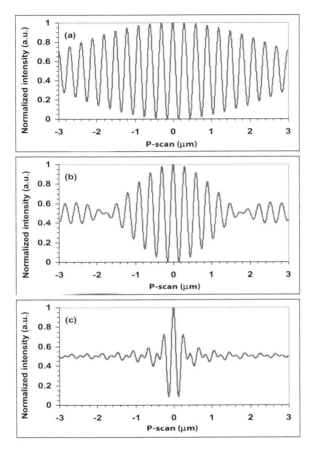

Figure 3.9 Interferograms obtained by scanning the path length (P-scan) with $\sigma = 1$, NA $= 0.95$, and different spectral bandwidths: (a) 580–620 nm, (b) 550–650 nm, and (c) 400–800 nm. This figure demonstrates that the P-scan interferogram is not affected by the objective NA; rather, it is determined only by the TC. The NA has an effect on the interferogram contrast only when the P-scan is performed at defocus in that the contrast decreases. Reproduced from Ref. [37].

3.7 Effects on Multilayer Sample Measurement

3.7.1 *LSC Effect on the Fringe Size*

A parameter of particular interest when working with monochromatic light is the fringe size, which is the distance between two maxima or two interference minima. Without the effect of the SC this is supposed to be $\lambda/2$, however due to SC it is decreased to $\lambda/2\eta$ where $\eta \geq 1$ is called the fringe size correction factor. This factor has seen a lot of attention since the early years of the 20th century [1, 63–69] till the 21st century [16, 36, 70–75] due to its importance in calibrating the path length scan. Only approximate expressions existed, which are valid mainly for low NA values, among which the best known is that of Ingelstam [67]: $\eta = 1/(1 - 0.25\,\mathrm{NA}^2)$. In previous works [36, 45] we have shown that η is influenced by the SC and that most of the experimental data fit very well the case with $\sigma = 0.5$. This effect is clearly seen based on Eq. 3.18, which yields the following expression for η if we ignore the effect of the two envelopes:

$$\eta = \frac{2}{1 + \cos\alpha_0 + \sigma^2 \sin^2(\alpha_0/2)}. \quad (3.24)$$

The relation by Ingelstam is derived easily from Eq. 3.24 if we take the case of $\sigma = 0$ and a low NA. In Fig. 3.10 we have plotted the exact correction factor determined numerically from Eq. 3.18, the one based on the Ingelstam expression and the one based on Eq. 3.24 for a coherence factor of $\sigma = 0.75$, together with our experimental results. It shows that to obtain good agreement with the experimental data one has to use $\sigma < 1$, most likely in the range of $\sigma \approx 0.5 - 0.75$, in particular for NA > 0.6. Note that the correction factor determined by Eq. 3.24 gives values closer to the exact ones than the one determined by the Ingelstam relation, which does not depend on σ. Equation 3.24 describes the fringe size correction factor quite nicely up to NA values as high as NA $= 0.9$ with less than 10% differences from the exact values. Although in a practical situation one may perform calibration of the fringe size, the derived analytic expression can be useful when calibration is not possible and it also helps understanding interference microscopy systems.

3.7.2 LSC Effect on the Thickness Determination When Imaging a Multilayered Sample

When relying on the LSC for imaging of a multilayered sample the relation between the axial Z-scan and the layer thickness is

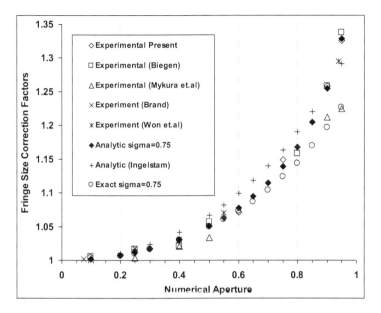

Figure 3.10 Fringe size correction factor versus the NA for different values of σ calculated both numerically and using Eq. 3.24, the Ingelstam relation, and some experimental data in the range of high NA. Reproduced from Ref. [37].

different from the case when relying on the low TC. To clarify this let us consider the propagation of two rays through the multilayer structure, as shown schematically in Fig. 3.11 for the two cases of which the top and bottom surfaces of the j-th layer are in focus. The translation of the objective focal plane from the top surface of layer j to the bottom surface is obtained by scanning the whole Linnik microscope by δz_j or equivalently the sample by δz_j. Each layer j is represented by the geometrical thickness d_j and the refractive index n_j. The propagation angles θ_j are related through Snell's law of refraction:

$$n_0 \sin\theta_0 = n_1 \sin\theta_1 = \ldots = n_j \sin\theta_j = NA. \qquad (3.25)$$

We start from the situation where the reference mirror is exactly in focus, although the formulation is straightforward even with the existence of a focus offset. Let the main objective be focused on the top surface of layer j. In this situation move the Linnik microscope

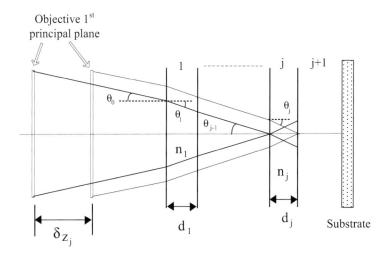

Figure 3.11 Trace of two ray paths through a multilayered object at two positions of the main objective, when focused on the top and bottom surfaces of layer j.

down toward the bottom surface until maximum contrast of the fringes is obtained. This will be the focus position because as can be seen from Eq. 3.18 the envelope of the interferogram does not depend on the phase. By equating the lateral shift of the ray in all layers, the following relation was found [35, 37]:

$$d_j tg\theta_j = \delta z_j tg\theta_0. \quad (3.26)$$

A combination of Eqs. 3.25 and 3.26 leads to the following relation between the layer thickness and the axial distance required to obtain the best contrast:

$$\delta z_j = d_j n_0 \cos\theta_0 / n_j \cos\theta_j. \quad (3.27)$$

With the approximation $\cos\theta_0 \approx \cos\theta_j$ we get the following:

$$\delta z_j = d_j n_0 / n_j. \quad (3.28)$$

Hence the correction factor for the thickness determination of layer j, $d_j = (n_j/n_0)\delta z_j$ using the Z-scan in FF-OCT, is different from that used with standard OCT systems that use collimated beams based on a short TC length where the more known relation is used: $d_j = \delta z_j/n_j$. This difference was confirmed experimentally recently by Safrani and Abdulhalim [76] and Eq. 3.28 was generalized to the case of multilayers.

3.7.3 LSC Effect on the Imaging Depth

Although the higher the NA in FF-OCT, the better the lateral and axial resolutions of the FF-OCT system, there is a limitation on the imaging depth when the TC length is not infinite. It was shown first by Abdulhalim [37] for the case of annular lenses that the interferogram contrast deteriorates as the layer thickness increases (Fig. 3.12). This undesirable effect is strongly evident when the illumination bandwidth is also wide, as confirmed by De Groot and

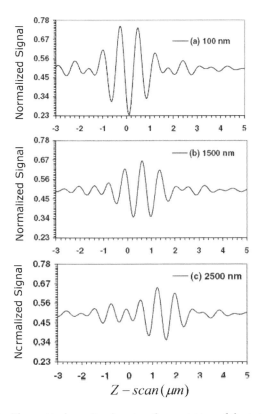

Figure 3.12 Theoretical results showing the variation of the interferogram from a Linnik microscope versus defocus using an annulus aperture with outer NA = 0.9, obstruction ratio $\varepsilon = 0.9$, and wavelength $\lambda = 500$ nm at different oxide layer thicknesses on a c-Si substrate: (a) 100 nm, (b) 1500 nm, and (c) 2500 nm. Note the deterioration in the contrast as the layer thickness increases. Based on Fig. 3.3 of Ref. [37].

Lega [77] using circular lenses. It becomes even worse when the space between the objective lens and the sample surface is not occupied with a suitable index-matching liquid, which in turn forces the application of a dynamic focusing compensation technique, as shown by Dubois et al. [78]. When using a suitable index-matching material the strength of the effect is significantly reduced; however, it does not completely vanish. In the context of biological tissue imaging using FF-OCT this physical phenomenon is more recognized since large imaging depths are attempted; however, it is also known in other imaging fields such as confocal microscopy, as it was recently observed [42, 79], and in interference confocal microscopy [80]. Hence under optically fixed conditions, the greater the width of the inspected sample the more severe is the erosion of the interference signal contrast in FF-OCT systems with finite TC length. A more detailed study of this phenomenon is published recently by Safrani and Abdulhalim [81].

To elucidate this effect further we consider the two cases sketched in Fig. 3.13, where in (a) a collimated beam is focused on the top surface of the first layer of a multilayered sample, while in (b) the sample moved up by δz, which is less than the thickness of the first layer d_1. As a result of the variation of the refractive index within the converging beam, the focal plane of the lens is shifted by δf, which is equivalent to say that the SC plane is shifted by δz_{SC}.

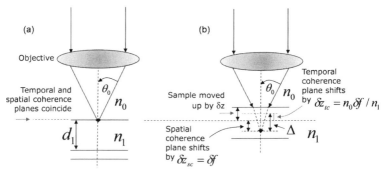

Figure 3.13 Illustration for estimating the TC versus SC mismatch effect when imaging deep samples and the medium above the sample has a refractive index different from that of the sample. (a) Upper sample surface is at focus so that both the TC and SC planes coincide. (b) The sample is moved up by δz and the mismatch occurs.

This shift can be found based on Eq. 3.29, which can now be written as

$$(\delta z + \delta f)tg\theta_1 = \delta z tg\theta_0. \tag{3.29}$$

This leads to

$$\delta z_{sc} = \left(\frac{tg\theta_0}{tg\theta_1} - 1\right)\delta z \approx \left(\frac{n_1}{n_0} - 1\right)\delta z. \tag{3.30}$$

On the other hand the TC plane has shifted backward by

$$\delta z_{tc} = \frac{n_1}{n_0}\delta z. \tag{3.31}$$

The shift between the SC and TC planes is then given by $\Delta = \delta z + \delta f - \delta z_{tc}$, which when combining Eqs. 3.29–3.31 leads to

$$\Delta = \left(\frac{tg\theta_0}{tg\theta_1} - \frac{n_0}{n_1}\right)\delta z \approx \left(\frac{n_1}{n_0} - \frac{n_0}{n_1}\right)\delta z. \tag{3.32}$$

The approximate signs in Eqs. 3.30 and 3.32 are valid for small angles. Hence the two coherence envelopes, the spatial and the temporal, are shifted one with respect to the other once the sample is scanned in z. Since the total interferogram envelope is approximately given by the product of the two envelopes multiplied by the infinite fringe pattern, as elucidated in Fig. 3.14, the end result is a deterioration of the interferogram contrast. The severity of the effect depends on the width of the TC envelope l_{tc}, which becomes very severe when the latter becomes less or comparable to the width of the SC envelope, that is, $l_{tc} \leq l_{sc}$.

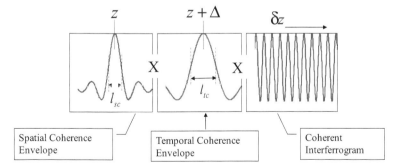

Figure 3.14 Illustration of the effect of the TC and SC mismatch on the image contrast when imaging deeper than the SC region.

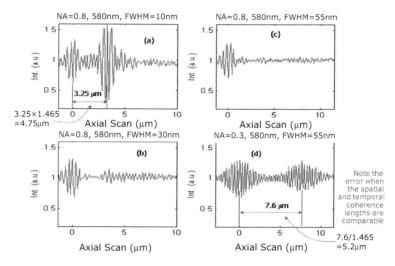

Figure 3.15 Experimental verification of two effects in FF-OCT: (1) The layer thickness determination in which the distance between the two interferograms obtained from the two interfaces of a 4.8 μm SiO_2 layer has to be multiplied by the index of refraction of the layer according to Eq. 3.28 (upper left), even for narrow-band illumination, and (2) the contrast degradation when a shorter TC length is used (b and c). In (d) the contrast is improved slightly because the SC length is increased as compared to (c). However, the layer thickness has a larger error from the nominal based on the division by the index of refraction. This means that in this case the NA of the objective has to be decreased further or the spectral bandwidth needs to be increased in order to rely on the TC imaging with negligible effect of the SC. Based on Fig. 2 of Ref. [80].

To compensate for this contrast deterioration Dubois et al. [78] used a dynamic focusing mechanism in which as the object is scanned for imaging at a depth δz, and the objective is moved by Δ for compensation. For the case when annular lenses are used Abdulhalim [35] proposed to use a compensation mechanism of the path length and from which the refractive indices and thicknesses of multilayers could be deduced. Wavefront correction approach to this defocusing effect was also proposed recently [82]. Note that when oil or water immersion microscope objectives are used at which point $n_0 \approx n_1$, the shift between the two coherence envelopes is minimized. Alternatively one can work with narrow-band illumination and obtain the high resolution depth imaging based on the short SC length, as demonstrated recently by Safrani et al. [81]. In Fig. 3.15 an experimental demonstration is presented for the two effects reported in this section and the previous one: (1) The layer thickness determination in which the distance between the two interferograms obtained from the two interfaces of a 4.8 μm SiO_2 layer has to be multiplied by the index of refraction of the layer according to Eq. 3.28 (Fig. 3.15a), even for narrow-band illumination, and (2) the contrast degradation when shorter TC length is used (Fig. 3.15b,c). In Fig. 3.15d the contrast is improved slightly because the SC length is increased as compared to Fig. 3.15c. However, the layer thickness has a larger error from the nominal based on the division by the index of refraction. This means that in this case the NA of the objective has to be decreased further (l_{sc} increased) or the spectral bandwidth needs to be increased (l_{tc} decreased) in order to rely on the TC imaging with negligible effect of the SC.

3.8 Conclusions and Future Trends

Spatiotemporal coherence effects are reviewed in this chapter in the context of FF-OCT and interference microscopy. There are several interesting and important phenomena that result from finite longitudinal coherence and TC lengths and the interplay between them. The main phenomena addressed can be summarized as follows:

(i) LSC takes an effect when the path length scan is performed with noncollimated beam such as defocus scan of a lens where the focal depth of the lens mainly determines the coherence region. The longitudinal coherence length is determined basically by the focal depth of the microscope objective (NA, λ, and the refractive index of the object space).
(ii) For each TC length there is a minimum NA value above which the interferogram starts to narrow due to the SC. In other words for each NA value there is a limit on the TC length above which its effect on the interferogram width starts to diminish.

(iii) In the latter case the fringe size for a monochromatic light is smaller than half the wavelength by a factor determined by the SC.

(iv) The lateral alignment between the main and reference wavefronts is crucial for the best performance of FF-OCT system as the lateral coherence region on the sample is comparable to the radius of the Airy function.

(v) When imaging a multilayered sample with FF-OCT the relation between the maximum contrast position and layer thickness is different from the case of collimated beam OCT that relies on the low TC (Eqs. 3.27 and 3.28).

(vi) When imaging multilayered samples the imaging depth is limited due to several factors:

1. Mismatch of the main and reference path lengths as the focused beam penetrates deeper into the layers as shown in Section 3.7.3.
2. Dispersion effects of the sample, which become more severe when the bandwidth is wide and the NA is high. Although results on the dispersion effects are not shown here, they were discussed heavily in Ref. [36], including the chromatic aberrations effects.
3. Multiple interference effects when the sample is nonscattering [83, 84].
4. Mismatch between the SC and TC interferograms, which becomes more severe when the LSC and TC lengths are short and comparable.

A promising high-contrast, high-resolution FF-OCT system for imaging of multilayered samples is one that uses narrow-band illumination and high NA objectives with an index-matching fluid. This is because wide-band illumination reduces the contrast of the interferogram from deeper layers, as discussed in Section 3.7.3. Other improvements are possible such as the use of phase and amplitude masks in the pupil plane to extend the depth of focus without deteriorating the resolution [85, 86] combined with frequency domain techniques and PSI and the use of a multilayered scattering reference mirror to minimize the effects of the mismatch [87]. This technique can become more efficient than confocal microscopy; however, it should be pointed out that the use of laser sources can be problematic because of their high lateral SC, which also implies high longitudinal coherence. Noncoherent narrow-band sources are recommended. For surface profiling these limitations are irrelevant, as already proved [88, 89], since no deep layer imaging is involved. For biomedical optics applications the use of near-infrared light is preferable as it penetrates deeper into tissues [90], which requires then the development of high-resolution infrared cameras. For frequency domain FF-OCT operation [91], tunable sources and femtosecond sources that can reveal the spectral interference images in a short time will be helpful.

Finally it should be mentioned that SC effects in other related fields such as interferometry in general, holography, surface profiling, and other applications have been an active field of research during the last two decades [92–97]. The group at Saratov University has been studying LSC extensively [98–103] from different aspects.

Acknowledgments

I am grateful to my students whose role was crucial in performing the research we conducted on this topic during the last few years: Ron Friedman, Rony Sharon, Lior Liraz, Ronen Dadon, Jenny Sokolovsky, and Avner Safrani. The majority of this chapter is based on the author's recent review article [104].

References

1. Krug W., Rienitz J., and Schultz G., eds (1964). *Contributions to Interference Microscopy* (Hilger and Watts, London).
2. Wilson T., and Shippard, C. J. R. (1984). *Theory and Practice of Scanning Optical Microscopy* (Academic, London).
3. Sheppard, C. J. R., and Zhou, H. (1997). Confocal interference microscopy. *Proc. SPIE*, **2984**, pp. 85–89.
4. Gu, M. (1996). *Principles of Three Dimensional Imaging in Confocal Microscopes* (World Scientific, Singapore).

5. Davidson M., Kaufman K., Mazor I., and Cohen, F. (1987). An application of interference microscopy to integrated circuit inspection and metrology. *Proc. SPIE*, **775**, pp. 233–247.
6. Chim, S. S. C., and Kino, G. S. (1990). Mirau correlation microscope. *Opt. Lett.*, **15**, pp. 579–581.
7. Chim, S. S. C., and Kino, G. S. (1992). Three-dimensional realization in interference microscopy. *Appl. Opt.*, **31**, pp. 2550–2553.
8. Lee, B. S., and Strand, T. C. (1990). Profilometry with a coherence scanning microscope. *Appl. Opt.*, **29**, pp. 3784–3788.
9. Danielson, B. L., and Boisrobert, C. Y. (1991). Absolute optical ranging using low coherence interferometry. *Appl. Opt.*, **30**, pp. 2975–2979.
10. Dresel, T., Hausler, G., and Venzke, H. (1992). Three-dimensional sensing of rough surfaces by coherence radar. *Appl. Opt.*, **31**, pp. 919–925.
11. Deck, L., and de Groot, P. (1994). High-speed noncontact profiler based on scanning white-light interferometry. *Appl. Opt.*, **33**, pp. 7334-7338.
12. Chen, S., Palmer, A. W., Grattan, K. T. V., and Meggitt, B. T. (1992). Digital signal-processing techniques for electronically scanned optical-fiber white-light interferometry. *Appl. Opt.*, **31**, pp. 6003–6010.
13. Rao, Y. J., Ning, Y. N., and Jackson, D. A. (1993). Synthesized source for white-light sensing systems. *Opt. Lett.*, **18**, pp. 462–464.
14. de Groot, P., and Deck, L. (1993). Three-dimensional imaging by sub-Nyquist sampling of white-light interferograms. *Opt. Lett.*, **18**, pp. 1462–1464.
15. Plissi, M. V., Rogers, A. L., Brassington, D. J., and Wilson, M. G. F. (1995). Low-coherence interferometric system utilizing an integrated optical configuration. *Appl. Opt.*, **34**, pp. 4735–4739.
16. Gale, D., Pether, M. I., and Dainty, J. C. (1996). Linnik microscope imaging of integrated circuit structures. *Appl. Opt.*, **35**, pp. 131–148.
17. Huang, D., Swanson, E. A., Lin, C. P., Schuman, J. S., Stinson, W. G., Chang, W., Lee, M. R., Flotte, T., Gregory, K., Puliafito, C. A., and Fujimoto, J. G. (1991). Optical coherence tomography. *Science*, **254**, pp. 1178–1181.
18. Bouma, B. E., and Tearney, G. J., eds. (2002). *Handbook of Optical Coherence Tomography* (Marcel Dekker, New York).
19. Wilkins, J. R., Puliafito, C. A., Hee, M. R., Duker, J. S., Reichel, E., Coker, J. G., Schuman, J. S., Swanson, E. A., and Fujimoto, J. G. (1996). Characterization of epiretinal membranes using optical coherence tomography. *Ophthalmology*, **103**, pp. 2142–2151.

20. Welzel, J., Lankenau, E., Birngruber, R., and Egelhardt, R. (1997). Optical coherence tomography of the human skin. *J. Am. Acad. Dermatol.*, **37**, pp. 958–963.
21. Beaurepaire, E., Boccara, A. C., Lebec, M., Blanchot, L., and Saint-Jalmes, H. (1998). Full -field optical coherence microscopy. *Opt. Lett.*, **23**, pp. 244–246.
22. Dubois, A., Vabre, L., Boccara, A. C., and Beaurepaire, E. (2002). High resolution full-field optical coherence tomography with a Linnik microscope. *Appl. Opt.*, **41**, pp. 805–812.
23. Dubois, A., Grieve, K., Moneron, G., Lecaque, R., Vabre, L., and Boccara, A. C. (2004). Ultrahigh-resolution full-field optical coherence tomography. *Appl. Opt.*, **43**, pp. 2874–2883.
24. Oh, W. Y., Bouma, B. E., Iftimia, N., Yun, S. H., Yelin, D., and Tearney, G. J. (2006). Ultrahigh-resolution full-field optical coherence microscopy using InGaAs camera. *Opt. Express*, **14**, pp. 726–735.
25. Laude, B., De Martino, A., Drevillon, B., Benattar, L., and Schwartz, L. (2002). Full-field optical coherence tomography with thermal light. *Appl. Opt.*, **41**, pp. 6637–6645.
26. Akiba, M., Chan, K. P., and Tanno, N. (2003). Full-field optical coherence tomography by two-dimensional heterodyne detection with a pair of CCD cameras. *Opt. Lett.*, **28**, pp. 816–818.
27. Yu, L., and Kim, M. K. (2004). Full-color three-dimensional microscopy by wide-field optical coherence tomography. *Opt. Express*, **12**, pp. 6632–6641.
28. Watanabe, Y., Hayasaka, Y., Sato, M., and Tanno, N. (2005). Full-field optical coherence tomography by achromatic phase shifting with a rotating polarizer. *Appl. Opt.*, **44**, pp. 1387–1392.
29. Moneron, G., Boccara, A. C., and Dubois, A. (2005). Stroboscopic ultrahigh-resolution full-field optical coherence tomography. *Opt. Lett.*, **30**, pp. 1351–1353.
30. Grieve, K., Dubois, A., Simonutti, M., Paques, M., Sahel, J., Le Gargasson, J. F., and Boccara, A.-C. (2005). In-vivo anterior segment imaging in the rat eye with high speed white light full-field optical coherence tomography. *Opt. Express*, **13**, pp. 6286–6295.
31. Bordenave, E., Abraham, E., Jonusauskas, G., Tsurumachi, N., Oberle, J., Rulliere, C., Minot, P. E., Lassegues, M., and Surleve Bazeille, J. E. (2002). Wide-field optical coherence tomography: imaging of biological tissues. *Appl. Opt.*, **41**, pp. 2059–2064.

32. Fercher, A. F., Hitzenberger, C. K., Sticker, M., Moreno-Barriuso, E., Leitgeb, R., Drexler, W., and Sattmann, H. (2000). A thermal light source technique for optical coherence tomography. *Opt. Commun.*, **185**, pp. 57–64.
33. Blazkiewicz, P., Gourlay, M., Tucker, J. R., Rakic, A. D., and Zvyagin, A. V. (2005). Signal-to-noise ratio study of full-field Fourier-domain optical coherence tomography. *Appl. Opt.*, **44**, pp. 7722–7729.
34. KLA-Tencor Corporation, 160 Rio Robles, San Jose, CA 95134-1809, USA.
35. Abdulhalim, I. (1999). Method for the measurement of multilayers refractive indices and thicknesses using interference microscopes with annular aperture. *Optik*, **110**, pp. 476–478.
36. Abdulhalim, I. (2001). Theory for double beam interference microscopes with coherence effects and verification using the Linnik microscope. *J. Mod. Opt.*, **48**, pp. 279–302.
37. Abdulhalim, I. (2001). Spectroscopic interference microscopy technique for measurement of layer parameters. *Meas. Sci. Technol.*, **12**, pp. 1996–2001.
38. Hopkins, H. H. (1953). On the diffraction theory of optical images. *Proc. R. Soc. London Ser. A*, **217**, pp. 408–432.
39. Hopkins, H. H. (1955). The frequency response of a defocused optical system. *Proc. R. Soc. London Ser. A*, **231**, pp. 91–103.
40. Karamata, B., Leutenegger, M., Laubscher, M., Bourquin, S., Lasser, T., and Lambelet, P. (2005). Multiple scattering in optical coherence tomography. II. Experimental and theoretical investigation of cross talk in wide-field optical coherence tomography. *J. Opt. Soc. Am. A*, **22**, pp. 1380–1388.
41. Grajciar, B., Pircher, M., Fercher, A. F., and Leitgeb, R. A. (2005). Parallel Fourier domain optical coherence tomography for in vivo measurement of the human eye. *Opt. Express*, **11**, pp. 1131–1137.
42. Sheppard, C. J. R., Roy, M., and Sharma, M. D. (2004). Image formation in low-coherence and confocal interference microscopes. *Appl. Opt.*, **43**, pp. 1493–1502.
43. Ryabukho, V., Lyakin, D., and Lobachev, M. (2005). Longitudinal pure spatial coherence of a light field with wide frequency and angular spectra. *Opt. Lett.*, **30**, pp. 224–226.
44. Ryabukho, V. P., Lyakin, D. V., and Lobachev, M. I. (2004). manifestation of longitudinal correlations in scattered coherent fields in an interference experiment. *Opt. Spectrosc.*, **97**, pp. 299–304.

45. Abdulhalim, I. (2006). Competence between spatial and temporal coherence in full field optical coherence tomography and interference microscopy. *J. Opt. A: Pure Appl. Opt.*, **8**, pp. 952–958.
46. Soroko, L. M. (1980). *Holography and Coherent Optics* (Plenum Press, New York).
47. Rosen, J., and Yariv, A. (1995). Longitudinal spatial coherence of optical radiation. *Opt. Commun.*, **117**, pp. 8–12.
48. Rosen, J., and Yariv, A. (1996). General theorem of spatial coherence: application to three-dimensional imaging. *J. Opt. Soc. Am. A.*, **13**, pp. 2091–2095.
49. Gokhler, M., and Rosen, J. (2005). General configuration for using the longitudinal spatial coherence effect. *Opt. Commun.*, **252**, pp. 22–28.
50. Zarubin, A. M. (1993). Three-dimensional generalization of the van Cittert–Zernike theorem to wave and particle scattering. *Opt. Commun.* **100**, pp. 491–507.
51. Ryabukho, V. P., Lyakin, D. V., and Lychagov, V. V. (2009). Longitudinal coherence length of an optical field *Opt. Spectrosc.*, **107**, pp. 282 287.
52. Ryabukho, V. P., Kal'yanov, A. L., Lyakin, D. V., and Lychagov, V. V. (2010). Influence of the frequency spectrum width on the transverse coherence of optical field. *Opt. Spectrosc.*, **108**, pp. 979–984.
53. Wada, K., Fujita, J., Yamada, J., Matsuyama, T., and Horinaka, H. (2008). Simple method for estimating shape functions of optical spectra. *Opt. Commun.* **281**, pp. 368–373.
54. Akcay, C., Parrein, P., and Rolland, J. P. (2002). Estimation of longitudinal resolution in optical coherence imaging. *Appl. Opt.*, **41**, pp. 5256–5262.
55. Abdulhalim, I. (2009). Coherence effects in applications of frequency and time domain full field optical coherence tomography to optical metrology. *J. Holography Speckle*, **5**, pp. 180–190.
56. Wolf, E. (1994). Radiometric model for propagation of coherence. *Opt. Lett.*, **19**, pp. 2024–2026.
57. Zeylikovich, I. (2008). Short coherence length produced by a spatial incoherent source applied for the Linnik-type interferometer. *Appl. Opt.*, **47**, pp. 2171–2177.
58. Biegen, J. F. (1994). Determination of the phase change on reflection-from two-beam interference. *Opt. Lett.*, **19**, pp. 1690–1692.
59. Born, M., and Wolf, E. (1993). *Principles of Optics*, 6th Ed. (Pergamon).

60. Denisyuk, Yu. N., Staselko, D. I., and Herke, R. R. (1970). On the effect of time and spatial coherence of radiation source on the image produced by a hologram. *Nouvelle Revue d'Optique Appliquée*, **1**, 4.
61. Abramson, N. H., Bjelkhagen, H. I., Caulfield, H. J. (1991). The ABCs of space-time-coherence recording in holography. *J. Mod. Opt.*, **38**, pp. 1399–1406.
62. Leith, E. N., Chien, W. C., Mills, K. D., Athey, B. D., and Dilworth, D. S. (2003). Optical sectioning by holographic coherence imaging: a generalized analysis. *J. Opt. Soc. A*, **20**, pp. 380–387.
63. Schulz, G., Deduction of theory, in reference 1, pp. 282–296.
64. Tolmon, F. R., and Wood, J. G. (1956). Fringe spacing in interference microscopes. *J. Sci. Instrum.*, **33**, pp. 236–238.
65. Gates, J. W. (1956). Fringe spacing in interference microscopes. *J. Sci. Instrum.*, **33**, pp. 507–507.
66. Bruce, C. F., and Thornton, B. S. (1957). Obliquity effects in interference microscopes. *J. Sci. Instrum.*, **34**, pp. 203–204.
67. Ingelstam, E. (1960). Problems related to the accurate interpretation of microinterferograms, in *Interferometry*, National Physical Laboratory Symposium No. 11 (Her Majesty's Stationery Office, London), pp. 141–163.
68. Mykura, H., and Rhead, G. E. (1963). Errors in surface topography measurements with high aperture interference microscopies. *J. Sci. Instrum.*, **40**, pp. 313–315.
69. Dowell, M. B., Hultman, C. A., and Rosenblatt, G. M. (1977). Determination of slopes of microscopic surface features by Nomarski polarization interferometry. *Rev. Sci. Instrum.*, **48**, pp. 1491–1497.
70. Biegen, J. (1998). Calibration requirements for Mirau and Linnik microscope interferometers. *Appl. Opt.*, **28**, pp. 1972–1974.
71. Creath, K. (1989). Calibration of numerical aperture effects in interferometric microscope objectives. *Appl. Opt.*, **28**, pp. 3335–3338.
72. Sheppard, C. J. R., and Larkin, K. G. (1995). Effect of numerical aperture on interference spacing. *Appl. Opt.*, **34**, pp. 4731–4734.
73. Brand, U. (1995). Comparison of interferometrical and stylus step height measurements on rough surfaces. *Nanotechnology*, **6**, pp. 81–86.
74. Dubois, A., Selb, J., Vabre, L., and Boccara, A.-C. (2000). Phase measurements with wide-aperture interferometers. *Appl. Opt.*, **39**, pp. 2326–2331.

75. Wan, D. S., Schmit, J., and Novak, E. (2004). Effects of source shape on the numerical aperture factor with a geometrical-optics model. *Appl. Opt.*, **43**, pp. 2023–2028.
76. Safrani, A., and Abdulhalim, I. (2011). Spatial coherence effect on layers thickness determination in narrowband full field optical coherence tomography. *Appl. Opt.*, **50**, pp. 3021–3027.
77. de Groot, P., and Colonna de Lega, X. (2004). Signal modeling for low-coherence height-scanning interference microscopy. *Appl. Opt.*, **43**, pp. 4821–4830.
78. Dubois, A., Moneron, G., and Boccara, C. (2006). Thermal-light full-field optical coherence tomography in the 1.2 μm wavelength region. *Opt. Commun.*, **226**, pp. 738–743.
79. Hell, S., Reiner, G., Cremer, C., and Stelzer, E. H. K. (1993). Aberrations in confocal fluorescence microscopy induced by mismatches in refractive index. *J. Microsc.*, **169**, pp. 391–405.
80. Izatt, J. A., Hee, M. R., Owen, G. M., Swanson, E. A., and Fujimoto, J. G. (1994). Optical coherence microscopy in scattering media. *Opt. Lett.*, **19**, pp. 590–592.
81. Safrani, A., and Abdulhalim, I. (2012). Ultra high resolution full field optical coherence tomography using spatial coherence gating and quasi monochromatic illumination. *Opt. Lett.*, **37**, pp. 458–460.
82. Labiau, S., David, G., Gigan, S., and Boccara, A. C. (2009). Defocus test and defocus correction in full-field optical coherence tomography. *Opt. Lett.*, **34**, pp. 1576–1578.
83. Abdulhalim, I., and Dadon, R. (2008). Multiple interference and spatial frequencies' effect on the application of frequency-domain optical coherence tomography to thin films' metrology. *Meas. Sci. Technol.*, **20**, 015108-19.
84. Roy, M., Cooper, I., Moore, P., Sheppard, C. J. R., and Hariharan P. (2005). White-light interference microscopy: effects of multiple reflections within a surface film. *Opt. Express*, **13**, pp. 134–170.
85. Abdulhalim, I., Friedman, R., Liraz, L., and Dadon, R. (2007). Full field frequency domain common path optical coherence tomography with annular aperture. *Proc. SPIE*, **6627**, 662719.
86. Zlotnik, A., Abraham, Y., Liraz, L., Abdulhalim, I., and Zalevsky, Z. (2010). Improved extended depth of focus full field spectral domain optical coherence tomography. *Opt. Commun.*, **283**, pp. 4963–4968.
87. Sharon, R., Friedman, R., and Abdulhalim, I. (2010). Multilayered scattering reference mirror for full field optical coherence tomography

with application to cell profiling. *Opt. Commun.*, **283**, pp. 4122–4125.
88. Rosen, J., and Takeda, M. (2000). Longitudinal spatial coherence applied for surface profilometry. *Appl. Opt.*, **39**, pp. 4107–4111.
89. Schmit, J., Reed, J., Novak, E., and Gimzewski, J. K. (2008). Performance advances in interferometric optical profilers for imaging and testing. *J. Opt. A: Pure Appl. Opt.*, **10**, 064001-8.
90. Oh, W. Y., Bouma, B. E., Iftimia, N., Yun, S. H., Yelin, D., and Tearney, G. J. (2006). Ultrahigh-resolution full-field optical coherence microscopy using InGaAs camera. *Opt. Express*, **14**, pp. 726–735.
91. Blazkiewicz, P., Gourlay, M., Tucker, J. R., Rakic, A. D., and Zvyagin, A. V. (2005). Signal-to-noise ratio study of full-field Fourier-domain optical coherence tomography. *Appl. Opt.*, **44**, pp. 7722–7729.
92. Duan, Z., Gokhler, M., Rosen, J., Kozaki, H., Ishii, N., and Takeda, M. (2002). Synthesis of longitudinal coherence functions by spatial modulation of an extended light source: a new interpretation and experimental verification. *Appl. Opt.*, **41**, pp. 1962–1970.
93. Duan, Z., Miyamoto, Y., and Takeda, M. (2006). Dispersion-free absolute interferometry based on angular spectrum scanning. *Opt. Express*, **14**, pp. 655–663.
94. Duan, Z., Miyamoto, Y., and Takeda, M. (2006). Dispersion-free optical coherence depth sensing with a spatial frequency comb generated by an angular spectrum modulator. *Opt. Express*, **14**, pp. 12109–12121.
95. Pavlíček, P., Halouzka, M., Duan, Z., and Takeda, M. (2009). Spatial coherence profilometry on tilted surfaces. *Appl. Opt.*, **48**, pp. H40–H47.
96. Liu, Z., Gemma, T., Rosen, J., and Takeda, M. (2010). Improved illumination system for spatial coherence control. *Appl. Opt.*, **49**, pp. D12–D16.
97. Lychagov, V. V., Ryabukho, V. P., Kalyanov, A. L., and Smirnov, I. V. (2012). Polychromatic low-coherence interferometry of stratified structures with digital interferogram recording and processing. *J. Opt. A: Pure Appl. Opt.*, **14**, 015702 (12pp).
98. Ryabukho, V. P., Lyakin, D. V., and Lobachev, M. I. (2004). The effects of temporal and longitudinal spatial coherence in a disbalanced-arm interferometer. *Tech. Phys. Lett.*, **30**, pp. 44–67.
99. Ryabukho, V., Lyakin, D., and Lobachev, M. (2004). Influence of longitudinal spatial coherence on signal of a scanning interferometer. *Opt. Lett.*, **29**, pp. 667–669.

100. Ryabukho, V. P., and Lyakin, D. V. (2005). The effects of longitudinal spatial coherence of light in interference experiments. *Opt. Spectrosc.*, **98**, pp. 273–283.
101. Ryabukho, V. P., Lychagov, V. V., Lyakin, D. V., and Smirnov, I. V. (2011). Effect of decoherence of optical field with broad angular spectrum upon propagation through transparent media interfaces. *Opt. Spectrosc.*, **110**, pp. 802–805.
102. Lyakin, D. V., and Ryabukho, V. P. (2011). Changes in longitudinal spatial coherence length of optical field in image space. *Tech. Phys. Lett.*, **37**, pp. 45–48.
103. Ryabukho, V. P., Lyakin, D. V., and Lychagov, V. V. (2007). What type of coherence of the optical field is observed in the Michelson interferometer. *Opt. Spectrosc.*, **102**, pp. 918–926.
104. Abdulhalim, I. (2012). Spatial and temporal coherence effects in interference microscopy and full-field optical coherence tomography. *Ann. der Physik*, **524**, pp. 787–804.

Chapter 4

Cross Talk in Full-Field Optical Coherence Tomography

Boris Karamata, Marcel Leutenegger, and Theo Lasser
Laboratoire d'Optique Biomédicale, École Polytechnique Fédérale de Lausanne, 1015 Lausanne, Switzerland
boris.karamata@a3.epfl.ch

The formation of cross-sectional images of biological tissues requires the discrimination between light conveying useful information—that is, propagating directly from object to image—from the abundant parasitic light caused by multiple scattering inherent to turbid media [1, 2]. In optical coherence tomography (OCT[a]) [3], selective detection of light undergoing a single backscatter event (reflection imaging) and rejection of multiply scattered light (MSL) has been successfully achieved by combining temporal coherence gating and confocal spatial filtering [4, 5].

---

[a] OCT systems incorporating sample objectives with relatively a large numerical aperture (NA) so as to provide enhanced en face optical sectioning are quite arbitrarily designated optical coherence microscopy (OCM) systems. Since our investigation and conclusions are independent of the NA, we use the term "OCT" throughout this chapter.

Handbook of Full-Field Optical Coherence Microscopy: Technology and Applications
Edited by Arnaud Dubois
Copyright © 2016 Pan Stanford Publishing Pte. Ltd.
ISBN 978-981-4669-16-0 (Hardcover), 978-981-4669-17-7 (eBook)
www.panstanford.com

In full-field OCT[b] (FF-OCT) [8–10], the rejection of MSL is even more challenging due to the large illumination field and the loss of confocal spatial filtering in a parallel detection scheme [1, 11]. As our investigation will reveal, despite temporal gating detection capabilities can be severely limited by cross talk, an unwanted signal contribution caused by MSL from the full-field illumination volume [11].

In this chapter, we develop a model accounting for multiple scattering in OCT and use it to predict cross talk effects in FF-OCT. As will be shown both theoretically and experimentally, the amount of cross talk strongly depends on the sample properties and system parameters and, above all, on the nature of the illumination. When the latter is spatially coherent, as obtained, for example, with a broadband laser, the cross talk contribution is generally a serious limitation to the method [11]. At the other extreme, spatially incoherent illumination (SII), as provided, for example, with a thermal light source, prevents the cross talk contribution [12].

Given this decisive advantage of SII, it would be interesting to understand the initial motivations for developing FF-OCT systems based on either type of illumination. In the early nineties, soon after the first developments of point-scanning OCT systems, two major development tasks were recognized: to improve the resolution and to increase the measurement speed. Image resolution at cellular level would secure sounder diagnosis and offer new applications in developmental biology [13], while faster measurement would allow the elimination of artefacts due to sample motions as well as observation of dynamic phenomena [14, 15].

On one side, FF-OCT systems with SII were primarily developed to obtain a very good resolution at minimum complexity and cost [8, 16, 17]. En face imaging inherent to FF-OCT allows the exploitation of thermal light sources, while maintaining a high enough signal-to-noise ratio (SNR) provided measurement time is sufficiently long. Since the naturally broad spectrum of thermal light sources yields very high longitudinal resolutions [18, 19], they are a valuable alternative to the sophisticated femtosecond

---

[b]With FF-OCT we implicitly consider full-field illumination. FF-OCT measurements for en face or three-dimensional imaging can also be performed with point-scanning [6] or line-scanning systems [7].

lasers used in point-scanning OCT, whose cost and complexity can be restrictive. Moreover, en face imaging is the configuration of choice with respect to transverse resolution [6, 8]. Surprisingly, a key advantage of the method—the cross talk rejection properties offered by thermal light sources—was not emphasized by most researchers with the exception of Fercher et al. [19]. However, as discussed further below, the weakness of FF-OCT systems relying on SII is first due to the low modal power of thermal light sources [19–21], and then to measurement speed limitations imposed by the insufficient performance of conventional detectors. This technological shortage, which obviously also affects FF-OCT with SCI, is even more important in this other method given its purpose explained hereafter.

On the other side, FF-OCT systems relying on spatially coherent illumination (SCI) were developed with the main aim to increase the measurement speed thanks to parallel acquisition [10, 22–24]. The idea was to exploit the broadband spatially coherent light sources developed for point-scanning OCT, that is, mainly superluminescent diodes (SLDs) and short-pulsed lasers. At that time, given the lack of appropriate quantitative knowledge, cross talk was mainly considered as a potential limitation to the method. The primary concern, when aiming at high measurement speed, was the technological barrier set by the too low readout speed and dynamic range of two-dimensional detectors such as CCDs [9, 15]. Indeed, to ensure a high enough SNR, while avoiding saturation of the detector due to the large reference signal component, it is necessary to average a relatively large collection of images at the same position [16, 18]. Moreover, the reconstruction of the interferometric OCT signal requires high sampling in the longitudinal dimension, which further increases the amount of images to be generated [16, 18]. Thus the image acquisition speed is limited by the time required for the accumulation of many images and the detector readout time. Thus requirement for high image acquisition rate–ultimately limited by the capabilities of the analog-to-digital converter,[c] considerably slows down the FF-OCT

---

[c]For a two-dimensional sensor such as CCDs, readout speed is determined by the capabilities of the analog-to-digital converter typically allowing for an acquisition speed of hundred frames per second, while the dynamic range available for signal capture is determined by the full well depth capacity of a pixel, which is typically

method. To address this crucial technological issue, the flexibility of design in CMOS detectors was exploited to develop arrays with a custom integrated circuit around each pixel performing the following operations: 1) high-pass filtering cancelling out the large reference signal, 2) amplification and rectification of the heterodyne signal, and, 3) low-pass filtering. The implementation of this parallel heterodyne detection scheme provides a higher dynamic range and requires a lower sampling rate to obtain the envelope of the OCT signal. Very high image acquisition rates were demonstrated with a prototype of such detectors [9, 22]. Presently, this can also be achieved with a commercially available custom camera (Heliotis AG) endowed with the same functionality and more pixels [25, 26].

Moreover, high-speed measurement with sufficient sensitivity for OCT biological imaging requires a spatially incoherent light source brighter than the commonly used thermal light sources, whose modal power is inherently low [19–21]. A few ways for creating such an ultrabright extended light source are briefly discussed in the conclusion of this chapter.

A new generation of FF-OCT systems incorporating an ultrabright spatially incoherent light source and a custom detector array may well compete with ultrafast Fourier domain OCT (FD-OCT) systems [14, 27] in terms of image acquisition speed. A parallel operation in the Fourier domain, which requires line illumination and a two-dimensional detector, provides longitudinal cross-sectional images without mechanical scanning [7]. Tremendous acquisition speeds without compromising on SNR and the number of pixels can be attained thanks to a more efficient use of light for building OCT signals yielding sensitivity superior to that of time domain OCT [27]. This remarkable feature allows the use of CCD detectors with good SNR while avoiding saturation by the large reference signal component. With potentially comparable acquisition speeds for en face cross-sectional images, the new generation of FF-OCT systems might become a valuable alternative to FD-OCT for fast three-

---

limited to a few hundred thousands photo-electrons. The corresponding dynamic range (without cooling system)—given by the square root of this number of photo-electrons, is then only of two orders of magnitude.

dimensional measurements, as briefly discussed in the conclusion of this chapter.

In Section 4.1 we will define cross talk noise in FF-OCT and provide a deeper analysis of its coherence properties. From there we will first derive the key assumptions on which an OCT model accounting for multiple scattering should be built, and second explain how cross talk can be suppressed. Our model is presented in Section 4.2 and the results of experimental validation are reported in Sections 4.3 and 4.4. In Section 4.4 we investigate the role of important sample properties and of the key system design parameters in relation to cross talk-generated noise for FF-OCT with SCI. The interest of this comprehensive study is twofold. First, it allows testing the validity of our theoretical model, which rests on assumptions fundamentally different from those on which other existing models are based; and, second, it offers a method to determine the quantitative contribution of cross talk in FF-OCT relying on SCI. In Section 4.5 we will present experiments that reveal how SII enables FF-OCT imaging free of cross talk noise. Finally, we will discuss our theoretical and experimental investigation and draw general conclusions. In particular, we will discuss the more complex models based on the extended Huygens–Fresnel principle [28], which rest on assumptions fundamentally different from ours.

Our study provides a deeper understanding of the role played by multiple scattering in coherence based detection methods and allows to better evaluate limitations in FF-OCT.

4.1 Optical Cross Talk in FF-OCT

In full-field optical coherence tomography (FF-OCT), the abundant amount of multiply scattered light (MSL) generates optical cross talk between the parallel detections channels. Despite temporal coherence gating, a fraction of such cross talk light usually generates a coherent cross talk noise contribution to the OCT signals. In Sections 4.1.1 and 4.1.2 we define more precisely cross talk noise. We analyze the coherence properties of cross talk light and determine the degree of correlation with the OCT reference signal,

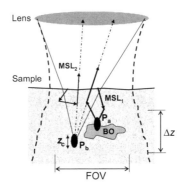

Figure 4.1 Scheme of an interferometer arm with a diffuse sample showing different optical paths yielding cross talk in FF-OCT: full-field illumination beam (dashed lines), field of view (FOV), ideal probe volume (P_b), virtual probe volume (P_a), backscattering object (BO), longitudinal scan range (Δz), longitudinal resolution (z_c), forward multiply scattered light from BO (MSL$_1$), and MSL from the scattering medium alone (MSL$_2$).

that is, its propensity to generate a detectable noise. In Section 4.1.3 we explain how cross talk-generated noise can be suppressed by exploiting the spatial coherence properties of the OCT light source.

4.1.1 Origin and Definition of Cross-Talk Noise

To define more precisely cross talk-generated noise that can occur in typical FF-OCT interferometers operated in the time domain (see Section 4.3.1), we will consider Fig. 4.1, which schematically represents the interferometer sample arm with several multiple scattering trajectories within the sample. The latter consists of a backscattering object (BO) embedded in a homogeneous scattering medium.

The full-field illumination beam diffusing through the sample determines the overall scattering volume. The ideal measurement volume is determined by the field of view (FOV) of the optical system and by the range (Δz) of the scan depth. The resolution of a given OCT system defines the size of a probe volume (P_n). With an appropriate design, the tranverse resolution and the depth resolution are determined by the numerical aperture (NA) of the sample arm objective and by the source coherence length l_c, respectively. A probe volume has a longitudinal dimension $z_c = l_c/2n$, where n is the sample refractive index. For each probing depth, a collection of probe volumes are defined in the same transverse plane (perpendicular to the optical axis). Each of these probe volumes P_n is imaged on a specific portion of the detector array (group of pixels), designated detector C_n.

Ideally, for a depth z, defined by the reference mirror position, only light originating from a given probe volume P_n should be detected by its conjugated detector C_n. Consider the two probe volumes P_a and P_b conjugated to the detectors C_a and C_b, respectively (see Fig. 4.1). Propagation of MSL across the whole scattering volume biases the ideal one to one correspondence between a given probe volume and its conjugated detector. The cross talk-generated noise on a detector can be defined as the total coherent signal contributions brought by light originating from all probe volumes in the measurement volume, with the exception of the probe volume conjugated to the detector at a given depth z. A coherent contribution may occur only if the random paths taken by MSL are equally long (within the distance of correlation determined by the source autocorrelation function g_0) to the ideal ballistic paths set by the reference mirror position. Figure 4.1 illustrates how forward MSL originating from the probe volume P_a (MSL_1), or from the homogeneous scattering medium (MSL_2), can reach the detector C_b conjugated to probe volume P_b, while taking identical path lengths.

4.1.2 Qualitative Analysis of the Cross-Talk Contribution

We propose to examine under what circumstances cross talk light—which necessarily consists of MSL—can interfere with the reference field (see Fig. 4.2). In other words, we need to determine the degree of correlation between MSL and the reference field in OCT.

In a typical FF-OCT setup with a spatially coherent light source, such as illustrated in Fig. 4.2, the reference field consists of a planar wavefront (RW) at the two-dimensional detector array, whereas the light scattered by a diffuse sample produces a distorted random wavefront (SW). The degree of correlation

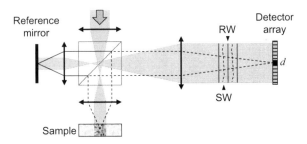

Figure 4.2 Scheme of an interferometer showing the distorted wavefronts from the sample (SW) and the plane wavefronts from the reference mirror (RW). The speckle size, corresponding to the Airy disc size d, is shown at the detector array.

between these two interfering fields depends both on their spatial and temporal coherence properties. It seems to be widely accepted that the interference process at the detector results from complex phenomena difficult to model. The analysis presented here below reveals that this is actually not the case.

4.1.2.1 Spatial coherence of cross talk light

With SCI the reference field is obviously spatially coherent. Let us examine the spatial coherence properties of the sample field. With a spatially coherent light source and a diffuse sample, the light backscattered from the sample generates a speckle pattern on the detector array. The distorted wavefront corresponding to this pattern interferes with the reference field producing a speckle pattern whose intensity obeys a known statistical distribution [29, 30]. More importantly the speckle size roughly corresponds to the Airy disc size d as determined by the objective NA of the detector arm (see Fig. 4.2). By definition, phase difference across the Airy disc is negligible. In an appropriate design, the pixel size of the detector array should be no more than half the size of the Airy disc so as to fulfil Shannon's condition for optimal spatial sampling. This means that, independent of the random orientation of the wavefront reaching the detector array (RW), the phase difference is negligible across a single pixel detector, which is exactly the condition for spatial coherence [31]. Thus, any portion of the distorted wavefront viewed by a single pixel detector can be but spatially coherent.[d]

Therefore, we can conclude that, in an FF-OCT design with adequate spatial sampling, despite distortion of the backscattered sample field due to multiple scattering, the portion of light collected by a single detector is spatially coherent. This means that any potential reduction of correlation between the reference and sample fields cannot be attributed to a loss of spatial coherence in the sample field.

4.1.2.2 Interference with multiply scattered light

Conservation of spatial coherence of light detected from the sample arm by a single detector is a necessary but not a sufficient condition to ensure the correlation between the reference and the sample field.

As explained in Section 4.1.1, at a given probing depth set by the reference mirror position, with MSL, equivalent optical path lengths can exist in the sample arm for different scattering paths outside of the ideal conjugated probe volume to be imaged. Obviously, if the random path lengths exceed the distance of correlation determined by the source autocorrelation function g_0, no interference can occur. Now, the following question becomes crucial: can the correlation be degraded by the random scattering events with MSL for path lengths falling within the coherence length set by g_0? Should this be the case, a reduction of correlation can be caused only by a temporal stochastic process such as Brownian motion generating a random phase relationship between the sample and reference fields. Actually, although a totally motionless biological sample is seldom met in practice, this assumption is generally valid in the context of OCT detection. In this case, a sample can be considered as locally motionless provided that fluctuations of the interference fringes obtained with the sample and reference fields are negligible during the measurement time interval necessary for recording one fringe period. This condition is fulfilled as shown in another

---

[d] Numerous studies on the degree of spatial coherence implicitly deal with areas much larger than the speckle size. For instance, for a large sampling area, Yang et al. showed how the loss of spatial coherence of forward-scattered light propagating in a turbid medium is related to the number of scattering events [32].

both theoretical and experimental study of the unfavorable case of an aqueous suspension of microspheres where the interference fringes are shown to be fully stationary (frozen) during the relevant measurement time in OCT [33].

Thus, contrary to a widespread belief, it turns out that, despite the random scattering events along various paths taken by MSL, the latter remains correlated to the reference field when falling within the source coherence length, exactly like for ballistic light. Under these conditions, the degree of correlation depends only on the source autocorrelation function g_0. It follows that MSL can strongly contribute to the OCT signal in the form of a coherent noise, which cannot be discriminated from the useful OCT signal (ballistic light) despite temporal coherence gating. This unwanted signal contribution exhibits strong speckle fluctuations as soon as the random phase delays of MSL—caused by a spatial stochastic process—fall within the range of the source's central wavelength. Agreement between theory and measurements regarding the statistical distribution of the speckle intensity strongly supports the view that MSL interference with the reference signal can be maximal ($g_0 = 1$) [33].

We conclude that MSL is fully correlated to the reference field for equivalent optical path lengths in both interferometer arms. The average magnitude of OCT signals at a given position depends only on the amount of light—ballistic or multiply scattered—taking an optical path equivalent to the one in the reference arm as set by the mirror position. Multiple scattering is responsible for a coherent noise contribution in the form of a speckled component, but not for a loss of correlation between the reference and sample fields.

4.1.2.3 Questions raised by our analyses

The above analyses provide a qualitative insight into cross talk-generated noise and raise two important questions:

(1) How important is the cross talk noise contribution relative to the useful OCT signal?
(2) How can cross talk noise be suppressed?

To answer the first question we present in Section 4.2 an OCT model which accounts for MSL and allows quantifying the cross talk noise contribution as a function of various relevant physical parameters. Regarding the second question, we will explain in the next section how cross talk-generated noise can be drastically reduced by exploiting the spatial coherence properties of the light source.

The above analyses also raise questions relative to OCT models built on assumptions incompatible with our findings. Indeed, sophisticated models based on the extended-Huygens principle and the use of mutual coherent functions (MCFs) rest on both spatial and temporal statistical averaging [34, 35]. Thus, they inherently assume a large sampling area when calculating the MCFs, as well as a long enough measurement time per probe volume, which would lead to some statistical averaging inducing a decorrelation between the reference and sample fields. The reduced degree of correlation obtained with such complex calculations leads to the prediction of lower cross talk noise contributions. Interested readers can find a more thorough discussion of this topic in Ref. [33].

4.1.3 Cross-Talk Noise Suppression with Spatial Coherence Gating

Generally, with full-field illumination large amounts of MSL are generated and collected (cross talk light). Therefore, cross talk-generated noise can be suppressed only by preventing the interference between MSL and the reference field. As explained hereafter, this can be achieved thanks to a low degree of spatial coherence within the full-field as generated by a spatially incoherent light source. The idea is to create, for each parallel detection channel, an effect equivalent to confocal spatial filtering by exploiting spatial coherence effects. This will provide a "spatial coherence gating" which one can figure out in a relatively straightforward and intuitive manner.

The principle of spatial coherence gating relies on the creation of mutually incoherent probe volumes within the sample. The probe volumes occupy adjacent positions with center coordinates $x_S y_S$ in the sample transverse cross section. In the interferometer, such probe volumes are duplicated in the reference arm at twin positions $x_R y_R$ on the reference mirror. When a reference field

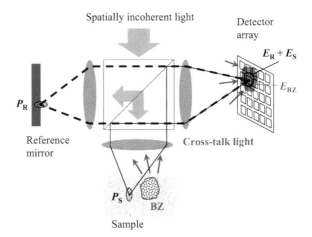

Figure 4.3 Schematic illustration of the spatial coherence gating concept. With spatially incoherent illumination, light from the probe volume P_S can interfere only with light from its replica P_R. Cross-talk light from the backscattering zone (BZ) produces a field that does not contribute to the OCT interference signal.

(E_R) and a sample field (E_S) recombine on the detector plane, interference can occur only if images of the corresponding twin probe volumes P_R and P_S are superimposed as illustrated in Fig. 4.3. The reference field, which is not perturbed by scattering, acts as a spatial coherence gate allowing interference only with light originating from the corresponding sample probe volume.

Assume cross talk noise generated by the sample backscattering zone (BZ) reaching the detector on which E_R and E_S overlap (see Fig. 4.3). An additional field contribution E_{BZ} due to cross talk is created on this detector yielding a total field $E_T(x, y) = E_R(x_R, y_R) + E_S(x_S, y_S) + E_{BZ}(x_{BZ}, y_{BZ})$. Since, by design, $E_{BZ}(x_{BZ}, y_{BZ})$ does not share the same spatial coherence properties as $E_R(x_R, y_R)$, interference cannot occur and the last term $E_{BZ}(x_{BZ}, y_{BZ}) = 0$. This selective interference allows cross talk noise suppression in full-field illumination.

Optimal cross talk noise rejection, that is, an amount of coherent noise owing to MSL as low as in point-scanning OCT, is obtained provided there is a good overlap between probe and coherence volumes $V_c = l_c A_c$, as defined by L. Mandel and E. Wolf [36]. Therefore, coherence volumes matching the probe volumes should be created. Generally, the longitudinal dimension of both coherence and probe volumes is the same in OCT since both are determined by the source coherence length l_c. Therefore, it suffices to match the cross-sectional area of a probe volume A_p with the coherence area A_c of the coherence volume. Since the Airy disc and coherence area share the same physical properties, namely, phase fluctuations significantly lower than 2π [31], the critical condition $A_p \approx A_c$ is quite naturally met in a practical case, given that the same sample arm is used both for illumination and detection. Thus, achieving an efficient spatial coherence gating requires only the pupil of the sample arm to be filled with spatially incoherent light. This is generally the case in a conventional optical setup used for OCT, where a spatially incoherent light source is imaged onto the sample [8, 12]. Some mathematical insights into the equivalence $A_p \approx A_c$, as well as more formal design guidelines for optimal coherence gating and power throughput can be found in Refs. [12, 20].

We would like to point out that spatial coherence gating in a parallel detection scheme offers even more than an effect equivalent to confocal spatial filtering for each channel. Indeed, besides rejection of MSL, spatial coherence gating yields an enhanced longitudinal and tranverse resolution identical to that of confocal imaging systems. This extremely interesting property, obtained whenever an extended source is combined with an interferometric detection process, was described by several authors in various contexts (Davidson et al. [37], Lee et al. [38], and more recently by Somekh et al. [39]). In a closely related context, Sun and Leith showed the equivalence of extended-source image plane holography and the confocal imaging process [40].

4.2 Theory and Model

A comprehensive model of OCT requires modeling both the light propagation in random media and the interference process, which depends on coherence properties of the sample field. Light propagation in random media, including temporal aspects, can be described quite successfully by the time-resolved diffusion theory [41, 42]. However, this theory is based on the diffusion approximation, which becomes valid only after a few scattering events [43] yielding delays of a few picoseconds [42, 44]. Such delays, which largely exceed typical source coherence times in OCT, correspond to considerable path lengths, that is, in the order of one millimeter. Therefore, models based on the diffusion approximation fail to properly describe MSL distribution in the range of interest met in OCT where relevant path lengths can also be one to three orders of magnitude lower. A calculation performed by a Monte Carlo simulation can provide the spatiotemporal distribution of light within the entire range of interest [42, 44]. Several existing models of OCT are based on Monte Carlo simulation, for example [45–47]. However, in principle, the latter cannot account for the coherence properties of MSL, whose knowledge is indispensable in the context of coherent detection techniques.

Here, we present a comprehensive model of OCT accounting for multiple scattering. The model rests on the two important assumptions derived in Section 4.1:

(i) The portion of the backscattered sample field collected by the detector (by each detector if there is more than one like in FF-OCT) is spatially coherent (see Section 4.1.2.1)
(ii) The interference fringes signal is stationary (frozen) during the time for recording one fringe period (see Section 4.1.2.2).

4.2.1 Mathematical Description

The mathematical description of interfering fields in an amplitude-splitting interferometer where both ballistic light backscattered once and MSL are collected in the sample arm (see Fig. 4.4), is greatly simplified with the above two assumptions. Indeed, assumption (i) implies that the phase of the electrical field is constant across the detector area. Thus, the description of the interference process does not require cross-correlation functions across the detector plane, leaving only its perpendicular dimension z to be considered in the model. Assumption (ii) allows a description of the interference process with a sample field altered by spatial

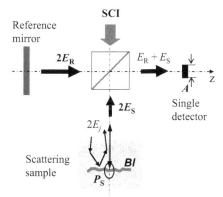

Figure 4.4 Scheme of an amplitude splitting interferometer with broadband spatially coherent illumination (SCI). E_R and E_S are the reference and sample fields, respectively. E_j are components of E_S corresponding to ballistic light backscattered once from backscattering interface BI (double line) and MSL (single line). Interference between E_R and E_S is considered on a single detector A—whose size is smaller than the Airy disc—located at a conjugated distance with the sample probe volume P_S. The beam splitter equally divides light. The full optical setup is shown in Fig. 4.5.

stochastic processes constant over the time for building one fringe period. Thus, calculation of the cross-correlation function between the sample and reference fields requires only accounting for spatial stochastic processes brought by the sample, but not for temporal stochastic processes.

Let us first calculate the intensity for a sample consisting of a simple plane mirror. Consider a broadband source power spectrum $I(k)$ where $k = 2\pi/\lambda$ is the wavenumber. Assuming 100% reflectivity of the reference mirror and a reflectivity $r(k)$ for the sample mirror, the electrical fields E_R and E_S reflected from the reference and from the sample arm, respectively, can be expressed for a given wavenumber as

$$E_R = \text{Re}\left\{\sqrt{I(k)}\exp\left(i(kz - \omega t)\right)\right\} \quad (4.1)$$

$$E_S = \text{Re}\left\{r(k)\sqrt{I(k)}\exp\left(i(\varphi(k) - \omega t)\right)\right\}, \quad (4.2)$$

where z is the reference mirror position and $\phi(k)$ a phase argument depending on the position of the sample mirror.

When the waves recombine in the interferometer, the total resulting intensity as a function of the wavenumber is

$$I_T(k) = \langle |E_R(k) + E_S(k)|^2 \rangle$$
$$= I(k)\left[1 + r^2(k) + 2r(k)\text{Re}\left\{\exp\left(i(\varphi(k) - kz)\right)\right\}\right], \quad (4.3)$$

where the brackets denote averaging over a time interval much longer than the time of an optical cycle.

The total intensity on the detector is obtained by integrating over the whole spectrum.

$$I_T = \int_0^\infty I_T(k)dk = \int_0^\infty I(k)\left(1 + r^2(k)\right)dk$$
$$+ 2\text{Re}\left\{\int_0^\infty I(k)r(k)\exp\left(i\varphi(k)\right)\exp\left(-ikz\right)dk\right\}. \quad (4.4)$$

The first term, independent of z, is a constant intensity I_0, while the second one corresponds to the Fourier transform of the power spectrum weighted by the sample spectral reflectivity. Therefore, Eq. 4.4 is equivalent to

$$I_T = I_0 + 2\text{Re}\left\{\mathbb{F}\left[I(k)r(k)\exp\left(i\varphi(k)\right)\right]\right\}. \quad (4.5)$$

Let us now consider the more general case met in OCT where both ballistic light backscattered once and MSL contributions are present. Here, the sample field E_S is the sum of many contributions $E_j = \text{Re}\{u_j(k)\exp(ikL_j)\}$, each corresponding to a light ray having undergone a random path owing to interactions with the scattering medium. The length L_j is the additional geometrical path length accumulated by a multiply scattered photon (double path), relative to ballistic photons, and $u_j(k)$ is a field-weighting coefficient proportional to the field magnitude. The total sample field is

$$E_S(k) = \text{Re}\left\{\sum_{j=1}^N E_j(k)\right\} = \text{Re}\left\{\sqrt{I(k)}\sum_{j=1}^N u_j(k)\exp\left(i(kL_j - \omega t)\right)\right\}. \quad (4.6)$$

Repeating for a scattering sample the calculation's steps that led to Eq. 4.5, and assuming the coefficients $u_j(k)$ to be independent of the wavelength (i.e., either no, or constant absorption over the spectrum), we obtain

$$I_T = I_0 + 2\text{Re}\left\{\mathbb{F}\left[I(k)\sum_{j=1}^{N}u_j\exp\left(ikL_j\right)\right]\right\} \quad (4.7)$$

and by virtue of the convolution theorem

$$I_T = I_0 + 2\text{Re}\left\{\mathbb{F}\left[I(k)\right]\otimes\sum_{j=1}^{N}u_j\mathbb{F}\left[\exp\left(ikL_j\right)\right]\right\}. \quad (4.8)$$

According to the Wiener–Khinchin theorem $\mathbb{F}[I(k)]$ is the source autocorrelation function $g(z)$. Defining $I'(\Delta k) = I(k)$, where $\Delta k = k - k_0$, and $\lambda_0 = 2\pi/k_0$ is the central wavelength of the light source, $g(z)$ can be expressed as follows

$$g(z) = \mathrm{F}\left[I(k)\right] = \mathbb{F}\left[I'(\Delta k)\right] = \exp\left(ik_0 z\right)\mathbb{F}\left[I'(k)\right] = \exp\left(ik_0 z\right)g_0(z), \quad (4.9)$$

where $g_0(z)$ is a complex function whose argument and amplitude vary slowly relative to $g(z)$ and whose module is the envelope of $g(z)$. The second Fourier transform in Eq. 4.8 corresponds to a delta function.

$$\mathbb{F}\left[\exp\left(ikL_j\right)\right] = \delta(z - L_j) \quad (4.10)$$

After substitution of Eqs. 4.9 and 4.10 into Eq. 4.8, one obtains

$$I_T = I_0 + 2\text{Re}\left\{\sum_{j=1}^{N}g_0(z)\exp\left(ik_0 z\right)\otimes u_j\delta(z - L_j)\right\}. \quad (4.11)$$

This expression reveals that the nature of the signal consists of a convolution of the autocorrelation function with randomly distributed delta functions, located at random L_j positions. Exploiting the shift properties of the delta function, we can simplify Eq. 4.11 by

$$I_T = I_0 + 2\text{Re}\left\{\exp\left(ik_0 z\right)\sum_{j=1}^{N}u_j g_0(z - L_j)\exp\left(-ik_0 L_j\right)\right\}. \quad (4.12)$$

This equation provides the intensity detected as a function of the reference mirror position z. In OCT that relies on heterodyne detection, the reference mirror is scanned at constant velocity inducing a modulation of the signal at the Doppler frequency. The effective OCT signal is the electrical current i_D obtained after bandpass filtering at the Doppler frequency and subsequent envelope demodulation. The first signal processing operation suppresses the constant term I_0 and the second operation removes the carrier modulation $\exp(ik_0 z)$ and leaves a signal proportional to the module of the filtered signal. Such signal processing yields the following detected current:

$$i_D(z) \propto \left| \sum_{j=1}^{N} u_j g_0(z - L_j) \exp\left(-ik_0 L_j\right) \right|. \quad (4.13)$$

This equation reveals that, despite multiple scattering, all the light detected in OCT interferes coherently within the coherence length around a position z imposed by the reference mirror. Moreover, for each position z, the sum of factors with random arguments leads to speckle formation, which accounts for the randomness of the measured signals.

4.2.2 Calculation of the Mean Signal

Since we are interested in determining the mean contribution of multiple scattering in OCT, we will calculate the mean of the random signal detected. The summation in expression (4.13) corresponds to a sum of random phasors, different around each position z, which results in a single random phasor with amplitude $A(z)$ and phase $\Phi(z)$. Assuming the argument of $g_0(z)$ to vary slowly relative to $\exp(-ik_0 L_j)$, one can write

$$\sum_{j=1}^{N} u_j g_0(z - L_j) \exp\left(-ik_0 L_j\right) \approx \sum_{j=1}^{N} u_j \left| g_0(z - L_j) \right| \exp\left(-ik_0 L_j\right)$$

$$= \sum_{j=1}^{N} \alpha_j(z) \exp\left(-i\theta_j\right) = A(z) \exp\left(i \Phi(z)\right). \quad (4.14)$$

Classical results from statistical optics derived by Goodman can now be exploited. His calculations of the statistical distribution of a sum of random phasors for various cases rest on two important initial assumptions [31]:

(i) The phases θ_j are uniformly distributed over the interval $[0, 2\pi]$.

(ii) The amplitude α_j and phase θ_j of the jth elementary phasor are statistically independent of each other, as well as of the amplitudes and phases of all elementary phasors.

Because we have $L_{max} >> 2\pi/k_0$ in all experiments and because we can assume the paths L_j to be randomly distributed in the interval $[0, L_{max}]$, the phase argument $\theta_j = k_0 L_j$ is uniformly distributed between 0 and 2π, as required by the first assumption.

The second assumption implies that both $g_0(z)$ and u_j are independent of L_j. In the first approximation, the independence between $g_0(z)$ and L_j is verified if the module of $g_0(z)$ varies slowly with L_j relative to the function $\exp(-ik_0 L_j)$. This condition is met for $\Delta \lambda << 1/k_0$, which is usually the case in OCT. However, u_j and L_j are generally interdependent. Fortunately, as shown in Appendix A, it is sufficient to verify independence locally, that is, for a relatively short path length interval ΔL in the order of the source coherence length l_c. Such local independence is generally achieved in OCT, as in our further experiments, and we can therefore assert the second assumption.

Therefore, the results derived by Goodman can be directly applied to the general case of OCT in which both ballistic light backscattered once and MSL contributions are present. The random phasor sum described by Eq. 4.14 obeys a probability density function, whose mean and variance are [31]

$$\overline{A(z)} = \sqrt{\frac{\pi}{2}} \sigma(z) \qquad (4.15)$$

and

$$\sigma^2(z) = \frac{\overline{\alpha_j^2(z)}}{2}, \qquad (4.16)$$

respectively. $\overline{A(z)}$ is directly proportional to the mean value of $i_D(z)$ in Eq. 4.13, i.e., to the mean amplitude of the random OCT signal. Such mean signal can be roughly measured by averaging demodulated signals as explained in Section 4.3.1. Applying Eq. 4.16 to the random phasor sum described by Eq. 4.14 yields

$$\sigma^2(z) = \frac{1}{2}\overline{u_j^2 |g_0(z-L_j)|^2} = \frac{1}{2N} \sum_{j=1}^{N} U_j \left(g_0(z-L_j)\right)^2, \qquad (4.17)$$

leading to

$$\sigma(z) = \sqrt{\frac{1}{2N} U_j \otimes g_0^2(z)}, \quad (4.18)$$

where $U_j = (u_j)^2$ is the intensity coefficient corresponding to light traveling a path length L_j.

$\overline{A(z)}$ is directly proportional to the mean value of $i_D(z)$ in Eq. 4.13, that is, to the mean amplitude of the random OCT signal. Such mean signal can be roughly measured by averaging demodulated signals as explained in Section 4.3.1.

Thus the mean signal detected in OCT can be calculated by combining Eqs. 4.15 and 4.18. However, to perform this calculation, one still needs to know the coefficients U_j, which are proportional to the intensity $I_S(L_j)$ measured in the sample at depth $L_j/2$ from the sample mirror. $I_S(L_j)$ corresponds to the spatiotemporal distribution of the intensity, that is, of photons. The calculation of such a distribution lends itself very well to a Monte Carlo simulation [42, 44]. Therefore, our model combined with a Monte Carlo simulation, allows one to calculate OCT signals in accounting for ballistic (backscattered once) and MSL. The important features of the Monte Carlo simulation further used in our model are described in Section 4.3.4.

Our model was developed for a backscattering interface covered with scattering medium. This corresponds to most practical cases of interest such as biological interfaces, generally made of densely packed scatterers [48]. Even ballistic light backscattered once by these complex submicrometric structures reaches the detector with random phase delays and gives rise to speckle formation. The samples used in all further experiments are made of a mirror covered with a scattering solution. In this case, all ballistic light is reflected by a mirror and reaches the detector with the same phase argument. Thus, the sample field distribution actually corresponds to a deterministic phasor plus a random phasor sum, instead of a purely random phasor sum. The term "deterministic phasor" signifies here that the phasor's argument is not randomized owing to multiple scattering. It is important to thoroughly treat this case in order to be able to reliably model the samples used in our experiments. The calculations for this case, provided in Appendix B, yield a slightly more complex solution, whose implementation is less straightforward.

4.2.3 Integration of Monte Carlo Results

Practical integration of Monte Carlo simulation's results into our model requires the following analysis. A Monte Carlo simulation provides a photon distribution whose density is proportional to the mean intensity I_v in a vth sampling volume determined by the detector size and temporal distribution ($\Delta t \propto \Delta L$ range). The resolution determines an average path of length L_v of all path lengths L_j falling into the vth sampling volume. To express Eq. 4.18 as a function of I_v, the sum operation is distributed into V sampling volumes, each containing $m_{v+1} - m_v$ of the N elements. This yields

$$\sigma^2(z) = \frac{1}{2N} \sum_{j=1}^{N} U_j \left(g_0(z - L_j)\right)^2$$

$$= \frac{1}{2} \sum_{v=1}^{V} \frac{1}{m_{v+1} - m_v} \sum_{j=m_v+1}^{m_{v+1}} U_j \left(g_0(z - L_j)\right)^2$$

$$= \frac{1}{2} \sum_{v=1}^{V} I_v \left(g_0(z - L_v)\right)^2 \qquad (4.19)$$

leading to

$$\sigma(z) = \sqrt{\tfrac{1}{2} I_v \otimes g_0^2(z)} \qquad (4.20)$$

and

$$\overline{A(z)} = \sqrt{\tfrac{\pi}{2}} \sigma(z) = \sqrt{\tfrac{\pi}{4} I_v \otimes g_0^2(z)}. \qquad (4.21)$$

Practical implementation details for the calculation of I_v are provided in Section 4.3.4.

4.3 Method for Cross-Talk Investigation

In this section we describe the setup (Section 4.3.1) and the samples (Section 4.3.2) used for the measurement of cross talk in FF-OCT with SCI. Explanations on the presentation of our experimental results and simulation details are provided in Sections 4.3.3 and 4.3.4, respectively.

4.3.1 Setup

The experimental setup used for FF-OCT is illustrated in Fig. 4.5. A broadband spatially coherent light source, coupled to a single mode fiber, is launching light into a free space Michelson interferometer. Light is equally divided by a nonpolarizing beam splitter into the reference and the sample arm containing identical microscope objectives with a × 10 magnification (L_3 and L'_3) and a numerical aperture (NA) of 0.25. Lenses L_1, L_2 and L_3 are positioned so as to obtain a collimated beam illuminating the sample with a 420 μm Gaussian intensity profile measured at full-width at half-maximum (FWHM). The sample is imaged with a × 30 magnification by lenses L_3 and L_4, which form a microscope as illustrated in the sample arm in Fig. 4.5.

The source is an SLD. Its spectrum, centered on 810 nm with 17 nm bandwidth at FWHM, corresponds to a coherence length (l_c) of 34 μm at FWHM. The SLD (Superlum 381-HP2) delivers a power of around 1 mW onto the sample.

Interference between light backscattered from the sample and the reference mirror (RM) can occur only when the optical path length difference lies within the source coherence length. The

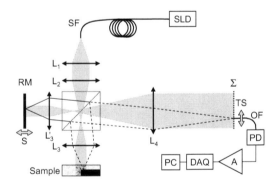

Figure 4.5 Scheme of an FF-OCT setup. Superluminescent diode (SLD); single-mode fiber (SF); reference mirror (RM); voice coil scanner (S); achromatic lenses (L_1, L_2, and L_4); microscope objective × 10, NA = 0.25 (L_3 and L'_3); translation stage (TS); 50 μm core optical fiber (OF); photodiode (PD); preamplifier (A); data acquisition card (DAQ); personal computer (PC).

reference mirror is mounted on a voice coil scanner (S) and scanned longitudinally at constant velocity over a depth of 750 μm at a frequency of 2 Hz with a 90% duty cycle. The resulting 6.7 kHz Doppler frequency modulation permits heterodyne detection.

To dispose of large dynamical range, avoid potential electronic cross talk and gain flexibility, a single movable detector is used instead of a detector array. The detection is made with a single photodiode (PD) to which light is delivered by an optical fiber (OF) moved in 120 μm steps across the image plane Σ with a motorized translation stage (TS). Each step corresponds to 4 μm on the sample side, a distance larger than the microscope objective resolution, which is around 2 μm. Such spatial undersampling allows for a larger measurement range while maintaining a reasonably low acquisition time. Note, however, that the 50 μm core of the optical fiber roughly matches the resolution of the microscope objective, which is around 60 μm after magnification with L_4.

The photocurrent produced by the photodiode is amplified and high-pass filtered by a low noise pre-amplifier (A). The signal is then digitized with a 12 bit analog-to-digital converter on a data acquisition card (DAQ) and numerically processed in a personal computer (PC). With the SLD, the signal processing consists of 0.6 kHz band-pass filtering around the 6.7 kHz Doppler frequency with a second-order Chebyshev filter, followed by envelope reconstruction with the Hilbert transform. The DAQ is triggered by the voice coil scanner at each new depth scan. An experimental sensitivity of −75 dB is obtained with the SLD when 25 demodulated signals are averaged.

Samples consisting of scattering solutions, such as used in our studies (see Section 4.3.2), undergo Brownian motion, that is, the scatterers are randomly changing their positions and inducing time-varying signals. Thus, a different random signal, modulated by speckle noise, is obtained for each OCT measurement. Taking the average of the envelopes a few OCT signals measured sequentially allows reducing speckle noise. Thanks to this operation—possible owing to the dynamic nature of our samples—one can easily obtain OCT measurements representative of the average cross talk signals, which is the value of interest in our study. Relevant speckle statistics and temporal behavior are thoroughly investigated in Ref. [33]. For instance, averaging 25 and 50 demodulated signals as in our further experiments, reduces the speckle contrast by 80% and 85%, respectively.

Some experiments require comparison with a confocal system, as used in point scanning OCT, without need for tranverse scanning. A confocal configuration with properties equivalent to point-scanning OCT can be obtained by simply removing lens L_2. The monomode fiber is then imaged onto the sample, providing a confocal spot illumination. The confocal configuration leads to an experimental sensitivity of −96 dB when 25 demodulated signals are averaged.

4.3.2 Sample

The sample consists of a cleaved GaAs edge coated with gold and embedded in a scattering solution maintained in a cell. The latter is limited by a 150 μm thick cover glass on the objective side. The distance Δz between the reflecting edge and the cover glass can be accurately varied. The scattering solution consists of monodisperse polystyrene microspheres in de-ionized water. Ultrasound shaking is applied for distributing the microspheres homogeneously. Tests revealed that, despite sedimentation, the sample optical properties remain stable during at least 10 minutes.

The absorption being negligible at the wavelengths used, the solution can be considered as purely scattering. Independent of the type of solution used (size of microspheres), the concentration is adjusted separately to a scattering coefficient $\mu_S = 6.2$ mm^{-1} using the collimated transmission method given in Ref. [49]. The cell inner thickness Δz determines the number of scattering mean free paths, that is, the sample optical density (OD) defined as $2\Delta z\mu_S$ when accounting for double optical path. The source wavelength ($\lambda \cong 810$ nm in air), as well as the size and refractive index ($n = 1.59$) of the microspheres suspended in water, yield an anisotropy parameter g defined as the average cosine of the scattering angles, that can be calculated with the Mie theory [50]. We use three different sizes of microspheres, that is, anisotropies g, in our experiments (see Section 4.4.1).

Depending on the experiment, the sample is moved in the transversally so as to present in the detection plane Σ either a full-mirror interface or half a mirror with the cleaved edge positioned on

(a) 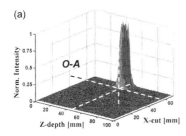 (b)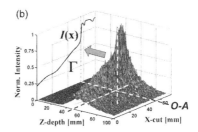

Figure 4.6 Cross-sectional FF-OCT image of a cleaved mirror with the edge break on the optical axis $(O - A)$: (a) in water; (b) in scattering solution (8 OD, $g = 0.93$); $I(x)$: projection of the maximum intensity profile in the plane Γ.

the optical axis. In this contribution, these positions are designated "full-mirror sample" and "half-mirror sample," respectively.

4.3.3 Presentation of Experimental Results

The graphical representation of the experimental results presented further in this chapter is not trivial and requires detailed explanations.

Consider the measurement performed with the FF-OCT system and the half-mirror sample, as described in the previous two sections. The three-dimensional plots shown in Fig. 4.6 are the cross-sectional images of the half-mirror sample—with the normalized intensity represented along the vertical axis—obtained in water and in a scattering solution, respectively. The plots correspond to the average cross talk signal obtained from a few tens of demodulated signals acquired at different time intervals. As explained in Section 4.3.1, random time-varying signals are obtained due to speckle fluctuations caused by the Brownian motion of the scatterers. In water, the half-mirror interface is clearly resolved in both dimensions, whereas with the scattering solution the cross talk spreads into the whole sample, degrading both the longitudinal and the tranverse resolution. Practically, a quantitative comparison between theory and experiment in a three-dimensional representation is difficult. Therefore, cross talk effects are investigated independently for the tranverse and the longitudinal dimensions with the half- and the full-mirror samples, respectively.

The tranverse cross talk extent is represented by the maximum intensity profile $I(x)$ corresponding to a projection of the three-dimensional intensity plot along the optical axis $(O-A)$ onto plane Γ, as shown in Fig. 4.6b. The edge break is always located on the optical axis. To reduce image acquisition time, we use measurements that cover 325 µm, from which 75 µm are on the cleaved mirror side and 250 µm are on the other side. $I(x)$ plots are obtained by averaging 25 demodulated signals so as to reduce the speckle contrast by 80% (see Section 4.3.1). To allow comparison with the ideal case of full cross talk rejection, we illustrate in all graphs the corresponding projection of the cleaved edge intensity profile measured in water.

The extent of longitudinal cross talk is measured along the optical axis with the full-mirror sample. The plots are normalized and shown as a function of the reference mirror displacement. The longitudinal plots correspond to the average of 50 demodulated signals yielding 85% lower speckle contrast (see Section 4.3.1). To allow comparison with the ideal case of full cross talk rejection, we illustrate the envelope of the autocorrelation function in all graphs.

Measurements are represented with corresponding theoretical results obtained with our model. Comparison of theoretical and experimental results for the tranverse dimension is less straightforward than for the longitudinal dimension, where all curves are normalized. Normalizing to the maximum intensity would not be reliable since the signals measured on the cleaved mirror side are very noisy (see results in Section 4.4) owing to rippled illumination profile, coating damage, and possibly microsphere aggregates sticking to the surface of the mirror. Therefore, experimental data are adjusted with a multiplicative factor K_f so as to obtain a least-square-fit difference with the theoretical curves. The adjustment is performed on data measured on the opposite side of the cleaved mirror.

On the practical side, between each measurement with a new parameter (see Section 4.4), we cleaned the sample cell with ethylene. Before introducing the scattering solution, we filled the sample cell with water to allow an accurate mirror positioning into focus.

4.3.4 Monte Carlo Simulation Details

The Monte Carlo simulation, which accounts for the sample properties and optical parameters, provides the spatiotemporal distribution of light collected from the sample arm. The algorithm used in our Monte Carlo simulation is described elsewhere, for example, by Wang et al. [51]. We used a Mie scattering distribution for unpolarized light at each particle interaction [50]. In our practical case we ignored it because absorption is negligible for polystyrene microsphere.

The position and the angle of a photon at the sample interface (after crossing the sample), determine on which detector it will fall. According to the rules of geometrical optics, photons collected by a detector originate from a sampling area whose points are conjugate to the detector area. For a scattering sample, the position and size of the detector actually determine a sampling area from which photons virtually originate. The accumulated path length delay (L_j) of a photon defines its position along the depth axis. The sampling volume introduced in Section 4.2.3 is then determined by the dimensions of the sampling area and by the longitudinal sampling resolution, which amounts to 1 μm in our simulation. The sampling area, which corresponds to the tranverse resolution of our system, is 4 μm.

For our specific case (mirror embedded in scattering solution), the intensity distribution $I_v(L_j)$, proportional to the number of photons m_v in a sampling volume v, is treated as follows. First, the ballistic component (interaction with the mirror only), here responsible for the major part of the high peak located at L_0 (sample mirror position), is removed from the distribution so that $\sigma(z)$ can be calculated from Eq. 4.22 in Appendix B. The mean OCT signal can then be derived from Eqs. 4.23–4.25 in Appendix B. More details of such calculation are shown for a case study in Ref. [33]. Note that, for the more general case involving backscattering interface with random microstructures (unlike mirror), the treatment is far more straightforward, since Eq. 4.21 can be directly applied to $I_v(L_j)$.

Particular care must be taken to implement the correct physical scaling factors into the simulations. Since L_j represents the geometrical path length, the corresponding length scale was multiplied by the water refractive index ($n = 1.33$) and divided by two, so as to account for the optical path length and reference mirror displacement, respectively. In addition, the width of g_0 must be divided by n.

4.4 Experimental and Theoretical Results for Cross-Talk Noise

In this section we will investigate how cross talk signals are correlated to the main properties of the sample, namely the optical density (OD) and the anisotropy (g) of the scattering solution (Section 4.4.1). The cross talk signals are measured with the setup described in Section 4.3.1. For each parameter, theoretical and experimental results are plotted on the same graph. In another experiment presented in Section 4.4.2, the intensity of the cross talk signal relative to that of the useful OCT signal (ballistic light) will be investigated as a function of the probing depth. At the end of this section, relevant additional results are briefly mentioned before the conclusions (Section 4.4.3).

4.4.1 Dependence on Sample Properties

Cross-talk noise dependence on the anisotropy of the scattering solution is investigated first. Results with three different anisotropy parameters are shown in Fig. 4.7a and Fig. 4.7b for the full-mirror sample and for the half-mirror sample, respectively (see Section 4.3.2). Microsphere diameters smaller ($d = 350$ nm), roughly equal ($d = 750$ nm) and larger ($d = 2050$ nm) than the illumination wavelength ($\lambda = 810$ nm in air, that is, 610 nm in water) were used, yielding the anisotropy parameters $g = 0.55$, $g = 0.85$, and $g = 0.93$, respectively [50]. For the three solutions the scattering coefficients were adjusted to $\mu_S = 6.2$ mm^{-1} as explained in Section 4.3.2, leading to OD $= 8$ for $\Delta z = 650$ μm. Note that the chosen anisotropy coefficients are representative of biological tissues, typically lying between 0.7 and 0.99 [52].

Let us first comment on the results obtained with the full-mirror sample. A wide-angle scattering solution is obtained with

Experimental and Theoretical Results for Cross-Talk Noise | 159

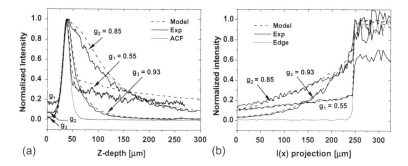

Figure 4.7 Experimental (Exp) and theoretical (Model) results obtained in FF-OCT for various anisotropies (g) with OD = 8, D = 420 μm, and NA = 0.25: (a) correlogram envelopes for the full-mirror sample, and envelope of the source autocorrelation function (ACF) given for reference; (b) projections of maximum intensity profiles obtained with the half-mirror sample, and the corresponding profile in water given for reference (Edge).

$g = 0.55$, resulting in a nearly flat cross talk signal extending over a long distance, while a large peak emerges at the mirror interface. This peak, whose width corresponds to the envelope of the source autocorrelation function (ACF), is caused by the ballistic light reflected by the mirror. The contrast, rather than the longitudinal resolution, is reduced. With anisotropy $g = 0.85$ moderate forward scattering is obtained giving rise to a long tail dramatically reducing the longitudinal resolution. With anisotropy $g = 0.93$ relatively high forward scattering is obtained resulting in limited MSL delays and in a signal width around 50% broader than the ACF envelope at FWHM.

With the half-mirror sample, cross talk effects are more pronounced with $g = 0.93$ showing that the moderately delayed MSL spreads quite significantly into the tranverse dimension. This study tends to show that, in our experimental conditions, the most deleterious cross talk effects, in terms of large noise contribution spreading far from the ideal probe volume, are obtained for microsphere diameters approaching the source central wavelength. In principle, for a larger NA, more wide-angle scattered light is collected and the worst cross talk figure is obtained for smaller microsphere diameters. Our model allows precisely determining the ratios λ/d yielding the best or the worse figure for each specific case.

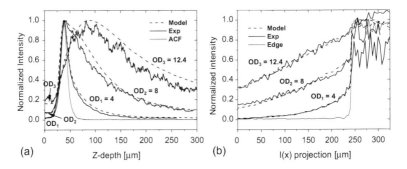

Figure 4.8 Experimental (Exp) and theoretical (Model) results obtained in FF-OCT for various optical densities (ODs) with $\mu_S = 6.2$ mm^{-1}, $g = 0.85$, $D = 420$ μm, and NA $= 0.25$: (a) correlogram envelopes for the full-mirror sample, and envelope of the source autocorrelation function (ACF) given for reference; (b) projections of maximum intensity profiles obtained with the half-mirror sample, and the corresponding profile in water given for reference (Edge).

Cross-talk noise dependence on the sample optical density (OD) is investigated here below. The cross talk signals measured for three different ODs with $g = 0.85$ are shown in Fig. 4.8a and Fig. 4.8b for the full- and the half-mirror sample, respectively. The different OD values are obtained by varying the cell thickness Δz while maintaining the mirror edge in focus, as explained in Section 4.3.2. With a constant $\mu_S = 6.2$ mm^{-1}, the thicknesses $\Delta z = 320$ μm, $\Delta z = 650$, and $\Delta z = 1000$ correspond to OD $= 4$, OD $= 8$, and OD $= 12.4$, respectively. As expected, the measurements clearly show that the smaller the OD of the sample, the less cross talk is generated, and the better is the resolution. With the thickest sample, the peak signal is well behind the mirror interface showing the dominant role played by delayed MSL relative to ballistic light.

The corresponding theoretical results obtained with the model presented in Section 4.2 are in very good agreement with experimental data for nearly all cases investigated. However, as can be observed with the full-mirror sample, the model systematically leads to slightly higher amplitudes for cross talk signals relative to the intensity peak. This means that the MSL contribution is slightly overestimated. One possible explanation for this small systematic difference could be the omission of polarization effects in our theoretical model that could lead to overestimation of the multiply scattered component relative to the ballistic component. This hypothesis is supported by the results obtained when varying the anisotropy of the scattering solution (see Fig. 4.7a) where the systematic difference is the highest for the lowest microsphere diameter (i.e., highest anisotropy) and vice and versa. Indeed, according to Mie theory, polarization effects increase inversely with the microsphere diameter. The interested reader can find a more comprehensive discussion on minor discrepancies between our theoretical and experimental results [11].

4.4.2 Cross-Talk Contribution Relative to the Useful Signal

The above presented results reveal how major sample properties affect the cross talk noise and, in turn, the longitudinal and tranverse resolution. However, the normalization procedure used for the presentation of the results is hiding an important consequence of cross talk. Indeed, in FF-OCT, the significant MSL contribution causes a signal enhancement highly dependent on the sample properties.

To gain more insight into the contribution of MSL responsible for cross talk-generated noise relative to the ballistic light component (useful signal), we will investigate the signal attenuation in the sample as a function of the probing depth and compare it to the ideal single-backscattering model used in point-scanning OCT [53]. In this model, which accounts for ballistic light only (backscattered once), the amplitude follows a negative exponential decrease according to Lambert–Beer's law. The significant MSL contribution in FF-OCT causes a signal enhancement that is highly dependent on the sample properties. The deviation from Beer's law caused by the MSL contribution is investigated in the experiment described here below.

The full-mirror sample was used with a scattering solution adjusted to $\mu_S = 6.2$ mm^{-1}, first for a relative forward scattering solution ($g = 0.93$). We varied the OD by increasing the sample thickness (Δz) in regular steps while maintaining the reflecting interface in focus, according to the procedure described in Section 4.3.3. The signal was measured for each thickness in both full-field and confocal configurations (see Section 4.3.1). The exponential intensity decrease, proportional to exp($-\alpha \Delta z$) predicted by Beer's

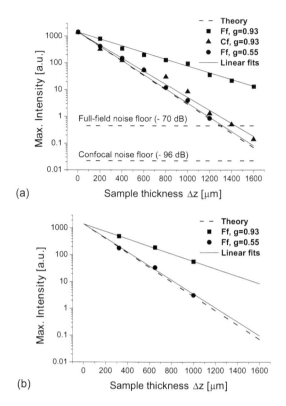

Figure 4.9 Maximum intensities of OCT signals obtained with the full-mirror sample for $\mu_S = 6.2$ mm^{-1} versus sample thickness for different anisotropies (g) and configurations. The decrease in intensity, for full-field configuration (Ff) with $D = 420$ μm, NA $= 0.25$, and confocal configuration (Cf) with NA $= 0.25$, is compared with the exponential decrease predicted by Lambert–Beer's law (Theory). Linear fits are illustrated for all cases. (a) Experimental results. (b) Theoretical results obtained with our model.

law ($\alpha = \mu_S = 6.2$ mm^{-1}), is plotted in Fig. 4.9. This figure shows as well the noise floors attained with the interferometer in the full-field configuration (sensitivity of -75 dB) and in the confocal configuration (sensitivity of -96 dB). The optical depth of our standard condition (OD $= 8$) corresponds to $\Delta z = 650$ μm.

With the full-field configuration (Ff), the signal attenuation is considerably lower than predicted by the Beer's law. Even at an OD $= 20$ ($\Delta z \approx 1600$ μm) the amplitude of the signal backscattered by the reflecting interface remains high. A linear fit (least squares) through the experimental data leads to the attenuation constant $\alpha = 2.9$ mm^{-1}. Thus, the attenuation constant $\alpha = 3.0$ mm^{-1} predicted by our model is in excellent agreement with experimental data. As shown in Fig. 4.9b, our theoretical value was obtained by a linear fit between three points.

With the confocal configuration (Cf), the linear fit through the experimental data yields $\alpha = 5.7$ mm^{-1}, corresponding to an attenuation slightly lower than given by Beer's law ($\alpha = 6.2$ mm^{-1}). The lower experimental coefficient is due to the detection of MSL residuals, in agreement with other studies [53, 54]. In addition, these results demonstrate the significant rejection of MSL obtained in point scanning OCT.

To further test our model and investigate the influence of the sample properties, we repeated the same experiment with a widely scattering solution ($g = 0.55$). In full-field configuration we obtained $\alpha - 6.1$ mm^{-1}, that is, nearly no signal enhancement as compared with Beer's law predictions. Again, our model proves to be in excellent agreement with this value, since it predicted an attenuation constant $\alpha = 6.0$ mm^{-1}. As shown in Fig. 4.9b, this theoretical value was obtained by a linear fit between three points. Since there is no significant deviation from Beer's law with the wide-angle scattering solution, we conclude that, in our case, only the forward-scattered light accounts for signal enhancement.

The significant deviation from Beer's law in FF-OCT for forward-scattered light reveals that the contribution of MSL relative to ballistic light dominates already at moderate probing depths. In the investigated case (Ff, $g = 0.93$), at a depth corresponding to one scattering mean free path ($\Delta z = 200$ µm), the ballistic and MSL components of the OCT signal are approximately equivalent. In point-scanning OCT, MSL contribution typically dominates after several mean free paths [55]. Last but not least, the excellent agreement between the experimentally determined coefficients α and the theoretically predicted values reveal the propensity of our model to provide an accurate extrapolation of tissue properties from OCT measurements. This important application of OCT is the object of intensive research [53, 56, 57].

4.4.3 Additional Results and Conclusions

The dependence of cross talk noise on some important design parameters of the optical system was investigated in a previous study [11]. Both experimental and theoretical results were reported for full-field illumination diameters (D = 420 µm, 210 µm and 85 µm), numerical apertures of the sample objective (NA = 0.25, 0.1 and 0.05), as well as the source coherence length (l_c = 34 µm and 15 µm). Experimental and theoretical results were in very good agreement.

The above referred results and those presented in Section 4.4.1 led us to the conclusion that cross talk increases with the full-field diameter, numerical aperture, source coherence length and sample optical density, and strongly depends on sample anisotropy. Therefore, design guidelines for minimal cross talk contribution, would generally recommend reducing the full-field diameter, the numerical aperture and the source coherence length down to a value depending on the sample properties. When striving to reduce the numerical aperture, a trade-off between the highest transverse resolution and the lowest cross talk must be found. A broader source spectrum only marginally favors cross talk rejection. Another evidence of the strong cross talk contribution is provided by our observations of the significant deviation from Lambert–Beer's law, revealing that the transition from single scattering to diffuse light regime happens for much lower optical depths in FF-OCT than in point scanning OCT.

4.5 Cross-Talk Suppression

The results presented in the previous section reveal to what extent cross talk-generated noise can limit FF-OCT. We will now experimentally investigate if the spatial coherence gating, introduced in Section 4.1.3, can be exploited for cross talk suppression. To prove our concept, we will investigate the cross talk rejection capability of spatial coherence gating by comparing spatially coherent to SII, under the same experimental conditions (setup and sample). We will show that rejection of cross talk by spatial coherence gating is comparable to that achieved in point-scanning OCT. To further investigate rejection performance, we will compare confocal illumination with SII.

4.5.1 *Setup and Sample*

The optical setup (Fig. 4.10) consists of a Linnik interferometer with a flip mirror (FM) allowing selecting either spatially coherent or SII. In both configurations, the image of the sample is obtained by lenses L_4 and L_5 forming a $\times 30$ magnifying microscope. The imaging optics and image construction method, which consists of a translation stage (TS) moving an optical fiber (OF) with appropriate core size, are the same as in our previous experimental setup. Except for the bandpass filter, which is increased to 1.4 kHz to account for the broader optical source bandwidth (see below), the same signal detection and processing (SDP) scheme, common to both configurations in Fig. 4.10, was employed in our previous experimental setup (see details in Section 4.3.1).

SCI is obtained with FM at $45°$ so as to select the interferometer source arm with the single mode fiber (SF). Details on the working principle of this configuration are provided in Section 4.3.1. Here the source is a mode-locked Ti:Sapphire laser (MLTS) whose spectrum is centered around 800 nm with a 70 nm bandwidth (FWHM).

SII is obtained with FM vertical so as to select the interferometer source arm with the multimode fiber (MF). The light from a 100 W Hg arc lamp injected (thermal light source) into MF allows delivering highly spatially incoherent light into the interferometer. The use of a multimode fiber provides a uniformly bright (equivalent to Koehler illumination) and easy to handle light source. In addition, the well-defined source size and geometry facilitate system design. In our case, the 550 μm fiber core, positioned at the focal plane of L_1, is directly imaged on the focal plane of objective L_4, leading to a 400 μm diameter full field.

In our design the sample objective aperture, which corresponds to the entrance pupil of the illumination system, is filled with spatially incoherent light. As explained in Section 4.3.1, the fulfillment of this requirement provides a design optimal for spatial coherence gating, that is, cross talk noise rejection capabilities.

Moreover, with a fiber NA of 0.21 and an objective NA of 0.25, the power throughput is nearly optimal since the spatial extent on the source side nearly corresponds to that on the sample side ($0.21^2 \times 550 \cong 0.25^2 \times 400$).

For a relevant comparison between spatially coherent and SII, full-field diameter, temporal coherence gating, and power must be the same in both configurations. The full-field intensity profile depends on the illumination. In SCI, a Gaussian beam is collimated onto the sample while, in SII, a top hat intensity profile is obtained by imaging the multimode fiber core. The waist (at 1/e) of the Gaussian illumination profile matches the 400 µm large incoherent full-field illumination (see Fig. 4.10, bottom inset). We attribute the relatively noisy top hat profile to insufficient polishing of our multimode fiber.

To obtain a similar spectrum with the mode-locked Ti:sapphire laser and the Hg light source, a smooth portion of the latter's spectrum—between 750 nm and 850 nm—is filtered through a combination of high- and low-pass interference filter (IF). As illustrated in the top inset of Fig. 4.10, very similar autocorrelation function envelopes are obtained for both sources. The corresponding measured longitudinal resolutions at FWHM in air are 4 and 4.5 µm, with the Hg and pulsed laser sources, respectively. The side lobes present in the filtered Hg source and in the mode-locked fs Ti:sapphire laser autocorrelation functions, are due to a nearly squared spectrum (resulting from the spectral filtering) and to polarization effects in the 100 meter long single mode fiber, respectively.

To illuminate the sample with the same power as with the Hg source, the mode-locked Ti:sapphire laser power was attenuated to 1 mW with a neutral density filter placed between L_2 and L_3.

The sample is a US air force reflecting resolution target covered with a scattering solution. The latter, consists of 2050 nm polystyrene microspheres diluted in de-ionized water, yielding an anisotropy parameter $g = 0.93$. A cell, with inner thickness of 650 µm, delimited by the resolution target and a 150 µm thick cover glass, was filled either with water or with the scattering solution. Full-field illumination was centered on element 2, group 6, of the resolution target (71.8 lines/mm) whose longitudinal position

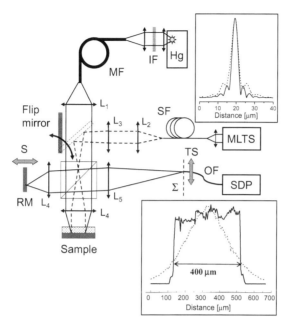

Figure 4.10 Scheme of an FF-OCT setup with spatially coherent (flip mirror at 45°) and spatially incoherent (flip mirror vertical) illumination: mercury arc lamp (Hg); mode-locked fs Ti:sapphire laser (MLTS); single-mode fiber (SF); multimode fiber (MF); interference filters (IF); reference mirror (RM); achromatic lenses (L_1, L_2, L_3, and L_5); microscope objectives $\times$ 10, NA = 0.25 (L_4); scanner (S); translation stage (TS); optical fiber (OF); signal detection and processing (SDP). Top inset: autocorrelation function envelopes measured in air for Hg (solid curve) and for MLTS (dashed curve) sources. Bottom inset: normalized full-field illumination profiles obtained with spatially coherent (dashed curve) and spatially incoherent illumination (solid curve).

was adjusted, with the cell filled with water, to form an image in the focal plane of L_5. Water was then replaced by the scattering solution whose scattering coefficient, measured independently with the method in Ref. [49], corresponds to 8 optical depths in the cell (two ways).

Note that, unlike for the half-mirror sample used for our quantitative investigation in Section 4.4, the glass interface between the reflecting stripes of the resolution target contributes to the OCT signal. This is not a problem in this comparative study.

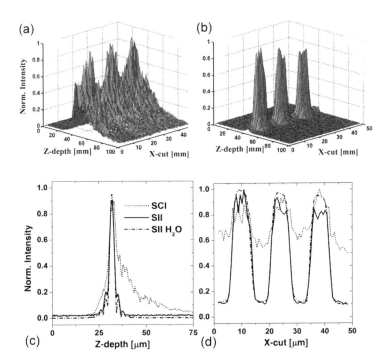

Figure 4.11 Cross-sectional FF-OCT image of a US Air Force resolution target covered with a scattering solution (OD = 8, $g = 0.93$): (a) spatially coherent illumination (SCI); (b) spatially incoherent illumination (SII); (c) cuts along the z axis of Fig. 4.2a (across the middle stripe), (d) cuts along the x axis of Fig. 4.2b (across the reflecting target interface). The curves labeled "SII H_2O" correspond to the same cuts of the same image acquired in water with SII. All curves are normalized and correspond to an average of 25 demodulated signals.

4.5.2 Results

Figures 4.11a and 4.11b show a three-dimensional representation of OCT cross-sectional images of the pattern's stripes—with the normalized intensity on the vertical axis—obtained with spatially coherent and SII, respectively. Despite short temporal coherence gating, SCI generates a considerable amount of cross talk noise, degrading both longitudinal and tranverse resolutions (Fig. 4.11a). The plots correspond to the average cross talk signal of 25 demodulated OCT signals (see Section 4.3.1). With full-field SII, the spatial coherence gating provided by our optimal design strongly reduces cross talk noise, and in turn yields a good resolution of the three stripes (Fig. 4.11b).

For a quantitative estimate of spatial coherence gating efficiency, longitudinal cuts of the SCI and SII three-dimensional plots are compared. Cuts along the z axis (across the middle stripe), and those along the x axis (across the sample interface), are shown in Fig. 4.11c and Fig. 4.11d, respectively. It appears that the cross talk-generated noise is responsible for a loss of contrast and resolution, which can be restored with SII.

To further illustrate the spatial coherence gating efficiency, the same cross-sectional image was acquired in water with SII. The two normalized longitudinal cuts displayed in Fig. 4.11c and Fig. 4.11d correspond to plots obtained with SII in water (dash-dot curves) and in the scattering solution (solid curves). Except slightly higher amplitude side lobes and a more rippled profile in the scattering solution, their near identity reveals that in our case, longitudinal and transverse resolutions are fully restored. The slight differences observed can be attributed to a noisy illumination intensity profile as well as to possible particle aggregates of the scattering solution at the resolution target interface.

For the same experimental setup, comparative results of spatially coherent with incoherent illumination for an ex vivo tooth have been reported [12]. SII has provided a more accurate image revealing new structures.

Comparison of confocal illumination with SII revealed that spatial coherence gating achieves cross talk suppression to the same extent as point-scanning OCT. As mentioned in Section 4.3.1, confocal configuration with properties equivalent to point-scanning OCT is achieved by simply removing lens L_3 in our setup (see Fig. 4.10). The monomode fiber is then imaged onto the sample providing a confocal spot illumination. The latter is attenuated with neutral density filters so as to obtain a sensitivity of 75 dB in both configurations. We use the full-mirror sample (see Section 4.3.2), with scattering solution with optical density OD $= 8$ and anisotropy $g = 0.85$. The correlograms obtained reveal that in both configurations cross talk rejection is equally efficient along the longitudinal dimension (Fig. 4.10). Differences near the side lobes are likely due to scanner instabilities as well as to residual MSL.

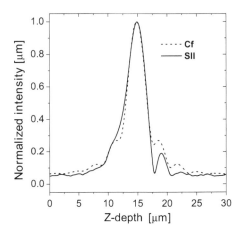

Figure 4.12 Comparison of FF-OCT with spatially incoherent illumination (SII) to point-scanning OCT with confocal configuration (Cf) for a mirror covered with the scattering solution (OD = 8, $g = 0.85$).

Our results demonstrate that cross talk noise generated in FF-OCT by SCI can be suppressed with SII provided by a thermal light source. With SII, the performance of FF-OCT is comparable to that of point-scanning OCT. However, despite the fact that FF-OCT requires less brightness than point-scanning OCT, the power per spatial mode radiated by thermal light sources is too low to permit a high SNR while maintaining a fast acquisition speed. This issue is discussed in the conclusions of this contribution.

We would like to emphasize that, besides allowing for spatial coherence gating, a thermal light source has a naturally broad spectrum, offering a very high longitudinal resolution at lower complexity and cost.

4.6 Conclusions and Discussions

4.6.1 Full-Field OCT

The theoretical and experimental investigations of FF-OCT presented in this chapter reveal the crucial role played by the nature of the light source. Despite temporal coherence gating, SCI generates considerable optical cross talk noise, which prevents shot-noise-limited detection and diffraction-limited imaging in scattering samples.

Cross-talk dependence on several important parameters of the optical system and on sample properties was investigated in a comprehensive study. The results have brought quantitative knowledge of cross talk noise contribution in FF-OCT relying on SCI, which has in turn allowed the elaboration of elementary design guidelines (see Section 4.4.3). The significant cross talk noise contribution drastically reduces the scope of practical applications in biological imaging. In principle, FF-OCT incorporating SCI is restricted to very moderately scattering samples. The method is suitable for topographic measurements.

By taking a closer look at the coherence properties of MSL in the context of OCT detection methods, we were able to explain how cross talk can be suppressed by exploiting spatially incoherent light. We have shown experimentally how SII realized with a thermal light source permits cross talk suppression, that is, rejection of MSL, to a level comparable to that of point-scanning OCT. This outstanding feature, combined with the naturally broad spectrum the thermal light source, allows the measurement of en face cross-sectional images with high quality and resolution, at minimum cost and complexity.

The main weakness of this method—as so far implemented— is its relatively low measurement speed, which does not allow observation of dynamic phenomena or in vivo imaging without artefacts due to sample motions. Thus, FF-OCT incorporating SII is incontestably a valuable method for obtaining high quality en face cross-sectional images, but remains mainly restricted to the observation of in vitro specimen.

The cause of limitation in measurement speed lies in (i) the insufficient performance of conventional detectors, and (ii) the relatively low brightness of thermal light sources. As reported in the introduction, the first limitation, of technological nature, can be overcome with presently available CMOS detectors endowed with custom functionality for OCT imaging (Heliotis AG). The second limitation, of fundamental nature, comes from the power per spatial mode radiated by thermal light sources that remains too weak to allow a sufficient sensitivity while maintaining a fast acquisition speed.

To overcome this limitation, an extremely bright spatially incoherent light source is required. Increasing the temperature of a thermal light source above the 6000 K offered by arc lamps (e.g., with a plasma) is not trivial at all. An alternative might be the creation of a so-called pseudothermal light source [58] obtained by destroying the spatial coherence of a powerful broadband laser source, that is, by creating a speckle field with very low degree of correlation between the speckles. This can be achieved by generating fast random fluctuations, for instance, with a rotating diffusor [20] or dynamical mode mixing in a multimode fiber [59]. However, it is not trivial to generate fluctuations much shorter than the actual OCT measurement time per probe volume. This is a requirement for obtaining sufficient statistical averaging to get a low enough degree of correlation for cross talk suppression [20]. Alternatively, a recently investigated solution to create an uncorrelated speckle field relies on the injection of light from a broadband laser into a multimode fiber [21]. Setting up appropriate length and diameter of the fiber, delays exceeding the source temporal coherence time between modes propagating within the fiber can be generated so as to obtain a low degree of spatial coherence at the fiber exit. However, we believe that this method should be carefully investigated prior to implementation.^{e} Thus, the creation of a pseudothermal light source suitable for FF-OCT does not seem trivial. A viable solution for creating a spatially incoherent light source might be an extended superluminescent light source based on the broadband fluorescence of a $Ti:Al_2O_3$ waveguide crystal [15].

Note that ANSI norms regarding the maximum exposure time of biological samples limit the increase of source power, which may in turn ultimately set a limitation to the method.

---

^{e}The following two effects should be considered carefully. First, the intercoupling between modes propagating into the multimode fiber with different path lengths may alter the source autocorrelation function, causing problems such as deleterious harmonics. Second, the delays between modes are altered—and in the worst case fully compensated—by the random propagation of MSL in the sample. Should the worst case prevail, conditions for cross talk noise generation would be restored.

The en face image of biological structures obtained with FF-OCT is of particular interest since it may complement information provided by the longitudinal image obtained with FD-OCT [6, 18]. Moreover, the interpretation of en face views is often more familiar to specialists, such as ophthalmologists. A new generation of FF-OCT systems incorporating an ultrabright light source and a custom detector array may well compete with ultrafast FD-OCT systems (see introduction) regarding image acquisition speed. Should the sample be scanned in the longitudinal dimension or dynamical focusing be implemented [60, 61], the new generation of FF-OCT systems may also become a valuable alternative for ultrafast three-dimensional measurements. Actually, it may provide the most accurate three-dimensional measurements thanks to its inherently superior transverse resolution.

Alternatively, a very interesting modality for ultrafast three-dimensional measurements based on FD-OCT operation allows directly obtaining en face images thanks to a swept-laser source [59]. However, to achieve cross talk suppression, the spatial coherence of the swept-laser source beam must be destroyed so as to create a pseudothermal light source, what is not trivial as explained above.

It should not be forgotten that, since this new generation of fast measuring FF-OCT systems relies on unconventional detectors and sources, the access to measurement speeds permitting reliable in vivo measurements can only be gained at the expense of cost and simplicity, which are major attributes of the previous generation. However, like with the sophisticated femtosecond laser sources, novel detectors and sources for FF-OCT should become more commonplace.

4.6.2 Modeling Multiple Scattering in OCT

We have conducted a comprehensive study of multiple scattering effects in FF-OCT realized with SCI. The agreement between theoretical and experimental results for a wide range of different parameters was very good, confirming the validity of our model for MSL in OCT and, implicitly, the relatively simple assumptions on which it rests. Thus, the role of multiple scattering in OCT does not seem to be as complex as so far suggested by several studies and models. Indeed, to the contrary of the widespread belief that a relevant OCT model should account for a partial correlation between interfering fields, we have explained why, in the context of OCT detection methods, multiple scattering actually induces neither a loss of spatial coherence of the sample field nor a reduction of the correlation between the latter and the reference field. This means that, for path length differences between the reference and sample arms falling within the source coherence length, the reference and sample fields interfere with a contrast determined uniquely by the source autocorrelation function. Multiple scattering generates a speckle noise contribution to the ideal OCT signal, but does not reduce the interference contrast.

Based on this important result, we have developed a comprehensive model of OCT where the signal is modeled as a sum of stationary random phasors and treated as a statistical signal. The mean intensity of this random signal, which is usually the variable of interest, can be calculated thanks to classical results of statistical optics and to a Monte Carlo simulation. This approach is very different from that of other models based on a Monte Carlo simulation, which take into account the effect of multiple scattering such as, for example, Ref. [62]. In the latter model the phase information, which depends on the history of scattering events, is recorded for each photon, represented by a plane wavelet. Thus, a simulation yields a sum of randomly delayed wavelets, which generate speckles. To obtain the mean signal, essentially the quantity of interest, many of such complicated and lengthy simulations must be run. In our model, the mean and the variance can be directly calculated thanks to the assumption that photons can undergo random phase excursions in the scattering medium of at least 2π.

The Monte Carlo simulation, necessary for the calculation of the raw spatiotemporal distribution of light, makes our model semianalytical. It is the prize one has to pay for accessing ranges where the diffusion approximation is not verified. Although our model is not as elegant as a fully analytical model, it offers the advantage of high versatility. Indeed, unlike fully analytical models, it is neither restricted to strongly simplified media nor limited by initial assumptions, such as the diffusion approximation or the small angle approximation.

As briefly discussed in Section 4.1.2.3, other existing models of OCT accounting for multiple scattering, are based on the extended Huygens–Fresnel principle and the use of mutual coherence functions MCFs [34, 35]. The calculation of MCFs relies on some spatial and temporal statistical averaging, generally yielding a reduction of correlation between the reference and sample fields. Thus, these models are incompatible with the two fundamental assumptions on which our model rests. Indeed, as emphasized in Section 4.1.2, statistical averaging is not applicable because of the high degree of spatial coherence of the measured sample field and of the short measurement time per probe volume. Therefore, our model implicitly puts in question other existing models accounting for multiple scattering in OCT. Interested readers can find a more thorough discussion of this topic in Ref. [33].

We do hope that our study and model will help to fully appreciate the role of multiple scattering in OCT and stimulate the research in this field. Future work of interest with our model could be a study of multiple scattering effects in point-scanning OCT in highly diffuse media such as skin. This would be particularly relevant for FD-OCT, in which confocal spatial filtering is inherently limited. Investigations of multiple scattering in OCT could be of particular interest when accounting for real tissues properties in the Monte Carlo simulation. Conclusions of our study may well concern a wider class of imaging methods limited by cross talk, such as parallel time-resolved detection [63], laser Doppler imaging [64] or holography with a broadband light source [65].

Appendix A: Independence of the Parameters u_j and L_j

The derivation of Eq. 4.18 requires assumption (ii) in Section 4.2.2. More specifically, this implies that u_j and L_j are independent, which is generally not the case. However, for our specific OCT case, it suffices to verify this condition only "locally," that is, for a relatively short path length interval ΔL in the order of the source coherence length l_c. Indeed, in the first approximation, the module of the source autocorrelation function $g_0(z)$ is null outside l_c. Thus, only the interval determined by l_c, needs to be considered. Independence between u_j and L_j requires first that the depth of field set by the sample objective NA is larger than the longitudinal resolution set by the source coherence length, which is usually the case in OCT. Second, the interval $\Delta L \approx l_c$ must contain path lengths L_j having accumulated the same number of scattering events. Indeed, the field coefficients u_j also depend on the number of scattering events due to loss of energy in scattering and absorption. This second condition is more restrictive and needs to be discussed. With the anisotropy g, corresponding to the mean scattering angle cosine, $\Delta L \approx (1-g)\mu_S$, where μ_S is the sample scattering coefficient in inverse millimeters. Thus, the larger g and μ_S, the shorter is ΔL_j; that is, the more difficult it is to fulfill the condition $\Delta L_j \geq \Delta L \approx l_c$. For our preliminary experiment with $l_c = 15$ μm and sample scattering properties ($\mu_S = 1/OD = 1/8$, $g = 0.85$) yielding $\Delta L_j \approx 19$ μm, we have $\Delta L > l_c$. Since OCT is generally not applied to $\mu_S > 10$ mm^{-1} (skin properties), a large g is in principle more critical. However, with increasingly shorter coherence lengths in present day OCT systems, our second condition is generally satisfied.

Note also that, without absorption, our second condition can be significantly relaxed since backscattering losses are negligible; that is, the influence on the number of scattering events is negligible on u_j coefficients. Indeed, Mie functions reveal that for $g > 0.7$ (as met in most biological samples) nearly all light energy is scattered under $2\pi\,sr$ in the direction of the wave propagation.

Appendix B: Sum of a Deterministic and a Random Phasor

To model the practical case corresponding to our experiments (mirror embedded in a scattering solution), it is necessary to account for the deterministic phase of light reflected by the mirror without scattering events (see Section 4.2.2). To account for this ballistic contribution, a deterministic phasor sum with amplitude S is added to the random phasor sum of Eq. 4.14,

$$S(z) + Q(z) \exp(-i\Phi(z)) = u_0 |g_0(z - L_0)| \exp(-ik_0 L_0) \\ + \sum_{j=1}^{N} u_j |g_0(z - L_j)| \exp(-ik_0 L_j),$$

(4.22)

where L_0 is the sample mirror position. Adopting the convention $L_0 = 0$, the additional path lengths are defined relative to the sample mirror interface leading to $S(z) = u_0\, g_0(z)$.

The phasor amplitude described by Eq. 4.22 obeys a Rician probability density function, whose mean is given by [31]

$$\overline{A'(z)} = \sqrt{\frac{\pi}{2}}\sigma(z)\exp\left(-\frac{\beta^2(z)}{4}\right)$$
$$\times \left[\left(1+\frac{\beta^2(z)}{2}\right)I_0\left(\frac{\beta^2(z)}{4}\right) + \frac{\beta^2(z)}{2}I_1\left(\frac{\beta^2(z)}{4}\right)\right], \quad (4.23)$$

where

$$\beta(z) = \frac{S(z)}{\sigma(z)} = \frac{u_0 g_0(z)}{\sigma(z)} \quad (4.24)$$

and where $\sigma(z)$ is the distribution's standard deviation. I_0 and I_1 are modified Bessel functions of the first kind with order zero and one, respectively. According to our analysis in Section 4.2.3, expressing the coefficient u_0 in $\beta(z)$ as a function of I_v for practical implementation of a Monte Carlo simulation yields

$$u_0 = \sqrt{U_0} = \sqrt{I_B}, \quad (4.25)$$

where I_B is the intensity of the ballistic light, determined either theoretically (Lambert–Beer's law) or by the Monte Carlo simulation.

Thus, the mean signal detected from a sample such as in our preliminary experiment, i.e the mean value of $i_D(z)$ in Eq. 4.13, is proportional to $\overline{A'(z)}$ and can be calculated by combining Eqs. 4.21 and 4.23–4.25. Note however that, even in the presence of a relatively large deterministic phasor, Eq. 4.21 may lead to an excellent approximation. Indeed, with the examples presented in Section 4.4, we obtained very similar results in nearly all cases, when using either Eq. 4.21 or the full treatment to account for the deterministic phasor.

References

1. Dunsby, C., and French, P. M. W. (2003). Techniques for depth-resolved imaging through turbid media including coherence-gated imaging. *J. Phys. D*, **36**, pp. R207–R227.

2. Rudolph, W., and Kempe, M. (1997). Topical review: trends in optical biomedical imaging. *J. Mod. Opt.*, **44**, pp. 1617–1642.
3. Fercher, A. F., Drexler, W., Hitzenberger, C. K., and Lasser, T. (2003). Optical coherence tomography: principles and applications. *Rep. Prog. Phys.*, **66**, pp. 239–303.
4. Izatt, J. A., Hee, M. R., Owen, G. M., Swanson, E. A., and Fujimoto, J. G. (1994). Optical coherence microscopy in scattering media. *Opt. Lett.*, **19**, pp. 590–592.
5. Kempe, M., Genack, A. Z., Rudolph, W., and Dorn, P. (1997). Ballistic and diffuse light detection in confocal and heterodyne imaging systems. *J. Opt. Soc. Am. A*, **14**, pp. 216–223.
6. Podoleanu, A. Gh., and Rosen, R. B. (2008). Combinations of techniques in imaging the retina with high resolution. *Prog. Retinal Eye Res.*, **27**, pp. 464–499.
7. Grajciar, B., Pircher, M., Fercher, A., and Leitgeb, R. (2005). Parallel Fourier domain optical coherence tomography for in vivo measurement of the human eye. *Opt. Express*, **13**, pp. 1131–1137.
8. Vabre, L., Dubois, A., and Boccara, A. C. (2002). Thermal full-field optical coherence tomography. *Opt. Lett.*, **27**, pp. 530–532.
9. Bourquin, S., Seitz, P., and Salathé, R. P. (2001). Optical coherence tomography based on two-dimensional smart detector array. *Opt. Lett.*, **26**, pp. 512–514.
10. Akiba, M., Chan, K. P., and Tanno, N. (2003). Full-field optical coherence tomography by two-dimensional heterodyne detection with a pair of CCD cameras. *Opt. Lett.*, **28**, pp. 816–819.
11. Karamata, B., Leutenegger, M., Lambelet, P., Laubscher, M., Bourquin, S., and Lasser, T. (2005). Multiple scattering in optical coherence tomography. Part II: experimental and theoretical investigation of cross talk in wide-field optical coherence tomography. *J. Opt. Soc. Am A*, **22**, pp. 1380–1388.
12. Karamata, B., Laubscher, M., Lambelet, P., Salathé, R. P., and Lasser, T. (2004). Spatially incoherent illumination as a mechanism for cross talk suppression in wide-field optical coherence tomography. *Opt. Lett.*, **29**, pp. 736–738.
13. Drexler, W., Morgner, U., Kärtner, F. X., Pitris, C., Boppart, S. A., Li, X. D., Ippen, E. P., and Fujimoto, J. G. (1999). In vivo ultrahigh-resolution optical coherence tomography. *Opt. Lett.*, **24**, pp. 1221–1223.
14. Potsaid, B., Gorczynska, I., Srinivasan, V. J., Chen, Y., Jiang, J., Cable, A., and Fujimoto, J. G. (2008). Ultrahigh speed Spectral/Fourier domain OCT

ophthalmic imaging at 70,000 to 312,500 axial scans per second. *Opt. Express*, **16**, pp. 15149–15169.
15. Sacchet, D., Brzezinski, M., Moreau, J., Georges, P., and Dubois, A. (2010). Motion artifact suppression in full-field optical coherence tomography. *Appl. Opt.*, **49**, pp. 1480–1488.
16. Beaurepaire, E., Boccara, A. C., Lebec, M., Blanchot, L., and Saint-Jalmes, H. (1998). Full-field optical coherence microscopy. *Opt. Lett.*, **23**, pp. 244–246.
17. Oh, W. Y., Bouma, B. E., Iftimia, N., Yun, S. H., Yelin, R., and Tearney, G. J. (2006). Ultrahigh-resolution full-field optical coherence microscopy using InGaAs camera. *Opt. Express*, **14**, pp. 726–735.
18. Dubois, A., Grieve, K., Moneron, G., Lecaque, R., Vabre, L., and Boccara, C. (2004). Ultrahigh-resolution full-field optical coherence tomography. *Appl. Opt.*, **43**, pp. 2874–2883.
19. Fercher, A. F., Hitzenberger, C. K., Sticker, M., Moreno-Barriuso, E., Leitgeb, R., Drexler, W., and Sattmann, H. (2000). A thermal light source technique for optical coherence tomography. *Opt. Commun.*, **185**, pp. 57–64.
20. Karamata, B. (2004). *Multiple scattering in wide-field optical coherence tomography*, Ph.D. thesis EPFL **3001** (École Polytechnique Fédérale de Lausanne, Switzerland).
21. Dhalla, A. H., Migacz, J. V., and Izatt, J. A. (2010). Crosstalk rejection in parallel optical coherence tomography using spatially incoherent illumination with partially coherent sources. *Opt. Lett.*, **35**, pp. 2305–2307.
22. Laubscher, M., Ducros, M., Karamata, B., Lasser, T., and Salathe, R. (2002). Video-rate three-dimensional optical coherence tomography. *Opt. Express*, **10**, pp. 429–435.
23. Bordenave, E., Abraham, E., Jonusauskas, G., Tsurumachi, N., Oberl, J., Rullière, C., Minot, P. F., Lassègues, M., and Surlève Bazeille, J. E. (2002). Wide-field optical coherence tomography: imaging of biological tissues. *Appl. Opt.*, **41**, pp. 2059–2064.
24. Miller, D. T., Qu, J., Jonnal, R. S., and Thorn, K. (2003). Coherence gating and adaptive optics in the eye. *Proc. SPIE*, **4956**, pp. 65–72.
25. Lambelet, P. (2011). Parallel optical coherence tomography (pOCT) for industrial 3D inspection. *Proc. SPIE*, **8082**, p. 80820X.
26. Sinclair, L. C., Cossel, K. C., Coffey, T., Ye, J., and Cornell, E. A. (2011). Frequency comb velocity-modulation spectroscopy. *Phys. Rev. Lett.*, **107**, p. 093002.

27. Leitgeb, R., Hitzenberger, C. K., and Fercher, A. F. (2003). Performance of fourier domain vs. time domain optical coherence tomography. *Opt. Express*, **11**, pp. 889–894.

28. Yura, H. T. (1979). Signal-to-noise ratio of heterodyne lidar signal systems in the presence of atmospheric turbulence. *Opt. Acta*, **26**, pp. 627–644.

29. Karamata, B., Hassler, K., Laubscher, M., and Lasser, T. (2005). Speckle statistics in optical coherence tomography. *J. Opt. Soc. Am. A*, **22**, pp. 593–596.

30. Schmitt, J. M., Xiang, S. H., and Yung, K. M. (1999). Speckle in optical coherence tomography. *J. Biomed. Opt.*, **4**, pp. 95–105.

31. Goodman, J. W. (1985). *Statistical Optics* (Wiley Classics Library), pp. 44–45.

32. Yang, C., An, K., Perelman, L. T., Dasari, R. R., and Feld, M. S. (1999). Spatial coherence of forward-scattered light in a turbid medium. *J. Opt. Soc. Am. A*, **16**, pp. 866–871.

33. Karamata, B., Lambelet, P., Laubscher, M., Leutenegger, M., Bourquin, S., and Lasser, T. (2005). Multiple scattering in optical coherence tomography. Part I: investigation and modeling. *J. Opt. Soc. Am. A*, **22**, pp. 1369–1379.

34. Schmitt, J. M., and Knuettel, A. (1997). Model of optical coherence tomography of heterogeneous tissue. *J. Opt. Soc. Am. A*, **14**, pp. 1231–1242.

35. Thrane, L., Yura, H. T., and Andersen, P. E. (2000). Analysis of optical coherence tomography systems based on the extended Huygens-Fresnel principle. *J. Opt. Soc. Am. A*, **17**, pp. 484–490.

36. Mandel, L., and Wolf, E. (1995). *Optical Coherence and Quantum Optics* (Cambridge University Press, UK).

37. Davidson, M., Kaufman, K., Mazor, I., and Cohen, F. (1987). An application of interference microscopy to integrated circuit inspection and metrology. *Proc. SPIE*, **775**, pp. 233–341.

38. Lee, B. S., and Strand, T. C. (1990). Profilometry with a coherence scanning microscope. *Appl. Opt.*, **29**, pp. 3784–3788.

39. Somekh, M. G., See, C. W., and Goh, J. (2000). Wide field amplitude and phase confocal microscope with speckle illumination. *Opt. Commun.*, **174**, pp. 75–80.

40. Sun, P. C., and Leith, E. N. (1994). Broad-source image plane holography as a confocal imaging process. *Appl. Opt.*, **33**, pp. 597–602.

41. Ishimaru, A., Kuga, Y., Cheung, R., and Shimizu, K. (1978). Diffusion of a pulse in densely distributed scatterers. *J. Opt. Soc. Am. A*, **68**, pp. 1045–1050.
42. Patterson, M. S., Chance, B., and Wilson, B. C. (1989). Time resolved reflectance and transmittance for the non-invasive measurement of tissue optical properties. *Appl. Opt.*, **28**, pp. 2331–2336.
43. Ishimaru, A., Kuga, Y., Cheung, R., and Shimizu, K. (1983). Scattering and diffusion of a beam wave in randomly distributed scatterers. *J. Opt. Soc. Am. A*, **73**, pp. 131–136.
44. Jacques, S. L. (1989). Time resolved propagation of ultrashort laser pulses within turbid tissues. *Appl. Opt.*, **28**, pp. 2223–2229.
45. Pan, Y., Birngruber, R., Rosperich, J., and Engelhardt, R. (1995). Low-coherence optical tomography in turbid tissue: theoretical analysis. *Appl. Opt.*, **34**, pp. 6564–6574.
46. Smithies, D. J., Lindmo, T., Chen, Z., Nelson, J. S., and Milner, T. E. (1998). Signal attenuation and localization in optical coherence tomography studied by Monte Carlo simulation. *Phys. Med. Biol*, **43**, pp. 3025–3044.
47. Yao, G., and Wang, L. V. (1999). Monte Carlo simulation of an optical coherence tomography signal in homogeneous turbid media. *Phys. Med. Biol.*, **44**, pp. 2307–2320.
48. Schmitt, J. M., and Kumar, G. (1996). Turbulent nature of refractive index variations in biological tissues. *Opt. Lett.*, **16**, pp. 1310–1312.
49. Van Staveren, H. J., Moes, C. J., van Marle, M. J., Prahl, S. A., and van Gemert, M. J. C. (1991). Light scattering in Intralipid-10% in the wavelength range of 400-1100 nm. *Appl. Opt.*, **30**, pp. 4507–4514.
50. Bohren, C. F., and Huffman, D. R. (1983). *Absorption and Scattering of Light by Small Particles* (Wiley-Interscience), p. 72.
51. Wang, L. H., Jacques, S. L., and Zheng, L.-Q. (1995). MCML: Monte Carlo modeling of photon transport in multi-layered tissues. *Comput. Methods Prog. Biomed.*, **47**, pp. 131–146.
52. Cheong, W. F., Prahl, S. A., and Welsh, A. J. (1990). A review of the optical properties of biological tissues. *J. Quantum. El.*, **26**, pp. 2166–2185.
53. Schmitt, J. M., Knuettel, A., and Bonner, R. F. (1994). Measurement of optical properties of biological samples in low-coherence reflectometry. *Appl. Opt.*, **32**, pp. 6032–6042.
54. Yadlowsky, M. J., Schmitt, J. M., and Bonner, R. F. (1995). Multiple scattering in optical coherence microscopy. *Appl. Opt.*, **34**, pp. 5699–5707.

55. Bizheva, K. K., Siegel, A. M., and Boas, D. A. (1998). Path-length-resolved dynamic light scattering in highly scattering random media: the transition to diffusing wave spectroscopy. *Phys. Rev. E*, **58**, pp. 7664–7667.
56. Faber, D. J., van der Meer, F. J., Aalders, M. C., and van Leeuwen, T. G. (2004). Quantitative measurement of attenuation coefficients of weakly scattering media using optical coherence tomography. *Opt. Express*, **12**, pp. 4353–4365.
57. Levitz, D., Thrane, L., Frosz, M. H., and Andersen, P. E. (2004). Determination of optical scattering properties of highly scattering media in optical coherence tomography. *Opt. Express*, **12**, pp. 249–259.
58. Martienssen, W., and Spiller, E. (1964). Coherence and fluctuations in light beams. *Am. J. Phys.*, **32**, p. 919.
59. Považay, B., Unterhuber, A., Hermann, B., Sattmann, H., Arthaber, H., and Drexler, W. (2006). Full-field time-encoded frequency-domain optical coherence tomography. *Opt. Express*, **14**, pp. 7661–7669.
60. Lexer, F., Hitzenberger, C. K., Drexler, W., Molebny, S., Sattmann, H., Sticker, M., and Fercher, A. F. (1999). Dynamic coherent focus OCT with dept-independent transversal resolution. *J. Mod. Opt.*, **46**, pp. 541–553.
61. Botcherby, E. J., Juskaitis, R., Booth, M. J., and Wilson, T. (2007). Aberration-free optical refocusing in high numerical aperture microscopy. *Opt. Lett.*, **32**, pp. 2007–2009.
62. Lu, Q., Gan, X., Gu, M., and Luo, Q. (2004). Monte Carlo modeling of optical tomography imaging through turbid media. *Appl. Opt.*, **43**, pp. 1628–1636.
63. Hebden, J. C., Kruger, R. A., and Song, K. S. (1991). Time resolved imaging through a highly scattering medium. *Appl. Opt.*, **30**, pp. 788–794.
64. Leutenegger, M., Martin-Williams, E., Harbi, P., Thacher, T., Raffoul, W., André, M., Lopez, A., Lasser, P., and Lasser, T. (2011). Real-time full field laser Doppler imaging. *Biomed. Opt. Express*, **2**, pp. 1470–1477.
65. Leith, E., Chen, H., Chen, Y., Dilworth, D., Lopez, J., Masri, R., Rudd, J., and Valdmanis, J. (1991). Electronic holography and speckle methods for imaging through tissue using femtosecond gated pulses. *Appl. Opt.*, **30**, pp. 4204–4210.

Chapter 5

Signal Processing Methods in Full-Field Optical Coherence Microscopy

Igor Gurov

Computational Photonics and Videomatics Department, ITMO University,
49 Kronverksky ave., Saint Petersburg, 197101, Russia
gurov@mail.ifmo.ru

Signal formation and processing methods are considered taking into account the requirements of high resolution and processing speed that are very important for full-field optical coherence microscopy. Special attention is devoted to the problem of information capacity of optical and electronic channels, where information is acquired and transmitted. Possibility to decrease the amount of data that has to be processed using signal downsampling, that is, missing uninformative video frames, is considered and discussed. Modern recurrence algorithms of signal processing in state space in dynamic mode are presented, and examples of experimental applications of high-speed processing algorithms are given.

5.1 Introduction

Evaluation of objects using light has been attractive for a long time, starting from visual observation and then using technical means due

---

Handbook of Full-Field Optical Coherence Microscopy: Technology and Applications
Edited by Arnaud Dubois
Copyright © 2016 Pan Stanford Publishing Pte. Ltd.
ISBN 978-981-4669-16-0 (Hardcover), 978-981-4669-17-7 (eBook)
www.panstanford.com

to the noncontact approach providing high resolution determined by short wavelength of light.

Coherence properties of light are widely used in many applications, including low-coherence interferometry (see, for example, Refs. [1, 2]). When observing micro-objects, interference microscopy methods are applied. Starting from Ref. [3] devoted to noninvasive cross-sectional imaging in biological objects, such an approach is widely entitled optical coherence tomography (OCT). The OCT systems utilizing high-aperture objectives to provide high lateral resolution are designated as optical coherence microscopy (OCM) systems. Full-field OCM (FF-OCM) is an interferometric technique that utilizes array detection for obtaining wide-field high-resolution microscopic images within scattering specimens under study [4–8]. Due to the fact that there is no fixed sharp border between OCT and OCM, and many interferometric signal processing methods are applicable to OCT and OCM, we use below both these abbreviations.

To extract information about an investigated object, one should obtain its image. In the so-called active optical systems utilizing artificial light sources, laser illumination is widely used, and light radiation is often represented by a scalar monochromatic plane wave

$$E_0(\mathbf{r}, \mathbf{k}; t, \nu) = A \exp[j2\pi(\mathbf{kr} + \nu t)] = A \exp(j2\pi \mathbf{kr}) \exp(j2\pi \nu t), \quad (5.1)$$

where A is the complex amplitude and $\mathbf{r}$, $\mathbf{k}$ and t, ν are spatial and temporal coordinates and frequencies, respectively.

When illuminating random tissue, the light wave (Eq. 5.1) generally suffers reflection, scattering, and absorption, and complex amplitude at an observation point $P(x, y, z)$ can be represented as a multitude of de-phased contributions from N different scattering regions of object in the form [9]

$$A(x, y, z) = \sum_{n=1}^{N} \frac{1}{\sqrt{N}} a_n(x, y, z) = \frac{1}{\sqrt{N}} \sum_{n=1}^{N} |a_n| \exp(j\phi_n). \quad (5.2)$$

It is clearly seen in Eq. 5.2 that it is impossible to find the parameters a_n and ϕ_n characterizing, correspondingly, the degree of reflection and position of each n-th region of an object having a single measurement of $A(x, y, z)$. To resolve this ambiguity, one should have at least $2N$ mutually independent equations. One of the

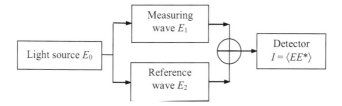

Figure 5.1 Schematic diagram of interferometric system.

possible approaches consists of the use of necessary number of light wavelengths in an interferometric system.

It is well known that an electric field of measuring and reference waves in interferometer (see Fig. 5.1) can be expressed in the form

$$E_1(v, t) = \sqrt{\alpha} E_0(v, t), \tag{5.3}$$

$$E_2(v, t, \tau) = \sqrt{\beta} E_0(v, t) \exp(-j2\pi v \tau), \tag{5.4}$$

where α and β are the intensity reflection/transmission coefficients in the beam splitter, $E_0(v, t) = A(v)\exp(-j2\pi vt)$ represents optical oscillations of light source radiation with amplitude spectrum $|A(v)|$, $\tau = \Delta/c$ is delay time of the reference wave with respect to the measuring one with optical path difference (OPD) $\Delta = 2n_s z$, n_s is refractive index, and c is the speed of light.

Fringe intensity is defined as

$$I(v, \tau) = \langle EE^* \rangle = \langle |E_1^2| \rangle + \langle |E_2^2| \rangle + 2\mathrm{Re}\langle E_1 E_2^* \rangle, \tag{5.5}$$

where $E = E_1 + E_2$, and angle brackets denote the time average value. Due to utilization of a practically available photodetector, the time averaging interval is much higher than the optical oscillation period $1/v$, and one obtains from Eqs. 5.3–5.5

$$I(v, \tau) = I_1(v) + I_2(v) + 2\sqrt{I_1(v)I_2(v)} \cos 2\pi v \tau, \tag{5.6}$$

where

$$I_1(v) = \langle A_1(v) A_1^*(v) \rangle, \quad I_2(v) = \langle A_2(v) A_2^*(v) \rangle, \tag{5.7}$$

$A_1(v)$ and $A_2(v)$ are complex amplitudes of interference waves at the output of interferometer. Equation 5.6 reflects the known spectral interference law (see, for example, Ref. [2]) in simplified form.

The first and the second terms in Eq. 5.6 correspond to the light source intensity spectrum, and the third term expresses the autocorrelation function of a harmonic component of optical oscillation with frequency v. In such a consideration, the light source radiation is interpreted as consisting of harmonic oscillations with (known) amplitudes $|A(v)|$.

Due to the light source spectrum being known a priori, it is possible to find the useful third term $\tilde{I}(v, \tau)$ in Eq. 5.6 in the form

$$\tilde{I}(v, \tau) \propto |A(v)|^2 \cos 2\pi v \tau. \qquad (5.8)$$

In the case of N reflections, this equation can be rewritten as

$$\tilde{I}(v, \tau_n) \propto |A(v)|^2 \cos 2\pi v \tau_n. \qquad (5.9)$$

This means a possibility to find each of τ_n when obtaining the intensities $\tilde{I}(v_k, \tau_n)$ at different wavelengths $\lambda_k = c/v_k, k = 1, \ldots, K$, $K > 2N$.

When interacting with a real object, the measuring wave Eq. 5.3 changes and takes the form

$$E_1(v, t) = \sqrt{\alpha} E_0(v, t) H(v, t), \qquad (5.10)$$

where $H(v)$ is a complex frequency characteristic of a sample under investigation. This changes Eq. 5.6, namely

$$I(v, \tau) = I_1(v)|H(v)|^2 + I_2(v) + 2\sqrt{I_1(v)I_2(v)}\operatorname{Re}\{H(v)\exp(j2\pi v \tau)\}, \qquad (5.11)$$

and causes the appearance of two problems: The first term in Eq. 5.11 depends now on an investigated object, and only the real part of the complex frequency characteristic of a sample is available for observation.

5.2 Basics of Fringe Signal Registration

The fringe intensity Eq. 5.11 is transformed into a photoelectric signal s that is then processed, $s = \mu I$, where μ is a coefficient. When a photodetector registers the intensity at all K wavelengths simultaneously, the fringe signal can be represented as

$$s(\tau) = \mu \sum_{k=1}^{K} I(v_k, \tau) = B + 2\mu \sqrt{\alpha \beta} \operatorname{Re} V(\tau), \qquad (5.12)$$

where

$$B = \mu \sum_{k=1}^{K} I_1(\nu_k) |H(\nu_k)|^2 + \mu \sum_{k=1}^{K} I_2(\nu_k) \quad (5.13)$$

is background component and

$$V(\tau) = \sum_{k=1}^{K} |A(\nu_k)|^2 H(\nu_k) \exp(j 2\pi \nu_k \tau) \quad (5.14)$$

corresponds to fringe visibility.

Equation 5.14 is the inverse discrete Fourier transformation (FT) of a product of two functions. Following the well-known convolution theorem, this equation can be rewritten in the time domain as

$$V(\tau) = R(\tau) * h(\tau), \quad (5.15)$$

where $R(\tau)$ is the autocorrelation function of light source radiation,

$$R(\tau) = \sum_{k=1}^{K} |A(\nu_k)|^2 \cos 2\pi \nu_k \tau, \quad (5.16)$$

and

$$h(\tau) = \sum_{k=1}^{K} H(\nu_k) \exp(j 2\pi \nu_k \tau) \quad (5.17)$$

is the impulse response of an object under study. If one uses a broadband light source, that is, a source with many wavelengths and a narrow autocorrelation function (Eq. 5.16), then the convolution (Eq. 5.15) gives $V(\tau) \cong h(\tau)$, and fringe visibility represents the impulse response $h(\tau)$.

To characterize an investigated object by its impulse response (Eq. 5.17), it is necessary to find the fringe visibility (Eq. 5.14) through the object depth by varying step by step the depth position of en face layer, within which the OPD in the interferometer does not exceed the coherence length of light l_c, where interference fringes exist. It is why this approach is known as time domain OCT (TD-OCT). Peculiarities of TD-OCT are considered in detail in many papers (see, for example, Ref. [10]).

When a photodetector registers the intensity obtained at each wavelength separately, the fringe signal can be represented as

$$s(\nu_k, \tau_n) = B(\nu_k) + 2\mu \sqrt{\alpha\beta} |A(\nu_k)|^2 \sum_{n=1}^{N} \text{Re}\{ H(\nu_k) \exp(j 2\pi \nu_k \tau_n) \},$$

$$(5.18)$$

where

$$B(v_k) = \mu I_1(v_k) |H(v_k)|^2 + \mu I_2(v_k) \qquad (5.19)$$

is the background component dependent on the temporal frequency v_k of light oscillations corresponding to the wavelength λ_k. Summation in Eq. 5.18 is caused by N simultaneous reflections within the investigated object and can be rewritten in the form

$$\sum_{n=1}^{N} \text{Re}\{H(v_k)\exp(j2\pi v_k \tau_n)\} = \sum_{n=1}^{N} |H(v_k)|\cos(j2\pi v_k \tau_n + \delta_k),$$
$$(5.20)$$

where $H(v_k) = |H(v_k)|\exp(j\delta_k)$, and δ_k are the phase shifts introduced by object at wavelengths λ_k. If these phase shifts are small, it is possible to interpret Eq. 5.20 as a cosine transformation of the impulse response $h(\tau_n)$ of the object.

Due to registration of the signal at different frequencies (wavelengths) such an approach is usually entitled as frequency domain OCT (FD-OCT) (see, for example, Ref. [11]).

A comparison of TD-OCT and FD-OCT is given in many publications (see, for example, Refs. [12, 13]). It has been shown that FD-OCT provides higher speed of tissue evaluation with respect to TD-OCT due to simultaneous registration of all reflections through a tissue depth A-scan in the former technique, while TD-OCT evaluates a tissue layer by layer that is accompanied by energy losses and takes more time.

The consideration given above allows distinguishing also the following important properties.

Influence of the light source wavelength number K is expressed in TD-OCT and FD-OCT in a different manner. In TD-OCT, this number determines the width of the autocorrelation function (Eq. 5.16), that is, the light coherence length, where interference fringes exist. The delay time τ is considered within this coherence length only. In FD-OCT, interference fringes exist at each wavelength for any delay time caused by object in interferometer. The number of wavelengths has to satisfy the condition $K > 2N$, where N is the number of (equidistant) reflections within the object corresponding to the delay time set τ_n.

In TD-OCT, the background component (Eq. 5.13) is constant and independent on the delay time τ (however, it usually depends on the depth coordinate z because of variable layer reflectivity), while in FD-OCT this component (Eq. 5.19) is essentially variable due to influence of the object frequency characteristic and nonuniformity of the intensity spectrum of the light source. These properties have to be taken into account when processing OCT signals.

5.3 TD-OCT and FD-OCT Information Capacity

It is interesting to compare the 3D tissue evaluation speed by TD-OCT and FD-OCT systems that can be considered from the viewpoint of information capacity following the general approach suggested by Wagner and Häusler [14].

From the viewpoint of information theory, a sample under study can be interpreted as a source of communications conducting information in series through optical and electronic subchannels. Communications are interpreted here as information about the local value of the reflection coefficient from each voxel of tissue.

The probability to obtain the reflection coefficient value r_i at a depth point z_j can be expressed as

$$P(r_i, z_j) = P(r_i)P(z_j)$$

taking into account the mutual statistical independence of the reflection coefficient and selection of a point coordinate. In this case, following Shannon's information theory, entropy of the information source is equal to

$$H(r, z) = H(r) + H(z) = -\sum_{i=1}^{m_r} P(r_i)\log_2 P(r_i) - \sum_{j=1}^{m_z} P(z_j)\log_2 P(z_j).$$
(5.21)

If the coefficient of reflection is represented, for instance, by 256 levels (m_r) and the depth uncertainty is equal to 4 μm within the range of 1 mm (the resolvable depth point coordinates number $m_z = 250$), the entropy of the information source (Eq. 5.21) is equal to about 2 bytes/voxel. This information must be transmitted through optical and electronic subchannels of the OCT system with maximal speed.

5.3.1 Information Channel and Capacity

The capacity C_T of an information channel is defined in the well-known form

$$C_T = B\log_2\left(1 + \frac{S_{in}}{N}\right) = B\log_2\left(\frac{S}{N}\right), \quad (5.22)$$

where B is the one-side temporal signal bandwidth, S_{in} is input power, N is the noise power mean value, and $S = S_{in} + N$ is the output power available for registration. Following the well-known Shannon's theorem (see, for example, Ref. [14]), one should provide the condition $C_T \geq H'$, where H' is productivity of information source (bits/sec). The amount of information transmitted during the time interval ΔT is

$$C = \Delta T B \log_2(S/N). \quad (5.23)$$

When evaluating optical channel information capacity, we need to consider 2D spatial space in lateral directions. This consideration is different in TD-OCT and FD-OCT. In the former case, an image is formed focusing the en face plane (see Fig. 5.2, left) coinciding with zero OPD in the interferometer, and it is possible to use an objective with an increased aperture. In the latter case, radiation reflected within the full range of tissue depth L (Fig. 5.2, right) should be directed simultaneously to the entrance slit of the spectrometer. This requires using a low-aperture objective.

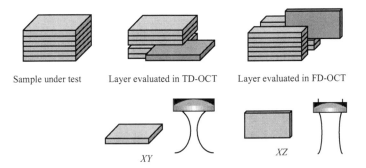

Figure 5.2 Comparison of evaluated layer geometry and measuring beam focusing in TD-OCT and FD-OCT [15].

The capacity of the optical channel in TD-OCT corresponds to the one of a conventional imaging system [14]:

$$C_{\text{OTD}} = 2\Delta X \Delta Y\, B_x B_y \log_2(S_0/N_0) = -4\Delta X \Delta Y\, B_x B_y \log_2 \mu, \tag{5.24}$$

where $\Delta X \times \Delta Y$ are the lateral extensions of the detector and

$$B_x B_y = \frac{\pi D^2}{16 \langle \lambda \rangle^2 f_{\text{TD}}^2} \tag{5.25}$$

is the bandwidth in the spatial frequency domain defined by a diffraction-limited lens system with diameter D and focal length f_{TD}, and a signal-to-noise ratio

$$\frac{S_0}{N_0} = \frac{\langle I^2 \rangle}{\sigma_I^2} \approx \frac{\langle I \rangle^2}{\sigma_I^2} = \frac{1}{\mu^2} \tag{5.26}$$

is defined mainly by speckle noise inherent in optical systems that use the radiation coherence property (see, for example, Ref. [9]). In Eq. 5.26, σ_I^2 is speckle intensity dispersion and μ is speckle contrast.

The capacity of the optical channel in FD-OCT can be defined as

$$C_{\text{OFD}} = 2\Delta X \Delta Z\, B_x B_z \log_2(S_0/N_0), \tag{5.27}$$

because the wavelength range in the spectrometer corresponds to the tissue depth range. Notice that the useful component of optical power S_0 is the same in both TD-OCT and FD-OCT because the phase spectrum in Eq. 5.11 is nonpower characteristic of the signal.

Taking into account optical conductivity of the system, the optical channel capacity can be expressed as [14]

$$C_0 = -\frac{\Sigma_p \Sigma_d}{\langle \lambda \rangle^2 f^2} \log_2 \mu, \tag{5.28}$$

where Σ_p is surface of the pupil and Σ_d is surface of the detector. To transmit the maximum amount of information about the object, largely incoherent illumination should be used (when $\mu \ll 1$) as well as a large observation aperture and a large detector array.

To compare the capacities of the optical channel in TD-OCT and FD-OCT, let us evaluate the ratio [7]

$$\frac{C_{\text{OTD}}}{C_{\text{OFD}}} = \gamma \left(\frac{f_{\text{FD}}}{f_{\text{TD}}}\right)^2 \approx \gamma \left(\frac{2L}{l_c}\right) \gg 1, \tag{5.29}$$

where

$$\gamma = \Sigma_{\text{pTD}}/\Sigma_{\text{pFD}} \approx \frac{\pi}{4}\frac{D}{\delta Y} > 1 \tag{5.30}$$

is the relative factor of using the square of the pupil in the optical system and δY is the lateral resolution of FD-OCT (see Fig. 5.2b). In derivation of Eq. 5.18, the value of focusing depth d_z of the objective along the z coordinate is calculated using the relation [16] $d_z = 1,8 \langle \lambda \rangle /(\text{NA})^2 = 1,8 \langle \lambda \rangle f^2/D^2$, where NA is the numerical aperture of the objective, which gives

$$\frac{d_{z\text{TD}}}{d_{z\text{FD}}} = \left(\frac{f_{\text{TD}}}{f_{\text{FD}}}\right)^2 \approx \frac{l_c}{2L}. \tag{5.31}$$

Equation 5.29 shows that the capacity of the optical channel in TD-OCT is essentially higher than in FD-OCT. It is explained by the two limits of FD-OCT: necessity to use a low-aperture objective and exploration of only part of the useful square of the pupil when transmitting a single lateral line of the observation field along the x coordinate to entrance the slit of the spectrometer.

The latter limit can be eliminated in the swept-source OCT (SS-OCT) in full-field mode when a light source illuminates the square of field of view entirely. However, this mode is not widely used due to observation of B-scans formed in SS-OCT by a laterally scanning measuring beam.

Let us consider the information capacity of an electronic subchannel.

Optical radiation is registered by a photosensitive matrix consisting of $P \times Q$ elements. This number is supposed to be the same in both TD-OCT and FD-OCT systems. Taking into account the Nyquist criterion, the transmission bandwidth along the x axis is defined as

$$B_x = \frac{P}{2\Delta X},$$

where ΔX is the discretization step. Using the same relation for another coordinate information capacity of the electronic channel can be expressed as

$$C_\text{E} = \frac{1}{2} P\, Q \log_2\left(\frac{S_\text{E}}{N_\text{E}}\right). \tag{5.32}$$

Optical power falling on a single sensitive element is small; it is why the noise term N_E in the electronic channel is determined mainly by photon noise with Poisson statistics. In this case, the noise dispersion σ_n^2 is equal to the mean value $\langle n \rangle$ of the number of registered photons that gives the signal-to-noise ratio

$$\frac{S_E}{N_E} = \frac{\langle n^2 \rangle}{\sigma_n^2} \approx \frac{\langle n \rangle^2}{\sigma_n^2} = \langle n \rangle. \tag{5.33}$$

Then Eq. 5.32 takes the form

$$C_E = \frac{1}{2} P\, Q \log_2 \langle n \rangle. \tag{5.34}$$

In correspondence with Eq. 5.34, the amount of information received by the photodetector increases linearly with increasing the number of photosensitive elements and logarithmically with the number of photons captured by each element.

It is important to note that useful information brings only $\langle n_c \rangle = \langle n \rangle\, l_c/2L$ photons within the light coherence length l_c in a TD-OCT system, while in FD-OCT this information is obtained using all $\langle n \rangle$ photons reflected through a tissue depth. Hence the ratio

$$\frac{C_{TDE}}{C_{FDE}} = \frac{\log_2 \langle n_c \rangle}{\log_2 \langle n \rangle} < 1. \tag{5.35}$$

Equation 5.35 shows that FD-OCT provides the gain in information capacity of the electronic channel in logarithmic dependence on relative resolving power in depth. Opposite, following Eq. 5.29 information capacity of the TD-OCT optical channel rises linearly with respect to FD-OCT with an increase of resolving power in depth. Therefore one has to conclude about the general advantage of TD-OCT in information capacity when evaluating a 3D microstructure entirely with high resolution, if the electronic subchannel in TD-OCT would not limit the total information capacity.

5.3.2 2D and 3D OCT Imaging

It is evident that 3D imaging of the internal microstructure of an object is more informative with respect to 2D imaging taking into account Eq. 5.21, which can be easily generalized to a multidimensional case. However, there appear two important questions: first, how to transmit and process the large amount of data acquired in 3D system, and second, how to use this 3D information.

Indeed, the human eyes and mind are able to observe and interpret 2D images mainly. It is why the majority of OCT systems are designed to visualize 2D images in the form of B-scans. This decreases the amount of information that has to be acquired and processed simultaneously and provides high-speed visualization.

Nevertheless, 3D information is very important when one can not surely evaluate an object with incompletely clear properties having a single B-scan. In this case, observer considers a few parallel B-scans to resolve ambiguity utilizing additional information.

From our point of view, 3D imaging is very important and a perspective for automatic diagnostics systems, which are able to capture 3D information simultaneously and provide reliable decisions about properties of an investigated object. Full-field OCT systems meet this property. It is reasonable to consider possibilities to increase their speed when acquiring 3D OCT data.

5.4 Signal Downsampling in Full-Field OCT

In full-field OCT, a sequence of video frames captured by video camera is recorded as illustrated in Fig. 5.3a.

In the case of TD-OCT, neighbor video frames have a mutual phase shift less than π that provides the Nyquist condition when sampling low-coherence interference fringes. This phase shift corresponds to the depth step less than $\langle \lambda \rangle /4$ and means a large

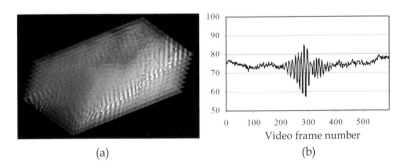

(a)   (b)

Figure 5.3 (a) Recorded video frame set and (b) example of low-coherence fringe signal (in arbitrary units) at single pixel through the video frame set.

total number of video frames. Due to variation of the fringe signal envelope slower than fringes, it is possible to miss a part of the frames without essential losses of useful information about the fringe envelope using the downsampling approach when the sampling frequency is set lower than the Nyquist limit with respect to the interference fringe frequency.

In Ref. [17], analysis of the downsampling conducted in detail with application to conventional interferometry, and possibility to overcome the Nyquist limit due to usage of a priori information about properties of interference fringes, is demonstrated. It has been shown in Ref. [18] that the downsampling allows minimizing the OCT signal processing root-mean-square (RMS) relative error up to a few percents under some conditions, providing a low-coherence fringe processing speed gain up to 1 order of magnitude.

It is reasonable to consider peculiarities of downsampling when recording low-coherence fringe signals and when using the asynchronous amplitude demodulation method applied to TD-OCT signals.

Amplitude demodulation of TD-OCT signal aims to evaluate low-coherence fringe envelope. There are well known two basic methods: synchronous demodulation (synchronous detection) where carrier fringe frequency should be known and stable and asynchronous detection of envelope that is free from the former requirement. The research results obtained before [18] show that the most preferable asynchronous amplitude demodulation method includes consequent procedures of suppressing the background OCT signal component by band-pass filtering, squaring of the useful fringe signal, and then low-pass filtering and calculation of the square root to recover the fringe envelope in the initial scale of magnitude. These procedures are illustrated in Fig. 5.4.

A low-coherence interferometric signal observed in an OCT system at a fixed point (x, y) in the lateral plane contains the useful component dependent on depth coordinate z

$$s(z) = A(z) \cos \Phi(z), \quad (5.36)$$

where the fringe envelope $A(z)$ contains useful information about the reflection degree through the depth of the investigated tissue and the fringe phase $\Phi(z)$ is defined as

$$\Phi(z) = 2\pi f_0 z + \phi + \delta\phi(z), \quad (5.37)$$

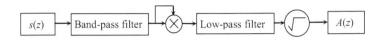

Figure 5.4 Diagram of the fringe processing algorithm.

where f_0 is the fringe carrier frequency, ϕ is initial phase at the point $z = 0$, which can be omitted without any decrease of consideration generality, and $\delta\phi(z)$ expresses random phase variations caused particularly by the influence of external factors. The fringe signal (Eq. 5.36) is distorted by observation noise $n(z)$ such that the real signal is expressed as

$$s(z) = A(z)\cos\Phi(z) + n(z). \tag{5.38}$$

The problem consists in obtaining the $A(z)$ estimates when processing the signal $s(z)$ with high resolution, noise immunity, and processing speed. The appropriate processing algorithm is illustrated by the diagram [19] shown in Fig. 5.4.

The signal $s(z)$ is processed by a band-pass filter to remove the fringe background component and high-frequency noise. The filter bandwidth is set wide enough to transmit fringe envelope variations with required resolution, taking into account possible variations of carrier frequency. The useful component (Eq. 5.36) of the interferometric signal is squared, namely

$$s^2(z) = \frac{1}{2}A^2(z)[1 + \cos 2\Phi(z)] = \frac{1}{2}A^2(z) + \frac{1}{2}A^2(z)\cos 2\Phi(z). \tag{5.39}$$

The second term in Eq. 5.39 changes faster in comparison to the first one taking into account (Eq. 5.37) and is removed by a low-pass filter with the cutoff frequency corresponding to the maximal frequency in the spectrum of the squared envelope signal. Calculation of the square root from the rest first term gives an estimate of the fringe envelope.

Advantages of the algorithm consist in its simplicity and absence of strong requirements to a priori information about fringe parameters. It surely provides good results, if the fringe carrier frequency is higher than the doubled spectral width of envelope variations and provides accurate results if the carrier frequency f_0 is twice higher than the Nyquist limit $f_N = 2f_M$, that is, $f_0 > 4f_M$, where f_M is maximal frequency in the envelope spectrum.

It is interesting to consider properties of the algorithm when processing a discrete signal samples series with different ratios of fringe and discretization frequencies, especially in the case of fringe signal downsampling with respect to the Nyquist limit.

5.4.1 Spectral Characteristics of Downsampled Signals

When sampling signal (Eq. 5.36), one obtains a discrete samples series

$$s(k) = A(k)\cos\Phi(k), \qquad (5.40)$$

where k corresponds to the discrete coordinate $z_k = k\Delta z$, $k = 0, 1, \ldots, K$, and Δz is the discretization step.

Consideration of continuous signal sampling depends on definition of this procedure. In sampling theory (see, for example, Ref. [20]), it is usually supposed that the discretization result is a product of a continuous signal and a set of delta functions with the step Δz. According to the well-known convolution theorem the sampled signal spectrum is represented in the frequency domain by a set of repeated initial signal spectra (called spectral orders) with the frequency interval between spectral order positions equal to the sampling frequency $f_s = 1/\Delta z$.

It is important to note that formation of such spectral orders takes place if the sample series a priori includes *zero values* of the signal between delta functions. If downsampling is conducted with all missing intermediate signal values between samples, then the spectral shift of the initial spectrum to a lower carrier frequency takes place without the formation of other spectral orders. This property is explained by the fact that in the latter case the timescale is defined by the sampling frequency but not by the initial signal.

In Fig. 5.5a,b, an example of a normalized TD-OCT signal as a samples serie $s(k)$ and the corresponding module of its spectrum are presented.

The signal contains six samples within the fringe period corresponding to the carrier frequency. The normalized value of the discretization step $\Delta z = 1$ is supposed, and the initial carrier frequency is equal approximately to 0.17. When modeling the fringe signal (Eq. 5.38) (Fig. 5.5) additive Gaussian noise had dispersion

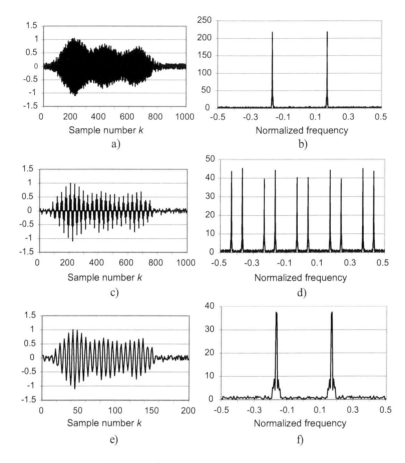

Figure 5.5 (a and b) Initial fringe signal (separate fringes are not seen) and its spectrum module, (c and d) sampled signal with zero values between samples and corresponding signal spectrum, and (e and f) downsampled signal and its spectrum.

$\sigma_n^2 = 0,01$, that is, its RMS value was equal approximately to 10%. The obtained values of the spectrum module in Fig. 5.5b are defined by the known Parseval's equation (the rule of energy conservation) in discrete form.

When downsampling the signal (taking each fifth sample in the considered example) and replacing all missed samples with zero value (see Fig. 5.5c), spectral orders appear (Fig. 5.5d). In the case of missing intermediate samples, the spectrum of the obtained signal (Fig. 5.5f) is generally similar to the initial spectrum (Fig. 5.5b).

The new frequencies f'_0 seen in Fig. 5.5d are caused by the effect of the so-called masking frequencies, which are defined by the known relation (see, for example, Ref. [21])

$$f'_0 = 2mf_N \pm f_0, \qquad (5.41)$$

where $f_N = f_s/2$ is Nyquist frequency with respect to sampling frequency f_s and $m = 1, 2, \ldots$

Let us return to the example in Fig. 5.5, where $f_0 = 0.17$, $f_s = 0.2$, and $f_N = 0.1$. When downsampling (Fig. 5.5c) with $m = 1$ the frequencies 0.03 и 0.37 appear. The frequencies 0.23 и 0.43 (third and fifth peaks from zero frequency) correspond to $m = 2$ and $m = 3$, that is, $2 \cdot 2f_N - f_0 = 2 \cdot 2 \cdot 0.1 - 0.17 = 0,23$; $2 \cdot 3f_N - f_0 = 2 \cdot 3 \cdot 0.1 - 0,17 = 0.43$.

It is seen that the spectrum in Fig. 5.5d is more complicated with respect to the one in Fig. 5.5f. This means that to avoid appearance of masking frequencies, it is reasonable to use the downsampling with missing all intermediate signal values (Fig. 5.5e,f) when the total number of samples essentially decreases.

Due to the value of the discretization step Δz of the signal in Fig. 5.5e being set as earlier equal to 1, and fewer number of samples being taken, this is equivalent to a shorter total length of the signal. In this case, the signal spectrum is determined by the known scaling property of FT, namely

$$\mathbb{F}\{s(\alpha k \Delta z)\} = (1/\alpha)S(f/\alpha), \qquad (5.42)$$

where $\mathbb{F}\{\cdot\}$ designates the Fourier transform operation, $S(f)$ is the spectrum of the initial signal, and α is a constant (in the example above, $\alpha = 5$). It is seen from Eq. 5.42 that downsampling causes spectrum widening proportional to the value α, as illustrated in Fig. 5.5f in comparison to Fig. 5.5b.

When squaring the signal following Eq. 5.39, the same effect of spectral widening appears (see Fig. 5.6).

It is seen in Eq. 5.39 that useful information about the signal envelope $A(k)$ is contained in the spectral component (Fig. 5.6d) near zero frequency. To extract this component, a low-pass filter is used with a cutoff frequency f_c corresponding to maximal frequency

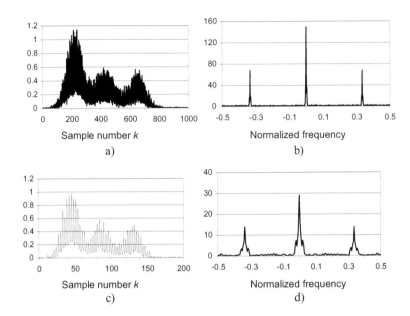

Figure 5.6 (a and b) Squared initial signal and spectrum module and (c and d) squared downsampled signal and spectrum module.

of this component. The correct choice of the f_c value taking into account the considered effect of the spectrum widening allows minimizing the envelope estimate error.

It is seen from Eqs. 5.39 and 5.42 that the cutoff frequency has to satisfy the condition

$$f_c = 2\alpha f_M, \quad (5.43)$$

where f_M is the maximal frequency in the initial signal envelope spectrum. Besides, to avoid masking frequencies, it is necessary to provide the carrier fringe frequency

$$f_0 \geq 4\alpha f_M. \quad (5.44)$$

Notice that quadratic transformation of noise when squaring the real signal (Eq. 5.38) gives a nonzero mean value. However, this does not influence essentially fringe envelope estimates.

Envelope estimate RMS errors obtained at the output of a low-pass filter with a different cutoff frequency applied to the signal presented in Fig. 5.6c are shown in Fig. 5.7. The curves correspond to

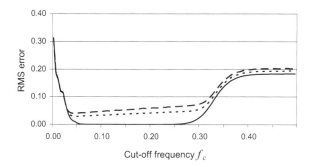

Figure 5.7 Dependence of envelope estimate RMS error on cutoff frequency of low-pass filter under different influence of additive noise (without noise, solid line; noise RMS equal to 4%, dotted line; noise RMS equal to 6%, upper dotted line).

the value of the parameter $\alpha = 5$. Calculations were conducted with the average over an ensemble of 100 OCT signal realizations.

It is clearly seen that too small values of the cutoff frequency cause essential distortions because of the spectral limit of the useful signal component near zero frequency. When increasing the cutoff frequency, the error decreases up to negligible values, if the noise influence is absent. Increase of the cutoff frequency leads to appearance of distortion again due to the influence of the second term in Eq. 5.39, which spectrum is represented in lateral peaks in Fig. 5.6d. Notice that increase of the noise level leads to a corresponding rise of the error due to the rest noise dispersion proportional to the bandwidth of the filter.

Experimental research of the method considered above shows that it allows obtaining a gain in processing speed typically up to 1 order of magnitude with a moderate RMS error up to a few percents if the carrier frequency of fringes is stable such that Eq. 5.44 is fulfilled.

Figure 5.8 illustrates the results of evaluation of a paper material relief using the considered method [22].

Figure 5.8a shows the reflection degree of paper material represented by recovered low-coherence fringe envelopes within the pixels' evolution in one line of a recorded frame set. There were 300 frames recorded with a height step of 64 nm, that is, the height

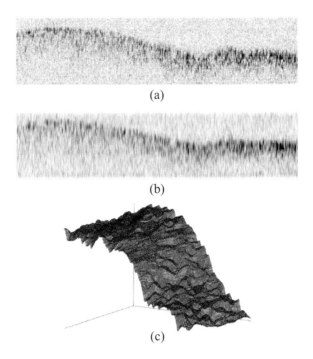

Figure 5.8 (a and b) Distribution of reflection degree of paper material (represented in inverse grey scale to be more visual) and (c) evaluated 3D surface relief.

range is equal to 19.2 µm. The lateral size of the observed surface area is equal to 200 µm. Figure 5.8a demonstrates high resolution of the system. The effect of partial light penetration into paper material is seen; that is why it is important to separate surface and subsurface reflection to characterize a pure surface relief.

The result illustrated in Fig. 5.8b relates to a high-speed registration mode, when 60 frames with a height step of 320 nm were recorded, that is, over the Nyquist sampling limit (equal to $\langle \lambda \rangle/4$ per step). A comparison of Figs. 5.8a and 5.8b shows that the frame speed gain in five times does not influence significantly the image quality.

The recovered surface relief inside the investigated area of 200 × 200 µm is illustrated in Fig. 5.8c. The statistical criterion of selecting fringe envelope peaks related to surface reflection was used to separate surface and subsurface reflection. The results obtained confirm high resolution and speed of the full-field low-coherence interferometric profilometer.

Thus, the method considered above is simple to use and insensitive to random variations of low-coherence fringe frequency and phase with moderate requirements to choice of the cutoff frequency of a low-pass filter. The method can be used in high-speed express evaluation of objects in OCT and OCM.

5.5 Dynamic Fringe Signal Evaluation in State Space

There are a few conventional fringe processing methods known to assess an envelope of low-coherence interference fringes. The local fringe amplitude can be evaluated using the well-known phase-shifting technique [23–25] or fringe amplitude demodulation by fringe signal squaring [26, 27]. Both these approaches are based on nonlinear transformations applied to a determinate model of interferometric data; that is why such methods are generally nonoptimal, especially when interference fringes are distorted under essential noise influence that takes place in OCM systems.

A new approach to interference fringe analysis based on recurrence calculations in state space was suggested [28] and investigated in detail in the last decade [7, 28–33]. In this approach, fringe processing is carried out using recurrence algorithms, in which the fringe signal value defined by the parametric stochastic model in state space is predicted to a following discretization step utilizing full information available before this step, and the fringe signal prediction error is used for step-by-step dynamic updating of the fringe parameters. Several versions of the recurrence fringe processing, in particular in the form of nonlinear Markov stochastic filtering [28], extended Kalman filtering [30, 31], and unscented Kalman filtering [32], were successfully applied to interference fringe analysis in rough surface profilometry and OCT [7, 19, 32, 33].

A commonly used mathematical model of interference fringes is expressed as

$$s(x) = B(x) + A(x)\cos\Phi(x), \qquad (5.45)$$

where $B(x)$ is the background term, $A(x)$ is the fringe envelope, $\Phi(x)$ is the fringe phase

$$\Phi(x) = \varepsilon + 2\pi f_0 x + \varphi(x), \quad (5.46)$$

ε is an initial phase at the point $x = 0$, f_0 is the fringe frequency mean value, and $\varphi(x)$ describes phase change nonlinearity. In the model in Eq. 5.45, the fringe background, envelope, and phase are usually supposed as belonged to a priori known kind of determinate function. Namely, the fringe envelope $A(x)$ and the background $B(x)$ are often considered as constants or slowly varied functions with respect to fringes [34–37]. Such assumption allows one to use processing algorithms applicable to high-quality fringes. In many cases, when evaluating real objects, fringe parameters can be essentially variable, that is, stochastic by nature, and determinate model is inappropriate.

Stochastic models are widely used in physics and engineering to describe random processes (see, for example, Refs. [21, 38]). Being applied to interferometry problems, stochastic models initially include possible random variations of fringe parameters [29]. In this way, the fringe signal is defined by a parametric model, that is,

$$s(x) = B(x) + A(x)\cos\Phi(x) = s(x, \theta); \; \theta = (B, A, \Phi, f)^T, \quad (5.47)$$

where θ is a vector of fringe parameters in the state space $\{\theta\}$. It allows introducing a priori knowledge about supposed variations of fringe parameters in explicit form.

One of the most productive approaches to describe stochastic system consists in usage of differential equations' formalism. Indeed, if the background component B, fringe amplitude A. and frequency f are supposed, for example, to be constants and the fringe phase Φ varies linearly in a given interferometric system, one can write

$$\frac{dB}{dx} = 0, \; \frac{dA}{dx} = 0, \; \frac{df}{dx} = 0, \; \frac{d\Phi}{dx} = 2\pi f_0. \quad (5.48)$$

The next step is to include stochastic properties in the model by modifying differential equations (Eq. 5.4) to the following form:

$$\frac{dB}{dx} = w_B(x), \; \frac{dA}{dx} = w_A(x), \; \frac{df}{dx} = w_f(x), \; \frac{d\Phi}{dx} = 2\pi f_0 + w_\Phi(x), \quad (5.49)$$

where $\mathbf{w} = (w_B, w_A, w_f, w_\Phi)^T$ is a random vector. The first-order Eq. 5.49 is a stochastic differential equation of the Langevin kind [38] and can be represented in vectorial form

$$\frac{d\Theta}{dx} = \mathbf{F}(x, \Theta) + \mathbf{w}(x), \qquad (5.50)$$

where the first term relates to determinate evolution of fringe parameters and the second one reflects their random variations. The a priori information about evolution of vector of parameters Θ is included by appreciable selecting a known vectorial function $\mathbf{F}$ and supposed statistical properties of "forming" noise $\mathbf{w}(x)$. It is important to emphasize that Eq. 5.50 defines as well nonstationary and nonlinear processes [38].

In a discrete case, Eq. 5.50 is rewritten in the form of a stochastic difference equation defining discrete series $s(k) = s(x_k)$ at the points $x_k = k\Delta x$, $k = 1, \ldots, K$, where Δx is the discretization step. The recurrence procedure in state space at the k-th step is based on the relation

$$\Theta(k) = \Theta(k/k-1) + \mathbf{w}(k), \qquad (5.51)$$

where $\Theta(k/k-1)$ is the predicted value from the $(k-1)$-th step to the k-th one, taking into account concrete properties of parameters determined by Eq. 5.50. The prediction in Eq. 5.51 contains an error, that is, difference between the a priori knowledge at the $(k-1)$-th step and actual information obtained at the k-th step. This difference is available for observation only as the fringe signal error. To obtain the a posteriori information about parameters at the k-th step, the signal error should be transformed to update fringe parameters, namely

$$\hat{\Theta}(k) = \Theta(k/k-1) + \mathbf{P}(k)\{\hat{s}(k) - s[k, \Theta(k/k-1)]\}, \qquad (5.52)$$

where $\mathbf{P}(k)$ is a vectorial function, which transforms the scalar difference between the observed signal sample value $\hat{s}(k)$ and the modeled (predicted) one $s(k, \Theta)$ to vectorial correction of fringe parameters for the current step.

Peculiarities of recurrence fringe processing algorithms based on the general formula Eq. 5.52 are considered in detail in the original paper [39] and in papers devoted to application of the recurrence algorithms in state space to low-coherence interferometry and OCT [19, 28, 29, 32].

5.5.1 Recurrence Computational Algorithms Applied to Fringe Processing

In recurrence computational algorithms based on Eq. 5.52 and applied to fringe processing, initial Eqs. 5.47 and 5.50 are considered as the observation equation and the system equation, respectively. These equations can be rewritten in discrete vectorial form as follows:

$$\mathbf{s}(k) = \mathbf{h}(k, \Theta(k)) + \mathbf{n}(k), \tag{5.53}$$

$$\Theta(k) = \mathbf{T}(k)\Theta(k-1) + \mathbf{w}(k), \tag{5.54}$$

where $\mathbf{h}(k, \Theta)$ is the vector containing nonlinear observation functions corresponding to Eq. 5.47, $\mathbf{n}(k)$ is the observation noise term, and $\mathbf{T}(k)$ is the transition matrix, which defines known determinate evolution of vector $\Theta(k)$.

Using Eq. 5.54, one can calculate the predicted vector of parameters at the k-th step as

$$\bar{\Theta}(k) = \Theta(k/k-1) = T(k)\hat{\Theta}(k-1) \tag{5.55}$$

and then assess the actual value of the vector of parameters following Eq. 5.52 in the form

$$\hat{\Theta}(k) = \bar{\Theta}(k) + \mathbf{P}(k)[\hat{\mathbf{s}}(k) - h(k, \bar{\Theta}(k))]. \tag{5.56}$$

The amplification factor $\mathbf{P}(k)$ in Eq. 5.56 transforms the difference (in square brackets) between the actual and the predicted signal value at the current sample k, taking into account the supposed fringe model (Eq. 5.45).

For instance, in the scalar case it has the form [29]

$$P_A(k) = (\sigma_A^2/\sigma_n^2) \cos\phi(k-1) \tag{5.57}$$

for the fringe envelope, and corresponding to Eq. 5.56 the dynamic envelope estimate is calculated as

$$\hat{A}(k) = A(k/k-1) + P_0 \cos\phi(k-1) \, [\hat{s}(k) - A(k/k-1)\cos\phi(k-1)], \tag{5.58}$$

where $P_0 = \sigma_A^2/\sigma_n^2$ is interpreted as (a priori evaluated) the signal-to-noise ratio in a concrete interferometric system.

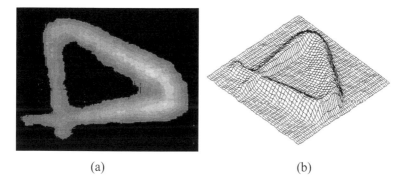

(a) (b)

Figure 5.9 (a) Initial 3D rough surface relief presented as a gray-level map and (b) 3D relief reconstruction shown as a convex form.

5.5.2 Application of Recurrence Processing Algorithm in State Space to Low-Coherence Profilometry

Application of the recurrence processing algorithm considered above was carefully studied in OCM mode, and corresponding results are presented in Ref. [29]. The input data $z = z(x, y)$ about the rough surface relief of a metal specimen had been used, with the surface deformed by a stamp in the form of the digit 4. The initial surface relief is presented in Fig. 5.9a as a gray-level map and in Fig. 5.9b as a 3D plot. The lateral dimensions of the useful surface area are equal to 3×2 mm, and the surface relief deviation range is equal to 180 μm. The surface relief deviations are presented in Fig. 5.9b for a convex relief form that is more visual in illustration.

It has been shown that the recurrence algorithm considered above when processing a set of video frames in dynamic mode provides the relief reconstruction with the estimate RMS error close to 1 μm.

5.5.3 Application of Recurrence Processing Algorithm in State Space When Evaluating Random Tissues

Nondestructive optical testing techniques are widely used in the field of painting diagnostics because of their effectiveness and safety. At present, many techniques for nondestructive investigations of paintings are available. The OCT techniques allow evaluating multilayer tissues. Being applied to painting diagnostics, OCT gives a possibility to measure the actual varnish thickness that is very important in painting restoration by the cleaning process. Because of the complicated local structure of layers and light scattering, noise-immune signal processing methods should be used. In Ref. [40], the Kalman filtering method involving the random fringe model applied to OCT signals is investigated and compared with the conventional fringe amplitude demodulation method.

The commonly used signal amplitude demodulation method based on signal rectification and subsequent low-pass filtering is well known. After subtraction of the average so that one finds a zero offset, the signal absolute value is taken to have only positive values. Finally, the data are resampled (by taking an average of the appropriate number of adjacent data points) so that the final number of depth points is adjusted (in our experiment 1 point per micron).

The method has the advantage in its simplicity but generally does not provide accurate discrimination between the fringe carrier signal and incoherent signal variations within a spectral band of the low-pass filter. Apart from it, the filter spectral band must be set wide enough to transmit a signal with the extended spectral band caused by signal envelope variations and by possible fringe frequency and phase variations. It may decrease the signal-to-noise ratio under the influence of the observation noise of uniform spectral density. If the filter spectral band is adjusted narrow, it may cause a decrease in the resolution of evaluated fringe envelope variations. To overcome these difficulties, the discrete Kalman filtering method was applied to process OCT signals.

The Kalman filtering processing algorithm described above has been used when processing OCT signals and recovering tomograms of paintings' layer structures. Figure 5.10 shows an example of the experimental tomograms represented in a logarithmic gray-level inverse scale to be more visual.

The initial recorded data contained 201 depth scans for 1 mm extension. Each depth scan signal includes 5000 signal samples.

It is seen in Fig. 5.10 that the Kalman filtering method provides better resolution. In spite of similar visual views of Fig. 5.10a,b, the

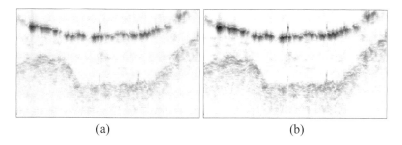

Figure 5.10 Tomograms recovered using (a) conventional fringe amplitude demodulation and (b) Kalman filtering method.

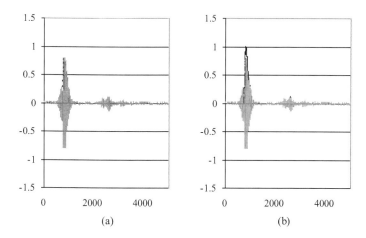

Figure 5.11 Evaluation of the OCT depth scan signal envelope using (a) conventional demodulation technique and (b) Kalman filtering method.

difference between signal envelopes in an axial scan is noticeable in Figs. 5.11 and 5.12.

Figure 5.12 shows the first left peak of the signal in Fig. 5.11 in enlarged scale. A comparison of Figs. 5.11a,b and Figs. 5.12a,b demonstrates higher resolution of the Kalman filter, but the signal envelope in Fig. 5.12b includes a small ripple-like noise component. The relation between resolution and noise can be selected by appropriate choice of the Kalman filter amplification factor.

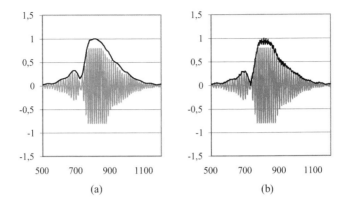

Figure 5.12 Evaluation of the fringe signal peak envelope using (a) conventional demodulation technique and (b) Kalman filtering method.

5.5.4 Combined Use of Signal Downsampling and Kalman Filtering

The method of sub-Nyquist sampling is considered in Ref. [18] that was used to decrease sampling speed a few times for narrow-band OCT signals. To provide high noise immunity and stability when processing OCT signals with randomly variable parameters in real time, the Kalman filtering method has been applied for evaluating sub-Nyquist-sampled OCT signals.

Figure 5.13 shows cross-sectional images (B-scans) of 3D tomograms of layered tissue recovered by the ordinary envelope evaluation method [40] and when applying sub-Nyquist sampling and the Kalman filtering method considered above. Tomograms are represented in the gray-level inverse scale to be more visual.

Tomograms in Fig. 5.13 include 200 vertical depth scans with a lateral step of 10 μm between them. A comparison of Figs. 5.13a and 5.13b confirms high accuracy of the suggested method, which provides significant processing speed gain almost without losses in resolution.

The advantage of the recurrence filtering algorithm consists mainly in high processing speed and adaptive tuning to signal carrier and sampling frequency variations that provide processing gain for typical OCT signals up to 10 times with a small resolution decrease

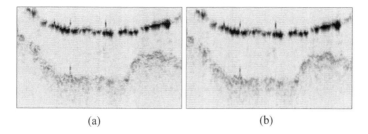

(a) (b)

Figure 5.13 (a) Initial layered tissue tomogram recovered by ordinary fringe envelope evaluation method and (b) the tomogram extracted from the sub-Nyquist sampled OCT signal by Kalman filtering method with processing speed gain of 4.5 times [18].

of a few percents. The advantage in processing speed is especially useful for full-field OCT and OCM systems to evaluate 3D tomograms in real time.

5.5.5 Recurrence Filtering Modifications and Processing Speed

There are known several modifications of recurrence stochastic filtering methods that can be applied to process interferometric signals in OCT and OCM [32], including Markov stochastic filtering and unscented Kalman filtering.

Markov stochastic filtering (considered in detail, for example, in Refs. [28, 41]) is the most general method, which provides optimal dynamic evaluation of fringe parameters, including filtering of locally variable fringe frequency and phase.

One of the problems in stochastic filtering consists in nonlinear transformation of the Gaussian random variable initially supposed as the prediction error. Such nonlinearity may lead to an essential difference with the true fringe model that causes instability of filtering. The unscented Kalman filtering method allows overcoming this drawback by calculating the mean value and covariance of the random variable passed through the nonlinear system (see, for example, Ref. [42]).

An example of the unscented Kalman filtering application to low-coherence fringe analysis [32] is presented in Fig. 5.14.

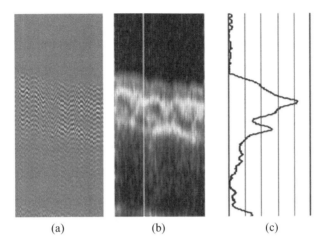

Figure 5.14 (a) Initial set of low-coherence interference fringes, (b) B-scan evaluated by the unscented Kalman filter, and (c) A-scan for the marked line.

The processing speed of recurrence algorithms was investigated in Ref. [32], taking into account a quantity of elementary computational primitives as well as averaged computational primitive speed for different modifications of the stochastic filtering procedures. It has been found that processing of the scalar fringe signal, including 512 signal samples, with evaluation of its envelope using conventional computational means takes 3–5 ms. Due to the possibility of parallel calculations provided by modern graphics processing units (GPUs) adapted to matrix calculations, it is possible to increase the computational speed of the stochastic filtering algorithms, providing high stability and accuracy of fringe processing and visualization of 2D and 3D microstructures of the investigated object.

5.6 Recurrence Processing of Swept-Source Spectral Interference Fringes

Conventional methods to extract useful information in spectral OCT involve FT of spectral fringes. Due to availability of light sources with controllable variation of wavelength in a wide spectral range it becomes possible to register time-variable spectral fringes appearing at different wavelengths, and it is needed to provide high-speed processing of spectral data in the time domain.

Consider the mutual interference of two beams from a light source with variable wavelength. When the optical path difference (OPD) Δ between the beams is adjusted in the interferometer, two-beam interference is governed at the wavelength λ by the spectral interference law (see, for example, Ref. [2]) that is expressed in simplified form omitting light dispersion when interacting with the object as follows:

$$I(\sigma) = G(\sigma)[1 + V(\sigma)\cos(2\sigma z)], \quad (5.59)$$

where $\sigma = 2\pi/\lambda$ is the wavenumber, $G(\sigma)$ is the spectral power density of light, $V(\sigma)$ is spectral fringe visibility, and OPD $\Delta = 2z$. After normalizing Eq. 5.58 relative to the light source spectrum, the interferometric signal is determined as

$$S(\sigma) = 1 + V(\sigma)\cos(2\sigma z). \quad (5.60)$$

The conventional method to find the distance z consists in applying FT to the second term in Eq. 5.59, taking into account that the spectral fringe frequency is obviously proportional to this distance. In OCT, there are many reflections from different layers of the object under study at distances z_i, such that the spectral fringe signal Eq. 5.59 contains components with different frequencies, and the FT procedure allows estimating these frequencies, that is, distances z_i.

It is important to note that FT allows obtaining results independently on a mathematical model of spectral fringes. However, the spectral fringe signal can be initially considered as a set of harmonic components defined by amplitude, frequency, and initial phase. In this case, the problem is considered as parametric identification of harmonic components of the signal Eq. 5.59 and can be solved in dynamic mode using Kalman filtering considered in previous sections of this chapter.

Suppose there is a sequence of values $S(n)$, $n = 0 \ldots N - 1$, representing the second term in Eq. 5.59, and the problem of identification of a harmonic component with a fixed frequency f contained in this sequence is solved. Here n is a current number of the wavenumber σ value within a sequence of wavenumbers set in the spectral interferometry system when tuning the wavelength of monochromatic light. A harmonic component is defined by the model $H(n) = a\cos(2\pi f n\Delta\sigma)$, where a is amplitude and $\Delta\sigma$ is the wavenumber step within the range of wavenumbers, that is, the overall wavelength tuning range. The recurrence algorithm of identification of the harmonic component consists in prediction of its amplitude at the step n, $\bar{a}(n) = \gamma a(n-1)$, where coefficient γ meets the linear law of amplitude change between the neighboring samples (it should be set as $\gamma = 1$ in the case of constant amplitude), and the a posteriori correction of amplitude, taking into account the difference between predicted and observed data values in the form

$$\tilde{a}(n) = \bar{a}(n) + P(n)[S(n) - \bar{a}(n)C(n)]. \qquad (5.61)$$

In Eq. 5.61, the transition factor $C(n)$ is defined as a derivative of the model $H(n)$ by parameters

$$C(n) = H'_a(n) = \cos(2\pi f n\Delta\sigma), \qquad (5.62)$$

$$P(n) = RC(n)[RC^2(n) + R_n]^{-1} \qquad (5.63)$$

is amplification factor (see Eq. 5.52), R is the error dispersion of a priori estimate of the parameter at the step $(k-1)$, and R_n is dispersion of observation noise. Thus, the algorithm of the identification of the harmonic component is reduced to calculation of the transition factor $C(n)$ and the amplification factor $P(n)$. If the wavenumber step $\Delta\sigma$ equal to 1 is supposed, then the Kalman filtering algorithm for evaluating amplitude of the harmonic component is described at each discrete wavelength step n by the following recurrence equation [33]:

$$\tilde{a}(n) = \bar{a}(n) + RC(n)[S(n) - \bar{a}(n)C(n)][RC^2(n) + R_n]^{-1}, \quad (5.64)$$

The values of R and R_n in Eq. 5.4 are defined by a priori information about the signal prediction error and observation noise statistical properties, respectively. The initial value of amplitude $\tilde{a}(0)$ can be set arbitrarily to a certain degree (as square root from supposed dispersion of possible amplitude values) since an error in the initial condition influences only the algorithm convergence speed and does not influence the final result. It is important to note that the nonzero initial phase of the harmonic component also has no effect on the amplitude estimate but influences convergence speed. When estimating amplitude only, the initial phase of the harmonic component can be neglected.

Kalman filtering allows involving an arbitrary number of processing steps using a stop criterion, that is, a limit of the number of spectral samples when sufficient resolution for a particular object is reached. The stop criterion can be based on calculating amplitude estimates at the current n-th and the previous $(n-1)$-th steps and then comparing the amplitude difference with a certain threshold value ε under the condition $\tilde{a}(n) - \tilde{a}(n-1) < \varepsilon$.

In Fig. 5.15a an example of a 2D object is shown. Such a model is applicable, for example, to describe borders of a blood vessel. Figure 5.15b represents spectral fringes corresponding to this object, and Fig. 5.15c,d illustrates the evolution of estimates $\hat{z}_{50}(n)$ and $\hat{z}_{250}(n)$ of the object borders along the lateral positions marked in Fig. 5.15a by dotted lines (with discrete positions 50 and 250 in the lateral direction; the total number of lateral positions is equal to 512).

Figures 5.15e–h show recovered B-scans obtained when using a different number of wavelengths. This example clearly demonstrates the possibility to apply the Kalman filtering algorithm to obtain tomograms with variable resolution dependent on the number of wavelengths that can be set in correspondence with particular properties of the investigated object.

The suggested algorithm was applied to investigate the internal microstructure of biological tissues. Examples of the B-scan of the nose area of the pool frog *Pelophylax lessonae* obtained using 113 spectral samples (wavelengths) by the Kalman filtering algorithm and using 1024 samples with application of fast Fourier transformation (FFT) are shown in Fig. 5.16. It is seen that the tomogram in Fig. 5.16a has better quality in spite of essentially fewer number of wavelengths with respect to FFT (Fig. 5.16b) because of the absence of the border effect (artifacts) inherent in FT when a harmonic component includes a noninteger number of periods along the signal length.

Dynamic data processing by the suggested algorithm in application to spectral interferometry systems with a tunable wavelength reduces the number of samples at different wavelengths, while maintaining high resolution. The algorithm allows identifying

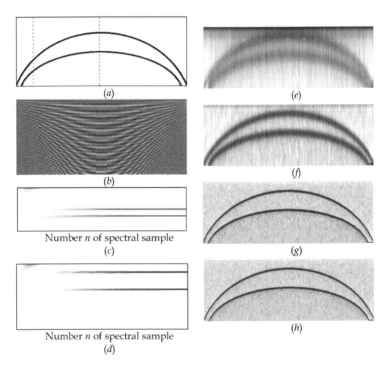

Figure 5.15 (a) The modeled cross section of an object, (b) spectral fringes, (c and d) evolution of estimates of the object borders $\hat{z}_{50}(n)$ and $\hat{z}_{250}(n)$, and (e–h) recovered tomograms (B-scans) with 30, 40, 50, and 255 wavelengths, respectively.

Figure 5.16 B-scans of nose area of pool frog *Pelophylax lessonae* obtained by (a) Kalman filtering and (b) the FFT algorithm.

harmonic components with a given set of frequencies and can be implemented in multidimensional mode of parallel calculations [33]. It should be noted that the considered algorithm is not sensitive to the initial phase of the signal and does not require an integer number of signal periods to avoid computational artifacts that presents an advantage compared to conventional FFT algorithms.

5.7 Discussion and Conclusions

Signal processing methods applied to FF-OCM have to provide, first of all, high resolution and processing speed. Due to high resolution and evaluation of the 3D microstructure of the investigated object, a large amount of data is registered, transmitted, processed, analyzed, and visualized. That is why information capacity of the FF-OCM systems considered above has to be taken into account to optimize procedures of image formation and evaluation.

In FF-OCM, a light detectors' array is used, and a decrease of the amount of data that has to be transmitted and processed plays an important role and can be achieved when utilizing downsampling of signals, that is, by missing an uninformative part of video frames that allows increasing processing speed. However, such a procedure is applicable to signals that meet the requirements considered in the chapter. Correct application of the downsampling procedure can give a speed gain typically up to 1 order of magnitude with moderate signal distortions about a few percents. Downsampling is especially useful in the express evaluation mode of an object.

Another way to increase processing speed consists in application of effective dynamic processing algorithms providing high noise immunity and possibility of parallel calculations. The signal processing algorithms in state space based on vector parametric stochastic models correspond to this requirement. However, such algorithms need correct a priori information about signal properties and initial conditions *before calculations*. This requirement does not present a drawback of recurrence algorithms, because one should take into account the same properties in conventional algorithms, however, *after calculations*. In some cases, posterior analysis can be more difficult with respect to accurate preparation of the experiment.

Because of space limitations, some important problems of signal processing in OCM have not been considered in the chapter. Possible methods of speckle noise reduction are presented, for example, in Refs. [43, 44]. Peculiarities of signal formation and registration in a few wavelengths (spectral regions) simultaneously are considered in Refs. [45–47]. The useful postprocessing procedures of image segmentation are discussed in Refs. [48–50]. In some cases, image deconvolution [51] and halftone histogram modifications [52] are able to enhance image quality and increase their contrast.

It is important to note that many approaches and methods of signal processing mentioned above and others are compatible with recurrence computational algorithms in state space that can present the direction of further research.

Acknowledgments

This work was partially supported by the Ministry of Education and Science of Russian Federation and by the Government of Russian Federation, Grant 074-U01.

References

1. Born, M., Wolf, E. (1968). *Principles of Optics* (Pergamon Press, Oxford).
2. Mandel, L., Wolf, E. (1995). *Optical Coherence and Quantum Optics* (Cambridge University Press, Cambridge).
3. Huang, D., Swanson, E. A., Lin, C. P., Schuman, J. S., Stinson, W. G., Chang, W., Hee, M. R., Flotte, T., Gregory, K., Puliafito, C. A., Fujimoto, J. G. (1991). Optical coherence tomography. *Science*, **254**, pp. 1178–1181.
4. Beaurepaire, E., Boccara, A. C., Lebec, M., Blanchot, L., Saint-Jalmes, H. (1998). Full-field optical coherence microscopy. *Opt. Lett.*, **23**, pp. 244–246.
5. Dubois, A., Vabre, L., Boccara, A. C., Beaurepaire, E. (2002). High-resolution full-field optical coherence tomography with a Linnik microscope. *Appl. Opt.*, **41**, pp. 805–812.
6. Dubois, A., Grieve, K., Moneron, G., Lecaque, R., Vabre, L., Boccara, C. (2004). Ultrahigh-resolution full-field optical coherence tomography. *Appl. Opt.*, **43**, pp. 2874–2883.

7. Gurov, I., Karpets, A., Margariants, N., Vorobeva, E. (2007). Full-field high-speed optical coherence tomography system for evaluating multilayer and random tissues. *Proc. SPIE*, **6618**, p. 661807.
8. Oh, W. Y., Bouma, B. E., Iftimia, N., Yun, S. H., Yelin, R., Tearney, G. J. (2006). Ultrahigh-resolution full-field optical coherence microscopy using InGaAs camera. *Opt. Express*, **14**, pp. 726–735.
9. Goodman, J. W. (1975). *Laser Speckle and Related Phenomena*, ed. Dainty, J. C., Chapter 2, Statistical properties of laser speckle patterns (Springer-Verlag, NY), pp. 9–77.
10. Schmitt, J. M. (1999). Optical coherence tomography: a review. *IEEE J. Sel. Top. Quantum Electron.*, **5**, pp. 1205–1215.
11. Tomlins, P. H., Wang, R. K. (2005). Theory, developments and applications of optical coherence tomography. *J. Phys. D: Appl. Phys.*, **38**, pp. 2519–2535.
12. De Boer, J. F., Cense, B., Park, B. H., Pierce, M. C., Tearney, G. J., Bouma, B. E. (2003). Improved signal-to-noise ratio in spectral-domain compared with time-domain optical coherence tomography. *Opt. Lett.*, **28**, pp. 2067–2069.
13. Leitgeb, R., Hitzenberger, C. K., Fercher, A. F. (2003). Performance of fourier domain vs. time domain optical coherence tomography. *Opt. Express*, **11**, pp. 889–894.
14. Wagner, C., Häusler, G. (2003). Information theoretical optimization for optical image sensors. *Appl. Opt.*, **42**, pp. 5418–5426.
15. Dubois, A. (2007). Private communication.
16. Pawley, J. B., ed. (1995). *Handbook of Biological Confocal Microscopy* (Plenum Press, NY).
17. Greivenkamp, J. E. (1987). Sub-Nyquist interferometry. *Appl. Opt.*, **26**, pp. 5245–5258.
18. Gurov, I., Taratin, M., Zakharov, A. (2006). High-speed signal evaluation in optical coherence tomography based on sub-Nyquist sampling and Kalman filtering method. *Proc. 5th Int. Workshop WIO'06, AIP Conf. Proc.*, **860**, pp. 146–150.
19. Alarousu, E., Gurov, I., Kalinina, N., Karpets, A., Margariants, N., Myllylä, R., Prykäri, T., Vorobeva, E. (2008). Full-field high-resolving optical coherence tomography system for evaluating paper materials. *Proc. SPIE*, **7022**, p. 702212.
20. Max, J. (1981). *Methodes et techniques de traitement du signal et applications aux measures physiques* (Masson, Paris).

21. Bendat, J. S., Piersol, A. G. (1971). *Random Data: Analysis and Measurement Procedures* (Wiley-Interscience, NY).
22. Alarousu, E., Gurov, I., Kalinina, N., Karpets, A., Margariants, N., Myllylä, R., Vorobeva, E. (2007). Evaluation of surface relief of paper materials by high-resolving full-field interferometric profilometer, presented at the 3rd Russian-Finnish Meeting, Photonics and Laser Symposium PALS 2007, Moscow, Russia, June 14–17, 2007.
23. Chang, S., Sherif, S., Mao, Y., Flueraru, C. (2008). Large area full-field optical coherence tomography and its applications. *Open Opt. J.*, **2**, pp. 10–20.
24. Deck, L., De Groot, P. (1994). High-speed noncontact profiler based on scanning white-light interferometry. *Appl. Opt.*, **33**, pp. 7334–7338.
25. Dresel, T., Häusler, G., Ventzke, H. (1992). Three-dimensional sensing of rough surfaces by coherence radar. *Appl. Opt.*, **31**, pp. 919–925.
26. Wang, Q., Ning, Y. N., Palmer, A. W., Grattan, K. T. V. (1995). Central fringe identification in a white light interferometer using a multi-stage-squaring signal processing scheme. *Opt. Commun.*, **117**, pp. 241–244.
27. Yoshimura, T., Kida, K., Masazumi, N. (1995). Development of an image-processing system for a low coherence interferometer. *Opt. Commun.*, **117**, pp. 207–212.
28. Gurov, I., Sheynihovich, D. (2000). Interferometric data analysis based on Markov nonlinear filtering methodology. *J. Opt. Soc. Am. A*, **17**, pp. 21–27.
29. Gurov, I., Ermolaeva, E., Zakharov, A. (2004). Analysis of low-coherence interference fringes by the Kalman filtering method. *J. Opt. Soc. Am. A*, **21**, pp. 242–251.
30. Gurov, I., Zakharov, A. (2004). Analysis of characteristics of interference fringes by nonlinear Kalman filtering. *Opt. Spectrosc.*, **96**, pp. 175–181.
31. Gurov, I., Volynsky, M., Zakharov, A. (2007). Evaluation of multilayer tissues in optical coherence tomography by the extended Kalman filtering method. *Proc. SPIE*, **6734**, p. 67341P.
32. Gurov, I., Volynsky, M. (2012). Interference fringe analysis based on recurrence computational algorithms. *Opt. Lasers Eng.*, **50**, pp. 514–521.
33. Gurov, I., Volynsky, M. (2013). Recurrence signal processing in Fourier-domain optical coherence tomography based on linear Kalman filtering. *Proc SPIE*, **8792**, p. 879203.
34. Bruning, J. H., Herriott, D. R., Gallagher, J. E., Rosenfeld, D. P., White, A. D., Brangaccio, D. J. (1974). Digital wavefront measuring interferometer for testing optical surfaces and lenses. *Appl. Opt.*, **13**, pp. 2693–2703.

35. Creath, K. (1988). Phase measurement interferometry techniques. *Prog. Opt.*, **26**, pp. 349–393.
36. Kreis, T. (1986). Digital holographic interference-phase measurement using the Fourier-transform method. *J. Opt. Soc. Am. A*, **3**, pp. 847–855.
37. Takeda, M., Ina, H., Kobayashi, S. (1982). Fourier transform method of fringe-pattern analysis for computer-based topography and interferometry. *J. Opt. Soc. Am.*, **72**, pp. 156–160.
38. Van Kampen, N. G. (1984). *Stochastic Processes in Physics and Chemistry* (North-Holland Physics Publishing, Amsterdam).
39. Kalman, R. E. (1960). A new approach to linear filtering and prediction problems. *Trans. ASME, J. Basic Eng.*, **82**, pp. 35–45.
40. Bellini, M., Fontana, R., Gurov, I., Karpets, A., Materazzi, M., Taratin, M., Zakharov, A. (2005). Dynamic signal processing and analysis in the OCT system for evaluating multilayer tissues. *Proc. SPIE*, **5857**, pp. 270–277.
41. Moura, J. M. F. (1987). Linear and nonlinear stochastic filtering. In *Signal Processing*, eds. Lacoume, J. L., Durrani, T. S., Storo, R. (North-Holland), pp. 205–276.
42. Wan, E., van der Merwe, R. (2001). The unscented Kalman filter. In *Kalman Filtering and Neural Networks*, ed. Haykin, S. (John Wiley & Sons, NY).
43. Hangai, M., Yamamoto, M., Sakamoto, A., Yoshimura, N. (2009). Ultrahigh-resolution versus speckle noise reduction in spectral-domain optical coherence tomography. *Opt. Express*, **17**, pp. 4221–4235.
44. Ozcan, A., Bilenca, A., Desjardins, A. E., Bouma, B. E., Tearney, G. J. (2007). Speckle reduction in optical coherence tomography images using digital filtering. *J. Opt. Soc. Am. A*, **24**, pp. 1901–1910.
45. Hendargo, H. C., Zhao, M., Shepherd, N., Izatt, J. A. (2009). Synthetic wavelength based phase unwrapping in spectral domain optical coherence tomography. *Opt. Express*, **17**, pp. 5039–5051.
46. Sacchet, D., Moreau, J., Georges, P., Dubois, A. (2008). Simultaneous dual-band ultra-high resolution full-field optical coherence tomography. *Opt. Express*, **16**, pp. 19434–19446.
47. Yu, L., Kim, M. K. (2004). Full-color three-dimensional microscopy by wide-field optical coherence tomography. *Opt. Express*, **12**, pp. 6632–6641.
48. Garvin, M. K., Abràmoff, M. D., Wu, X., Russel, S. R., Burns, T. L., Sonka, M. (2009). Automated 3-D intraretinal layer segmentation of macular spectral-domain optical coherence tomography images. *IEEE Trans. Med. Imag.*, **28**, pp. 1436–1447.

49. Kajić, V., Považay, B., Hermann, B., Hofer, B., Marshall, D., Rosin, P. L., Drexler, W. (2010). Robust segmentation of intraretinal layers in the normal human fovea using a novel statistical model based on texture and shape analysis. *Opt. Express*, **18**, pp. 14730–14744.
50. Lu, S., Cheung, C. Y., Liu, J., Lim, J. H., Leung, C. K., Wong, T. Y. (2010). Automated layer segmentation of optical coherence tomography images. *IEEE Trans. Biomed. Eng.*, **57**, pp. 2605–2608.
51. Liu, Y., Liang, Y., Mu, G., Zhu, X. (2009). Deconvolution methods for image deblurring in optical coherence tomography. *J. Opt. Soc. Am. A*, **26**, pp. 72–77.
52. Liu, Y., Liang, Y., Tong, Z., Zhu, X., Mu, G. (2007). Contrast enhancement of optical coherence tomography images using least squares fitting and histogram matching. *Opt. Commun.*, **279**, pp. 23–26.

PART II
TECHNOLOGICAL DEVELOPMENTS

Chapter 6

High-Speed Image Acquisition Techniques of Full-Field Optical Coherence Tomography

Byeong Ha Lee,[a] Woo June Choi,[b] Kwan Seob Park,[b] and Gihyeon Min[c]

[a] *School of Information and Communications, Gwangju Institute of Science and Technology (GIST), 261 Cheomdangwagi-ro, Buk-gu, Gwangju 500-712, Republic of Korea*
[b] *Department of Bioengineering, University of Washington, 3720 15th Avenue NE, Seattle, Washington 98195, USA*
[c] *Energy System Research Section, Electrics and Telecommunications Research Institute (ETRI), 176-11 Cheomdangwagi-ro, Buk-gu, Gwangju, Republic of Korea*
leebh@gist.ac.kr, wjchoi78@uw.edu, kspark81@uw.edu, ghmin@etri.re.kr

6.1 Introduction

We review image acquisition strategies and techniques studied and developed to achieve high-speed imaging of full-field optical coherence tomography (FF-OCT). Usually, the time domain FF-OCT imaging has employed conventional phase-shifting interferometry in which at least three interference images, for example with quadrant phase shifts, were needed to extract one coherence-gated tomogram image [1–6]. Such imaging process may be time

Handbook of Full-Field Optical Coherence Microscopy: Technology and Applications
Edited by Arnaud Dubois
Copyright © 2016 Pan Stanford Publishing Pte. Ltd.
ISBN 978-981-4669-16-0 (Hardcover), 978-981-4669-17-7 (eBook)
www.panstanford.com

consuming and not suited for in vivo imaging of mobile biological samples. Recently, many techniques to improve the FF-OCT imaging speed have been proposed, and they suggested great potentials for the in vivo imaging of living specimens. In this chapter, we will review the FF-OCT imaging schemes in terms of principles and processes. For effective reviewing, it is categorized into two main sections: (1) high-speed 2D FF-OCT imaging and (2) high-speed 3D FF-OCT imaging. For each section, various recent efforts for enhancing the imaging rate will be presented and analyzed in detail.

6.2 High-Speed 2D Imaging of Full-Field Optical Coherence Tomography

6.2.1 Phase-Stepping Method Based on Hilbert Transformation

In general, interference signals in a FF-OCT system are recorded by a 2D image sensor array such as a charge-coupled device (CCD) camera. The light intensity delivered to each pixel at a point (x, y) on the CCD plane can be expressed as [7]

$$I_0(x, y) = \bar{I}_{DC}(x, y) + A(x, y)\cos\phi(x, y). \quad (6.1)$$

Here $\bar{I}_{DC}(x, y)$ denotes the average light intensity, $A(x, y)$ is the envelope function of the interference signals affected by the reflectance of a sample, and $\cos\phi(x, y)$ is the fast-varying phase modulation resulted from the interference. Phase-shifting interferometry (PSI) has been developed to extract the coherent envelope signal from the interference signals, for which at least three interference images with certain phase delays were required. In this section, we describe the PSI techniques developed to reduce the number of required interference images, ensuring faster image acquisition than the conventional one.

In 2008, Na et al. [8] has proposed the mathematical Hilbert transformation method based on the phase-stepping PSI for high-speed FF-OCT imaging. In their method, the background intensity term was eliminated by taking the difference between two nearby phase-shifted interference images. The phase factor in the interference term could be removed by taking the mathematical operation of Hilbert transformation [9] with the background-free interference signal, which allowed getting only the coherent envelope function as the sample information.

Theoretically, when the reference arm of an OCT interferometer undergoes the optical path length (OPL) change within the coherence gating length, the interference signal of Eq. 6.1 is varied as

$$I_\alpha(x, y) = \bar{I}_{DC}(x, y) + A(x, y) \cos(\phi(x, y) - \alpha). \tag{6.2}$$

Here α is the phase shift induced by the OPL difference between the reference and the sample arms. Difference of the original interference signal (Eq. 6.1) and the phase-shifted signal (Eq. 6.2) is taken to eliminate the background intensity term $\bar{I}_{DC}(x, y)$, which gives the background-free interference signal S_1 defined as

$$S_1 \equiv I_\alpha - I_0 = 2A(x, y) \sin\left(\frac{\alpha}{2}\right) \sin\left(\phi(x, y) - \frac{\alpha}{2}\right). \tag{6.3}$$

This signal can be expressed in a succinct form of

$$S_1 = A'(x, y) \sin(\Phi(x, y)), \tag{6.4}$$

where the envelope function and the phase factor are redefined as

$$A'(x, y) \equiv 2A(x, y) \sin\left(\frac{\alpha}{2}\right), \tag{6.5}$$

$$\Phi(x, y) = \phi(x, y) - \frac{\alpha}{2}. \tag{6.6}$$

The quadrature function of this DC-free function S_1 is mathematically obtained by applying the Hilbert transformation (denoted by H) as

$$S_2 = H\{S_1\} = -A'(x, y) \cos(\Phi(x, y)). \tag{6.7}$$

Then the newly defined envelope function is simply derived by taking the square sum of two quadrature functions S_1 and S_2:

$$A'(x, y) = \sqrt{S_1^2 + S_2^2}. \tag{6.8}$$

This equation shows that the phase value $\Phi(x, y)$ in the interference term is suppressed and the envelope signal $A'(x, y)$ is extracted. It is noteworthy that the envelope signal $A'(x, y)$ can be extracted regardless of the exact amount of the phase shift. However, the

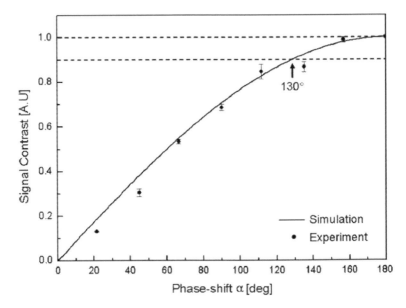

Figure 6.1 OCT signal contrast against the applied phase shift α. The solid line is the theoretical result. The signal contrast increases with the phase shift and has a maximum at the phase shift of 180°. At even 130°, only a signal reduction of 10% from its maximum is shown (dotted line). The measurements were made for step increment of phase shift by 22.5°. From Ref. [8]. © 2008 Optical Society of America.

magnitude of the extracted envelope signal is influenced by the amount of the phase shift α.

The proposed scheme was applied to a conventional FF-OCT system based on a Linnik-type interferometer and a visible CCD camera. A description of the system is detailed in the Na's paper [8]. To evaluate the effect of the induced phase shift on the magnitude of the OCT signal, a series of FF-OCT imaging of a flat mirror were performed with gradual increment of the phase shift. The intensities of FF-OCT images taken for some phase values are plotted in Fig. 6.1. The solid line in the figure is a simulation curve calculated with Eq. 6.5. We can see that the maximum signal contrast occurred at $\alpha = 180°$, and the signal contrast is remained more than 90% of its maximum value at even $\alpha = 130°$ (see horizontal dotted lines in Fig. 6.1). This indicates that the proposed scheme is able to provide the OCT signal despite of appreciable phase variation. In practice, environmental perturbations such as vibration and system drift can cause unwanted residual fringe patterns on the en face OCT image, but the proposed one may be insensitive to them. The discrepancy between the simulation and the experiment may be due to the hysteresis of the piezoelectric transducer (PZT) used to induce the phase shift.

The imaging feasibility of the proposed method was tested with FF-OCT imaging of a biosample. Figure 6.2 shows the Hilbert transformation based FF-OCT imaging of a garlic. Figures 6.2b and 6.2d are en face FF-OCT images of the garlic taken at the top surface and 40 μm below the surface, respectively. The corresponding optical microscope images are shown in Figs. 6.2a and 6.2c. The image size is 433 μm $(X) \times$ 325 μm (Y) for all figures. In the OCT images, individual hexagonal epithelium cells and cell walls are

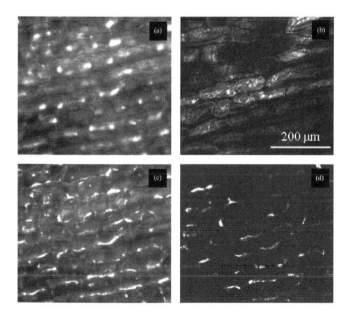

Figure 6.2 FF-OCT imaging of a garlic. (a and b) An optical microscope image and the corresponding en face OCT image taken at the top surface. (c and d) An optical microscopic image and the corresponding en face OCT image taken at 40 μm below the surface. From Ref. [8]. © 2008 Optical Society of America.

clearly resolved in depth, whereas their presence is not obvious in the microscopic views.

Current limitation of this method is in a rather long computation time for the Hilbert transformation. The application of the mathematical Hilbert transformation on the 2D image needed an appreciable calculation time. In this work, the acquisition time required for getting one en face OCT image was ~1 s, in which most of the time was cost to Hilbert transformation. Development of a digital signal processing (DSP) hardware such as a field-programmable gate array (FPGA) integrated circuit or a graphic processing unit (GPU) [10, 11] can be useful to significantly reduce the computation time, providing real-time high-speed FF-OCT imaging.

6.2.2 Integrating Bucket Method Based on Hilbert Transformation

The mathematical Hilbert transformation has been utilized to a continuous temporal PSI so-called the integrating bucket method by Lee et al. for the high-speed OCT imaging [12]. Unlike the phase-stepping method, the interference signal intensity is accumulated while the phase is under continuous shifting. This approach may be less sensitive to external noises compared to the phase-stepping method [1]. In this work, the sinusoidal phase modulation is introduced to the OCT interferometer for the phase shifting, similar to the conventional four-bucket integrating method [1]. Where, the interference signal of Eq. 6.1 is integrated over each half of the modulation period T as [12]

$$E_m = \int_{(m-1)T/2}^{mT/2} \{\bar{I}_{DC}(x, y) + A(x, y)\cos(\phi(x, y) + \Psi\sin(\omega t + \theta))\} dt, \quad \text{with} \quad m = 1, 2 \quad (6.9)$$

where Ψ and θ are the modulation amplitude and the initial phase of the sinusoidal phase modulation.

To extract the coherent envelope function $A(x, y)$, firstly, the difference between the two integrated intensity signals (E_1, E_2) are taken, which removes the background intensity signal as

$$S'_1(x, y) = E_1(x, y) - E_2(x, y) = \frac{4T}{\pi} A(x, y) \Gamma \sin \phi$$
$$= CA(x, y) \sin \phi, \quad (6.10)$$

with

$$\Gamma \equiv \sum_{n=0}^{\infty} \frac{J_{2n+1}(\Psi)}{2n+1} \cos((2n+1)\theta) \quad \text{and} \quad C \equiv \frac{4T}{\pi} \Gamma. \quad (6.11)$$

Here $J_n(\Psi)$ is the nth-order first-kind Bessel function given as a function of the phase modulation amplitude Ψ. By applying the Hilbert transformation to the background-free signal $S'_1(x, y)$ of Eq. 6.10, its quadrature function $S'_2(x, y)$ is obtained as

$$S'_2(x, y) = H\{S'_1\} = -CA(x, y) \cos \phi. \quad (6.12)$$

Accordingly, the resulting envelope function $A(x, y)$ is simply extracted by taking the squared sum of both quadrature functions (S'_1, S'_2) as

$$\sqrt{S'^2_1 + S'^2_2} = CA. \quad (6.13)$$

With Eq. 6.11, we can see that the extracted signal of Eq. 6.13 is maximized when Γ is maximum. The maximum of Γ was found when $\Psi = 1.98$ and $\theta = 0$ which were numerically calculated as shown in Fig. 6.3 [12]. These optimal phase modulation parameters could be adjusted in practice.

The imaging capability of the Hilbert transform method based on the integrating bucket PSI algorithm has been evaluated with the same FF-OCT setup of Section 6.2.1 with a biological sample. Figure 6.4 shows the depth-resolved en face FF-OCT images of an insect wing ex vivo (gold beetle). In Figs. 6.4a and 6.4b, several punctures on the surface of the hardened forewing (elytron) are observed as a typical feature of the beetle wing [13]. Figure 6.4e shows a hind wing structure folded in the hardened forewing. At 60 µm below the surface of the hardened forewing, an abdomen of the gold beetle is observed, as shown in Fig. 6.4f.

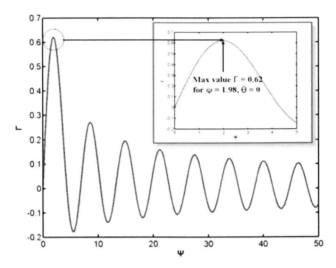

Figure 6.3 Numerical simulation for the optimal phase modulation parameters Ψ and θ. The maximum value of Γ for $\Psi = 1.98$ and $\theta = 0$ is determined, as indicated with a dotted circle and enlarged in the inset. Reproduced from Ref. [12] with permission from *The Review of Laser Engineering*, 2008.

6.2.3 Quadrature Fringes Wide-Field Optical Coherence Tomography

Quadrature fringes wide-field optical coherence tomography (QF WF-OCT) has been proposed by Sato et al. in 2007 [14]. WF-OCT is similar to FF-OCT in the en face imaging geometry. However, FF-OCT is based on a Linnik-based Michelson interferometer with a pair of high numerical aperture (NA) objectives in both arms of the interferometer, whereas WF-OCT uses a single lens in front of the image sensor to image a large sample area (a few tens of mm^2), compromising the lateral resolution [15–17]. In this OCT technique, two phase-quadrature interference images are produced by a series of polarization optics and then simultaneously captured by a single CCD camera (thus called the optical Hilbert transformation, as in Ref. [14]). Schematic of the system is illustrated in Fig. 6.5.

In brief, the output beam of a superluminescent diode (SLD) light source (1309 nm center wavelength and 26 nm spectral bandwidth)

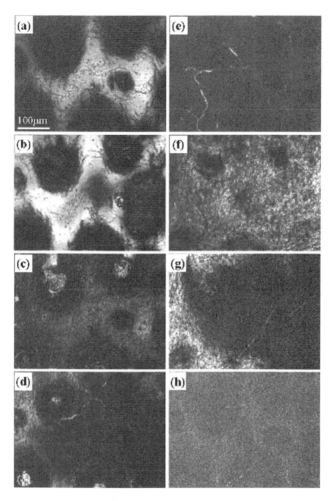

Figure 6.4 FF-OCT imaging of an insect wing ex vivo. (a–h) The images were taken at 10–80 μm below the surface of the wing, respectively. The image size of each figure was 470 μm × 350 μm. Reproduced from Ref. [12] with permission from *The Review of Laser Engineering*, 2008.

was collimated and linearly polarized at 45° and then split into the sample arm and the reference arm by a nonpolarizing beam splitter (NPBS). In the reference arm, the beam became circularly polarized through an achromatic quarter-wave plate and a linear polarizer. In the sample arm, a linear polarizer (45°) was placed to select

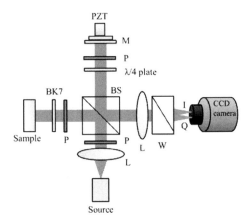

Figure 6.5 Schematic illustration of quadrature fringes wide-field optical coherence tomography. BS, beam splitter; P, polarizer (45°); M, mirror; L, lens; W, Wollaston prism. From Ref. [14]. © 2007 Elsevier.

the backscattered light of identical 45° linear polarization from the sample. A flat BK7 plate was placed in the sample beam path in order to compensate for the material dispersion caused by the quarter-wave plate in the reference arm. The two beams returning from the sample and the reference arms were recombined at the BS and interfered with each other. The interference light field was spatially separated into two orthogonal linearly polarized light fields by a Wollaston prism, and then captured by a single CCD camera.

The simultaneously captured pair of interference images have a quadratic phase difference, because the two orthogonal (horizontal and vertical) components of the circularly polarized beam from the reference arm have a 90° phase delay to each other, whereas the two orthogonal components of the 45° linearly polarized beam from the sample arm are in phase. Therefore, the two interference images captured through the Wollaston prism, but located side by side on a single CCD frame, are in phase (horizontal component) and quadratic phase (vertical component), giving two phase-quadrature interference images. Consequently, a single CCD frame gives the two interference images of

$$I_I(x, y) = \bar{I}_{DC}(x, y) + A(x, y) \cos\phi(x, y), \qquad (6.14)$$

$$I_Q(x, y) = \bar{I}_{DC}(x, y) - A(x, y) \sin\phi(x, y). \qquad (6.15)$$

One mechanical phase modulation (180°) on the reference arm produces an additional set of two phase-quadrature interference images of

$$I'_I(x, y) = \bar{I}_{DC}(x, y) - A(x, y) \cos \phi (x, y), \quad (6.16)$$

$$I'_Q(x, y) = \bar{I}_{DC}(x, y) + A(x, y) \sin \phi (x, y). \quad (6.17)$$

Therefore, the coherent envelope signal $A(x, y)$ can be extracted from the four phase-quadrature CCD images as

$$A = \frac{1}{2}\sqrt{(I_I - I'_I)^2 + (I_Q - I'_Q)^2}. \quad (6.18)$$

With the QF WF-OCT system, a healthy human fingernail was imaged in vivo. Figure 6.6a shows the en face OCT image taken at a 1.03 mm depth from the surface of the nail plate. The white ellipsoid pattern in the figure means that the coherence-gated plane is located below the curved (convex) nail surface. Figure 6.6b is the en face OCT image taken at a 1.64 mm depth from the surface of the nail plate. The white circle area shows the half-moon region at the bottom of the nail plate. Figure 6.6c is the XZ cross-sectional image reconstructed along the white dotted line in Fig. 6.6a. The dorsal nail plate, eponychium, nail root, and half-moon are visible.

Although the method has successfully performed en face OCT imaging with only two acquisitions as being achieved by the

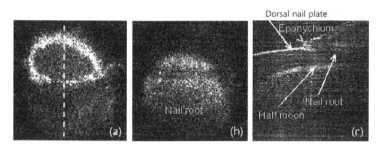

Figure 6.6 QF WF-OCT images of an in vivo human nail: (a) en face OCT image at 1.03 mm below the nail surface, (b) en face OCT image of the half-moon at 1.64 mm below the nail surface, and (c) XZ cross-sectional image reconstructed along the dotted line in (a). The depth-scanned range is 4 mm with steps of 10 μm. This depth corresponds to about 3.1 mm with a refractive index of 1.3. Irradiation power and exposure time were 506 μW and 1 ms. From Ref. [14]. © 2007 Elsevier.

mathematical Hilbert transformation approach, there are some technical issues: Detection of two interference images with a single CCD at a time reduces the effective pixel number of the obtained OCT image by half, leading to degradation of its lateral resolution. Moreover, precise alignment between the Wollaston prism and the CCD camera is required for the interpixel matching between the spatially divided interference images.

6.2.4 Single-Shot Phase-Stepping Methods

The Hilbert transformation approaches described above need two CCD data acquisition processes, including one mechanical movement of the reference mirror (RM) for the suppression of background intensity in the interference images. However, it may be still subject to system vibration or sample movement. Because a series of phase-shifted images are taken at different times, the external perturbation can cause an additional unwanted phase drifts between the successive images.

Unlike the temporal phase-shifting method, the spatial phase-shifting method acquires three or four phase-shifted images simultaneously; thus it can minimize the measurement errors. This scheme of simultaneous measurement of three or four phase-shifted signals using the spatial phase-shifting method is called the single-shot method. Various types of the spatial phase-shifting techniques have been proposed [18–20], and several single-shot approaches have been utilized for FF-OCT [15, 21–24].

In 2003, Dunsby et al. [21] demonstrated single-shot FF-OCT for the first time. Figure 6.7 shows the system setup, which had two distinct parts, a low-coherence polarizing Michelson interferometer and a four-channel polarization phase stepper (FC-PPS). In the polarizing Michelson interferometer, however, the two beams did not interfere with each other, but the reference beam P_R and the sample beam P_O were assigned with orthogonal linear polarizations. Each of P_R and P_O was equally divided into two optical paths, A and B, by the non-polarizing beam splitter (NPBS) in the FC-PPS part. The two lights going to path A through the quarter-wave plate Q3, whose fast axis was oriented at 0°, gained a relative phase delay of 90° between P_R and P_O. Then, after passing through

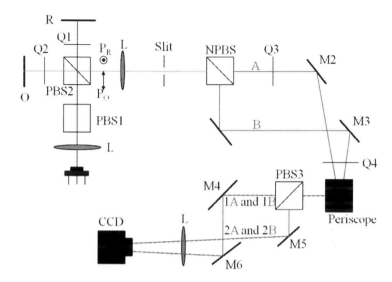

Figure 6.7 Single-shot wide-field optical coherence-gated imaging setup proposed by Dunsby et al. O, object; R, reference mirror; P_R, horizontally polarized light coming from the reference arm; P_O, vertically polarized light coming from the sample arm; L#, lenses; PBS#, polarizing beam splitter cube; M#, mirrors; Q#, quarter-wave plates. From Ref. [21]. © 2003 Optical Society of America.

the quarter-wave plate Q4, set to be oriented at 45°, P_O and P_R became left-hand circularly polarized light and right-hand circularly polarized light, respectively. The other two lights entering the optical path B also became left-hand circularly polarized light and right-hand circularly polarized light after the Q4. Finally, horizontal and vertical components of those circularly polarized lights were extracted by PBS3. Eventually the horizontal (1A) and the vertical (2A) components, passed through optical path A, had 0° and −180° phase delays, respectively. And the horizontal (1B) and the vertical (2B) polarization components coming from path B had 90° and −90° phase delays, respectively. This phase stepping is known as Pancharatnam's phase [22]. The four phase-quadrature images were captured by a single CCD camera at a time. The six mirrors in the polarization phase stepper were well adjusted to have each of the four interference images on each of the four quadrants of the CCD plane.

These rather long and complicated polarization coding processes can be easily understood by using the well-known Jones matrix. With the polarizing Michelson interferometer, the Jones vector of the input beam to the FC-PPS is made to have the complex object field E_O and the complex reference field E_R as

$$E_{in} = \begin{bmatrix} E_O \\ E_R \end{bmatrix}. \quad (6.19)$$

The Jones matrices for a quarter-wave plate with a horizontal fast axis Q_0 and the one of 45° fast axis Q_{45} are [21]

$$Q_0 = \begin{bmatrix} 1 & 0 \\ 0 & -i \end{bmatrix}, \quad Q_{45} = \frac{1}{\sqrt{2}} \begin{bmatrix} 1 & i \\ i & 1 \end{bmatrix}. \quad (6.20)$$

Therefore, the complex field of the beam passing through path A is simply given as

$$E_A = Q_{45} Q_0 \begin{bmatrix} E_O \\ E_R \end{bmatrix} = \frac{1}{\sqrt{2}} \begin{bmatrix} E_O + E_R \\ i(E_O - E_R) \end{bmatrix}. \quad (6.21)$$

It means that its horizontal component is the sum of the object and reference fields, and the vertical component is the difference between them with a common phase of 90° (representing i). These in-phase (0°) and out-of-phase (180°) interference images are separated by PBS3 and directed to the first two quadrants of the CCD. In the same way, the beam passing through path B is given by

$$E_B = Q_{45} \begin{bmatrix} E_O \\ E_R \end{bmatrix} = \frac{1}{\sqrt{2}} \begin{bmatrix} E_O + iE_R \\ i(E_O - iE_R) \end{bmatrix}. \quad (6.22)$$

We can see that the horizontal component is the sum of the object and reference fields with a 90° phase delay and the vertical one is with 270°. Of course these interference images are separated by PBS3 also and directed to the other two quadrants of the CCD. Therefore, we have the four phase-quadrature images simultaneously with a single CCD frame.

Image reconstruction were performed with the four phase-quadrature images by the well-known phase-shifting algorithm of

$$S = \sqrt{(I_\pi - I_0)^2 + (I_{3\pi/2} - I_{\pi/2})^2}. \quad (6.23)$$

However, unlike the temporal phase stepped method, the spatial phase stepped method involves additional calculation process such as geometric correction and nonuniformity compensation. Since four phase-shifted images are captured by a single CCD camera, each image should be geometrically separated in the CCD plane. Accordingly, the calculated image is easily affected by optical aberration, image projection, and pixel response of the CCD camera. Also, the measured value of a CCD pixel is different from the desired phase value due to imperfect polarization optics.

To demonstrate the feasibility of the system, a watch cog was imaged with a repetition rate of 16.5 Hz, which is the frame rate of the CCD camera as shown in Fig. 6.8. The sensitivity was 44 dB corresponding to the signal-to-noise ratio of 45.5 dB. In Fig. 6.9, as a biological sample, an onion was imaged with water immersion

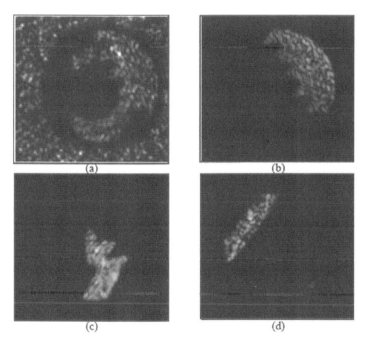

Figure 6.8 (a) Direct image of a watch cog; (b–d) en face (XY) WF-OCT images obtained at depths of $z = 0.55$ mm, 0.97 mm, and 2.54 mm respectively, relative to the front surface of the watch cog. The exposure time was 1 ms and the frame rate was 16.5 Hz using 2 × 2 binning. From Ref. [21]. © 2003 Optical Society of America.

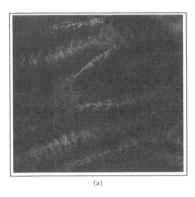
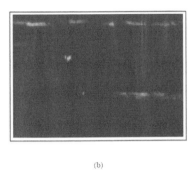

(a) (b)

Figure 6.9 (a) En face (XY) WF-OCT image at top surface of an onion: field of view 270 μm (X) × 250 μm (Y). (b) Representative cross-sectional (XZ) image taken from the 3D OCT data volume of the onion: field of view 270 μm (X) × 210 μm (Z) From Ref. [21]. © 2003 Optical Society of America.

objectives, which reduced the strong surface reflection at the air–cell boundary and enabled deeper penetration into the sample.

A more compact single-shot scheme for the 3D shape measurement was demonstrated in 2008 [23]. The proposed system was based on a polarizing Michelson interferometer configuration, which used a paired wedge prism (PWP) and a combined-wave plate (CWP), as shown in Fig. 6.10.

The four-channel phase stepper, which spatially separates each phase-shifted image, has been simplified by use of the newly designed polarization optics, PWP and CWP. The PWP was fabricated using two wedge prisms with a deviation angle of 1° to vertically split into two beams in the XZ plane. CWP consisted of two quarter-wave plates, QWP1 (10 × 20 mm^2) with its fast axis oriented at 0° and QWP2 (10 × 10 mm^2) oriented at 45°, and a half-wave plate (HWP; 10 × 10 mm^2) oriented at 22.5° with respect to the X axis. The up-directed beam from the PWP passed through QWP1, and the down-directed beam passed through QWP2 and the HWP. The four phase-shifted images were then simultaneously captured by a single CCD frame.

To check the feasibility of this scheme, topographic images of a 5 yen Japanese coin were obtained. Figure 6.11 shows a reconstructed 3D volume rendered image of the 5 yen Japanese coin.

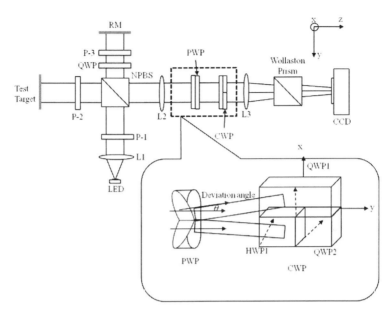

Figure 6.10 Spatial phase-shifting interferometer using paired wedge prism and combined-wave plate. P#, polarizer; L#, lenses; QWP, quarter-wave plate; HWP, half-wave plate; RM, reference mirror; PWP, paired wedge prism; CWP, combined-wave plate. From Ref. [23]. © 2008 Elsevier.

In the following year, Hrebesh et al. [24] introduced a new design of phase-stepper optics to supplement the existing one, which had a phase error problem due to the bonding between different polarization optics when fabricating the PWP and the CWP. Figure 6.12 shows a newly designed phase stepper, allowing the system to be more compact and the sensitivity of the system to be improved. The light emerging from the polarizing Linnik interferometer consisted of two orthogonally polarized lights. Two polarized lights passed through a half-wave plate and a dual-channel parallel beam splitter (DCPBS) one by one. Since the half-wave plate was oriented at 22.5° to the X axis, two orthogonally polarized lights had the same polarization direction but the phase delay between them was 180°. Then, the lights from reference and sample arms were equally divided into two channels (B1 and B2) by an NPBS. In channel B1, optical phases of 0° and 180° remained as they were. In

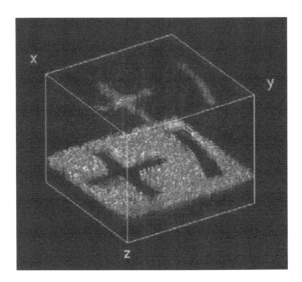

Figure 6.11 The 3D rendered image of a coin reconstructed with a set of 160 en face OCT images. From Ref. [23]. © 2008 Elsevier.

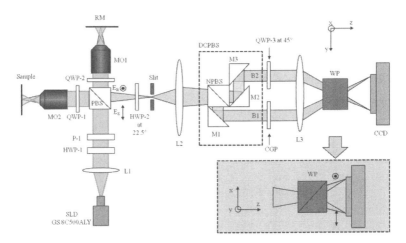

Figure 6.12 Single-shot FF-OCT system based on the DCPBS. E_S, Signal light; E_R, reference light; HWP#, half-wave plates; L#, lenses; P#, polarizers; PBS, polarizing beam splitter; QWP#, quarter-wave plates; MO#, microscope objectives; RM, reference mirror; NPBS, nonpolarizing beam splitter; M#, mirrors; CGP, compensation glass plate; WP, Wollaston prism; CCD, charge coupled device. From Ref. [24]. © 2009 Elsevier.

channel B2, since the quarter-wave plate QWP3 was positioned so as to generate a 90° delay between horizontal and vertical polarization lights, the optical phases of two lights become 90° and 270°. At the Wollaston prism, the horizontal and vertical polarization components having four different phases were directed upward and downward. Finally, the four phase-shifted images were spatially separated and simultaneously captured by a single CCD camera.

Image extraction was performed by a well-known four-phase-stepped algorithm. With this scheme, a beating internal structure of a diaptomus as an in vivo biological sample was successfully imaged at a frame rate of 28 fps in real time without pixel binning and image averaging, as shown in Fig. 6.13. The sample was trapped in a water chamber and partially sedated by adding watery ethyl alcohol in the chamber for immobilization of the specimen. The sensitivity of the system was measured to be 63–65 dB. The sensitivity of single-shot FF-OCT is mainly determined by the full-well capacity (FWC) of the

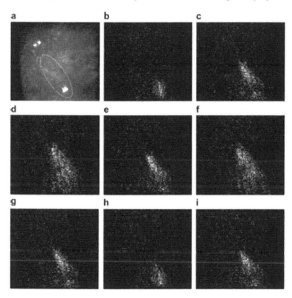

Figure 6.13 (a) In vivo microscopic image of the diaptomus. A white oval mark shows a measured site in the diaptomus. (b~i) In vivo en face OCT images of the moving internal structure at a depth of 170 μm from the dorsum with a time interval of 2 s. The field of view is 280 μm × 320 μm. From Ref. [24]. © 2009 Elsevier.

CCD camera like FF-OCT using the temporal phase-shifting method [15].

6.3 High-Speed 3D Imaging of Full-Field Optical Tomography

In the previous sections, various image acquisition techniques for high-speed 2D FF-OCT imaging were reviewed. In this section, we review several volumetric or high-speed 3D acquisition techniques in FF-OCT or WF-OCT imaging. Fast 3D imaging is essential for advanced understanding of dynamics in biological microstructures in clinical or laboratory studies. For most of FF-OCT imaging, however, its 3D image acquisition time has been thoroughly limited by the mechanical slow step scan in the depth direction (Z axis), which has restricted its applications especially when the sample is mobile. Methods to be introduced in this section are able to overcome the issue in the imaging speed and provide a great potential in fast 3D OCT imaging. We also discuss the achievement of high-speed 3D image acquisition of FF-OCT in the time domain.

6.3.1 *2D Heterodyne Detection with a Pair of CCD Cameras*

For high-speed 3D image acquisition, fast longitudinal scan of the reference mirror (RM) or the sample position is obviously necessary. Since the transverse (XY) cross-sectional image of a specimen is obtained at a time, the 3D image construction is possible by successively capturing a series of en face interference images of the specimen during a single longitudinal scan. With the scan, the coherence-gated imaging sheet travels into the depth of the specimen, which is similar to the heterodyne detection technique of typical time domain OCT [25].

In 2001 and 2002, Bourquin et al. [26] and Laubscher et al. [27], respectively, have demonstrated the parallel heterodyne detection schemes using smart sensor arrays. Each pixel in the sensor array was composed of a photodiode for light detection and an electric circuitry for signal processing such as DC filtering and demodulation, allowing extraction of the envelope function (or sample reflectance) from the interference signal generated and modulated by the scanning RM. Using the array of 58 × 58 pixels, this technique has demonstrated 3D OCT imaging of a 0.21 × 0.21 × 0.08 mm^3 sample volume with a sensitivity of 76 dB and at a repetition rate of 25 Hz [27]. The acquisition speed corresponded to about 4.88 Mpixels/s. However, this approach had limitation of poor lateral resolution due to the small number of pixels and high cost in fabrication of the customized sensor array. For the low-cost and high-pixel counts, however, a commercially available CCD camera is particularly attractive as the 2D detector array for the fast 3D acquisition. Unfortunately, however, the read-out frequency of a commercial CCD is usually a few tens of Hz, far below the intermediate heterodyne frequency (also called the Doppler frequency) of a few kHz; $f_D = 2v/\lambda_0$ where v is the scan speed and λ_0 is the center wavelength of the OCT light source [28]. When the fast varying interference signal is accumulated during a relatively long exposure time of the CCD, the interference component of the signal can be completely washed out. Hence, some technical tricks had to be made in order to realize the heterodyne detection with the slow CCD camera.

Akiba et al. [29, 30] reported a CCD-based 2D heterodyne detection FF-OCT system using the frequency synchronous detection method. In this technique, both the reference beam and the sample beam were time-gated at the Doppler frequency Δf of the system with a duty cycle of 50% or less. This could be readily carried out by switching the light illumination on and off at the rate of Δf, yielding the intensity modulated beams in both arms. Subsequently, the heterodyne signal was sampled by the same on-off pulse train, as in Fig. 6.14 [30]. It is obvious that such sampling enables the CCD camera to pick up the partial phase of the interference components (construction, destruction, or intermediate phases) in the heterodyne signals, keeping survival of the coherent components in the CCD output.

Theoretically, provided that $m(t)$ is a rectangular sampling function of a duty of 50% and frequency of Δf, it can be expressed as a sum of Fourier series [29]

$$m(t) = \frac{1}{2} + \sum_{n=1}^{\infty} \frac{\sin(n\pi/2)}{n\pi/2} \cos(2n\pi \Delta f t + \theta). \qquad (6.24)$$

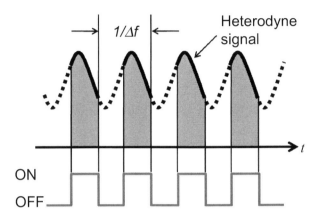

Figure 6.14 Time flow of a heterodyne signal sampled with a rectangular function (duty cycle of 50%) of frequency Δf, equal to that of the heterodyne signal (from Ref. [30]).

Here θ is a relative phase between the sampling function and the interference signal. Because the image read-out (or exposure) time of the CCD camera is much longer than the period $(1/\Delta f)$ of the signal, its output can be regarded as proportional to the time average of the light intensity incident on the CCD [29]

$$I = \left\langle \left\{ \frac{1}{2}A_s^2 + \frac{1}{2}A_r^2 + A_s A_r \cos(2\pi \Delta f t + \phi) \right\} \cdot m(t) \right\rangle, \quad (6.25)$$

where A_s and A_r are the amplitudes of the sample and the reference beams, respectively. ϕ is the initial phase of the heterodyne signal. The angled bracket $\langle \rangle$ denotes the time average during the CCD exposure time. Inserting the sampling function of Eq. 6.24 into Eq. 6.25 gives

$$I = \frac{1}{4}\left(A_s^2 + A_r^2\right) + \frac{1}{2}A_s^2 A_r^2 \langle \cos(2\pi \Delta f t + \phi) \rangle$$

$$+ \frac{1}{2}\left(A_s^2 + A_r^2\right) \sum_{n=1}^{\infty} \text{sinc}\left(\frac{n}{2}\right) \langle \cos(2n\pi \Delta f t + \theta) \rangle$$

$$+ A_s^2 A_r^2 \sum_{n=1}^{\infty} \text{sinc}\left(\frac{n}{2}\right) \langle \cos(2n\pi \Delta f t + \theta) \cos(2\pi \Delta f t + \phi) \rangle.$$

(6.26)

Since the CCD exposure time for each frame is much longer than the period $(1/\Delta f)$ of the signal, most of the terms are averaged out except the first DC term and the last term, giving

$$I \cong \frac{1}{4}\left(A_s^2 + A_r^2\right) + \frac{1}{2}A_s A_r \left(\frac{2}{\pi} - \frac{2}{3\pi} + \frac{2}{5\pi} - \ldots\right)\cos\psi. \quad (6.27)$$

The new phase parameter $\psi = \phi - \theta$ is defined as the phase difference between the heterodyne signal and the sampling function. Considered that the summation expansion in the second term of Eq. 6.27 is negligible except for the first $2/\pi$, the output signal of a CCD pixel is approximately given by

$$I \cong \frac{1}{4}\left(A_s^2 + A_r^2\right) + \frac{1}{\pi}A_s A_r \cos\psi. \quad (6.28)$$

Extraction of the sample reflectivity information A_s from the CCD interference signal is possible by removing the first term (background) and demodulating the second term of Eq. 6.28. For this, they have proposed a dual-channel detection scheme in which an interference image and its phase-quadrature interference image were captured with dual CCD cameras, thus eliminating the need for consecutive recording of multiple CCD frames [29].

The schematic of the system consisted of a bulk Michelson interferometer as shown in Fig. 6.15. The output beam from an SLD (center wavelength 826 nm and spectral bandwidth 18 nm) was collimated and linearly polarized. The polarized beam was launched into the interferometer and then split into the sample and the reference arms by a BS (BS1). With a translation stage, RM was longitudinally scanned, serving as a local oscillator to get a 3D en face image set. The beam reflected by the sample was recombined with the beam from the RM, forming an interference signal. Due to the longitudinal scan of the RM, the interference signal has the Doppler frequency, thus called the heterodyne signal. Dual-channel heterodyne detection was performed by splitting the heterodyne signal into two identical detection parts with another BS (BS2). Each detection part consisted of a liquid crystal shutter (LCS) and a CCD camera (DALSA CAD1, 256 × 256 pixels, 12 bit). The split heterodyne signals were sampled by LSCs operated at the Doppler frequency, but $\pi/2$ phase-shifted to each other by a phase shifter.

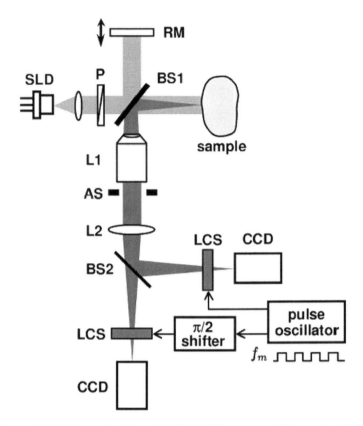

Figure 6.15 Schematic of a wide-field OCT system with a pair of CCD cameras for 2D heterodyne detection. SLD, superluminescent diode; P, polarizer; BS#, beam splitters; RM, reference mirror; L#, lenses; AS, aperture stop; LCSs, liquid crystal shutters. From Ref. [29]. © 2003 Optical Society of America.

From Eq. 6.28 the outputs from the two CCD cameras can be expressed as [29]

$$I_A = \frac{1}{4}\left(A_s^2 + A_r^2\right) + \frac{1}{\pi}A_s A_r \cos\left(\psi + \psi'\right)$$

$$I_B = \frac{1}{4}\left(A_s^2 + A_r^2\right) + \frac{1}{\pi}A_s A_r \sin\left(\psi + \psi'\right), \quad (6.29)$$

where ψ' is a random phase drift due to the external disturbance such as vibration. It is expected that both CCD outputs, I_A and I_B, are influenced by the same amount of ψ' due to the sharing of the detection path of the interferometer. Since the background intensity I_{DC} could be measured in advance by adjusting the optical path difference (OPD) between both arms beyond the coherence length of the light source, the interference signal intensity was obtained as

$$(I_A - I_{DC})^2 + (I_B - I_{DC})^2 = \frac{1}{\pi^2} A_s^2 A_r^2 \propto A_s^2. \qquad (6.30)$$

Because of its common path configuration in the detection channel and the lock-in process using the in-phase and the quadrature signals, the heterodyne detection scheme has the advantage of immunity to the random phase drift that can adversely affect the OCT image quality. Use of a pair of CCD cameras, however, is not cost effective. Further it requires pixel by pixel precise prepositioning between two cameras.

With the WF-OCT system, fast volumetric OCT imaging for a plant leaf has been demonstrated. En face OCT measurements were performed at a rate of 100 frames/s during a single longitudinal scan of RM at a scan speed of 0.6 mm/s, yielding an imaging depth interval of 6 μm and a Doppler frequency of 1.5 kHz. The imaging results of the back surface of the leaf are shown in Fig. 6.16a. The numbered tomographic images clearly show the inner structural variation within the plant leaf. The image size was 800 μm × 800 μm. The axial and the lateral resolutions were limited to 17 μm and 4 μm, respectively, because of the narrow bandwidth of the light source and the use of a low NA (0.14) lens. Figure 6.16b shows the 3D OCT image reconstructed with 60 consecutive en face OCT images of a vegetable seed (turnip) as another biological sample. The 3D image acquisition time was about 0.6 s/volume.

Despite of the rapid acquisition in 3D, its detection sensitivity was 85 dB, which is ~20 dB lower than that of the fiber-based spectral domain OCT [31–33]. This is mainly attributed to the FWC of the CCD (e.g., 200 Ke⁻ for this experiment) [34], the typical characteristic factor determining the sensitivity of CCD-based full-field or WF-OCT imaging. The sensitivity can be improved by image accumulation with a number of iterations, compromising the heavy measurement time. This low sensitivity limits the imaging depth of the current system to only a few hundreds of micrometers with most biological samples. However, it would be useful for the applications

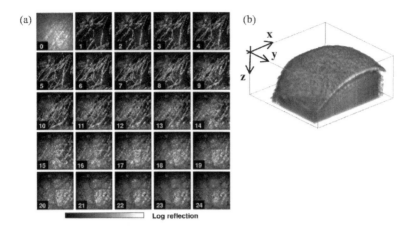

Figure 6.16 Fast 3D OCT imaging of biological samples by the 2D heterodyne detection scheme. (a) En face OCT images (800 μm × 800 μm) of a back surface of a plant leaf obtained at a rate of 100 frames/s, with a depth interval of 6 μm. For reference, image 0 shows a direct bright field view of the sample. (b) 3D OCT image (800 μm × 800 μm × 360 μm) of a vegetable seed reconstructed from 60 en face OCT images. All figures adapted from Ref. [29]. © 2003 Optical Society of America.

with optically transparent eye structures (anterior segments or retina) in vivo.

6.3.2 2D Heterodyne Detection with a Single CCD Camera

The 2D heterodyne detection scheme using a single CCD camera has been achieved also with a typical WF-OCT system, in which consecutive en face images could be acquired with a single axial scan of RM [35]. To avoid the averaging or washing out of interference signals by the slow response of CCD, the frequency synchronous detection technique has been utilized. Unlike the method presented in the previous section [29], the sampling of the heterodyne signal is performed by switching the illumination of the OCT light source on and off at the same frequency as the heterodyne signal. The heterodyne signals modulated by the flashed light illumination are integrated during the rather long exposure time of the CCD and then read out.

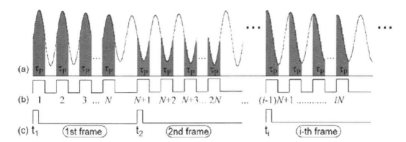

Figure 6.17 Time flow for 2D heterodyne signal detection. (a) Heterodyne signal; (b) rectangular signal with 50% duty cycle as an input for switching the light illumination. Its frequency is equal to that $(1/2\tau_p)$ of (a). (c) Synchronized read-out signal of the CCD camera. The sampled portions of the heterodyne signal are integrated in the $(t_{i+1}-t_i)$ period. Reprinted from Ref. [35], Copyright (2008), with permission from Elsevier.

Figure 6.17 illustrates the condition for the heterodyne signal detection [35]. The heterodyne signal (Fig. 6.17a) is modulated and sampled by the switched lights fed by a rectangular pulse train of the frequency same as the Doppler frequency of the heterodyne signal (Fig. 6.17b). The signal is recorded by the CCD whose read-out is synchronized with the modulation signal in Fig. 6.17c. To obtain the sample information, this technique employs squared difference of two sequential interference images. Since the sampling distance between the captured adjacent interference images is only a few micrometers and less than the coherence gating length, they can be treated as having almost identical background intensities. Thus, the squared difference of two images eliminates the background intensity and extracts only the interference component.

Numerical analysis for the overall detection process helps easy understanding. Initially, the light intensity of the interference signal at the image sensor plane is given as

$$I = \int_{-\infty}^{\infty} \frac{S(\omega)}{2\pi} \left(A_r^2 + A_s^2 + 2 A_r A_s \cos(2k\Delta z) \right) d\omega. \quad (6.31)$$

Here A_r and A_s are the light amplitudes reflected from the RM and the sample, respectively, and Δz is the path length difference between both arms of the OCT interferometer. We also assume that the OCT light source has a Gaussian power spectral density distribution of

$$S(\omega) = \left(\frac{2\pi}{\sigma_\omega^2}\right)^{1/2} e^{-\frac{(\omega-\omega_0)^2}{2\sigma_\omega^2}}, \qquad (6.32)$$

where ω_0 is the center frequency and σ_ω the standard deviation power spectral bandwidth (radians per second). Considering the axial scan motion of the RM, the interference light signal intensity is given as

$$I(t) = A_r^2 + A_s^2 + 2A_r A_s e^{-\frac{\Delta \tau_g^2}{2\sigma_\tau^2}} \cos(2\pi \Delta f t), \qquad (6.33)$$

where Δf is the Doppler frequency and σ_τ the standard deviation temporal width, inversely proportional to the σ_ω. We note that the Doppler frequency is proportional to the scan speed, and the group delay mismatch $\Delta \tau_g$ is a function of time through the scanning.

The signal of Eq. 6.33 is composed of DC and AC components as

$$I_{DC} \equiv A_r^2 + A_s^2, \qquad (6.34a)$$

$$I_{AC}(t) \equiv 2A_r A_s e^{-\frac{\Delta \tau_g^2}{2\sigma_\tau^2}} \cos(2\pi \Delta f t). \qquad (6.34b)$$

Since the rectangular pulse train for the modulation of the light source illumination is designed to have a 50% duty cycle and has the same frequency as the OCT signal, it can be expanded with a Fourier series similarly as in Eq. 6.24:

$$m(t) = \frac{1}{2} + \sum_{n=1}^{\infty} \operatorname{sinc}\left(\frac{n}{2}\right) \cos(2n\pi \Delta f t + \theta_m). \qquad (6.35)$$

where θ_m is the initial phase difference between the modulation function and the interference signal function. The CCD output signal is given by integrating the interference light intensity signal of Eq. 6.33, multiplied with the modulation function of Eq. 6.35, during the CCD accumulation time period T as

$$I_{CCD}(t) = \int_t^{t+T} (I_{DC} + I_{AC}) m(t) dt$$

$$= \frac{T}{2} I_{DC} + I_{DC} \sum_{n=1}^{\infty} \operatorname{sinc}\left(\frac{n}{2}\right) \int_t^{t+T} \cos(2n\pi f_D t + \theta_m) dt$$

$$+ A_r A_s \int_t^{t+T} e^{-\frac{\Delta \tau_g^2}{2\sigma_\tau^2}} \cos(2\pi f_D t) dt$$

$$+ 2 A_r A_s \sum_{n=1}^{\infty} \operatorname{sinc}\left(\frac{n}{2}\right) \int_t^{t+T} e^{-\frac{\Delta \tau_g^2}{2\sigma_\tau^2}} \cos(2\pi f_D t)$$

$$\times \cos(2n\pi f_D t + \theta_m) dt. \qquad (6.36)$$

In the equation, the second term can vanish when the accumulation period T is much longer than the oscillation period of the OCT signal, or $T \gg 1/\Delta f$. Moreover, the third term becomes negligible when the temporal width of the Gaussian envelope function of the OCT system is much wider than the signal period, or $\sigma_\tau \gg 1/\Delta f$. The same approximation can be made with the last term, which is expanded as

$$\sum_{n=1}^{\infty} \operatorname{sinc}\left(\frac{n}{2}\right) \int_{t}^{t+T} e^{-\frac{\Delta \tau_g^2}{2\sigma_\tau^2}} \cos(2\pi f_D t) \cos(2n\pi f_D t + \theta_m) \, dt$$

$$= \frac{1}{\pi} \int_{t}^{t+T} e^{-\frac{\Delta \tau_g^2}{2\sigma_\tau^2}} \{\cos(4\pi f_D t + \theta_m) + \cos\theta_m\} \, dt$$

$$- \frac{1}{3\pi} \int_{t}^{t+T} e^{-\frac{\Delta \tau_g^2}{2\sigma_\tau^2}} \{\cos(8\pi f_D t + \theta_m) + \cos(4\pi f_D t + \theta_m)\} \, dt + \cdots$$

(6.37)

The even terms regarding the summation index n were vanished due to the normalized sinc function, and the odd terms were decomposed by the trigonometric identity. With the condition of a wide envelope function of the OCT system, same as for the third term of Eq. 6.36, all terms of Eq. 6.37 become negligible except the term of

$$\frac{1}{\pi} \cos\theta_m \int_{t}^{t+T} e^{-\frac{\Delta \tau_g^2}{2\sigma_\tau^2}} \, dt, \qquad (6.38)$$

and the resulting CCD output is approximately given as

$$I_{\mathrm{CCD}}(t) \approx \frac{T}{2} I_{\mathrm{DC}} + \frac{2 A_r A_s}{\pi} \cos\theta_m \int_{t}^{t+T} e^{-\frac{\Delta \tau_g^2}{2\sigma_\tau^2}} \, dt. \qquad (6.39)$$

We can see that the CCD output signal does not depend on the Doppler frequency at all. The integrand can be considered as the sampled portions of the temporal Gaussian envelope function, which are accumulated during the integration period T of the CCD. Finally, the absolute difference of the two adjacent CCD frames gives an en face OCT image, by which the incoherent background intensity is almost completely removed as

$$|I_{\mathrm{CCD}}(t+T) - I_{\mathrm{CCD}}(t)|^2 \propto \left(\frac{2 A_r A_s}{\pi} \cos\theta_m\right)^2. \qquad (6.40)$$

However, the contrast of the CCD frame is degraded by the phase θ_m. The interference component of the signal can be even disappeared at the phase condition of $\theta_m = \pm 90°$. Especially, the effect of the residual phase is critical for imaging the object having discrete and flat multilayers, because the phase delay at a certain layer is the same across the entire plane of the layer, which leads to overall signal reduction in the en face OCT image. Therefore, timely flashing of the OCT light in harmony with the axial scanning is empirically required to obtain an improved 3D OCT image of the sample. This strict rule seems to be somewhat relaxed for biological samples. Because in general a biological sample consists of complex but rather continuous reflective layers, the initial phase is quite different for pixel by pixel. Therefore, even if a pixel has the phase of $\theta_m = \pm 90°$, giving a zero-interference image, the nearby pixels can have different phases which allows having nonvanishing signal of Eq. 6.40. Of course, it is unavoidable for the partial signal degradation due to nonperfect phase matching [35]. Despite of the drawback, this imaging manner takes advantage of cost-effectiveness and ease in system alignment compared to the 2D heterodyne detection technique using a pair of CCD cameras.

Figure 6.18 shows the experimental setup to achieve the 2D heterodyne detection using a single camera (here, a complementary metal–oxide semiconductor [CMOS] camera was used instead of a CCD camera) [35]. The system was based on a polarizing Michelson interferometer, which reduced the influence of the birefringence of the sample. The light source was an SLD with a central wavelength of 830 nm and a full-width at half-maximum (FWHM) spectral bandwidth of 26 nm, yielding the axial resolution of ~11.7 μm in air. The collimated output of the SLD was switched on and off using an acoustic optical deflector (AOD) and then divided with the polarizing beam splitter (PBS) into two orthogonally polarized beams. The polarizer allows the relative intensity of the beams in two arms of the interferometer to be adjusted to have maximum fringe contrast. The beams retroreflected from the RM and the sample were recombined by PBS, and then detected by the CMOS camera (512 × 512 pixels, 10 bit resolution, 1500 frames/s).

Heterodyne signal detection using the system was verified at first by measuring the interference signal with a photodetector (PD)

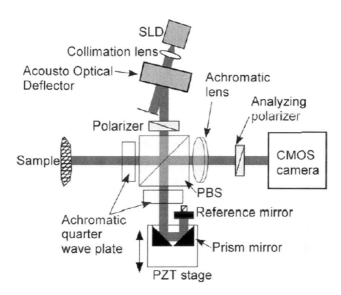

Figure 6.18 Schematic of the wide-field OCT system with a single CMOS camera. SLD, superluminescent diode; PBS, polarizing beam splitter; PZT, piezoelectric transducer. Reprinted from Ref. [35], Copyright (2008), with permission from Elsevier.

instead of the CMOS camera. Use of the PD having a MHz bandwidth enabled to record the change in the interference signal intensity at one particular pixel without signal blurring. Axial translation of the RM was performed at a scan speed of 2.4 mm/s, producing a Doppler frequency of 11.5 kHz. The SLD light was flashed with the same frequency (11.5 kHz) by the AOD. Figure 6.19 shows the electric waveforms of the interference signal detected by the PD and the AOD triggering signal. It is observed that the phases of the detected heterodyne signals are almost periodic and the signal frequency agrees with that of the input electric signal, which means that the sampling of the heterodyne signal is well operated. When the PD is replaced with the CMOS camera, the camera would accumulate four sampled components during a 1/3000 s exposure time. The difference between two sequential CMOS outputs produces one en face OCT image.

Further, the system sensitivity was measured by moving a −30 dB attenuation reflector with a step of 0.2 mm as shown in Fig. 6.20.

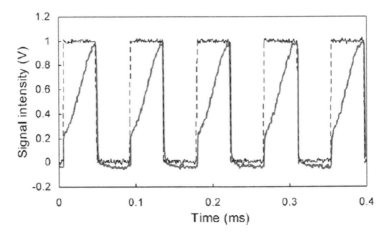

Figure 6.19 Heterodyne signals measured as an electric waveform with a photodetector (solid line) and overlapped input signal of the AOD (dashed line). Reprinted from Ref. [35], Copyright (2008), with permission from Elsevier.

For each measurement, 375 en face OCT images were extracted with a total of 750 en face interference images captured through a scan length of 1.2 mm, from which the averaged value by 5 × 5 pixel binning was plotted. In Fig. 6.20, the experimental sensitivity was measured to be about 73 dB, close to the theoretical value of 72.9 dB [35].

It is noted that the OCT signal intensities can be different at different positions along the depth. That is because the Doppler frequency is practically time variant due to may-existing nonlinear motion of the moving stage; thus, can be not equal to the constant frequency of the stroboscopic OCT illumination. Because of this frequency mismatch, the amount of the sampled interference component can be temporally random, leading to a change in the OCT signal intensities. It can be effectively compensated by utilizing a frequency follower with an active feedback loop, which will be discussed in the next section.

With the system, volumetric OCT imaging was performed with a healthy human finger in vivo. A 3D data set was acquired with a single axial scan with a time of 0.25 s (4 volumes/s) and used to extract a series of en face OCT images. Figure 6.21 shows en face OCT

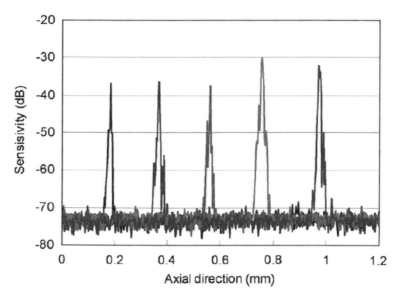

Figure 6.20 OCT signal intensities of the −30 dB attenuation mirror measured at different positions. Reprinted from Ref. [35], Copyright (2008), with permission from Elsevier.

images—2.6 mm (X) × 2.6 mm (Y)—of the human finger in depth of (a) 202 μm, (b) 227 μm, (c) 265 μm, and (d) 744 μm below the surface. Ridges of the fingerprint and sweat glands (SGs) (arrows in Figs. 6.21b and 6.21c) in the epidermis are clearly visible. A 3D structure—2.6 mm (X) × 2.6 mm (Y) × 1.2 mm (Z)—of the human finger was reconstructed from the 375 depth-resolved en face OCT images as Fig. 6.21e.

6.3.3 *2D Heterodyne Detection with a Single CCD Camera and an Active Feedback Loop*

As was mentioned in Section 6.3.2, though it gives rapid 3D volumetric imaging, the conventional single-camera heterodyne detection technique is subject to rather random phase variations due to the mismatch between the Doppler frequency of the OCT system the modulation frequency of the illumination beam. In the reference arm of the system, the translation of the axial scanner such

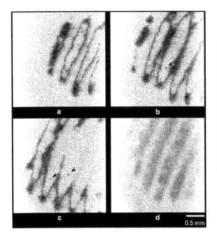
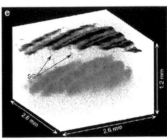

Figure 6.21 High-speed volumetric OCT imaging of a human finger in vivo. En face OCT images of the human finger pad at depth of (a) 202 μm, (b) 227 μm, (c) 265 μm, and (d) 744 μm, (e) 3D image of the human finger pad. The volume is 2.6 (X) × 2.6 (Y) × 1.2 (Z) mm^3. SG, sweat gland. Adapted from Ref. [35], Copyright (2008), with permission from Elsevier.

as a mechanical stage or a PZT actuator is actually not linear because the motion is easily affected by acceleration, friction, backlash of the mechanical stage, and hysteresis of the PZT. Therefore, the Doppler or beat frequency of the interference signal can be changed during the scanning.

In 2010, Choi et al. [36] has proposed an alternative method to compensate for the OCT signal degradation. They have added an auxiliary laser interferometer having a laser diode (LD) with the same central frequency as the OCT light source and a feedback loop system to a conventional OCT interferometer, where the reference arm of the OCT interferometer was shared with the laser interferometer. While translating the scanner in the reference arm, an OCT interference signal was generated. At the same time, the interference signal generated by the laser interferometer was converted into an electric pulse train and then used as an input for optical chopping of the OCT light source. As a result, the illumination on-off beam was completely frequency-locked with the heterodyne signal, allowing stable OCT signal detection. As shown in Fig. 6.22a, the main key of the proposed scheme was introducing the Doppler

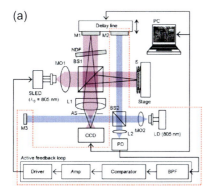 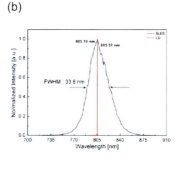

Figure 6.22 (a) Schematic of a wide-field OCT system with a frequency lock module. SLED, superluminescent emitting diode; BS#, beam splitters; MO#, micro-objectives; L#, lenses; S, sample; NDF, neutral density filter; M#, mirrors; PD, photodetector; LD, laser diode; AS, aperture stop; BPF, bandpass filter; Amp, amplifier. (b) Optical spectra of SLED (black line) and LD (red line). From Ref. [36] © IOP Publishing. Reproduced with permission. All rights reserved.

frequency lock module (indicated with the dotted red box) to a single-CCD-based WF-OCT system.

In brief, a 5 mW superluminescent emitting diode (SLED; center wavelength of 805 nm and spectral bandwidth of 33.6 nm at FWHM) was used as an OCT light source. The OCT interference signal was detected with a CCD camera (640 × 480 pixels, 8 bits, 200 frames/s). The auxiliary laser interferometer was based on a double-beam Michelson interferometer, which was designed to share its reference arm with the OCT interferometer. As a source of the laser interferometer, a 10 mW LD (wavelength of 805 nm) was used. By sharing the reference delay line and using the light sources of similar center wavelengths (see Fig. 6.22b), the interference signals produced at both interferometers equally undergo the same variation in the Doppler frequency during the scanning. The laser interference signal was detected by a single PD and its output was directed to electric circuits, where it was bandpass-filtered and converted into a DC-free pulse train with a 50% duty cycle, and finally fed to the current driver of the SLD. Through the feedback loop circuitry, the SLD beam was automatically fired at the identical frequency of the OCT operation. Finally, the sampled interference

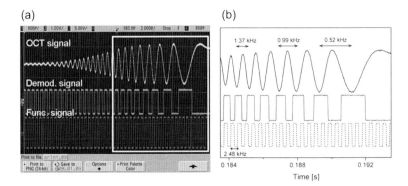

Figure 6.23 (a) (Top) Electric waveform of the OCT interference signal of the flat mirror surface detected by a single photodetector during axial scan in deceleration. (Middle) The OCT beam demodulation signal with the active feedback routine and without (bottom). (b) Enlarged view of the white box in (a). From Ref. [36]. © IOP Publishing. Reproduced with permission. All rights reserved.

components were accumulated during the exposure time of the CCD camera, and then a series of en face OCT images were extracted from the squared absolute difference between the consecutive CCD frames [36].

To identify the idea of active frequency locking with a real system, the OCT interference signal of a flat mirror was taken with a PD instead of the CCD camera. For better appearance of the variation in the Doppler frequency, the signal measurement was carried out when the axial scanning was in a deceleration state (see Fig. 6.23b). The axial scan speed was set to 0.99 mm/s, corresponding to the Doppler frequency of 2.48 kHz. Figure 6.23a shows an electric waveform (top) of the OCT interference signal detected by the PD and each electric pulse train (demodulation signal) produced with the feedback circuitry (middle) and with a conventional function generator (bottom). We can see that the interference signal was highly down chirped because of the deceleration of the scanner near the end of the motion span. The white box region of Fig. 6.23a is enlarged in Fig. 6.23b, showing that the frequency of the modulation signal (red solid line) from the feedback loop tends to follow nonlinear Doppler frequency of the interference signal, whereas the

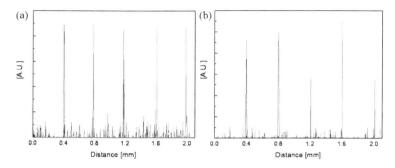

Figure 6.24 OCT signal intensities of the flat mirror surface at five different positions obtained from the proposed method (a) and the conventional one (b).

pulse train signal (blue dotted line) made by the function generator has a constant frequency of 2.48 kHz, same as the nominal Doppler frequency at the stable scanning. It is obvious that the OCT light flashing by the proposed scheme is able to be actively synchronized with any variation in the OCT Doppler frequency, which enables stabilized OCT signal for the 3D OCT imaging.

Comparison of OCT signal intensities of the flat mirror surface obtained at five different positions (with an increment of 0.4 mm) for the proposed scheme and a conventional one is shown in Fig. 6.24. A nominal scan speed of the delay line was set to 0.805 mm/s. For the proposed scheme (Fig. 6.24a), the magnitudes of OCT signal are less variable than those of the conventional one (Fig. 6.24b). Even with the proposed scheme, however, there are many factors that may affect the system accuracy or stability. For instance, the fraction of the Doppler cycle being integrated during the constant CCD exposure time might be different when the Doppler frequency is increased or decreased; therefore, the amplitude of the signal would be affected by the chirping.

Fast 3D OCT imaging performance of the proposed method has been demonstrated with a metallic spring watch consisting of multiple layers. Axial scanning was performed at a scan speed of 0.805 mm/s (Doppler frequency of 2 kHz), forming a 524 OCT image set with a volume size of 8 mm $(X) \times 6$ mm $(Y) \times 2.19$ mm (Z). The depth interval between adjacent OCT images was 4.02 μm. The 3D

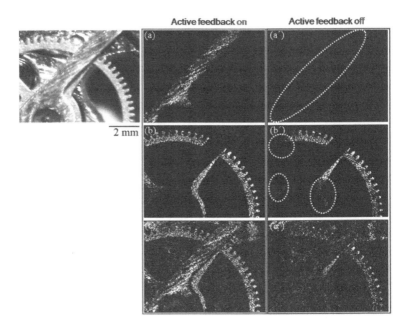

Figure 6.25 High-speed wide-field 3D OCT imaging of a metallic spring watch. With the active feedback on: (a) at the surface, (b) 1.054 mm below the surface, and (c) co-registered. With the active feedback off: (a') at the surface, (b') 1.054 mm below the surface, and (c') co-registered. 3D OCT data set consists of 640 (X) × 480 (Y) × 524 (Z) voxels corresponding to 8 mm (X) × 6 mm (Y) × 2.19 mm (Z). Total 3D data acquisition time was 2.7 s.

data acquisition time was 2.7 s in the one-way scan, corresponding to 0.3 volume/s. Figures 6.25a and 6.25b show the topographic images of the spring watch taken at the top and 1.054 mm below it, where a gear wheel was located. In Fig. 6.25c, both tomograms are co-registered to compare with the photograph at the left of the figure. The same measurements were made with the conventional method without the active feedback and presented with Figs. 6.25a'–c'. Where whole or partial loss of OCT signal indicated with the dashed circles was clearly observed, which may be due to the frequency mismatch. By stacking a total of 524 en face OCT images, a 3D structure of the spring watch could be reconstructed as in Fig. 6.26.

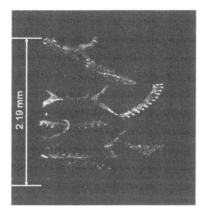

Figure 6.26 3D rendering of the spring watch.

6.4 Summary

We have reviewed the methods for fast image acquisition in time domain FF-OCT or WF-OCT. Technically, the methods have been described over two main categories, high-speed 2D en face OCT imaging and high-speed volumetric (or 3D) OCT imaging. In high-speed 2D en face OCT imaging, the phase-shifting techniques that had been developed to reduce the number of CCD images (up to even a single shot) required for getting one en face OCT image have been presented. Also, their rapid imaging feasibility has been demonstrated with various biological samples in vivo or ex vivo. In fast volumetric OCT imaging, the heterodyne signal detection techniques developed to extract the interference component (having the information of sample reflectivity) from the successive heterodyne signals generated by the longitudinal scan of the reference mirror have been presented. Especially, the WF-OCT systems using dual CCD cameras and a single CCD camera, respectively, have been explained in detail. The 3D OCT structure images of several turbid objects, like a human finger or a spring watch, have been achieved with the volumetric imaging within a few seconds. Despite a low sensitivity of less than 80 dB, the future potential of these FF-OCT imaging techniques is extremely promising. The high-speed image acquisition ability permits real-time tracking of specimens in two or three dimensions, enabling evaluation of temporal dynamics in the sample structures with a cellular resolution (e.g., cellular developmental process of in vivo embryo models in development biology [37]).

Acknowledgments

This work was supported by the "GIST Research Institute (GRI)" Project through a grant provided by GIST.

References

1. Dubois, A. (2001). Phase-map measurements by interferometry with sinusoidal phase modulation and four integrating buckets. *J. Opt. Soc. Am. A*, **18**, pp. 1972–1979.
2. Laude, B., De Martino, A., Drévillon, B., Benattar, L., and Schwartz, L. (2002). Full-field optical coherence tomography with thermal light. *Appl. Opt.*, **41**, pp. 6637–6645.
3. Choi, W. J., Jeon, D. I., Ahn, S.-G., Yoon, J.-H., Kim, S., and Lee, B. H. (2010). Full-field optical coherence microscopy for identifying live cancer cells by quantitative measurement of refractive index distribution. *Opt. Express*, **18**, pp. 23285–23295.
4. Akiba, M., and Chan, K. P. (2007). *In vivo* video-rate cellular-level full-field optical coherence tomography. *J. Biomed. Opt.*, **12**, p. 064024.
5. Choi, W. J., Park, K. S., Eom, T. J., Oh, M.-K., and Lee, B. H. (2012). Tomographic imaging of a suspending single live cell using optical tweezer-combined full-field optical coherence tomography. *Opt. Lett.*, **37**, pp. 2784–2786.
6. Min, G., Choi, W. J., Kim, J. W., and Lee, B. H. (2013). Refractive index measurements of multiple layers using numerical refocusing in FF-OCT. *Opt. Express*, **21**, pp. 29955–29967.
7. Dubois, A., Vabre, L., Boccara, A. C., and Beaurepaire, E. (2002). High-resolution full-field optical coherence tomography with a Linnik microscope. *Appl. Opt.*, **41** pp. 805–812.
8. Na, J., Choi, W. J., Choi, E. S., Ryu, S. Y., and Lee, B. H. (2008). Image restoration method based on Hilbert transform for full-field optical coherence tomography. *Appl. Opt.*, **47**, pp. 459–466.

9. Hahn, S. L. (1996). *Hilbert Transforms in Signal Processing*, 1st Ed. (Artech House, USA).
10. Choi, D.-H., Hiro-Oka, H., Shimizu, K., and Ohbayashi, K. (2012). Spectral domain optical coherence tomography of multi-MHz A-scan rates at 1310 nm range and real-time 4D-display up to 41 volumes/second. *Biomed. Opt. Express*, **3**, pp. 3067–3086.
11. Huang, Y., Liu, X., and Kang, J. U. (2012). Real-time 3D and 4D Fourier domain Doppler optical coherence tomography based on dual graphics processing units. *Biomed. Opt. Express*, **3**, pp. 2162–2174.
12. Lee, B. H., Choi, W. J., and Na, J. (2008). Full-field optical coherence tomography based on Hilbert transform. *Rev. Laser. Eng.*, **36**, pp. 1347–1350.
13. Jewell, S. A., Vukusic, P., and Roberts, N. W. (2007). Circularly polarized colour reflection from helicoidal structures in the beetle *Plusiotis boucardi*. *New J. Phys.*, **9**, p. 99.
14. Sato, M., Nagata, T., Niizuma, T., Neagu, L., Dabu, R., and Watanabe, Y. (2007). Quadrature fringes wide-field optical coherence tomography and its applications to biological tissues. *Opt. Commun.*, **271**, pp. 573–580.
15. Hrebesh, M. S. (2012). Full-field and single-shot full-field optical coherence tomography: a novel technique for biomedical imaging applications. *Adv. Opt. Technol.*, **2012**, Article ID 435408, pp. 1–26.
16. Bordenave, E., Abraham, E., Jonusauskas, G., Tsurumachi, N., Oberlé, J., Rullière, C., Minot, P. E., Lassègues, M., and Surlève Bazeille, J. E. (2002). Wide-field optical coherence tomography: imaging of biological tissues. *Appl. Opt.*, **41**, pp. 2059–2064.
17. Sato, M., Nomura, D., Tsunenari, T., and Nishidate, I. (2010). *In vivo* rat brain measurements of changes in signal intensity depth profiles as a function of temperature using wide-field optical coherence tomography. *Appl. Opt.*, **49**, pp. 5686–5696.
18. Nakadate, S., and Isshiki, M. (1997). Real-time vibration measurement by a spatial phase shifting technique with a tilted holographic interferograms. *Appl. Opt.*, **36**, pp. 281–284.
19. Asundi, A., Tong, L., and Boay, C. G. (2001). Dynamic phase-shifting photoelasticity. *Appl. Opt.*, **40**, pp. 3654–3658.
20. Love, G. D., Oag, T. j. D., and Kirby, A. K. (2005). Common path interferometric wavefront sensor for extreme adaptive optics. *Opt. Express*, **13**, pp. 3491–3499.

21. Dunsby, C., Gu, Y., and French, P. M. W. (2003). Single-shot phase-stepped wide-field coherence-gated imaging. *Opt. Express*, **11**, pp. 105–115.
22. Ferrari, J. A., Frins, E. M., and Perciante, C. D. (2002). A new scheme for phase-shifting ESPI using polarized light. *Opt. Commun.*, **202**, pp. 233–237.
23. Hrebesh, M. S., Watanabe, Y., and Sato, M. (2008). Profilometry with compact single-shot low-coherence time-domain interferometry. *Opt. Commun.*, **281**, pp. 4566–4571.
24. Hrebesh, M. S., Dabu, R., and Sato, M. (2009). *In vivo* imaging of dynamic biological specimen by real-time single-shot full-field optical coherence tomography. *Opt. Commun.*, **282**, pp. 674–683.
25. Fercher, A. F., Drexler, W., Hizenberger, C. K., and Lasser, T. (2003). Optical coherence tomography-principles and applications. *Rep. Prog. Phys.*, **66**, pp. 239–303.
26. Bourquin, S., Seitz, P., and Salathé, R. P. (2001). Optical coherence topography based on a two-dimensional smart detector array. *Opt. Lett.*, **26**, pp. 512–514.
27. Laubscher, M, Ducros, M., Karamata, B., Lasser, T., and Salathé, R. (2002). Video-rate three-dimensional optical coherence tomography. *Opt. Express*, **10**, pp. 429–435.
28. Schmitt, J. M. (1999). Optical coherence tomography (OCT): a review. *IEEE J. Sel. Top. Quant.*, **5**, pp. 1205–1215.
29. Akiba, M., Chan, K. P., and Tanno, N. (2003). Full-field optical coherence tomography by two-dimensional heterodyne detection with a pair of CCD cameras. *Opt. Lett.*, **28**, pp. 816–818.
30. Akiba, M., Chan, K. P., and Tanno, N. (2000). Real-time, micrometer depth-resolved imaging by low-coherence reflectometry and a two-dimensional heterodyne detection technique. *Jpn. J. Appl. Phys.*, **39**, pp. L1194–L1196.
31. Yun, S. H., Tearney, G. J. Bouma, B. E., Park, B. H., and de Boer, J. F. (2003). High-speed spectral-domain optical coherence tomography at 1.3 µm wavelength. *Opt. Express*, **11**, pp. 3598–3604.
32. Ju, M. J., Lee, S. J., Kim, Y., Shin, J. G., Kim, H. Y., Lim, Y. Yasuno, Y., and Lee, B. H. (2011). Multimodal analysis of pearls and pearl treatments by using optical coherence tomography and fluorescence spectroscopy. *Opt. Express*, **19**, pp. 6420–6432.
33. Choma, M. A., Sarunic, M. V., Yang C., and Izatt, J. A. (2003). Sensitivity advantage of swept source and Fourier domain optical coherence tomography. *Opt. Express*, **11**, pp. 2183–2189.

34. Dubois, A., Grieve, K., Moneron, G., Lecaque, R., Vabre, L., and Boccara, C. (2004). Ultrahigh-resolution full-field optical coherence tomography. *Appl. Opt.*, **43**, pp. 2874–2883.
35. Watanabe, Y., and Sato, M. (2008). Three-dimensional wide-field optical coherence tomography using an ultrahigh-speed CMOS camera. *Opt. Commun.*, **281**, pp. 1889–1895.
36. Choi, W. J., Na, J., Choi, H. Y., Eom, J., and Lee, B. H. (2010). Active feedback wide-field optical low-coherence interferometry for ultrahigh-speed three-dimensional morphometry. *Meas. Sci. Technol.*, **21**, p. 045503.
37. Syed, S. H., Larin, K. V., Dickinson, M. E., and Larina, I. V. (2011). Optical coherence tomography for high-resolution imaging of mouse development *in utero*. *J. Biomed. Opt.*, **16**, p. 046004.

Chapter 7

Toward Single-Shot Imaging in Full-Field Optical Coherence Tomography

Bettina Heise

Christian Doppler Laboratory for Microscopic and Spectroscopic Material Characterization (CDL-MS-MACH), Center for Surface and Nanoanalytics (ZONA), Institute for Knowledge-based Mathematical Systems (FLLL), Johannes Kepler University Linz, Altenberger Str. 69, A-4040 Linz, Austria
bettina.heise@jku.at

Full-field optical coherence tomography (FF-OCT) and full-field optical coherence microscopy (FF-OCM) are regarded from a rather abstract point of view: the number of interferometric images, captured per modulation period and required for reconstruction, and the way of their recording and processing are discussed. Furthermore, application examples for dual- and single-shot FF-OCT and FF-OCM imaging, illustrating the particular benefits of these methods, are provided.

7.1 Introduction

Full-field optical coherence tomography (FF-OCT) as well as full-field optical coherence microscopy (FF-OCM) are emerging

---

Handbook of Full-Field Optical Coherence Microscopy: Technology and Applications
Edited by Arnaud Dubois
Copyright © 2016 Pan Stanford Publishing Pte. Ltd.
ISBN 978-981-4669-16-0 (Hardcover), 978-981-4669-17-7 (eBook)
www.panstanford.com

nondestructive optical imaging techniques for the investigation of semitransparent, turbid materials. They have shown their potential in various application fields providing a visual characterization of the sample due to the general backscattering behavior of the specimen. In biology and material research alike, FF-OCT delivers information about the structural composition, tissue morphology or, in the context of material sciences, about the local arrangement of interfaces, fillers, inclusions, and defects distributed within the specimen.

Due to the enhanced lateral resolution and optical sectioning abilities achieved in FF-OCM [14]—a modification of FF-OCT by incorporating high numerical aperture optics—detailed structural information in the micron size range within the subsurface region of the sample can be gathered. In this way FF-OCT and FF-OCM provide new possibilities for supporting medical diagnosis [3, 34], or for inspection of technical materials [29, 52].

During the short period of recording the biological specimens or technical materials are frequently assumed to stay unaltered. But, increasingly FF-OCT imaging for in vivo visualization of specimens [39] or for monitoring formation and evolution of microstructures in material testing procedures [55] has come into focus of interest. The opening of novel fields of application for FF-OCT and FF-OCM—in biology but also for applications with a technical background [9]—requires an appropriate imaging and suitable image reconstruction methods adapted for fast and stable acquisition and processing. FF-OCT realizations performing well in reconstruction with a reduced number of interferometric en face images acquired at each depth position, as typical for dual-shot or single-shot approaches, are often the methods of choice for time- or stability-critical imaging tasks.

In the following we review acquisition concepts, stretching from temporal phase-shifting techniques toward single-shot approaches, reported in the low-coherence interferometric (LCI) application field. We will focus on dual- and single-shot FF-OCT and FF-OCM, discuss standard realizations and interesting modifications, and throw a glance on complementary single-shot imaging techniques. In addition, we will consider the advantages, restrictions, and challenges of pure single-shot FF-OCT schemes and illustrate signal processing techniques that perform for these measurement approaches.

7.2 From Phase Shifting toward Single-Shot Techniques

FF-OCT, as an LCI imaging technique, records en face (x, y) images of the specimen by scanning over depth (z), and can be realized under different geometries like Michelson, Mach-Zehnder, Linnik, or Mirau configuration [1, 40, 60]. In time domain FF-OCT, the optical path length (OPL) difference is varied by shifting the relative position of the sample with respect to the reference mirror. In this way the axial position of the coherence gate that moves through the specimen can be controlled. For each depth position, a series of phase-shifted or— in the discrete manner—phase-stepped interferometric images are recorded. The phase-shifting procedure itself is well established in conventional interferometry, too.

Applying the standard framework for interferometric signal description, an individual interferometric image $I(x, y)$ resulting from the superposition of sample and reference wave field $\mathbf{E}_S(x, y)$ and $\mathbf{E}_R(x, y)$ is expressed as

$$I(x, y) = |\mathbf{E}_S + \mathbf{E}_R|^2 = |\mathbf{E}_S|^2 + |\mathbf{E}_R|^2 + 2\,Re\left\{\mathbf{E}_S^*\mathbf{E}_R\right\}(x, y). \quad (7.1)$$

Renamed with $B(x, y) = |\mathbf{E}_S(x, y)|^2 + |\mathbf{E}_R(x, y)|^2$ denoting the background, $A(x, y) = 2\,|\mathbf{E}_S(x, y)|\,|\mathbf{E}_R(x, y)|$ describing the local amplitude, and $\phi(x, y)$ denoting the local phase, corresponding to the optical path length (OPL) difference between sample and reference arm, the interferometric image exhibits a form of

$$I(x, y) = B(x, y) + A(x, y)\cos(\phi(x, y)). \quad (7.2)$$

The first term in (7.2) describes the noninterfering part of the signal, corresponding to an incoherent background component. The second term corresponds to the interfering part representing the signal of interest.

By introducing additional phase shifts $\Delta\varphi_n$, a sequence of interferometric images (frames)

$$I_n(x, y) = B(x, y) + A(x, y)\cos(\phi(x, y) + \Delta\varphi_n) \quad (7.3)$$

can be recorded.

Local amplitude and phase map encoded in the fringe pattern of the interferograms can be extracted by an appropriate demodulation method like phase shifting or phase stepping. But also numerical approaches for quadrature and analytic signal calculation should be cited.

Referred to FF-OCT, for a fixed depth z and defined visibility $V(z)$, the individual en face interferometric image can be modeled in the way of (7.3). The local amplitude image expresses the (en face) reflectivity image, which provides insight into the structural and backscattering features of the sample. Additionally, the phase map may deliver information about internal optical inhomogeneities or the topography of the specimen within the considered coherence range.

For conventional LCI, various optical configurations that enable a temporal or spatial phase-shifted image acquisition have been reported, with applications mainly related to profilometry, optical inspection, or surveillance tasks [11, 20, 43]. In wide-field (WF)-OCT, FF-OCT, and FF-OCM imaging corresponding types of multi-, dual-, or single-shot acquisition techniques have been reported in the literature. We will review them briefly in the following.

7.2.1 Temporal Phase-Shifting Techniques

A time-sequential recording of phase-shifted interferometric images characterizes temporal phase-shifting methods. These techniques are frequently implemented in FF-OCT by means of a piezoelectric transducer (PZT) unit coupled with the reference mirror of the interferometric setting. The PZT moves the mirror in fractions of the wave length and causes its oscillating motion. The temporal modulation of the interference signal is obtained by a periodically alternating optical path length and phase difference, respectively, between sample and reference wave fields.

The mechanical displacement of the reference mirror is often limited in speed by the inertia of the PZT unit. In the case of high operational frequencies, the precision of a PZT-based mirror displacement can be reduced. So, an exact phase shifting and stable driving of the reference mirror might be technically demanding. The setup is easily influenced by instabilities, particularly for rapidly oscillating mirror units.

In most phase-shifting approaches, the PZT is driven by a sinusoidal or sawtooth-like function. Whereas sawtooth-like oscillations are rather restricted to lower driving frequencies, sinusoidal phase modulations have proven useful in ensuring stability at higher operational speed. In [15], a phase-shifting "integrating bucket" method is demonstrated for FF-OCT, there combined with a synchronized stroboscopic illumination allowing fast modulation schemes.

A further aspect to be noted for temporal phase shifting is the dependence of the phase shift introduced by OPL variations from the wave length. In particular, this issue arises for spectral broadband light sources, which are the usual case for illumination in FF-OCT imaging. This systematic error can lead to a reduced fringe contrast [49].

Geometric rather than temporal phase shifting may offer a solution. A geometric phase-shifting (GPS) approach has already been described in context of coherence probe microscopy (CPM) [49, 60]. CPM can be interpreted as an early version of the (FF-)OCT technique, with origin in the field of LCI for surface inspection and profilometry [11]. In the CPM setup a GPS unit is included, as schematically depicted in Fig. 7.1. The GPS unit consisting of a quarter-, a half-, and a quarter-wave plate (QWP, HWP, QWP) facilitates an achromatic phase shifting by rotation of the HWP. Recently, an achromatic phase shifting for FF-OCT realized with rotating HWP has again been presented in [70].

However, temporal phase-shifting or phase-stepping techniques, although relatively easily implemented, are less practicable for real time imaging of dynamic scenes. For the observed specimen a quasi-static behavior has to be supposed during recording the frame sequence. As the image acquisition speed is limited by the frame rate of the area detector, such as CCD (charged coupled device) or CMOS (complementary metal oxide semiconductor) cameras, and moreover, the achievable image reconstruction rate is still reduced by taking a time series of multiple interferograms per modulation cycle, the bounds for imaging of such dynamic scenes are predefined.

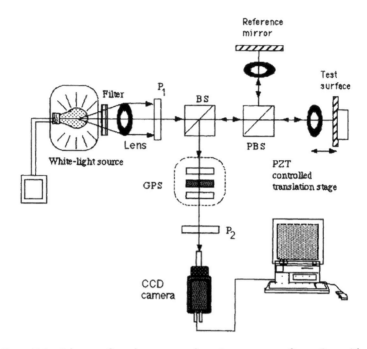

Figure 7.1 Scheme of a coherence probe microscopy configuration with a geometric phase-shifting (GPS) unit included. GPS unit in (QWP, HWP, QWP) combination; P: polarizer, PBS: polarizing beam splitter. Figure reprinted from [49], with kind permission from SPIE, © 1999 SPIE.

Additionally, the time-sequential capturing of interferograms is very sensitive to mechanical or thermal fluctuations in the measurement system. In particular, random phase drifts and jittering effects can reduce the performance. Artifacts and erroneous effects, which results in stripes, reduced contrast, or in objects multiply displayed, are reported [51].

Regarding fast dynamical processes, like transient formation of micropatterns, rapidly moving objects, or modifying biological structures, these applications require concepts for a simultaneous recording of phase-shifted interferograms, free of artifacts induced by time-sequential acquisition schemes. Such concepts attain their realization by simultaneous phase-shifting techniques, replacing the temporal shifting with a spatial splitting and stepping of phase.

7.2.2 Simultaneous Phase-Shifting Techniques

In interferometry different concepts for spatial splitting and phase shifting have been followed: configurations operating with reflective or refractive components, like beam splitters or Wollaston prisms, or solutions using diffractive elements, such as conventional or holographic gratings for beam division [41, 62]. Recently liquid crystal spatial light modulators (LC-SLMs), displaying the grating of interest, have gained importance [26].

Although LC-SLM-based methods are favored by their flexibility for introducing arbitrary phase patterns, it should be taken into account that diffractive solutions in general are well suited only for narrow-band light sources due to the dispersion artifacts caused. In particular for FF-OCT, where broadband light sources are usually applied, dispersion effects resulting from incorporated gratings need correction by expensive compensation methods. These compensation settings often compromise the simplicity of the optical setup [42, 63]. Therefore, a simultaneous phase shifting by various beam splitter arrangements represents a widespread base for single-shot FF-OCT applications.

With respect to the number of (temporal) acquisitions we will distinguish between dual-shot (or dual-step) and true single-shot techniques and survey historical developments and principles.

7.2.2.1 Dual-shot image acquisition

In [53] an approach for a quadrature fringe WF-OCT is suggested, based on an optical realization of the Hilbert transform to record four phase-shifted images by only two image acquisition steps. There, a *two-channel* configuration is realized, which operates with linearly and circularly polarized light and uses polarization optics elements (Wollaston prism) for splitting. In the first step two spatially divided interferograms are recorded by a single camera corresponding to the in-phase and in-quadrature component of the signal.

In the second acquisition step a phase shift of π relative to the initial phase position is introduced by the PZT. Combining the temporal and spatial phase shifting in this way, four interferograms

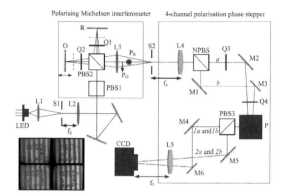

Figure 7.2 Experimental setup. O, object; R, reference mirror; L1-5, lenses; S1-2, slits; PBS1-3, polarizing beam splitter cube; NPBS, nonpolarizing beam splitter cube; P, periscope; M1-6, mirrors; Q1-4, quarter-wave plates. Inset shows a CCD image acquired with a USAF test chart. Figure reprinted from [16], with kind permission from OSA, © 2003 Optical Society of America.

are finally recorded. For the reported two-channel dual-shot approach a single indium-gallium-arsenide CCD camera at a frame rate of 30 fps has been used.

7.2.2.2 Single-shot image acquisition

The two-channel dual-shot concept for spatial multiplexing of interferograms can be continued toward a simultaneous recording of four spatially divided interferograms. Such a *four-channel* single-shot version has already been described for a WF-OCT system in [16], and has been demonstrated for FF-OCT in [32].

The particular feature of configuration consists in two main components: a polarizing interferometer and a four-channel polarization phase-stepper unit, as shown in Fig. 7.2 and Fig. 7.3. A polarization-optics-based splitting and shifting scheme is performed. It leads to four spatially separated, phase-stepped interferograms, which can be simultaneously obtained. They are recorded as subimages by a single CCD camera and provide the required information for reconstruction.

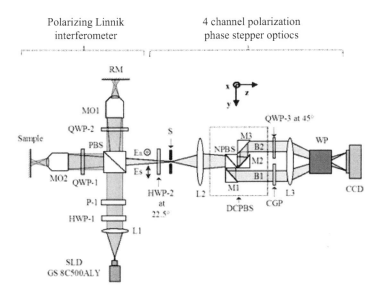

Figure 7.3 Experimental setup (lateral view). L1–3, lenses; HWP, half-wave plate; P, polarizer; PBS, polarizing beam splitter; QWP, quarter-wave plate; MO1–2, microscopic objectives; RM, reference mirror; S, rectangular slit; NPBS, nonpolarizing beam splitter; M1–3, mirrors; CGP, compensation glass plate; WP, Wollaston prism; CCD, charge-coupled device. Figure reprinted from [32], with kind permission Elsevier, © 2008 Elsevier.

This concept follows the classical reconstruction based on four phase-shifted interferograms. However, a careful registration and intensity adjustment for all four subimages is mentioned as prerequisites for the success of the method. Therefore, coming back to the two-channel techniques the registration efforts can be partly diminished. Before discussing reconstruction suited for these two-channel techniques, a short overview on further simultaneous phase-shifting schemes is given.

7.2.2.3 Polarization-based schemes

In [31] a single-shot two-channel FF-OCT realization is described following again the polarization optics concept as depicted in Fig. 7.4. This dual-channel method performs similar to the above-mentioned polarization-based four-channel technique but contains

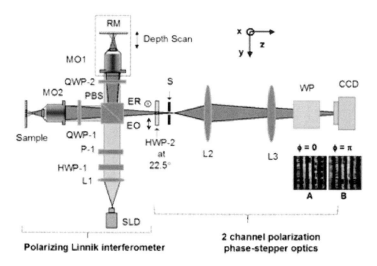

Figure 7.4 Single-shot FF-OCT: experimental setup. L1–3, lenses; HWP, half-wave plate; P, polarizer; PBS, polarizing beam splitter; QWP, quarter-wave plate; MO1–2, microscopic objectives; RM, reference mirror; S, rectangular slit; WP, Wollaston prism; CCD, charge-coupled device. Figure reprinted from [31], with kind permission from M. S. Hrebesh, © 2012 Hrebesh Molly Subhash.

only a dual-channel phase-stepping unit and is extended by a suitable processing afterwards.

7.2.2.4 Unpolarized instantaneous phase-shifting schemes

In case of birefringent specimens optical splitting and shifting techniques applying nonpolarizing elements are of advantage.

In [51] a single-shot FF-OCT system realized in a Linnik–Michelson configuration is cited. Schematically it is illustrated in Fig. 7.5. There, the two phase-opposed interferograms both generated at the central (nonpolarizing broadband) beam splitter are directed—by including two further beam splitters—onto two cameras. The demonstrated single-shot full-field approach should be regarded in relation to the balanced detection principle, which is well known for point-scanning OCT systems.

FF-OCT configurations realized in a Mach-Zehnder geometry offer benefits for incorporating additional functional components,

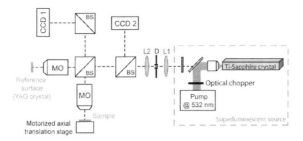

Figure 7.5 Schematic representation of the unpolarized instantaneous phase-shifting FF-OCT setup. BS, nonpolarizing broadband beam splitters; MO, microscope objectives; L1, L2, lenses; D, diaphragm; superluminescent source. Figure reprinted from [51], with kind permission from OSA, © 2010 Optical Society of America.

such as a Fourier plane filtering unit [55, 56]. Different contrast modifications can be performed by these filtering techniques. A spatial splitting and simultaneous phase-stepping scheme that is suitable for single-shot FF-OCT in Mach-Zehnder settings is illustrated in Fig. 7.6. The splitting unit generates two interferometric subimages, which are phase opposed. They are geometrically separated and recorded by a single camera. The reported camera system (Neo sCMOS) possesses beneficial properties for the chosen single-shot detection scheme and enables the observation of dynamical processes. A full frame rate of 100 fps has been achieved in the described setting.

7.2.2.5 Heterodyne detection schemes

A heterodyne single-shot FF-OCT solution operating with two cameras and two liquid crystal shutters in front is described in [2]. For acquiring a 3D data set in a single longitudinal depth scan a frequency-synchronous detection method is applied: The interference signal, modulated over time by the Doppler frequency f_D due to the mirror motion, is equally divided by a beam splitter and sent onto both high-speed shutters operating with a sampling frequency of f_M and $\pi/2$ phase difference to each other. The beating signal with $(f_D - f_M)$ can be recorded. After an additional

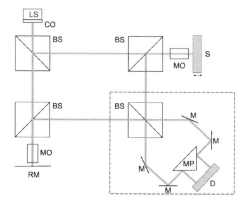

Figure 7.6 Single-shot FF-OCT scheme realized in a Mach-Zehnder geometry enabling a simultaneous phase shifting by a balanced detection-like configuration, indicated by the dashed line. LS, light source; CO, collimator; BS, beam splitter; MO, micro-objective; M, mirror; RM, reference mirror; S, sample; D, area detector; MP, mirrored prism. Figure kindly provided by CDL MS-MACH, JKU, Linz, Austria.

background estimation the in-phase and quadrature component can be simultaneously derived.

7.2.2.6 SLM-based schemes

Programmable optics devices, like by LC-SLMs or micromirror arrays, offer possibilities for a defined phase shaping of wave fields. Using these devices for spatial phase modulation of the reference wave field, the technique can be exploited for generating a defined spatial phase pattern. For FF-OCT this method has been reported in [46] as an interesting alternative to conventional splitting approaches. Addressing there a periodic, checkerboard-like spatial phase pattern on a micromirror array, spatially distributed zero- and π-phase shifts are introduced, as illustrated schematically in Fig. 7.7. Each mirror element forms a small image, called uniform area image, at the 2D detector. By a mathematical procedure, which takes into account neighborhood relations between uniform areas and the spatial periodicity of the phase pattern, (directional) differences between adjacent area signals are calculated. Applying the partial

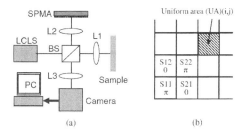

Figure 7.7 (a) Basic system of single-shot FF-OCT with a spatial micromirror array (SPMA), applied for spatial modulation of the reference wave field. (b) Uniform area (UA) on CCD camera. Figure reprinted from [46] with kind permission from Springer, © 2011 The Japan Society of Applied Physics.

Hilbert transform on the difference image, finally the FF-OCT image can be reconstructed in an approximated way.

7.2.3 Hyperspectral Single-Shot Techniques

The previously presented single-shot schemes all represent more or less modifications of various time domain FF-OCT concepts. A completely different approach toward a single-shot OCT is followed in [45]. There a 3D snapshot OCT is described, which seems more related to a single-shot Fourier domain OCT or hyperspectral single-shot LCI [69].

The method combines features of spectral domain OCT (spectral data), full-field imaging (full-field area illumination), and image mapping spectrometry (multiple slits interferometer for spatial splitting and multiplexing), as illustrated in the scheme of Fig. 7.8. In this way, a hyperspectral (x, y, λ) data cube is obtained within one camera integration time (snapshot). The full (x, y, z) structural information of the sample can be reconstructed by performing the 1D Fourier transform of the recorded data resampled with respect to the wavenumber. For the demonstrated snapshot 3D OCT system, an axial resolution of 16 µm and a lateral resolution of 13 µm for an imaging depth of about 400 µm has been achieved. In Fig. 7.9 the feasibility of the reported concept is illustrated, the 3D reconstruction shows a tape structure applied on a resolution test target that is imaged by a 3D snapshot OCT.

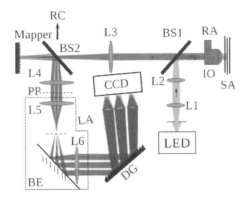

Figure 7.8 System layout for a 3D snapshot OCT. BS: beam splitter, IO: interferometry objective, LA: lenslet array, PP: pupil plane, RC: reference camera, RA: reference arm, SA: sample arm. Figure reprinted from [45], with kind permission from OSA, © 2013 Optical Society of America.

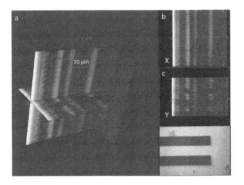

Figure 7.9 3D structure recorded in snapshot mode from the 3D OCT system. (a) Reconstructed structure of clear tape on USAF target. (b and c) Its xz and yz cross sections. (d) Camera image for reference. Figure reprinted from [45], with kind permission from OSA, © 2013 Optical Society of America.

With respect to single-shot interferometric full-field imaging techniques, the demand for a reduced number of acquisitions per cycle on one hand causes the need for suitable demodulation methods on the other hand. Signal processing techniques derived from a 2D quadrature signal model [18] for computing local amplitude and local phase should be pointed out as valuable methods extending the range of possibilities for reconstruction. Referred to FF-OCT they can extract the complete structural information for sufficiently modulated signals. In their results, analytic signal techniques are comparable to phase-stepping methods, as we will discuss in the following section.

7.3 Signal Processing Methods

Different signal processing algorithms for reconstruction of FF-OCT images have been designed, suited for the above-mentioned measurement procedures. In general, the methods base on using multiple phase-stepped and continuously phase-shifted (interferometric) FF-OCT images, respectively.

Concerning reconstructions based on a pair of (usually) two phase-opposed interferograms, subtraction methods represent a common technique in FF-OCT. Furthermore, mathematical approaches that contain the Hilbert transform as an essential element should be highlighted, having a distant counterpart in quadrature signal techniques in 1D.

Briefly, reconstruction by phase stepping is regarded, which represents a widespread technique and partly can act for comparison. Later, we focus in more detail on analytic signal-based approaches, which seem less known in their variety of 2D realizations.

7.3.1 *Phase-Stepping Approaches*

Mathematical fundamentals for phase-stepping approaches can be derived from least square solutions [24] and Fourier theory. In (equidistant) phase-stepping approaches, local amplitude and phase can be reconstructed by the complex summation of individual phase-stepped interferograms I_n, multiplied with the complex-valued phase shifts $\exp(i\varphi_n)$, $\varphi_n = 2\pi(n-1)/N$, $n = 1, \ldots, N$, which are introduced within the modulation period [41, 58]. Amplitude and phase appear as modulus and argument of the complex-valued sum

$$A(x, y) = abs\left(\sum_{n=1}^{N}(I_n \exp(i\varphi_n))\right)(x, y), \quad (7.4)$$

$$\phi(x, y) = arg\left(\sum_{n=1}^{N}(I_n \exp(i\varphi_n))\right)(x, y). \quad (7.5)$$

For the conventional case of four interferograms—each shifted by $\pi/2$—the amplitude yields

$$A(x, y) = ((I_3 - I_1)^2 + (I_4 - I_2)^2)^{\frac{1}{2}}(x, y), \quad (7.6)$$

and the wrapped phase is given as

$$\phi(x, y) = \arctan((I_4 - I_2)/(I_3 - I_1))(x, y). \quad (7.7)$$

7.3.2 The Subtraction Method

For the reconstruction of structural information in FF-OCT imaging several authors describe approaches, in which only the squared difference image $I_S(x, y)$ is calculated by subtracting two phase-shifted, consecutive interferometric frames and squaring the difference [25, 51, 68].

The resulting signal is in a manner of

$$I_S(x, y) \propto A^2(x, y) \cos^2(\phi(x, y)), \quad (7.8)$$

in which amplitude and phase terms are not separated.

The subtraction method allows a very fast and elegant reconstruction. It seems a sufficient procedure for real-time visualization of biological samples [50]. An explanation could be that these specimens frequently show a highly scattering behavior, which leads to a rather random-like appearance of microtextures in tissues. In analogy to image processing [22], visualization for such microtextures is not affected in a statistical sense by an additional random phase term. So also amplitude and phase need not be further separated.

However, regarding rather piece-wise continuous features, as appearing, for example, at inclined interfaces or elongated fiber structures in technical materials, the remaining phase modulation would lead to a degraded appearance and can influence a proper visualization. Additionally, fringe patterns can be introduced due to a minor tilt or a mismatch in curvature between the wave fronts. In these cases continuing the subtraction method toward an exact separation into amplitude and phase would be preferable.

In the following we will describe reconstruction methods, which can perform this separation. They can already be found in their origins in conventional interferometric imaging [7, 38] and are adapted to FF-OCT.

7.3.3 Analytic Signal–Based Reconstruction

In analogy to the calculation of the well-known 1D analytic signal by means of the 1D Hilbert transform [21], a generalized 2D analytic signal definition is related either to the 2D (hypercomplex) analytic signal [8] or to the 2D monogenic signal [17], which are based on the 2D Hilbert (HT) or 2D Riesz transform (RT) for their calculation. Briefly, the theoretical background will be sketched here:

In the 1D case, the HT of the 1D function $f(x)$ is mathematically described by convolving $f(x)$ with the kernel $h(x) = -\frac{1}{\pi x}$ [23], as

$$f_H(x) = -\frac{1}{\pi x} \otimes f(x). \tag{7.9}$$

In spatial domain, in the 2D case, the 2D HT of a 2D function $f(x, y)$ can be performed in a component-wise way as

$$f_{H_x}(x, y) = -\frac{\delta(y)}{\pi x} \otimes f(x, y), \tag{7.10}$$

$$f_{H_y}(x, y) = -\frac{\delta(x)}{\pi y} \otimes f(x, y),$$

$$f_{H_t}(x, y) = \frac{1}{\pi^2 x, y} \otimes f(x, y).$$

The terms f_{H_x}, f_{H_y}, f_{H_t} denote the partial (directional) and total HT components applied on $f(x, y)$ [27].

This notation delivers a 2D (quaternionic) analytic signal $\mathbf{f}_A(x, y)$, which can be expressed in the following way

$$\mathbf{f}_A(x, y) = f(x, y) + i f_{H_x}(x, y) + j f_{H_y}(x, y) + k f_{H_t}(x, y), \tag{7.11}$$

with (i, j, k) as quaternionic units. (Slightly simplified, $\mathbf{f}_A(x, y)$ could be read as a four-component vector.)

Local amplitude (representing the reflectivity image in FF-OCT context) and phase (describing a kind of topographic information) are defined in the 2D (quaternionic) analytic signal model as

$$A(x, y) = \sqrt{(f(x,y))^2 + (f_{H_x}(x,y))^2 + (f_{H_y}(x,y))^2 + (f_{H_t}(x,y))^2}, \quad (7.12)$$

$$\phi(x, y) = \arctan\left(\sqrt{\frac{(f_{H_x}(x,y))^2 + (f_{H_y}(x,y))^2 + (f_{H_t}(x,y))^2}{(f(x,y))^2}}\right). \quad (7.13)$$

For the RT-based approach the procedure is repeated in a modified way. The RT is rather related to polar coordinates (r, η) and therefore occasionally also called radial HT [12]. Performing now the RT, it yields

$$f_{R_x}(x, y) = \frac{\cos(\eta)}{2\pi r^2} \otimes f(x, y), \quad (7.14)$$

$$f_{R_y}(x, y) = \frac{\sin(\eta)}{2\pi r^2} \otimes f(x, y),$$

with the terms f_{R_x}, f_{R_y} corresponding to both components of the RT applied on $f(x, y)$.

The monogenic signal $\mathbf{f}_M(x, y)$ [17] is defined in analogy to (7.11) as

$$\mathbf{f}_M(x, y) = f(x, y) + i f_{R_x}(x, y) + j f_{R_y}(x, y), \quad (7.15)$$

with (i, j) as two generalized complex units.

Local amplitude and phase based on 2D monogenic signal model can be derived [17] as

$$A(x, y) = \sqrt{(f(x,y))^2 + (f_{R_x}(x,y))^2 + (f_{R_y}(x,y))^2}, \quad (7.16)$$

$$\phi(x, y) = \arctan\left(\sqrt{\frac{(f_{R_x}(x,y))^2 + (f_{R_y}(x,y))^2}{(f(x,y))^2}}\right). \quad (7.17)$$

Furthermore, the orientation $\beta(x, y)$ of the fringes is given by

$$\beta(x, y) = \arctan\left(\frac{f_{R_y}(x,y)}{f_{R_x}(x,y)}\right). \quad (7.18)$$

Here, with respect to image reconstruction for single-shot FF-OCT imaging, we will focus more on the aspects of practical realization than on deeper mathematical derivations. The theoretical concept for these methods can be found for more detail in [19, 27, 64].

In FF-OCT, the discussed approach starts from the difference image between two subsequent interferograms I_k, I_{k+1}; that is, for the abstract function $f(x, y)$ is chosen $I_D(x, y) = I_{k+1}(x, y) - I_k(x, y)$. Assuming both interferograms are phase-shifted by φ, the difference image, being virtually free of any background signal, can be described as a 2D amplitude-frequency-modulated (AM-FM) signal expressed by [44]

$$I_D(x, y) = 2\sin(\varphi/2) A(x, y) \sin\left(\phi(x, y) + \frac{\varphi}{2}\right). \quad (7.19)$$

For the usual case that of two exactly phase-opposed interferograms ($\varphi = \pi$) it leads to the well-known form

$$I_D(x, y) = 2A(x, y) \cos(\phi(x, y)). \quad (7.20)$$

It should be noted, 2D analytic signal-based demodulation techniques are applicable under the condition of a distinguished frequency support of AM and FM component in Fourier domain [4]. Conventionally, a slowly varying amplitude modulation (AM) and a highly varying (spatial) frequency modulation (FM) of the interference fringes, as introduced by tilts or at inclined structures, is supposed.

In [44] an image restoration technique for single-shot FF-OCT applying an HT-based method is suggested. There a partial 2D HT approach is applied, which performs the reconstruction in a directional way.

The partial HT-based approach, which is intrinsically 1D in its conception, can be extended toward a true 2D image reconstruction technique. For single-shot FF-OCT imaging it is successfully demonstrated in [31]. There, the reconstruction of the reflectivity image is performed based on the quaternionic analytic signal concept. Although the different HT and RT definitions and concepts are rather of mathematical interest, it should be noted that RT-based methods exhibit a more isotropic response characteristics, compared to their HT-based counterparts. This can be advantageous for smoother reconstruction results. Such an RT-based reconstruction is demonstrated in [55], there performed on both simulated and experimental data for FF-OCM.

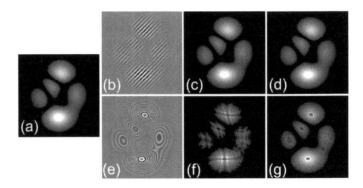

Figure 7.10 Demodulation performance of the analytic and monogenic approach on simulated FF-OCM raw data: (a) ground truth of the amplitude map, (b and e) simulated 2D AM-FM input data, (c and d) reconstructed 2D analytic and monogenic amplitude map (reflectivity image) in the case of linear fringes, and (b, f, and g) corresponding results for closed fringe structures (e). Figure reprinted from [55], with kind permission from OSA, © 2012 Optical Society of America.

Fig. 7.10 shows a comparison of the 2D analytic and monogenic method for reconstruction on simulated data. For closed fringe patterns, the directionally dependent artifacts in the image reconstructed by HT approach are visible. Similar erroneous features are also known from image processing, related there to RT and HT-based edge detection and orientation estimations [5].

Fig. 7.11 shows a comparison of phase stepping and 2D analytic signal–based reconstruction for experimental data. Two-dimensional analytic signal–based reconstruction methods closely match phase-stepping techniques in their results, as depicted in the detailed views (f2)-(f4). But it should be recalled, only a reduced number of interferograms is needed for achieving almost the same performance. Again, directional artifacts are better prevented in the monogenic approach than in the conventional 2D analytic reconstruction scheme. Furthermore, as local phase and orientation of the fringe pattern are also obtained, even these quantities can contribute to gain additional information, for example about inclination of objects observed within the coherence range.

In comparison with (mathematically) similar 2D reconstruction tasks in other coherent imaging applications, in [59] an RT-based

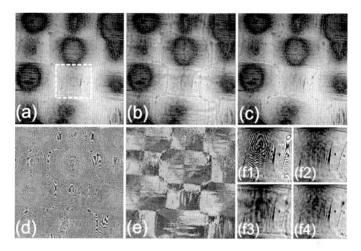

Figure 7.11 Reconstructed FF-OCM en face image of a fiber-reinforced polymer structure. Reflectivity image obtained with (a) conventional phase stepping (8 frames), (b) 2D analytic reconstruction (2 frames), (c) monogenic approach (2 frames), and (d and e) calculated monogenic phase and orientation map of the fringe pattern. Field of view 3 mm × 3 mm; (f1) magnified part of the raw fringe pattern, (f2), (f3), and (f4) magnified reconstruction details corresponding to (a), (b), and (c), as indicated in (a). Figure reprinted from [55], with kind permission from OSA, © 2012 Optical Society of America.

local demodulation for single-shot holographic image data is described. A multiresolution Riesz–Laplacian wavelet approach is presented in [67], amongst others, tested there for demodulation, too. Including elements of monogenic wavelet analysis in FF-OCT image reconstruction, this idea seems promising for enhancing particular scale-dependent features [48]. Fig. 7.12 shows the outcome of such a multiscale reconstruction approach for a technical material structure probed by FF-OCM. The reconstruction method applies a monogenic wavelet scheme [30] for scale-dependent decomposition into amplitude, phase, and orientation. The outcomes of the multiscale approach should be compared to the (conventional) reconstruction results that had been previously illustrated in Fig. 7.11.

The RT-based reconstruction technique can be successfully combined with the single-shot FF-OCT setup in Mach-Zehnder

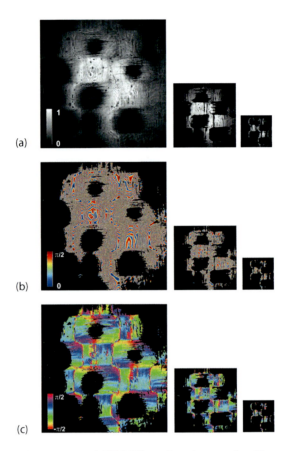

Figure 7.12 Reconstructed FF-OCM en face image of a fiber-reinforced polymer structure in multiscale representation. (a) Reflectivity image (monogenic amplitude), (b) monogenic phase image, and (c) monogenic orientation image, obtained by an RT isotropic wavelet-based reconstruction technique. Demodulation results are depicted at three subsequent scales. The phase and orientation map are overlaid with an reliability mask according to the reflectivity image. Figure kindly provided by Martin Reinhardt, TU Bergakademie Freiberg, Germany, [48].

geometry, as was sketched above in Fig. 7.6). An illustration for the outcome of this combined imaging and reconstruction approach is given in Fig. 7.13. A reflective, metalized Siemens Star resolution target acts as test object for a first prove of concept here.

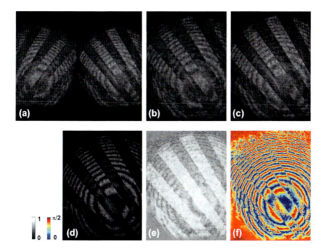

Figure 7.13 Illustration for single-shot FF-OCM reconstruction procedure: (a) two phase-opposed interferograms taken by one camera shot, (b and c) registered interferograms, (d) difference image, (e) reconstructed reflectivity map, and (f) reconstructed phase map. The procedure is exemplified at the surface of a Siemens Star test target. Figure kindly provided by CDL MS-MACH, JKU Linz, Austria.

7.4 Applications

Surveying current developments, meanwhile dual- or single-shot FF-OCT and FF-OCM techniques can be found in use for manifold imaging tasks coming from biology, medicine, or material research.

In biology, FF-OCT has been emerged as an technique of interest for a label-free in situ and in vivo imaging of tissues and living microorganisms.

FF-OCT imaging following a dual-shot concept for in vivo tissue imaging and visualization of fluctuations is demonstrated for example in [6], there applied for monitoring dynamics within the (sub)-surface blood vessels of a rat cortex tissue. For the reported FF-OCT system, complemented with an adaptive optics tool for correcting defocus aberrations, an in vivo imaging up to a penetration depth of several hundred microns at a frame rate of 33 Hz (or 6.6 Hz by averaging) has been shown. Its performance is illustrated for selected image frames in Fig. 7.14.

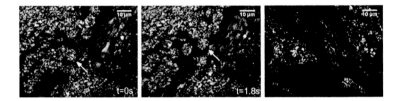

Figure 7.14 Single-frame excerpts produced from a 33 Hz FF-OCT video recordings of surface blood vessels in rat cortex. Left and center: Large vessel 6.6 Hz frame rate (by averaging 5 FF-OCT images each to increase signal to noise). Right: Junction of two blood vessels joining in the cortex (33 Hz frame rate, taken without any averaging). Figure reprinted from [6], with kind permission of OSA, © 2011 Optical Society of America.

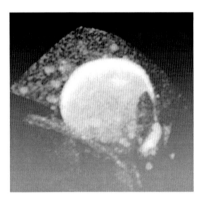

Figure 7.15 3D volumetric image of a tadpole eye recorded by single-shot FF-OCT and reconstructed by Riesz transform-based method. Image volume: $1 \times 1 \times 0.55$ mm^3. Figure is kindly provided by Hrebesh Molly Subhash, [31].

Even a complete sectioning of biological microstructures can be realized by a well-aligned single-shot FF-OCT system. As example the reconstructed 3D structure of a tadpole eye should be cited, shown in Fig. 7.15. The individual frames are recorded by a single-shot FF-OCT configuration, proving the ability of the polarization optics concept for single-shot imaging [31]. The en face images are reconstructed by a 2D analytic signal-based demodulation technique and finally composed to the 3D volumetric image.

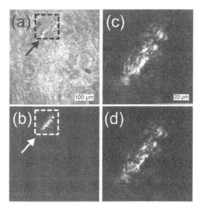

Figure 7.16 Single-frame excerpt of a living microorganism (*Paramecium*) produced from a video sequence. (a) Interferometric raw data. The arrow marks the specimen. (b) Corresponding demodulated FF-OCM image with living specimen and suppressed object structures. The dashed square marks the region of interest reconstructed in (c) with the 2D analytic HT-based and in (d) with the monogenic RT-based method. Figure reprinted from [55], with kind permission of OSA, © 2012 Optical Society of America.

The real-time imaging of a living paramecium microorganism surrounded by a liquid medium is reported in [55], there performed by a dual-shot FF-OCM technique. The time sequence of records for pursuing the moving specimen is taken at a chosen imaging depth. The individual reflectivity images are reconstructed by both HT- and RT-based methods. It is of interest that the contrast and detail enhancement achieved by the RT-based method in comparison to the 2D classical analytic signal method can be seen for this biological specimen in Fig. 7.16.

Single-shot approaches enable the observation of dynamic processes in the context of material sciences, too. Altering internal structures or material features can be directly monitored during treatment or testing procedures. Successful application of FF-OCM with reduced image acquisition number has been reported in [28, 57] for tensile or bending tests on blank plastics or fiber-reinforced composite materials. In [55] the FF-OCM imaging technique has been combined with monogenic signal analysis for the structural investigation of fiber-reinforced polymers in both

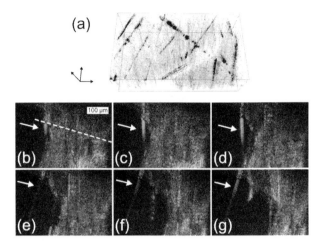

Figure 7.17 (a) 3D stack of fiber-filled polymer sample before testing (540 × 300 × 200 μm). (b–g) Extracted image sequence taken during tensile test with fracture front marked by dashed line and arrows indicating the position of an individual detaching fiber. Figure reprinted from [55], with kind permission of OSA, © 2012 Optical Society of America.

static and dynamic testing scenes. There, the temporal formation of the polymer matrix pattern during the stretching process is investigated by a dual-shot method. Results are shown in Fig. 7.17 for characteristic events like the pulling out of fibers and for monitoring the instantaneous position of fracture front.

Yet other applications for FF-OCT imaging can be found in microfluidics. There, the geometry and size of microchannels themselves can be retrieved by dimensional FF-OCM measurements [65]. Furthermore, single-shot FF-OCM imaging enables the depth-resolved localization of the microparticles that are transported through the channels. The local, static distribution of particles dispersed in the fluidic material can be analyzed.

FF-OCT imaging should also be mentioned for characterization of particle motion and fluidic behavior in microfluidic systems. But while conventional OCT mostly probes the sample by a cross-sectional raster scanning extracting axial phase information [66] or exploits related features like Doppler variance or cross-correlation [13], single-shot FF-OCT enables en face visualization of the fluid

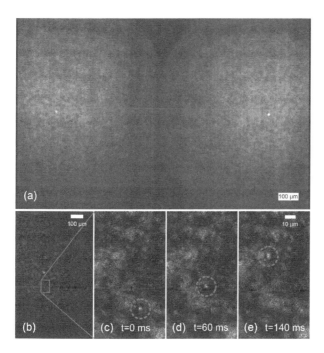

Figure 7.18 Monitoring dynamics within a particle-water suspension injected in a microfluidic channel by single-shot FF-OCT imaging: (a) interferometric single-shot image pair, (b) reconstructed reflectivity image, and (c) magnified reconstructed FF-OCM scans at different time states, corresponding to the region as indicated in (b). The displacement of the marked microparticle within the surrounding fluidic medium can be tracked over time. The sequence is monitored within the microchannel in a depth of 100 μm at a frame rate of 100 fps. Figure kindly provided by S. E. Schausberger, CDL MS-MACH, Austria, [54].

field and moving structures at a chosen depth of interest. By locally correlating subsequent reconstructed en face images a lateral motion analysis can be performed [28].

In Fig. 7.18 a demonstration of single-shot FF-OCT imaging in a microfluidics context is given, [54]. As illustrated there, the lateral displacement of the microparticles within the fluid could be determined from the subsequent frames. The technique uses a single-shot FF-OCT in a Mach-Zehnder configuration. The reflectivity images are reconstructed by the Riesz transform-based demodulation scheme.

Over the last decade, an increasing number of other, closely related single-shot imaging techniques has been published. With respect to the underlying principles or applications the borders are sometimes fuzzy and only distinguished by subtle differences. But the individual techniques can complement each other and offer a broad spectrum for nondestructive testing and 3D imaging of microstructures.

The abilities of single-shot FF-OCT also should be contrasted with techniques like digital holographic microscopy and regarded in the light of current trends, where single-shot holographic imaging comes together with novel computational reconstruction techniques [37] or with achromatic coherence-gated solutions [61]. Remarkable results for a lensless high resolution imaging by means of single-shot full-field holographic techniques have been described recently in [35, 36].

Further full-field imaging techniques should be named that can bridge the gap between FF-OCT imaging and phase microscopy: single-shot full-field reflection phase microscopy [71], diffraction phase microscopy [47], or Hilbert phase microscopy [33] represent related methods. The techniques mostly enable a quantitative phase characterization of the investigated specimens. As an example, the performance of single-shot full-field reflection phase microscopy is illustrated in Fig. 7.19.

Also noninteferometric single-shot techniques exploiting the transport of intensity equation for phase reconstruction should be recalled [72]. There, inclined optical components enable the spatial splitting; a SLM integrated in the Michelson interferometer performs the digital defocusing.

Full-field single-shot quantitative phase microscopy using dynamic speckle illumination is presented in [10]. Dynamics of small intracellular particles in live biological cells can be monitored in this way.

Although these methods are mostly well suited for in vivo imaging of cellular or other almost transparent structures, they seem easily affected when monitoring dynamics in turbid samples due to the distinct scattering characteristics and surface features. It

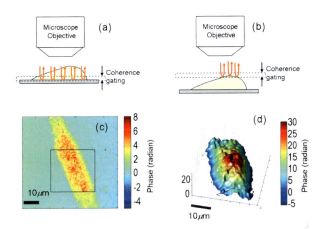

Figure 7.19 (a and b) Location of coherence gate for double-pass transmission and reflection phase imaging, respectively. (c) Double-pass transmission phase image of a HeLa cell and (d) single-shot reflection phase image of the region inside square box in (c). Figure reprinted from [71], with kind permission from OSA, © 2011 Optical Society of America.

should be remembered finally, that single-shot FF-OCT and FF-OCM show their particular advantage to cope the encountered problems and facilitate the observation of dynamics in scattering materials, too.

7.5 Summary and Outlook

Starting from traditional multiframe imaging techniques and continuing toward dual-shot and single-shot FF-OCT/FF-OCM realizations, these methods have been regarded and compared with respect to their benefits and challenges in this chapter.

The advantage of single-shot FF-OCT has been discussed especially in the context of monitoring dynamical processes. Single-shot imaging methods require extra efforts like a careful alignment and balancing of the intensities of the spatially multiplexed interferometric images. However, an improved mathematical processing facilitated by increasing computational power nowadays enables a fast registration and correction of misalignment.

The extra efforts for single-shot FF-OCT can be balanced out by the prospect of providing a valuable modification of FF-OCT suited for real-time imaging of dynamic processes without the restriction to 'freezing' the motion of specimens. Furthermore, by sophisticated imaging concepts, incorporating novel detectors like scientific CMOS or other high speed cameras into the optical setup, the trend toward a dynamic process monitoring by FF-OCT is additionally emphasized.

But also current developments in the field of programmable optics open interesting perspectives for single-shot full-field imaging. Amongst these possibilities the field of adaptive optics should be highlighted, enabled by devices as spatial light modulators or micromirror arrays in combination with computational techniques derived from the field of phase retrieval and compressed sensing. The potential of these approaches to contribute for advantages in single-shot FF-OCT as well should be envisaged in future.

Acknowledgments

The author thanks D. Stifter, S. E. Schausberger, M. Reinhardt, and B. Buchroithner for discussion and assistance. The financial support by the Federal Ministry of Economy, Family and Youth, the National Foundation for Research, Technology and Development, is gratefully acknowledged.

References

1. Abdulhalim, I. (2012). Spatial and temporal coherence effects in interference microscopy and full-field optical coherence tomography, *Ann. Phys.*, **524**, 12, pp. 787–804.
2. Akiba, M., Chan, K. P., and Tanno, N. (2003). Full-field optical coherence tomography by two-dimensional heterodyne detection with a pair of CCD cameras, *Opt. Lett.*, **28**, 10, pp. 816–818.
3. Assayag, O., Grieve, K., Devaux, B., Harms, F., Pallud, J., Chretien, F., Boccara, C., and Varlet, P. (2013). Imaging of non-tumorous and tumorous human brain tissues with full-field optical coherence tomography, *NeuroImage: Clinical*, **2**, 0, pp. 549– 57.

4. Bedrosian, E. (1963). A product theorem for Hilbert transform, **51**, pp. 868–869.
5. Bernstein, S., Bouchot, J.-L., Reinhardt, M., and Heise, B. (2013). *Generalized Analytic Signals in Image Processing: Comparison, Theory and Applications*, (TIM Birkhäuser), pages to appear.
6. Binding, J., Arous, J. B., Léger, J.-F., Gigan, S., Boccara, C., and Bourdieu, L. (2011). Brain refractive index measured in vivo with high-NA defocus-corrected full-field OCT and consequences for two-photon microscopy, *Opt. Express*, **19**, 6, pp. 4833–4847.
7. Bülow, T., Pallek, D., and Sommer, G. (2000). Riesz transform for the isotropic estimation of the local phase of moiré interferograms, in *Mustererkennung 2000, 22. DAGM-Symposium* (Springer-Verlag, London, UK), pp. 333–340.
8. Bülow, T., and Sommer, G. (2001). Hypercomplex signals: a novel extension of the analytic signal to the multidimensional case, *IEEE Trans. Signal Process.*, pp. 2844–2852.
9. Choi, W. J., Jung, S. P., Shin, J. G., Yang, D., and Lee, B. H. (2011). Characterization of wet pad surface in chemical mechanical polishing (cmp) process with full-field optical coherence tomography (ff-oct), *Opt. Express*, **19**, 14, pp. 13343–13350.
10. Choi, Y., Yang, T. D., Lee, K. J., and Choi, W. (2011). Full-field and single-shot quantitative phase microscopy using dynamic speckle illumination, *Opt. Lett.*, **36**, 13, pp. 2465–2467.
11. Davidson, M., Kaufman, K., Mazor, I., and Cohen, F. (1987). An application of interference microscopy to integrated circuit inspection and metrology, in *Microlithography Conference* (International Society for Optics and Photonics), pp. 233–249.
12. Davis, J. A., McNamara, D. E., Cottrell, D. M., and Campos, J. (2000). Image processing with the radial Hilbert transform: theory and experiments, *Opt. Lett.*, **25**, 2, pp. 99–101.
13. Dsouza, R. I., Zam, A., Subhash, H. M., Larin, K. V., and Leahy, M. (2013). In vivo microcirculation imaging of the sub surface fingertip using correlation mapping optical coherence tomography (cmOCT), *Proc. SPIE*, **8580**, pp. 85800M–85800M–5.
14. Dubois, A. (2012). Full-field optical coherence microscopy, in D. G. Liu (ed.), *Selected Topics in Optical Coherence Tomography* (Intech), pp. 3–20.
15. Dubois, A., Vabre, L., Boccara, A.-C., and Beaurepaire, E. (2002). High-resolution full-field optical coherence tomography with a Linnik microscope, *Appl. Opt.*, **41**, 4, pp. 805–812.

16. Dunsby, C., Gu, Y., and French, P. (2003). Single-shot phase-stepped wide-field coherence gated imaging, *Opt. Express*, **11**, 2, pp. 105–115.
17. Felsberg, M., and Sommer, G. (2001). The monogenic signal, *IEEE Trans. Signal Process.*, **49**, 12, pp. 3136–3144.
18. Felsberg, M., and Sommer, G. (2002). Image features based on a new approach to 2d rotation invariant quadrature filters, in A. Heyden, G. Sparr, M. Nielsen, and P. Johansen (eds.), *Computer Vision: ECCV 2002*, *LNCS*, Vol. 2350 (Springer, Berlin, Heidelberg), pp. 369–383.
19. Felsberg, M., and Sommer, G. (2003). The monogenic scale-space: a unifying approach to phase-based image processing in scale-space, *J. Math. Imaging Vision*, **21**, pp. 5–26.
20. Gabai, H., and Shaked, N. T. (2012). Dual-channel low-coherence interferometry and its application to quantitative phase imaging of fingerprints, *Opt. Express*, **20**, 24, pp. 26906–26912.
21. Gabor, D. (1946). Theory of communication, *J. Inst. Elect. Eng.*, **93**, pp. 429–457.
22. Galerne, B., Gousseau, Y., and Morel, J.-M. (2011). Random phase textures: theory and synthesis, *IEEE Trans. Image Process.* **20**, 1, pp. 257–267.
23. Granlund, G. H., and Knutsson, H. (1995). *Signal Processing for Computer Vision* (Kluwer Academic).
24. Greivenkamp, J. E. (1984). Generalized data reduction for heterodyne interferometry, *Opt. Eng.*, **23**, 4, pp. 350–352.
25. Grieve, K., Dubois, A., Simonutti, M., Paques, M., Sahel, J., Gargasson, J.-F. L., and Boccara, C. (2005). In vivo anterior segment imaging in the rat eye with high speed white light full-field optical coherence tomography, *Opt. Express*, **13**, 16, pp. 6286–6295.
26. Guo, C.-S., Rong, Z.-Y., Wang, H.-T., Wang, Y., and Cai, L. Z. (2003). Phase-shifting with computer-generated holograms written on a spatial light modulator, *Appl. Opt.*, **42**, 35, pp. 6975–6979.
27. Hahn, S. L. (1996). *Hilbert transforms in Signal Processing* (Artech House, Norwood, MA).
28. Heise, B., Schausberger, S. E., and David, S. (2013). Full field optical coherence microscopy: imaging and image processing for micro-material research applications, in M. Kawasaki (ed.), *Optical Coherence Tomography* (InTech), pp. 139–162.
29. Heise, B., Schausberger, S. E., Häuser, S., Plank, B., Salaberger, D., Leiss-Holzinger, E., and Stifter, D. (2012). Full-field optical coherence

microscopy with a sub-nanosecond supercontinuum light source for material research, *Opt. Fiber Technol.*, **18**, 5, pp. 403–410.

30. Held, S., Storath, M., Massopust, P., and Forster, B. (2010). Steerable wavelet frames based on the riesz transform, *IEEE Trans. Image Process.*, **19**, 3, pp. 653–667.

31. Hrebesh, M. S. (2012). Full-field and single-shot full-field optical coherence tomography: a novel technique for biomedical imaging applications, *Adv. Opt. Technol.*, **2012**, p. 26 page.

32. Hrebesh, M. S., Dabu, R., and Sato, M. (2009). In vivo imaging of dynamic biological specimen by real-time single-shot full-field optical coherence tomography, *Opt. Commun.*, **282**, 4, pp. 674–683.

33. Ikeda, T., Popescu, G., Dasari, R. R., and Feld, M. S. (2005). Hilbert phase microscopy for investigating fast dynamics in transparent systems, *Opt. Lett.*, **30**, 10, pp. 1165–1167.

34. Jain, M., Manzoor, M., Shukla, N., Mukherjee, S., and Nadolny, S. (2011). Modified full-field optical coherence tomography: a novel tool for rapid histology of tissues, *J. Pathol. Inform.*, **2**, 1, p. 28.

35. Jesacher, A., Harm, W., Bernet, S., and Ritsch-Marte, M. (2012). Quantitative single-shot imaging of complex objects using phase retrieval with a designed periphery, *Opt. Express*, **20**, 5, pp. 5470–5480.

36. Kanka, M., Riesenberg, R., Petruck, P., and Graulig, C. (2011). High resolution (NA=0.8) in lensless in-line holographic microscopy with glass sample carriers, *Opt. Lett.*, **36**, 18, pp. 3651–3653.

37. Khare, K., Ali, P. T. S., and Joseph, J. (2013). Single shot high resolution digital holography, *Opt. Express*, **21**, 3, pp. 2581–2591.

38. Larkin, K. G., Bone, D. J., and Oldfield, M. A. (2001). Natural demodulation of two-dimensional fringe patterns. I. General background of the spiral phase quadrature transform, *J. Opt. Soc. Am. A*, **18**, 8, pp. 1862–1870.

39. Latrive, A., and Boccara, A. C. (2011). In vivo and in situ cellular imaging full-field optical coherence tomography with a rigid endoscopic probe, *Biomed. Opt. Express*, **2**, 10, pp. 2897–2904.

40. Lu, S.-H., Chang, C.-J., and Kao, C.-F. (2013). Full-field optical coherence tomography using immersion Mirau interference microscope, *Appl. Opt.*, **52**, 18, pp. 4400–4403.

41. Malacara, D., Servin, M., and Malacara, Z. (2005). *Interferogram Analysis for Optical Testing* (CRC Press, Taylor & Francis).

42. Maurer, C., Jesacher, A., Bernet, S., and Ritsch-Marte, M. (2010). What spatial light modulators can do for optical microscopy, *Laser Photon. Rev.*, **5**, pp. 1863–8899.

43. Mehta, D. S., Inam, M., Prakash, J., and Biradar, A. M. (2013). Liquid-crystal phase-shifting lateral shearing interferometer with improved fringe contrast for 3D surface profilometry, *Appl. Opt.* **52**, 25, pp. 6119–6125.
44. Na, J., Choi, W. J., Choi, E. S., Ryu, S. Y., and Lee, B. H. (2008). Image restoration method based on Hilbert transform for full-field optical coherence tomography, *Appl. Opt.*, **47**, 3, pp. 459–466.
45. Nguyen, T.-U., Pierce, M. C., Higgins, L., and Tkaczyk, T. S. (2013). Snapshot 3D optical coherence tomography system using image mapping spectrometry, *Opt. Express*, **21**, 11, pp. 13758–13772.
46. Nugroho, W., Ito, Y., Hrebesh, M., and Sato, M. (2011). Basic characteristics of interference image obtained using spatially phase-modulated mirror array, *Opt. Rev.*, **18**, 2, pp. 247–252.
47. Popescu, G., Ikeda, T., Dasari, R. R., and Feld, M. S. (2006). Diffraction phase microscopy for quantifying cell structure and dynamics, *Opt. Lett.*, **31**, 6, pp. 775–777.
48. Reinhardt, M. (2012). *Wavelets in Bildgebung und Bildverarbeitung, MA Thesis* (TU Bergakademie Freiberg, Germany).
49. Roy, M., Cherel, L., and Sheppard, C. J. R. (1999). Geometric phase coherence probe microscope for surface profiling, *Proc. SPIE*, **3749**, pp. 462–463.
50. Sacchet, D. (2010). *Tomographie par coherence optique plein champ lineaire et non lineaire, These de Doctorat* (University Paris-Sud, France).
51. Sacchet, D., Brzezinski, M., Moreau, J., Georges, P., and Dubois, A. (2010). Motion artifact suppression in full-field optical coherence tomography, *Appl. Opt.*, **49**, 9, pp. 1480–1488.
52. Salbut, L., Tomczewski, S., and Pakula, A. (2014). VIS-NIR full-field low coherence interferometer for surface layers non-destructive testing and defects detection, in W. Osten (ed.), *Fringe 2013* (Springer, Berlin, Heidelberg), pp. 925–928.
53. Sato, M., Nagata, T., Niizuma, T., Neagu, L., Dabu, R., and Watanabe, Y. (2007). Quadrature fringes wide-field optical coherence tomography and its applications to biological tissues, *Opt. Commun.*, **271**, 2, pp. 573–580.
54. Schausberger, S. E. (2013). *Full Field Optical Coherence Microscopy with Fourier Plane Filtering*, PhD thesis, Johannes Kepler University, Linz, Austria.
55. Schausberger, S. E., Heise, B., Bernstein, S., and Stifter, D. (2012). Full-field optical coherence microscopy with Riesz transform-based demodulation for dynamic imaging, *Opt. Lett.*, **37**, 23, pp. 4937–4939.

56. Schausberger, S. E., Heise, B., Maurer, C., Bernet, S., Ritsch-Marte, M., and Stifter, D. (2010). Flexible contrast for low-coherence interference microscopy by fourier-plane filtering with a spatial light modulator, *Opt. Lett.*, **35**, 24, pp. 4154–4156.
57. Schausberger, S. E., Heise, B., and Stifter, D. (2013). Dynamic imaging by full-field optical coherence microscopy with a sCMOS detector and Riesz transform-based demodulation, *Proc. SPIE*, **8802**, pp. 88020A–88020A–5.
58. Schlager, V., Schausberger, S. E., Stifter, D., and Heise, B. (2011). Coherence probe microscopy imaging and analysis for fiber-reinforced polymers, in A. Heyden and F. Kahl (eds.), *Image Analysis, LNCS*, Vol. 6688 (Springer, Berlin, Heidelberg), pp. 424–434.
59. Seelamantula, C. S., Pavillon, N., Depeursinge, C., and Unser, M. (2012). Local demodulation of holograms using the Riesz transform with application to microscopy, *J. Opt. Soc. Am. A*, **29**, 10, pp. 2118–2129.
60. Sheppard, C. J. R. (2007). Low-coherence interference microscopy, in *Optical Imaging and Microscopy, Optical Sciences*, Vol. 87 (Springer Berlin Heidelberg), pp. 329–345.
61. Slabý, T., Kolman, P., Dostál, Z., Antoš, M., Lošťák, M., and Chmelík, R. (2013). Off-axis setup taking full advantage of incoherent illumination in coherence-controlled holographic microscope, *Opt. Express*, **21**, 12, pp. 14747–14762.
62. Song, Y., Chen, Y., Wang, J., Sun, N., and He, A. (2012). Four-step spatial phase-shifting shearing interferometry from moiré configuration by triple gratings, *Opt. Lett.*, **37**, 11, pp. 1922–1924.
63. Steiger, R., Bernet, S., and Ritsch-Marte, M. (2012). SLM-based off-axis Fourier filtering in microscopy with white light illumination, *Opt. Express*, **20**, 14, pp. 15377–15384.
64. Stein, E. M. (1970). *Singular Integrals and Differentiability Properties of Functions* (Princeton University Press).
65. Stifter, D. (2014). Optical coherence tomography: a novel technique for the characterisation of micro-parts and -structures, in Yi Qin, (ed.), *Micromanufacturing Engineering and Technology*, 2nd ed. (William Andrew Applied Science, Norwich).
66. Trantum, J. R., Eagleton, Z. E., Patil, C. A., Tucker-Schwartz, J. M., Baglia, M. L., Skala, M. C., and Haselton, F. R. (2013). Cross-sectional tracking of particle motion in evaporating drops: flow fields and interfacial accumulation, *Langmuir*, **29**, 21, pp. 6221–6231.

67. Unser, M., Sage, D., and Van De Ville, D. (2009). Multiresolution monogenic signal analysis using the Riesz-Laplace wavelet transform, *IEEE Trans. Image Process.*, **18**, 11, pp. 2402–2418.
68. Watanabe, Y., and Sato, M. (2008). Three-dimensional wide-field optical coherence tomography using an ultrahigh-speed CMOS camera, *Opt. Commun.*, **281**, 7, pp. 1889–1895.
69. Widjanarko, T., Huntley, J. M., and Ruiz, P. D. (2012). Single-shot profilometry of rough surfaces using hyperspectral interferometry, *Opt. Lett.*, **37**, 3, pp. 350–352.
70. Yang, Y.-L., Ding, Z.-H., Wang, K., Wu, L., and Wu, L. (2010). Full-field optical coherence tomography by achromatic phase shifting with a rotating half-wave plate, *J. Opt.*, **12**, 3, p. 035301.
71. Yaqoob, Z., Yamauchi, T., Choi, W., Fu, D., Dasari, R. R., and Feld, M. S. (2011). Single-shot full-field reflection phase microscopy, *Opt. Express*, **19**, 8, pp. 7587–7595.
72. Zuo, C., Chen, Q., Qu, W., and Asundi, A. (2013). Noninterferometric single-shot quantitative phase microscopy, *Opt. Lett.*, **38**, 18, pp. 3538–3541.

Chapter 8

Frequency Domain Full-Field Optical Coherence Tomography

Rainer A. Leitgeb,[a,b] Abhishek Kumar,[a,b] and Wolfgang Drexler[a]

[a]*Center for Medical Physics and Biomedical Engineering, Medical University Vienna, Waehringer Guertel 18-20/4L, A-1090 Vienna, Austria*
[b]*Christian Doppler Laboratory for Innovative Optical Imaging and its Translation to Medicine, Medical University Vienna, Vienna, Austria*
rainer.leitgeb@meduniwien.ac.at, wolfgang.drexler@meduniwien.ac.at, abhishek.kumar@meduniwien.ac.at

8.1 Introduction

In Fourier domain (FD) optical coherence tomography (OCT), the interference pattern is recorded as a function of light frequency which can either be encoded spatially with the help of a diffraction grating and a camera or be encoded in time using a sweeping frequency laser source. FD-OCT systems typically have a sensitivity advantage of about 20–30 dB over the conventional time domain (TD) OCT systems [1, 2]. The sensitivity advantage of FD-OCT and possibility to enhance imaging speed by parallelization has motivated researchers to develop line field (LF) FD-OCT and swept source (SS) full-field (FF) OCT imaging techniques. The LF-FD-OCT setup typically consists of a Michelson interferometer

---

Handbook of Full-Field Optical Coherence Microscopy: Technology and Applications
Edited by Arnaud Dubois
Copyright © 2016 Pan Stanford Publishing Pte. Ltd.
ISBN 978-981-4669-16-0 (Hardcover), 978-981-4669-17-7 (eBook)
www.panstanford.com

with an anamorphic illumination system using a broadband light source to produce a line illumination on the sample and the reference mirror. The light reflected from the sample and the reference arm is combined and dispersed using a diffraction grating, and subsequently detected by a 2D charge-coupled device (CCD)/complementary metal-oxide semiconductor (CMOS) camera. Thus, a spectrally resolved interference signal of each point on the line illuminating the sample is obtained in parallel. A simple 1D fast Fourier transformation (FFT) along the spectral dimension of the data (after $\lambda \rightarrow k$ interpolation) yields an XZ cross-sectional image. A working LF-FD-OCT system was first demonstrated by Zuluaga and Richard-Kortum using a technical sample [3]. This technique has been further developed by Grajciar et al. [4] and Nakamura et al. [5] for in vivo ophthalmic imaging. However, this technique still requires scanning along one lateral direction in order to acquire 3D volume image. On the other hand, SS-FF-OCT does not require scanning of the sample in any direction. The FF-SS-OCT setup usually consists of a free-space Michelson interferometer and involves area illumination with a sweeping wavelength laser source on the sample and the reference mirror. The detection of the generated interference signal is performed by a high-speed 2D camera. The 2D interference signal is recorded for each wavelength sequentially in time, and 1D FFT (after $\lambda \rightarrow k$ interpolation) is performed along the spectral dimension to get a 3D volume image with depth information of the sample.

The interference signal with varying wavelength can also be recorded in the Fresnel or the Fourier plane instead of the image plane, and digital holographic (DH) reconstruction can then be performed on the recorded data to get OCT equivalent images [6–9]. In early 2000, M. K. Kim demonstrated 3D imaging of a millimeter sized biological sample using wavelength scanning wide field digital interference holography [7]. A wavelength sweeping ring dye laser ($\lambda_0 = 601.7$ nm) was used with a narrow sweeping range of 3.08 nm. Thus, a low axial resolution of only 100 μm could be achieved [7].

Povazay et al. (2006) demonstrated an improved FF-SS-OCT system using a broad band Ti-sapphire laser ($\lambda_0 = 800$ nm, $\Delta\lambda = 100$ nm) with an acousto-optic frequency tunable element (linewidth <0.4 nm) [10]. The axial and lateral resolutions achieved were 3 µm and 4 µm respectively. The detected sensitivity was ~83 dB with average power per A-scan of about 20µW. An acoustic mode-mixer was used to average out the different speckle patterns formed due to different transversal mode of the illuminating multimode fiber during the exposure time. Due to the speed limitations only ex vivo images of a biological sample (fruit fly) were demonstrated [10]. Later in 2010, Bonin et al. demonstrated in vivo retinal imaging using FF-SS-OCT system equipped with ultrahigh speed CMOS camera (Y4, Redlake/IDT, USA), and a SS laser form Superlum (Broadsweeper BS840-01, $\lambda_0 = 850$ nm, $\Delta\lambda = 50$ nm) with an output power of 5mW [6]. An A-scan rate of 1.5 million A-lines/sec was shown for a sample volume of size 640 (x) × 24 (y) × 512 (z) pixels. The detected sensitivity was only ~72 dB with an equivalent power on sample limited to 92µW/pixel. It was noted that at least a 100 Hz sweep rate was required for in vivo imaging in order to avoid any significant sample motion blurring [6].

In this chapter we discuss the general concept of FD-FF-OCT, numerical focusing techniques used to achieve depth invariant lateral resolution, and challenges and a future outlook.

8.2 General Concept

8.2.1 *Setup and System Components*

Employing SS technology allows for a very compact OCT design, similar to TD-FF-OCT, but with the important sensitivity advantage of FD-OCT. The complexity of the system is shifted to the spectral tuning of the laser source. The source delivers ideally a spectral output over time that is linear in wavenumber k, $I(k(t))$ according to $k(t_n) = (k - \Delta k/2) + \delta k n$. Here, k is the central wavenumber, Δk is the spectral tuning bandwidth, δk is the instantaneous output linewidth $t_n = t + n\delta t$, with n being an integer and $1/\delta t$ being the sampling frequency. Fast sweeping lasers usually employ mechanical tuning elements that are driven with a sinusoidal signal. In this case the temporal dependence of the wavenumber is highly nonlinear, leading to a chirped interference fringe signal. The remapping to a

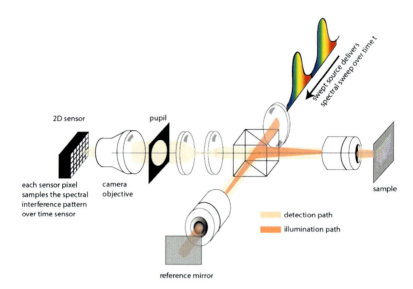

Figure 8.1 SS-FF-OCT setup employing microscope objectives in a Linnik-type interferometer. The illumination path allows for exposing a large sample area to a collimated beam. The detection path images the sample in the focus of the objective lens onto the sensor plane. The pupil in the Fourier plane defines the spatial frequency subtend and suppresses aliasing.

linear signal in k can be performed either by electronically sampling the signal at defined intervals δk_S or by recording a reference interference signal and resampling in postprocessing. Figure 8.1 depicts a schematic SS-FF-OCT system setup. The spectrum is then sampled in time at $N = \Delta k/\delta k$ points, within time $T = N\delta t$ at each pixel of a 2D sensor. The two coordinates of the 2D sensor then represent the lateral sample coordinates. The number of pixels in each direction defines spatial sampling. The distance between the sampling points is given by the magnification factor M from the sample to the sensor plane and depends on the optical setup. Alternatively, the sensor could be placed in the Fresnel domain, as opposed to the sample image plane. The reconstruction would then be performed similar to holography by applying Fresnel backpropagation or by the angular spectrum method. The latter approach, termed "holoscopy," will be explained in detail elsewhere in this book.

For most commercially available SSs, the A-scan time translates not directly to A-scan rate, via $1/T$. As mentioned before, fast tuning sources use sinusoidal driving signals, leading to two sweep directions during the driving period. The optimal laser performance is in general different for both directions, which is why some producers optimize the laser output only for a single sweep direction and suppress the output for the other direction. In that case, the duty cycle would be only 50%, or sometimes slightly more than 50% depending on the tuning function. The actual A-scan rate in this case is equal to the laser sweeping frequency f. In case both sweep directions are used, one achieves an A-scan rate of $2f$. Recent laser developments demonstrated the possibility to fully electronically control the laser sweep, avoiding mechanically frequency filtering, and leading to linear in k sweep with close to 100% duty cycle. Linear tuning reduces significantly the signal processing effort, as no interpolation of the recorded interference pattern during k-remapping is needed [11–15].

Rapid commercially available SSs already achieve sweep rates of 100 kHz and more. The concept of FD mode locking is particularly apt for high sweeping frequencies as high output power can be maintained when increasing the tuning rate. The laser resonator is extended by a long fiber, which is chosen such that the round trip time of the light matches the tuning period of the cavity Fabry–Perot filter. Hence each wavelength experiences the optimal resonance condition, leading to synchronous amplification of the full laser spectrum. For SS-FF-OCT however, the data recording speed is limited by the frame rate of the employed 2D sensor, which samples the spectral interference pattern over time. Affordable fast CMOS sensors achieve full frame rates of about 100 Hz. The speed can be improved by reducing the number of active detector pixels. Alternatively one could use line sensors that allow for several hundreds of kHz frame rate. Assuming now N sampling points for each recorded spectrum and a time τ as minimal exposure time of the sensor will then result in a recording time of a volume of $T = \tau N$. With a frame rate of 100 Hz or an exposure time of 10 ms, 1000 sampling points, we have 10 sec for a full volume. The sweep rate of the laser would then be 1/10 Hz, which is very low as compared to the fast tuning lasers mentioned before. In fact, point-scanning systems might easily achieve volume rates in the order of hertz. Recent work demonstrated several tens of Hz volume rate imaging with FDML lasers. Hence the advantage of Swept Source Full Field OCT (SS-FF-OCT) is currently not the speed but potentially the higher sensitivity due to the parallel recording. If one uses a line sensor, the recording speed could be faster. With a 100 kHz frame rate, the recording time of a single tomogram would be 100 Hz. But in order to create a typical 500(x)x500(y)x500(k) volume data stack, we need 500 samples in scanning direction. Then we come close to 1 Hz volume rate. In the following section we will, however, concentrate on a true SS-FF-OCT configuration, as it has the attractive capability to record full sample volumes without the need of mechanical scanning. This has the important feature of maintaining high lateral phase stability and correlation, which is advantageous for digital wavefront correction techniques (see Section 8.3). A more detailed discussion on its intrinsic sensitivity advantage is provided in Section 8.2.3.

8.2.2 Standard SS-OCT Signal Processing

As mentioned before, the 2D sensor records the spectral interference pattern over time for each spatial location. Let us assume that the spectra are recorded as function of wavelength, that is, $I(x, z, k)$. To reconstruct the sample structure a simple Fourier transform along k for each pixel location is needed. Assuming, for example, a simple layer model for the sample with reflectivities $R_{S,i}$ and a reflector with reflectivity R_r the spectral interference pattern at a given position (x, y) can be written as

$$I(k) = S(k) \left(R_r + \sum_i R_{S,i} \right) + 2S(k) \sum_{i>j} R_{S,i} R_{S,j} \cos\left[2k \left(z_{S,i} - z_{S,j}\right)\right]$$
$$+ 2S(k) \sum_i R_{S,i} R_r \cos\left[2k(z_{S,i} - z_r)\right], \quad (8.1)$$

with $S(k)$ the spectral output power of the light source, and $z_{S,i}$ and z_r the distances to the ith sample interface and to the reference mirror, respectively. Fourier transform of the expression for $I(k)$

yields

$$\tilde{I}(z) \equiv FT\{I(k)\} = \gamma\left(\frac{z}{c}\right) S_0 \left(R_r + \sum_i R_{S,i}\right)$$

$$+ \gamma\left(\frac{z}{c}\right) S_0 \otimes \left[\sum_{i>j} R_{S,i} R_{S,j} \delta\left(z_{S,i} - z_{S,j}\right)\right.$$

$$\left. + \sum_i R_r R_{S,i} \delta\left(z_r - z_{S,i}\right)\right] + c.c., \qquad (8.2)$$

where z is the depth coordinate, $\delta(.)$ the Kronecker delta function, S is the total output power of the source, and $\gamma\left(\frac{z}{c}\right)$ the temporal degree of coherence, with c being the speed of light. The temporal degree of coherence, which we will further on simply call the coherence function, is related to the spectral power density as $\gamma\left(\frac{z}{c}\right) = FT\{S(k)\}/S_0$. The first term in Eq. 8.2 is the direct current (DC) term, the second term, called the autocorrelation term, is located close to the DC and can be seen as coherent noise. Only the last term contains the actual sample structure. Since the spectrum $I(k)$ is real valued, Fourier transform results in a symmetric function $\tilde{I}(z)$ with complex conjugate (c.c.) mirror terms. By proper adjustment of the reference mirror position z_r the sample structure can be offset from the DC and autocorrelation terms, and mixing with the c.c. terms can be avoided. The Fourier analysis has to be repeated for each in a parallel recorded spectrum at a lateral position (x, y). The reconstruction steps are displayed in Fig. 8.2. The axial resolution of OCT is defined by the width of the round trip coherence function. Given the Fourier relation to the spectral power density, the axial resolution is inversely proportional to the spectral bandwidth of the source. In the case of Gaussian-shaped spectra, the axial resolution becomes $\delta z = 0.44 \lambda_0^2 / \Delta \lambda$, with $\Delta \lambda$ being the full-width at half-maximum of the source spectrum and λ_0 the central wavelength. If the spectrum is linearly sampled in $k = 2\pi/\lambda$ by N points, the spectral resolution is given as $\delta k_S = \Delta k/N$, with Δk being the recorded full spectral width, which is covered by one spectral sweep of the source. The maximum depth range is then calculated from the Nyquist condition $2\delta k_s \Delta z_{max} = \pi$ as $\Delta z_{max} = \pi N/(2\Delta k)$. In FD-OCT we need to consider an additional important feature that is related

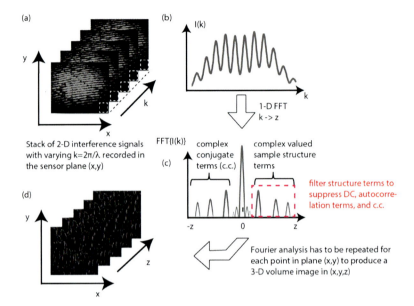

Figure 8.2 SS-FF-OCT data processing. (a) Recorded stack with sensor plane (x, y) and wavenumber k spectrum recorded in time for each sensor pixel. (b) Single spectrum $I(k)$ at indicated pixel in (a) (dashed box). (c) Result after FFT of (b); the structure terms are obtained by simple cropping (red dashed box). (d) Taking the FFT for each spectrum at each pixel in (x, y) results in a 3D image in (x, y, z).

to the finite sampling or resolution of the spectrum—the sensitivity roll-off with depth. In SS-OCT this roll-off is determined by the instantaneous linewidth of the source output δk, which defines the physical resolution of the spectral sampling. In mathematical terms, the sampling of the spectrum can be seen as convolution of the full spectrum with the instantaneous spectral line shape, of width δk. After Fourier transform the convolution becomes a multiplication of $\tilde{I}(z)$ with the coherence function of the instantaneous output spectrum. A very important parameter of the performance of swept-sources is therefore the width of this coherence function, which often is defined as the width at which the coherence function decreased by -6 dB.

Looking back to Eq. 8.2 we observe that the result of the Fourier transform along k is a complex valued function in z. This allows extracting the structural phase as function of (x, y, z) at any sample depth as $\varphi(x, y, z) = FT_{k \to z}\{I(x, y, k)\}$ with $\angle$ being the argument of a complex number. For scanning systems the phases at different lateral positions may be uncorrelated due to motion or mechanical jitter. This is different in the case of a parallel system, where a clear spatial phase relation is maintained. This intrinsic phase stability can, for example, be used for quantitative FF phase microscopy [16] and is the basis for the digital refocusing algorithm explained in Chapter 3.

8.2.3 Sensitivity

One of the important advantages of FD-OCT is the intrinsic sensitivity advantage as compared to TD-OCT. This has been described in 2003 by Leitgeb et al. on the basis of spectrometer-based FD-OCT [1] but has soon afterward been exemplified as well for SS-OCT [2]. In the following section we give a short explanation for this advantage and describe parameters to quantify OCT system performance.

Let us first define sensitivity and the signal-to-noise ratio (SNR). In OCT we define the SNR as the quotient of the average squared signal divided by the noise variance, that is, $\text{SNR} = <I^2>/\sigma_{\text{noise}}^2$. The angle brackets define the ensemble average, which in the case of ergodic signals, is equal to the time average. The sensitivity Σ is defined as the inverse of the smallest sample reflectivity, for which the SNR equals 1, that is, $\Sigma = 1/R_{s,\min}$.

Each OCT modality has its specific flexibilities and limitations. In TD-OCT we observe the interference signal at a single axial location or delay at a time. Hence the optics can be dynamically adapted to this location by dynamic focusing or by managing depth-dependent sample dispersion [17]. This flexibility comes, however, at the price of lower detection sensitivity: The detection system records at each axial location the full spectral bandwidth and power, whereas the actual OCT signal is present only within the temporal coherence gate. Thus only a small part of the deposited light energy actually contributes to the OCT signal, which fraction is given by the ratio of the round trip coherence length to the full-depth scanning range. In FD-OCT light backscattered from all depth regions interferes with

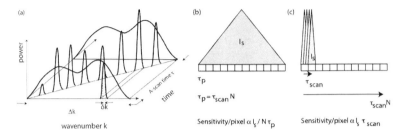

Figure 8.3 (a) Total energy in the case of SS-OCT versus sb-OCT. (b) Parallel detection during exposure time τ_P. (c) Scanning system with dwell time per sampling point τ_{scan}. A full scan of N pixels takes $N\tau_{scan}$ seconds. In contrast, for parallel OCT the time $N\tau_{scan}$ can be used for collecting light at all pixels in parallel.

the reference beam at the detector and contributes to the spectral interference pattern, which is recorded in spectrometer-based (sb) OCT across all detector pixels. Fourier analysis of the spectral interference pattern obtains the full axial structure in parallel. Thus, apart from the DC term, the recorded signal contributes in its entirety to the axial structure. This immediately translates into higher detection sensitivity as compared to the case of TD-OCT. The sensitivity advantage is also present in SS-OCT, where the interference pattern is recorded in time. As seen from Fig. 8.3 sb-OCT systems integrate the full spectral width Δk during exposure time τ, whereas in SS-OCT the detector is only exposed to the instantaneous line width δk during time $\tau \delta k/\Delta k$. If the detection is shot noise limited, the SNR is directly proportional to the number of recorded photo-electrons from the sample, that is, $SNR \propto I_s\tau$, with I_s being the power on the sample. A simplified comparison would lead then to $SNR_{sb\text{-}OCT}/SNR_{SS\text{-}OCT} = \Delta k/\delta k$, which seems like a sensitivity disadvantage for SS-OCT. However, we need to consider the actual total energy at the sample (Fig. 8.3a) [18]: The total energy during exposure τ is in fact much larger for sb-OCT than for SS-OCT assuming the same spectral power density of the illumination. The full spectrum contributes at each time instant. The energy at the sample for both modalities can in fact be made equal if the SS-OCT power per instantaneous linewidth δk is increased by a factor of $\delta k/\Delta k$. This cancels the factor of $\Delta k/\delta k$ in the expression for the SNR quotient, leaving finally the enhanced sensitivities of both FD-OCT modalities equal.

In addition to the inherited FD-OCT sensitivity advantage SS-FF-OCT exhibits the advantage of parallel detection as compared to a scanning OCT system. Assume first the same power I_S incident on the sample. In FF-OCT this power spreads over all N parallel pixels, leading to a power per pixel or lateral position of I_S/N (Fig. 8.3b,c). For point-scanning OCT, this power is present at each lateral position; however, the dwell time at each position is smaller than in FF-OCT. For a recording time of a full scan of all N lateral pixels or scanning locations, OCT needs N times the dwell time τ_{scan}. This time is effectively used in FF-OCT for collecting photoelectrons, thus compensating the spread of power across all parallel pixels. Thus, the sensitivity is actually the same in both cases assuming the same power at the sample. The parallel OCT sensitivity advantage comes in when considering the allowed power at the sample according to laser safety regulations [19]. Since in parallel OCT one has an extended light source that illuminates the sample, the allowed power is in general higher as for point-scanning OCT, even in the case of retinal imaging. Higher illumination power directly results in the shot noise limit in higher sensitivity.

8.3 Digital Refocusing

FF imaging enables collection of out-of-focus light beyond the Rayleigh range as it does not introduce confocal depth gating. If the information about the amplitude and phase of the out-of-focus signal is accessible, then numerical holographic reconstruction can be applied to get the focused image. Thus, a depth-invariant lateral resolution can be achieved over the whole imaging range. Hillmann et al. have proposed *holoscopy*, which combines wide-field digital lensless holography and SS-OCT to allow focusing beyond the Rayleigh range [8]. The optical setup consists of a simple lensless free-space Michelson interferometer with a convex mirror in the reference arm to produce a spherical reference wave. The holographic data was acquired with a sweeping laser source. For 3D image reconstruction, the recorded holographic image at each wavenumber was multiplied with the conjugate of the reference wave and propagated to a specific distance using the angular spectrum approach. Finally, the 1D FFT ($k \to z$) was calculated to yield an image with depth information. This process was repeated for different reconstruction distances to focus in different depths in the sample. The focused region in each reconstruction was filtered and stitched together to get an image with depth-invariant lateral resolution. An improvement of over 30 Rayleigh ranges was demonstrated using an iron oxide phantom sample [8]. Hillmann et al. further proposed an efficient holoscopy reconstruction method to obtain focused 3D volume images in a single step [9]. It involves interpolating the recorded 3D holographic data on a nonequidistance grid in a 3D spatial frequency space, similar to inverse scattering reconstruction called inverse synthetic aperture microscopy (ISAM) proposed for FF-SS-OCT [20] and then calculating the 3D FFT to get an OCT-equivalent image of the sample. However, unlike ISAM, this method does not calculate the regularized pseudo inverse of the forward-propagating kernel. Instead, the recorded object field is numerically propagated back into the sample volume. A decrease by a factor of 15 in the processing time was demonstrated for an imaging numerical aperture (NA) of 0.14 [9].

ISAM and holoscopy reconstruction provide 3D volume images focused over the whole imaging range. But, these reconstruction techniques work well only when the sample has a uniform refractive index, the imaging NA is low, and defocus is the major aberration [8, 9, 20]. However, at higher NA, the higher-order aberrations such as spherical, astigmatism, coma, etc., can affect the image quality. Also, proper holographic image reconstruction becomes difficult. A computational adaptive optics technique has been suggested to reduce higher-order aberration in the case of a phase-stable point-scanning FD-OCT system [21]. However, this technique is based on an optimization method that is iterative in nature and increases the computation cost. Numerical complexity increases if the sample is highly inhomogeneous, causing variable aberration with depth [21]. Kumar et al. (2013) proposed a digital aberration correction method, which is in fact a digital equivalent of a Hartmann sensor, to deal with variable higher-order aberrations in FF-SS-OCT [22]. In this method, a 2D FFT is performed on the aberrated image field corresponding to a layer in the sample to get to the spatial

Fourier plane, and then the Fourier data is segmented into $K \times K$ small subapertures. Two-dimensional IFFTs on these segmented Fourier data yield images that are shifted with respect to each other due to aberrations. Cross-correlating the intensity of these images with respect to a reference image, which is formed by the central subaperture, gives shift data from which local slope information of the wavefront error can be calculated as

$$s_{x_{p,q}} = \frac{\partial \phi_p}{\partial x} - \frac{\partial \phi_q}{\partial x} = \frac{\Delta m \pi}{M} \quad \text{and} \quad s_{y_{p,q}} = \frac{\partial \phi_p}{\partial y} - \frac{\partial \phi_q}{\partial y} = \frac{\Delta n \pi}{M}, \tag{8.3}$$

where $s_{x_{p,q}}$ and $s_{y_{p,q}}$ are x and y components of the relative slope between the pth and qth subaperture, Δm and Δn are the relative shifts in terms of pixels in the x and y directions, respectively, and M is the size of the image array in pixels. The above equation is derived assuming that in each subaperture the phase error can be approximated by a first-order Taylor approximation, and we can see that it is independent of any system parameters. Knowing the slope information, the wavefront error can be calculated using a suitable basis function such as Zernike's or Taylor monomials. Phase correction can then be applied digitally on the whole Fourier data and an aberration-free image is finally obtained in a single step by calculating the 2D IFFT. The schematic of the algorithm is shown in Fig. 8.4. This method assumes that the data is uniformly distributed in the spatial Frequency plane, that is, the sample gives

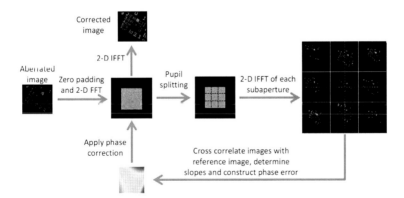

Figure 8.4 Schematic of the subaperture correlation-based phase correction method.

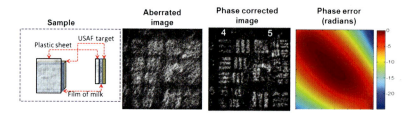

Figure 8.5 Result of applying subaperture-based digital adaptive optics phase correction to an aberrated en face image of an RTT. The resolution improvement is clearly visible in the image after phase correction.

diffuse reflection on illumination, and also the spatial coherence is maintained over the field of view. This condition is usually fulfilled when imaging tissue with a spatially coherent laser light source.

The proof of the principle of the subaperture-based digital adaptive optics was shown using an en face OCT image of a technical sample, which consisted of a layer of a nonuniform plastic sheet, a thin film of dry milk, and a resolution test target (RTT). The plastic sheet was used to introduce random aberrations, whereas the film of milk produced scattering. Figure 8.5 shows the original aberrated en face image of the RTT layer, showing the IV and V group elements, obtained by the FF-SS-OCT system and the image obtained after phase correction using 5×5 subapertures. We can clearly appreciate the improvement in resolution of the image after phase correction where we can resolve horizontal bars up to element (5, 4). This corresponds to the improvement in resolution by a factor of 2 over the aberrated image where the highest resolvable element is (4, 3). The improvement in resolution is calculated using the relation $2^{\Delta m/6}$, where Δm is the improvement in elements.

For demonstrating the method on a biological sample, the defocus correction using just two nonoverlapping subapertures was applied to the 3D volume image of a grape [22]. An extended depth of field was achieved using the digital defocus correction. The first layer of the grape was at the focal plane of the microscope objective (NA = 0.15). The layers that are out of the depth of focus of 130 μm (assuming a refractive index of 1.5) get blurred. Figure 8.6a shows a tomogram of the grape sample with an arrow indicating a layer at the depth of 424.8 μm from the focal plane. Figure 8.6c shows

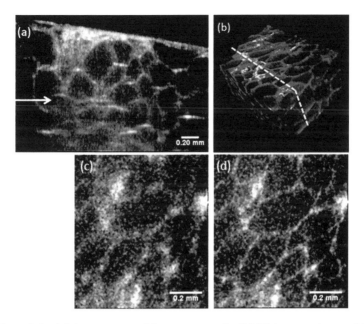

Figure 8.6 (a) A tomogram of the grape sample, (b) 3D image volume with dotted line showing location of the tomogram shown in (a), (c) en face image obtained at the depth of 424.8 μm in the grape sample indicated by arrow in (a), and (d) digitally focused image of (c).

an en face view of that layer and Fig. 8.6d shows the defocus-corrected image, where we can appreciate the improvement in lateral resolution as the cell boundaries are now clearly visible.

8.4 Outlook

A critical issue in SS-FF-OCT is the inherent problem of cross-talk or speckle noise caused by multiple scattering. This multiple scattering cross-talk is decreased in TDFF-OCT due to the tight coherent gate that is scanned along the depth. SS-FF-OCT on the other hand collects the light from within a large coherence range at each time instant. This makes it highly sensitive to multiple scattering, which is why this method might be optimal for the assessment of low-scattering tissue, such as the retina and the cornea, or for microscopic phase samples.

The unique possibility of FF-OCT to reconstruct a fully focused and aberration-corrected image is of particular interest in retinal imaging but also for high-resolution microscopy. The digital refocusing approach can of course be applied to any complex-valued FF-OCT data that preserves the lateral phase correlation.

The currently tightest bottleneck of SS-FF-OCT is certainly the available and affordable high-speed sensor technology. To be competitive with ultrahigh-speed point-scanning OCT systems [23], there is a need for fast 2D sensors. For example, a volume rate of 10 Hz would need a full frame rate of 5–10 kHz, depending on the number of spectral sampling points. Such rates are already achieved with CMOS sensors when reducing the number of active pixels. Ultrahigh-speed sensors achieve already much faster frame rates but at a price that reduces its their commercial potential, at least today [6]. Hence there is hope for SS-FF-OCT systems in that this bottleneck might be overcome in the near future by new silica-based sensor design. For longer wavelengths InGaAs sensors are needed, which currently offer much lower scan rates than CMOS sensors based on silicon.

Another technical limitation today is the low output power of available SSs. To exploit the sensitivity advantage of parallel OCT, the full power according to the maximum permissible sample exposure should be applied. For this task it would be desirable to have sources with 50 mW output power and more.

The rise and fall of FF-OCT is therefore strongly dependent on the technological developments of fast sensors and powerful light sources in the next few years. Given the rapid development of OCT technology during the last two decades there is enough reason to believe that parallel SS-OCT with its intrinsic advantages and the attractive potential of digital adaptive imaging will play an important role in the near future.

Acknowledgments

This work is supported by the Medical University of Vienna, the European projects FAMOS (FP7 ICT 317744) and FUN OCT (FP7 HEALTH 201880), the Macular Vision Research Foundation (MVRF,

USA), the Austrian Science Fund (FWF) project number S10510-N20, and the Christian Doppler Society (Christian Doppler Laboratory "Laser Development and Their Application in Medicine").

References

1. Leitgeb, R., Hitzenberger, C., and Fercher, A. (2003). Performance of fourier domain vs. time domain optical coherence tomography. *Opt. Express*, **11**, pp. 889–894.
2. Choma, M., Sarunic, M., Yang, C., and Izatt, J. (2003). Sensitivity advantage of swept source and Fourier domain optical coherence tomography. *Opt. Express*, **11**, pp. 2183–2189.
3. Zuluaga, A. F., and Richards-Kortum, R. (1999). Spatially resolved spectral interferometry for determination of subsurface structure. *Opt. Lett.*, **24**, pp. 519–521.
4. Grajciar, B., Pircher, M., Fercher, A., and Leitgeb, R. (2005). Parallel Fourier domain optical coherence tomography for in vivo measurement of the human eye. *Opt. Express*, **13**, pp. 1131–1137.
5. Nakamura, Y., Makita, S., Yamanari, M., Itoh, M., Yatagai, T., and Yasuno, Y. (2007). High-speed three-dimensional human retinal imaging by line-field spectraldomain optical coherence tomography. *Opt. Express*, **15**, pp. 7103–7116.
6. Bonin, T., Franke, G., Hagen-Eggert, M., Koch, P., and Hüttmann, G. (2010). In vivo Fourier-domain full-field OCT of the human retina with 1.5 million A-lines/s. *Opt. Lett.*, **35**, pp. 3432–3434.
7. Kim, M.-K. (2000). Tomographic three-dimensional imaging of a biological specimen using wavelength-scanning digital interference holography. *Opt. Express*, 7, pp. 305–310.
8. Hillmann, D., Lührs, C., Bonin, T., Koch, P., and Hüttmann, G. (2011). Holoscopy: holographic optical coherence tomography. *Opt. Lett.*, **36**, pp. 2390–2392.
9. Hillmann, D., Franke, G., Lührs, C., Koch, P., and Hüttmann, G. (2012). Efficient holoscopy image reconstruction. *Opt. Express*, **20**, pp. 21247–21263.
10. Povazay, B., Unterhuber, A., Hermann, B., Sattmann, H., Arthaber, H., and Drexler, W. (2006). Full-field time-encoded frequency-domain optical coherence tomography. *Opt. Express*, **14**, pp. 7661–7669.

11. Changho, C., Morosawa, A., and Sakai, T. (2008). High-speed wavelength-swept laser source with high-linearity sweep for optical coherence tomography. *IEEE J. Sel. Top. Quantum Electron.* **14**, pp. 235–242.
12. Eigenwillig, C. M., Biedermann, B. R., Palte, G., and Huber, R. (2008). K-space linear Fourier domain mode locked laser and applications for optical coherence tomography. *Opt. Express,* **16**, pp. 8916–8937.
13. Drexler, W., and Fujimoto, J. G. (2008). *Optical Coherence Tomography: Technology and Applications* (Springer Berlin Heidelberg).
14. Bonesi, M., Minneman, M. P., Ensher, J., Zabihian, B., Sattmann, H., Boschert, P., Hoover, E., Leitgeb, R. A., Crawford, M., and Drexler, W. (2014). Akinetic all-semiconductor programmable swept-source at 1550 nm and 1310 nm with centimeters coherence length. *Opt. Express,* **22**, pp. 2632–2655.
15. Huo, T., Zhang, J., Zheng, J.-G., Chen, T., Wang, C., Zhang, N., Liao, W., Zhang, X., and Xue, P. (2014). Linear-in-wavenumber swept laser with an acousto-optic deflector for optical coherence tomography. *Opt. Lett.,* **39**, pp. 247–250.
16. Choma, M. A., Ellerbee, A. K., Yang, C., Creazzo, T. L., and Izatt, J. A. (2005). Spectral-domain phase microscopy. *Opt. Lett.,* **30**, pp. 1162–1164.
17. Schmitt, J. M., Lee, S. L., and Yung, K. M. (1997). An optical coherence microscope with enhanced resolving power in thick tissue. *Opt. Commun.,* **142,** pp. 203–207.
18. Leitgeb, R. A. (2011). Current technologies for high-speed and functional imaging with optical coherence tomography. In *Advances in Imaging and Electron Physics,* Hawkes, P. W. (ed.), Vol 168 (Elsevier Academic Press San Diego).
19. American National Standards Institute (2000). *American National Standards for Safe Use of Lasers. American National Standards Institute, ANSI Z.136.1-2000* (Laser Institute of America, Orlando, FL).
20. Marks, D. L., Ralston, T. S., Boppart, S. A., and Carney, P. S. (2007). Inverse scattering for frequency-scanned full-field optical coherence tomography. *J. Opt. Soc. Am. A,* **24**, pp. 1034–1041.
21. Adie, S. G., Graf, B. W., Ahmad, A., Carney, P. S., and Boppart, S. A. (2012). Computational adaptive optics for broadband optical interferometric tomography of biological tissue. *Proc. Natl. Acad. Sci. U S A,* **109**, pp. 7175–7180.

22. Kumar, A., Drexler, W., and Leitgeb, R. A. (2013). Subaperture correlation based digital adaptive optics for full field optical coherence tomography. *Opt. Express,* **21**, pp. 10850–10866.
23. Klein, T., Wieser, W., Reznicek, L., Neubauer, A., Kampik, A., and Huber, R. (2013). Multi-MHz retinal OCT. *Biomed. Opt. Express,* **4**, pp. 1890–1908.

Chapter 9

Full-Field OCM for Endoscopy

Anne Latrive and Claude Boccara
Institut Langevin, ESPCI-ParisTech, 1 rue Jussieu, Paris, 75005, France
claude.boccara@espci.fr

Endoscopy has become a valuable tool to diagnose and help to treat pathologies in a minimally invasive way. Optical imaging techniques such as confocal microscopy or conventional OCT have already proven their utility for external examination and are therefore being implemented for endoscopy in vivo. They are beginning to be used in clinical domains such as cardiology, pulmonology, gastroenterology, and dermatology.

The full-field approach of OCT gives images of a quality approaching that of histopathological sections thanks to its micrometer-scale resolution. An endoscopic FF-OCM system would allow access to this histological information in vivo and in situ. While no such industrial system has been developed yet, a few experimental setups have already been demonstrated.

We will discuss in this chapter the technical approaches and challenges of implementing FF-OCM in systems with flexible and rigid endoscopic probes.

9.1 Introduction

9.1.1 Overview of Endoscopy

Endoscopy has existed since the early 19th century; but it is only since the 20th century that it is widely used in the medical and also the industrial world, as a mean to explore otherwise inaccessible cavities. An endoscopic probe can be described in the most general way as a tube provided with an optical illumination system. It performs imaging either directly with an integrated miniature camera or by carrying an image from one end to the other with an assembly of optical components. This probe can access internal organs either by natural means (respiratory and aerodigestive tracts) or through a small incision (cardiovascular system, brain). Endoscopy allows the physician to perform diagnosis, treatment, and even surgery in the least invasive way. In addition to the imaging channel, some endoscopes also have operating channels in which the surgeon can insert the tools during microsurgical interventions, which is called interventional endoscopy.

For each clinical application, specific technical characteristics are required, such as size and flexibility of the probe, field of view, resolution, imaging speed, etc. The endoscopic probe itself can be designed in different ways [1]:

- Rigid approach.
  The optical components are miniaturized microlenses or graded refractive index (GRIN) arranged in relays and objectives with diameters less than 1 mm. Such a rigid probe allows for imaging of areas such as the skin (also possible with a portable imaging head) or the head and neck, but also brain, breast, liver, and other internal organs when used as a penetrating needle.

- Flexible approach.
  - With a single optical fiber (not adaptable for full field). Distal scanning is performed by electronical and optical miniaturized components, often microelectromechanical systems (MEMS) [2]. A 1D image can also be obtained in some cases without scanning by using spectral encoding

of depth. The technical difficulty is the miniaturization of scanning components, but such probes can reach diameters as small as 500 μm.
- With a bundle of optical fibers. The fiber bundle directly carries a 2D image; it can be used directly in a full-field approach or with proximal scanning of each of the fiber core. Industrial bundles have up to 100,000 fiber cores in diameters less than 1 mm. However, the imaging quality is degraded by the pixelation effect of the cores.

The flexible probe allows for imaging of the vascular system, the respiratory tract, and the aerodigestive tract.

9.1.2 Endoscopic OCT

Conventional optical coherence tomography (OCT) systems are often based on a fibered interferometer so that the adaptation for endoscopy is quite straightforward and the first endoscopic systems designed for in vivo endoscopic use appeared only a few years after the invention of OCT [3]. Very diverse methods have been demonstrated, using flexible or rigid probes designed with optical fibers, arrays of microlenses or GRIN lenses [4].

9.1.2.1 OCT for flexible endoscopy

A major application of endoscopic OCT is intravascular imaging, especially for the characterization of atherosclerotic plaques and diagnosis of cardiovascular diseases [5]. Commercial systems (LightLab Imaging, St. Jude Medical) offer a transverse resolution of the order of 5 to 10 microns and an axial resolution around 20 to 25 microns, with diameters smaller than 1 mm that allow its integration into current catheters. The optical fiber probe works with side viewing (field of view parallel to the direction of the probe) using a fast 2D scan combined with a circular movement of withdrawal called pull-back. Thus the volume of the full artery can be reconstructed in only a few seconds while blood circulation is stopped. OCT can also be coupled to echography within an intravascular ultrasound probe to combine the deeper millimeter-scale information of ultrasound with the micrometer scale of OCT [6].

Similar probes can be used for imaging of the respiratory and aerodigestive tracts [7]. Other probes are designed for imaging with front viewing (field of view perpendicular to the direction of the probe), especially in optical coherence microscopy (OCM), which achieves resolutions from 2 to 4 microns [8].

9.1.2.2 OCT for rigid endoscopy

Since the penetration depth of OCT does not exceed a few millimeters, even a flexible endoscope does not allow for imaging inside the organ. For this purpose it is necessary to have a rigid probe that acts as a piercing needle to penetrate the tissue, or that is inserted in the channel of a biopsy needle. Such a system was first demonstrated in 2000 [9], and ever since various probes have been proposed that can reach diameters less than 0.5 mm, in particular to guide biopsies in the breast or in the lung [10–12]. In such a probe OCT can also be coupled with elastographic methods in order to combine the tissue mechanical elasticity map with the optical morphological image [13].

Another application of rigid endoscopic OCT is to use a front viewing hand-held probe during surgery on small animals [14] or humans, for example, in the brain to identify blood vessels ahead of the probe and avoid cutting them during the operation [15].

9.1.3 *Problematic of Endoscopic Full-Field OCM*

The benchtop full-field optical coherence microscopy (FF-OCM) system is based on a Linnik interferometer with strictly identical arms. It may seem that a simple endoscopic adaptation would be to integrate the endoscopic probe directly in the object arm of the interferometer. However, it would be mandatory to compensate for the endoscope optical path by integrating exactly the same optical path in the reference arm. This configuration is possible if the endoscope is made of classic reproducible optics such as microlenses, but not when using GRIN lenses or fiber bundles. Moreover, all available commercial fiber bundles are multimode. During in vivo use of

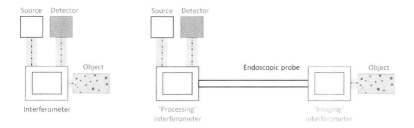

Figure 9.1 Principle of OCT with one and two interferometers.

the system, the probe would be subjected to various movements, bends, and twists, which alter the propagation of light in a different way in each fiber core, making it extremely difficult to compensate the optical paths. The resulting interferometric signal would be greatly degraded. It is therefore preferable for in vivo and in situ clinical use that the probe would be as insensitive as possible to the environment, and therefore not be part of an interferometer arm.

A solution is to use an OCT system with not one but two interferometers, which was first proposed in 2005 [16] (see Fig. 9.1). The first interferometer, called processing, is external to the probe and is illuminated by a broadband spatially incoherent source. The second interferometer, called imaging, is situated at the distal end of the probe and collects light backscattered by the tissue. The camera detects the OCT signal as a superposition of the signals from each interferometer.

In this configuration the main part of the endoscopic probe is not part of any interferometer; it only transports a 2D image. It is thus entirely passive and insensitive to the environment.

9.2 Theory of Full-Field OCM with Two Coupled Interferometers

9.2.1 *Dual-Interferometer OCT Signal*

The OCT signal detected by the camera in a two-interferometer setup is different from the classical OCT signal. It is the superposition of the interferometric signals coming from two incoherently coupled interferometers.

The processing interferometer is formed by two reflective surfaces illuminated by a broad spectrum source. This results in a modulation of the spectrum by a sine function [17]:

$$I_c(\sigma) \propto I(\sigma)[r_{c1}(\sigma) + r_{c2}(\sigma)\cos(2\pi\delta_c\sigma)]$$

where σ is the wavenumber $\sigma = 1/\lambda$ (cm^{-1}), $I_c(\sigma)$ is the output spectrum of the processing interferometer, $I(\sigma)$ the spectrum of the source, r_{c1} and r_{c2} the reflectivity coefficients of each arms (often $r_{c1} = r_{c2}$), and δ_c the path length difference of the processing interferometer.

The resulting spectrum is called channeled spectrum. Its envelope is the source spectrum, the fringe period depends on the path length difference: it is precisely equal to $1/\delta$. The processing interferometer is directed set by the user, who defines the control path length difference, and therefore the processing channeled spectrum.

This channeled spectrum light is injected into the endoscopic probe and illuminates the second imaging interferometer. In the latter two beams interfere: the light backscattered by the object and the reference light, created in the interferometer by a reflective surface. This interferometer therefore creates an additional modulation of the spectrum, similar to the processing interferometer, at a given frequency given by d. In the considered spectral range and small path length differences we can neglect the effects of dispersion between the two interferometers. The spectrum at the output of the imaging interferometer is:

$$I_i(\sigma) \propto I_c(\sigma)[r_i(\sigma) + R(\sigma)\cos(2\pi\delta_i\sigma)]$$

With δ_i the path length difference of the imaging interferometer, r_i the reflectivity of the reference surface, and $R(\sigma)$ the reflectivity from the object. $R(\sigma)$ is the term that contains the information about the object.

The detector receives light that emerges from the imaging interferometer after passing back through the endoscopic probe. The resulting spectrum is the superposition of the two spectra:

$$I_{\text{detector}}(\sigma) \propto I(\sigma)[(r_{c1}(\sigma) + r_{c2}(\sigma)\cos(2\pi\delta_c\sigma))(r_i(\sigma) \\ + R(\sigma)\cos(2\pi\delta_i\sigma))]$$

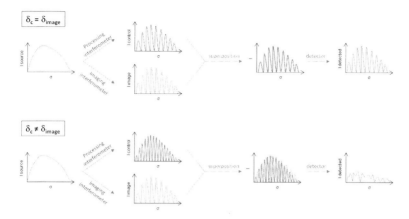

Figure 9.2 Signal resulting from the superposition of two channeled spectra.

Interferences occur only for matched paths within the coherence length. In the total spectrum, with $r_{c1}(\sigma) = r_{c2}(\sigma)$, the resulting interference term is [16]:

$$I_{\text{interference}}(\sigma) \propto I(\sigma)\left[R(\sigma)\cos(2\pi\delta_i\sigma) + r_i(\sigma)\cos(2\pi\delta_c\sigma)\right]$$

The sensor integrates the spectrum over a given range of wavenumbers. If the fringe frequencies of the processing and the imaging spectra correspond, that is, if the path length differences are equal, there is a perfect overlapping of the spectra: the output is a maximum signal. If the frequencies do not match, the overlay is poor, and the resulting signal is low, which is illustrated in Fig. 9.2.

Figure 9.3 shows an example of the resulting signal obtained on a single object, a mirror placed in front of the probe, when scanning the path length difference of the processing interferometer. We see that the signal peaks when the path length difference of the processing interferometer is equal to that of the imaging interferometer.

Let us note that if the object is flat, like a mirror, light is reflected at a single path length difference, and there is therefore a single frequency in the spectrum. If the object exhibits a complex structure, such as biological tissue, light is backscattered from several depths and there are several frequencies in the spectrum. On scanning the

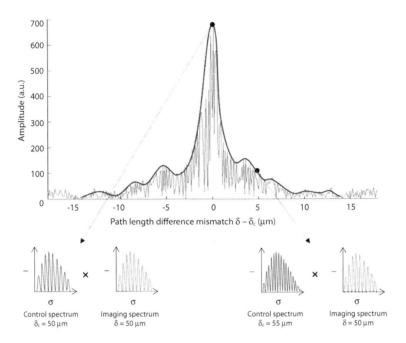

Figure 9.3 OCT signal with a two-interferometer setup, recorded on a mirror placed 25 μm ahead of the probe.

control path length difference, a peak signal appears at each depth that has a value proportional to the backscattered intensity of light, thus reconstructing the axial profile of the object (A-scan).

In summary, the control path length difference defines the corresponding imaging depth in a plane perpendicular to the optical axis. The 2D detector receives an en face FF-OCM image.

9.2.2 Performances

9.2.2.1 Sensitivity

As in nonendoscopic FF-OCM the signal-to-noise ratio is increased by extracting the interferometric signal from the incoherent background using a two-step or four-step phase modulation. This method can be implemented by introducing a small displacement element such as a piezoelectric into the processing interferometer

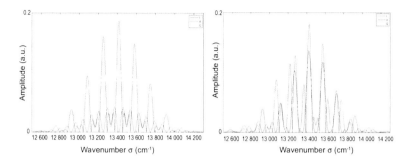

Figure 9.4 Spectrum modulation on the detector.

so as to be able to vary the path length difference by a fraction of the wavelength. Let us note that this modulation could also be performed into the imaging interferometer; however this would require to add moving parts at the end of the endoscope and would increase the complexity of the probe.

In a simple and faster two steps phase modulation scheme, the detector receives a first image I_1, then a second image I_2 obtained with δc shifted by $\lambda/2$. The computer unit calculates the difference $I = I_1 - I_2$, which is the en face FF-OCM image. A calculation of the spectra thus obtained is shown in Fig. 9.4.

Still, dual-interferometer FF-OCM systems have reduced sensitivity compared to classical FF-OCM setups [16].

9.2.2.2 Transversal resolution

The transversal resolution of the system depends on several parameters: the optical components of the probe tip, the sampling on the camera chip, and the characteristics of the fiber bundle if used.

The optics at the distal end of the probe defines a diffraction limit resolution, which is thus the highest value achievable by the system. Typically, at a wavelength of 800 nm, for a numerical aperture (NA) = 0.1, $R_{min01} = 4.8$ µm; NA = 0.2, $R_{min02} = 2.4$ µm; NA = 0.3, $R_{min03} = 1.6$ µm.

The camera chip is a 2D pixel array that performs discrete spatial sampling of the image. The well-known Shannon's theorem tells that a signal of frequency ν_s can be properly reconstructed only if sampled at a frequency at least twice: $\nu_{\text{sampling}} \geq 2\nu_s$. Applying this general principle to our system, to obtain a resolution on the final digital image of R μm, then R μm in the object plane must represent at least 2 pixels of the detector. Since the detector size is limited (chips are often square with typically 1024 × 1024 pixels), there is a trade-off between field of view and transversal resolution. More precisely: $R.\text{FOV} \geq 2/N$, where N is the number of pixels of the detector (if the detector is not square then consider each x or y direction and the corresponding resolutions independently). For a camera chip with 1024 × 1024 pixels, a resolution of 2 microns gives a maximum field of 1 × 1 mm and a resolution of 5 microns a maximum field of 2.5 × 2.5 mm. These parameters are tuned by changing the magnification on the camera, that is to say the focusing optics, often a doublet lens.

The bundle induces a pixelation effect which can degrade the resolution. Each fiber core carries one effective pixel of information of the image, while it is itself imaged on several pixels on the camera. Therefore it reduces the number of effective pixels in the final image and introduces an additional sampling frequency, as the inverse of the distance between adjacent cores of the bundle. Typically for a bundle of 30,000 fibers having a diameter of 0.8 mm, a core diameter is about 2 μm and intercore distance 3.5 μm. Eventually, the transversal resolution can be of a few micrometers.

9.2.2.3 Axial resolution

There is no simple analytical solution to the equation describing the signal of dual-interferometer OCT that would allow for a simple expression of the axial resolution of the system. However, the resulting OCT signal can be computed numerically for given parameters and the axial resolution can be calculated as shown in Fig. 9.5.

It demonstrates that the broader the source spectrum, the higher the resolution, as in classical OCT. This fact can be intuitively understood by considering that a broad spectrum improves the efficiency of the channeled spectra comb principle. The broader the wavenumber range, the better the efficiency of the superposition gets blurred. This is linked to the concept of temporal coherence:

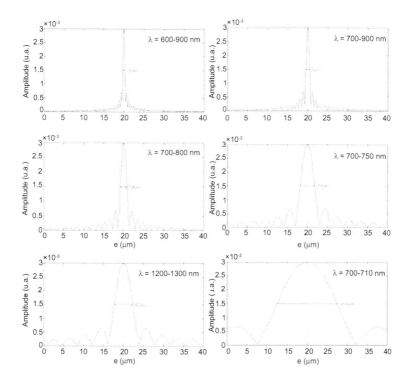

Figure 9.5 OCT signal from two coupled interferometers calculated for different wavelength ranges, as a function of the distance *e* between the path length differences of the processing and imaging interferometers.

more wavenumbers means more interference signals from each wavenumber so that the signal blurs rapidly when one moves away from the zero path length difference.

Typically, for a system working in a wavelength range between 700 and 900 nm, the axial resolution is around 1 μm, which is comparable to nonendoscopic FF-OCM.

9.2.3 Conclusion

A dual-interferometer FF-OCM system is well suited to endoscopic in vivo and in situ use in a clinical environment thanks to an entirely passive endoscopic probe.

This system can be implemented in multiple ways depending on the choice of the interferometers and of the probe. The processing interferometer can be bulky such as a Michelson or a Linnik, whereas the imaging interferometer at the tip of the probe has to be kept simple and miniaturized. The endoscopic probe can be flexible, based on a fiber bundle, or rigid, based on an assembly of lenses (classical or GRIN).

9.3 Problematics of Endoscopic Imaging with a Flexible Fiber Bundle

In flexible endoscopy, the use of a fiber bundle is the only solution to perform full-field imaging and record a 2D en face image without scanning. There are only a few types of bundles commercially available, either leached (Schott Inc, USA) or fused (Sumitomo Electrics, Japan, Fujikura Ltd., Japan). A typical bundle of highest resolution has 30,000 fiber cores of diameters around 2 μm for a total bundle diameter around 0.8 mm (Sumitomo GN08/30, Fujikura FIG-30-850N); but it can goes up to 100,000 fiber cores and 2 mm diameter, r down to 10,000 fiber cores and 0.5 mm diameter.

However, the fiber bundle reduces the final image quality in two ways: by the image pixelation effect and by introducing stray interferometric signal from cross-talk.

9.3.1 Pixelation Effect Removal

Each fiber core in the fiber bundle represents one pixel of information. Thus the image on the camera suffers from additional pixelation effect. Endoscopic optical microscopy systems therefore include a real-time image processing treatment to remove this artifact. A wide variety of methods can be used.

9.3.1.1 Fourier domain filters

The fiber core have irregular shapes but relatively similar sizes, with diameters varying of about 10% [18]. The intercore distance defines therefore a rather well-defined narrow range of spatial frequencies,

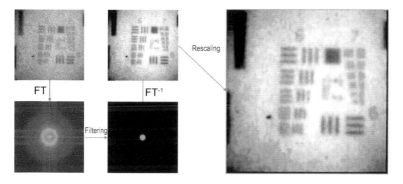

Figure 9.6 Image of a USAF target through a fiber bundle and pixelation removal by Fourier domain filtering.

which can be found on the 2D Fourier transform of the image (Fig. 9.6). A well-adjusted filter in the Fourier domain will eliminate the core frequencies, and the inverse Fourier transform gives a non pixelated image.

Although fast and simple, this technique is not entirely satisfactory. Some artifacts are left on the image because of the Fourier-space filtering: some intercore frequencies are still present while frequencies in the filtered range corresponding to OCT signal details are lost. Moreover, the results vary with each image, they are correct on well-contrasted objects as a grid, but much less for biological tissues because of the plurality of spatial frequencies.

9.3.1.2 Deconvolution

The pixelated image can be convolved with a well-chosen mask, so that the pixels inside of the cores are amplified and those of the cladding are offset (Figs. 9.7 and 9.8). The mask can for example have a simple binary distribution, or more efficiently a Gaussian decay profile.

The processed images are well depixelated (Fig. 9.7). However, the processing time is of the order of several seconds, too long for real-time treatment [17].

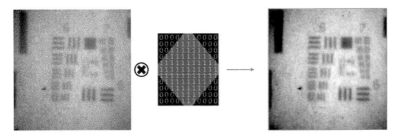

Figure 9.7 Image of a USAF target through a fiber bundle and pixelation removal by deconvolution.

9.3.1.3 Interpolation between the cores

A more complex technique is to extract the positions of the core centers, that is about 30,000 points, and interpolate the signal between these points to reconstruct a typical image of 1024 × 1024 pixels [17]. This method is implemented in come commercial confocal systems [19].

There are two distinct stages that can be realized in different ways: a first step of mapping of the fiber core centers and a second step of image reconstruction.

The position of the fiber core centers can for example be obtained by searching the local maxima of the image, supposing that the spatial repartition of light in the cores is homogeneous enough. The resulting data matrix forms a discrete and irregularly sampled grid, as the fibers are often themselves irregularly arranged in the bundle. There are many interpolation functions to reconstruct the image from this matrix, some of them already available in the library of programming languages such as Matlab. The Gaussian algorithm uses an interpolation function with a Gaussian decay profile [20]. In Shepard's algorithm the interpolated value is function of the inverse of the distance to the reference core [21, 22]. Delaunay triangulation uses a linear or cubic function (in Matlab's gridData function). Interpolation by multilevel B-spline functions is another method, which is more complex but shows improved accuracy [23].

Figure 9.8 compares results obtained with the Gaussian algorithm and the multilevel B-splines interpolation.

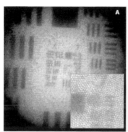 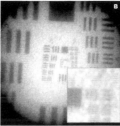 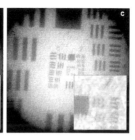

Figure 9.8 Image of a USAF target through a fiber bundle (A) and pixelation removal by a Gaussian algorithm (B) and multilevel B-splines interpolation (C).

9.3.2 Cross-Talk Interferences

9.3.2.1 Core-to-core coupling

In an imaging fiber bundle, cores are close (about 1 µm radius, separated by $d = 3.5$ µm in Sumitomo IGN 08/30) and immersed in a common sheath. The cores can then be coupled, that is to say that light passes from one core to the other, via the evanescent waves associated to propagating modes. This is a well-known phenomenon called cross-talk [24].

The coupling equations can only be solved analytically in some simple cases (low number of fibers, identical fibers of perfectly cylindrical section throughout the bundle, single mode) or in a numerical manner, by a finite difference method in particular [25].

Coupling is more important for small core radius, high wavelength, high index contrast between core and sheath, and of course low intercore distance. It is characterized by a coefficient C, in m^{-1}. $1/C$ can be considered as the characteristic length after which the whole signal has passed from one fiber to the other [25].

For example, for two fiber cores of type SMF28, with a radius $a = 4$ µm, the coupling coefficient C for $\lambda = 1.5$ µm is 10^{-4} µm^{-1} for intercore distance $d = 2a$, 10^{-6} µm^{-1} for $d = 4a$ (situation of the imaging bundle), which means that the characteristic coupling length is of the order of 1 m. In such a situation it is impossible to correctly transport an image through a bundle.

However, it was shown numerically that coupling efficiency is reduced in imaging fiber bundles due to the nonuniformity of cores

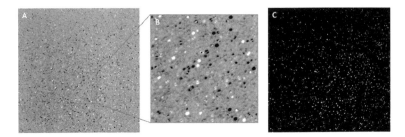

Figure 9.9 Cross-talk interferometric signal recorded on a dual-interferometer FF-OCM setup with an imaging fiber bundle Sumitomo IGN08/30. (A and B) Tomographic images with phase modulation and (C) absolute value.

of different sizes and different shapes that vary along the length the bundle. In a case similar to an imaging bundle (Fujikura FIGH 10-350S), with $a = 1$ µm, $d = 3.2$ µm, and $\lambda = 0.6$ µm, a numerical model with seven cores gives a coupling length of 1 cm and a coupling amplitude of only 10% of the total signal in a core [18].

In an imaging bundle with a high number of pixels, with a length of typically 1 m, all or part of the signal passes thus from one to another core, and may optionally return to the starting core. With about 30,000 fibers, it creates a large number of different optical paths inside the bundle, and thus a parasitic interferometric signal inherent to the bundle.

Figure 9.9 shows images of this stray signal on the camera [17], and Fig. 9.10 profiles of cross-talk signals with depth [26].

However the use in FF-OCM of a source of low spatial coherence considerably reduces crosstalk. Light injected into one core cannot interfere with that coming from a neighboring core, because they are not coherent. While a considerable amount of light is still lost in passing through the bundle, there is cross-talk interference only from signals that came from the same core and recombined.

Moreover, in fiber bundles of leached type, fiber cores are fused together at each end but not inside the bundle. Thus, the fibers keep their own sheaths that are not fused together. In such a device coupling between the cores is reduced [17, 24]. However the available bundles have fewer pixels (Schott SAP 1136247) and are less easy to use for endoscopy as they are completely shapeless.

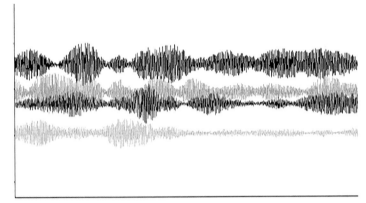

Figure 9.10 Intensity modulation of the signal from individual fibers when the processing interferometer is scanned.

Each bundle fiber exhibits a singular behavior, with cross-talk signals arising at different depths in a determinist manner. However, the stray cross-talk signal can also be considered globally by considering a mean on the highest cross-talk signals of the image at each depth. The obtained parasitic signal exhibits an interesting decay with depth, that is to say with the control path length difference δ_c (Fig. 9.11) [17].

Indeed, considering a given δ_c means considering interference between signals that have made a given number of coupling events from core to core, each coupling event creating a small path length difference. The global cross-talk signal is thus proportional to the probability of said path. Because only around 10% of the signal is coupled from one core to another, the signal loses a factor 10 with each coupling event. Thus the paths with higher numbers of coupling events are less probable.

To understand this intuitively, one can consider a section of the bundle where the signal would follow a random walk in two dimensions, each step being a coupling event. δ_c represents the number of coupling events, that is, the number of steps. The intensity of the cross-talk signal represents the number of optical paths of said length δ_c.

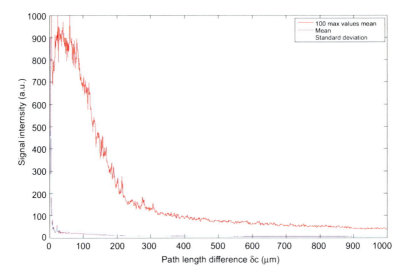

Figure 9.11 Profile of the cross-talk interferometric signal as a function of the control path length difference (mean on the 100 highest values).

9.3.2.2 Mode-to-mode coupling

In addition to core-to-core coupling, cross-talk can occur inside a single core from mode-to-mode coupling [24].

Indeed the imaging fiber bundles are usually multimode in the visible and near-infrared (NIR) range. The monomodal or multimodal state of an optical fiber is characterized by the normalized frequency V defined by:

$$V = \frac{2\pi a}{\lambda} \text{NA} \quad \text{with} \quad \text{NA} = \sqrt{n_{core} - n_{sheath}}$$

with a the core radius and NA the numerical aperture. The fiber has a single propagating mode if $V \leq V_c$ with $V_c = 2.405$. If $V > V_c$ then several propagating modes coexist [25]. The number of modes can under some conditions be approximated by $V^2/2$. In the case of the 30,000 fiber bundle, $a \simeq 1$ µm, NA $\simeq 0.35$ µm, $\lambda = 0.8$ µm, and $V \simeq 2.75$. There are approximately four modes of propagation in each core.

Thus there is modal dispersion in every core: each mode is characterized by a propagation constant, often called β, and propagates at a different speed [27]. These modes can be coupled

Figure 9.12 Cross-talk signal from mode-to-mode coupling observed directly on the camera (A and B) and on the processed tomographic signal (C).

to one another within a single core, for example by surface roughness. This phenomenon also induces a distribution of path length differences in the bundle and participates in the formation of the parasitic cross-talk signal.

Figure 9.12 shows typical images interferometric signal arising from mode-to-mode coupling in an endoscopic FF-OCM setup. The cores exhibit characteristic spatial patterns of propagation modes.

9.3.3 Conclusion

The use of a fiber bundle in an endoscopic FF-OCM setup leads to degraded image quality in terms of resolution because of the pixelation effect and in terms of signal-to-noise ratio because of cross-talk phenomena. However, some working setups have already been demonstrated, which will be detailed in Section 9.4.

Endoscopic FF-OCM systems with a rigid probe avoid those problems and exhibit better imaging results, as will be shown in Section 9.5.

9.4 Full-Field OCM with a Flexible Endoscope

The previous section showed that an FF-OCM setup with a fiber-bundle-based endoscope would suffer from high sensitivity loss due to the dual-interferometer scheme and the cross-talk parasitic signals. Nonetheless a few experimental setups have already

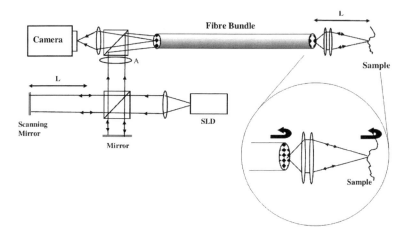

Figure 9.13 Configuration for 3D OCT employing a fiber imaging bundle. SLD, superluminescent diode; A, optional focusing lens controlling illumination onto bundle face. The average path length imbalance L is identical in both Fizeau and Michelson interferometers. The inset circle shows where the two reflections occur in the Fizeau interferometer [26].

been demonstrated; their characteristics and performances will be discussed and compared here.

9.4.1 FF-OCM System with Simple Imaging Interferometer and SLD Source

Tatam et al. demonstrated in 2007 an FF-OCM setup with a SLD source at 830 nm, Michelson processing interferometer, and Fizeau imaging interferometer as represented in Fig. 9.13 [26]. The resolution is 30 µm axial and 30 µm transversal; the sensitivity is 40 dB for 1 image without averaging.

The advantage of the system is the very simple common-path imaging interferometer: the reference beam is generated by the reflection at the end of the fiber bundle. However, because of the use of a SLD source which is spatially coherent the level of cross-talk is quite high and dramatically reduces the sensitivity and image quality (Fig. 9.10). Thus, only images of reflective surfaces can be made, not of biological tissues.

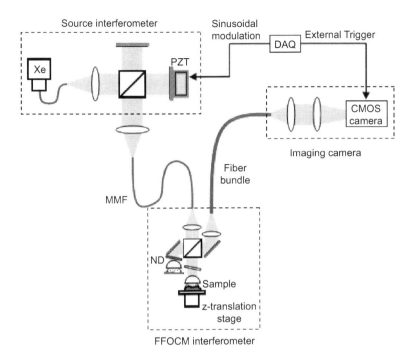

Figure 9.14 Schematic of the experimental setup. Xe, xenon arc lamp; ND, neutral density filter; GP, glass plate; MMF, multimode fiber; DAQ, data acquisition board in computer [28].

9.4.2 FF-OCM System with Complex Imaging Interferometer and Xenon Arc Source

Oh et al. demonstrated in 2006 a setup with an incoherent Xenon arc lamp, a Michelson processing interferometer, and a Linnik imaging interferometer as shown in Fig. 9.14 [28]. The resolution is 1.1 μm axial and 4.1 μm transversal, sensitivity is 76.8 DB when averaging over 800 images.

The setup presents a higher signal-to-noise ratio that allows for imaging of biological tissues. Moreover, the use of a Linnik imaging interferometer allows to scan the sample in depth by moving the coherence plane and focal plane at the same time.

Figure 9.15 shows images obtained ex vivo on a Xenopus laevi embryo heart with a bundle Schott of 17,000 fibers of core diameter

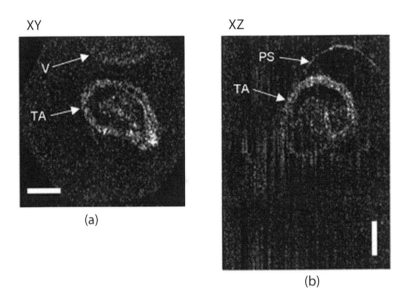

Figure 9.15 Images of Xenopus laevi embryo heart, ex vivo, obtained through a multimode fiber optic imaging bundle. (a) En face image. (b) Reconstructed cross-sectional image. PS, pericardial sac; V, ventricle; TA, truncus arteriosus. Scale bar: 100 μm [28].

12.7 μm for a total diameter 1.93 mm. The en face images and cross-sectional reconstructions reveal the different structures such as the pericardial sac, the truncus arteriosus, ventricle, and atrium. The pixelation artifacts are removed by convolution with a smoothing disk filter.

The drawback of the setup is the use of a Linnik imaging interferometer which in spite of having no moving parts is too bulky to be adapted for clinical endoscopic use. Moreover the imaging time is high: 8 seconds for one image, mainly because each image is averaged 800 times to improve the signal-to-noise ratio.

The signal-to-noise ratio of this setup is higher than that of the previous system. One of the causes is that the path length difference of the interferometers is set to 600 μm, so the cross-talk phenomena are less intense in this range, as explicated previously with Fig. 9.11.

9.4.3 FF-OCM System with Simple Imaging Interferometer and Xenon Arc Source

A third solution was proposed in 2012 to achieve at the same time a sufficient signal-to-noise ratio with very low cross-talk signal, and a miniature simple imaging interferometer suitable for endoscopy [17, 29].

The light source is a Xenon arc lamp, the processing interferometer is a Linnik, the imaging interferometer is a common-path Fizeau.

The path length differences of the interferometers are set very high, of the order of a few millimeters, in order to avoid cross-talk; but the imaging depth inside the sample is kept at less than 100 μm by introducing a GRIN relay lens at the tip of the bundle as the optical cavity of the imaging interferometer. The same optic is introduced in the processing interferometer to compensate for dispersion.

The resolution is 1.8 μm axial and 4.9 μm transversal, sensitivity is 70 dB when averaging over 50 images.

Figure 9.16 shows en face images of a phantom made of polyurethane with micrometer-scale TiO_2 beads, where individual beads and clusters of beads are clearly seen. The fiber bundle is Sumitomo IGN08/30 of 30,000 pixels with a diameter of 2 μm for a total diameter of 0.8 mm.

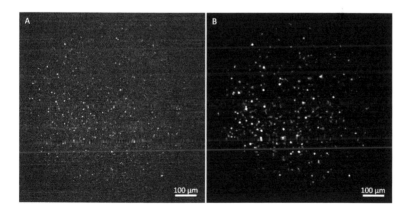

Figure 9.16 En face endoscopic FF-OCM image of a phantom acquired at a depth of 50 μm, without fiber bundle pixelation removal treatment (A) and with Gaussian interpolation reconstruction (B) [17].

The advantage of this setup is the very simple imaging interferometer which makes the endoscopic miniaturization straightforward. The drawback of this simplicity is that the focal plane cannot be moved inside the sample at the same time as the coherence plane, so that the imaging depth is limited by the depth of field of the optics, typically of the order of 100 μm. The sensitivity is, however, too low to really allow imaging of biological tissues. It can be increased by averaging over more images, but this would also increase the acquisition time over 1 second.

9.4.4 Discussion

Adapting FF-OCM for endoscopy with a fiber bundle is challenging. The interferometers have to be carefully chosen in order to counterbalance the loss of sensitivity due to the dual-interferometer scheme and to the crosstalk signal. There is a strong constraint on the choice of the imaging interferometer that has to be kept as simple as possible, without moving parts, to allow for easy integration. Eventually there is a trade-off between miniaturization of the probe and sensitivity of the FF-OCM system.

Still, systems may in the future be able to achieve in vivo imaging capabilities thanks to novel faster cameras that would solve the problems of both speed and sensitivity, allowing for averaging a greater number of frames.

9.5 Full-Field OCM with a Rigid Endoscope

Compared to flexible endoscopy, rigid endoscopy seems to give a direct noninvasive access to a more limited number of areas: skin, head & neck, urogenital tract, or internal organs during open surgery operations. However, if a flexible endoscope can indeed reach the organ, it can only image its surface because of the limited depth of OCT. To perform OCT imaging inside the organ one must mechanically penetrate the tissue with a needle or a needle-like tool. An imaging system with a rigid probe will therefore be able to image areas inaccessible even by flexible endoscopy, such as inside the breast, brain, or kidney.

9.5.1 Choice of the FF-OCM Setup

Section 9.2 explains the principle and advantages of using an FF-OCM setup with two interferometers for endoscopic purpose. This system can be implemented with any type of endoscope, either flexible or rigid. However, a more classical FF-OCM setup with one interferometer could also be used for rigid endoscopy. Indeed, the question of the integrity of the interferometric signal by passing through the optical fiber bundle does not arise with a rigid probe. A possible setup could be an interferometer such as a Linnik with two identical rigid endoscopic probes in the two arms.

Yet this solution would be more difficult to implement for clinical applications. Indeed such a setup would be bulkier due to the presence of two endoscopes. Furthermore, a system with removable endoscope is needed in order to decontaminate, sterilize, or replace the endoscope between each use. Since the imaging endoscope has to be exactly the same as the endoscope in the reference arm, changing the imaging endoscope could not be done without degrading the system performances.

For these reasons, an FF-OCM system with two interferometers is to be preferred even for rigid endoscopy.

9.5.2 FF-OCM Setup with a Rigid Probe

The rigid probes can be an assembly of conventional microlenses mounted within a support tube, or GRIN lenses which already have a cylindrical shape. Until now, FF-OCM imaging with a rigid probe has only been demonstrated in one paper, where GRIN lenses are used [30].

9.5.2.1 Setup and performances

The processing interferometer is of Michelson type, while the imaging interferometer is very simple and common path, using the reflection at the tip of the probe as a reference beam.

The light source is a Xenon arc lamp, which is both spatially and temporally incoherent, with passband filters from 700 to 800 nm.

The camera has a complementary metal-oxide semiconductor (CMOS) sensor of 1024 × 1024 pixels and can operate at up to 160 Hz.

Results were obtained with a rigid probe based on a GRIN lens assembly with a diameter of 2 mm, a length of 150 mm, and a numerical aperture of 0.1 (GRINTECH GmBH, Germany).

With such a probe, the transverse resolution of the system was measured to be 3.5 µm, and the axial resolution 1.8 µm in water. Sensitivity was experimentally evaluated using a partial reflector with −33 dB reflectivity. A value of −80 dB was obtained when averaging 20 images during about 1 second.

9.5.2.2 Imaging results

The rigid probe is applied in direct contact with the sample, which can be covered in water-based gel (Ultrasonic) to prevent reflection from air bubbles.

Proof-of-principle was conducted on a phantom made of polyurethane with micrometer-scale TiO_2 beads and on optical cleaning paper, as shown in Fig. 9.17. One can distinguish each individual bead as well as clusters of beads in the phantom, and the fibers of the optical paper.

Ex vivo imaging was performed on fixed unstained human breast tissue samples Examples of imaging results are presented on Fig. 9.17. One can see strong backscattering connective tissue as well as adipocytes easily recognizable as black round-shaped cells.

In vivo imaging was performed on the skin of a healthy volunteer. No clearing agents were used, although it has been previously demonstrated that it could quite enhance the quality of images and increase depth penetration of light [31]. Figure 9.18 shows results obtained on normal forearm skin and on a mole, at depths around 20 µm. Normal tissue exhibits wrinkles, which have a typical depth of 100 µm and a width between 20 and 50 µm, and regular epithelial cells with a typical diameter of 30 µm. Structures are totally different on mole tissue: only a few epithelial cells are present, and the tissue exhibits large-scale fibered structures.

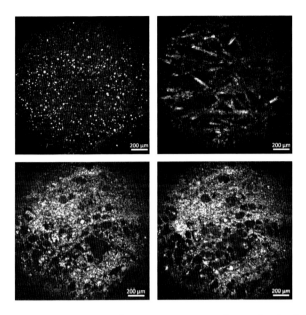

Figure 9.17 En face images of a phantom acquired at a depth of 100 μm (A), of optical paper at a depth of 50 μm (B), of fixed human breast tissue at depths of 20 μm (C), and 30 μm (D). Field of view is 1.5 mm × 1.5 mm.

9.5.3 Discussion

The FF-OCM setup with rigid probe has proven its ability to image biological tissues ex vivo as well as in vivo and reveal fine tissue structures thanks to its micrometer-scale resolution. The simple design of the probe allows diameters ranging from less than 1 mm to a few mm, making it suitable for in situ imaging.

Nonetheless, several points have to be improved for future development of a medical device.

The acquisition time of one 2D image was typically 1 second which is quite long, so that motion artifacts can be a problem for in vivo imaging. Speed can be increased by using a more powerful light source and a faster 2D camera, and by improving the mechanical setup to increase the sensitivity of the system, thus reducing the number of accumulations needed to have a good image quality.

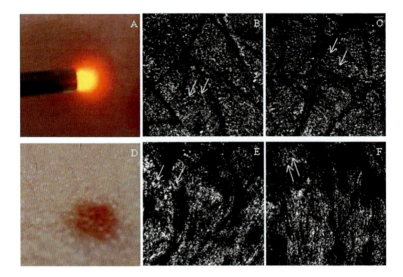

Figure 9.18 En face images of in vivo human skin. (A) The probe is applied on forearm tissue: healthy tissue at 20 μm under the surface (B and C) showing wrinkles and epithelial cells (arrows). (C) The probe is applied on a mole: mole tissue at 20 μm under the surface (E and F) showing only a few epithelial cells (arrows) and long fibered structures. Field of view is 1 mm × 1 mm.

Another limitation is that although the imaging depth can be easily scanned through the tissue, the focal depth stays fixed at a given depth. The imaging range is therefore limited by the depth field of the probe optics. Since it is typically 50 μm it is only possible to image tissues up to depths around 100 μm. Dynamic focusing could be introduced relatively easily thanks to the use of GRIN lenses. Indeed one can displace the focal plane at the exit of a GRIN lens by displacing the focal plane at the entrance, for example, using a microscope objective or lens system mounted on a linear motor [32].

Although comprising an endoscopic probe, the setup is still bulky and has to be designed into a hand-held probe in order to be able to access areas of interest on the patient's body. Such a hand-held probe could be used in dermatology for imaging of the skin; it could also guide the surgeon during tumor removal operations and guide biopsy procedures, for example, in the brain or the breast.

9.6 Conclusion

While endoscopic OCT is already implemented in commercial systems, the full-field approach is more difficult to adapt for endoscopy. Indeed, integrating an endoscope capable of transporting a 2D image into an interferometric setup is an optical challenge. To ensure the probe being insensitive to the perturbations of the environment, an interferometric setup with two coupled interferometers should be preferably used. While the external processing interferometer can be bulky and freely chosen, the imaging interferometer located at the distal extremity of the probe has to be kept simple and miniature, a common-path design being the best solution.

In flexible endoscopy FF-OCM has been demonstrated in several setups with different interferometers: Michelson or Linnik for processing, Linnik or Fizeau for imaging. The use of a fiber bundle degrades the image quality because of the pixelation and cross-talk effects. Ex vivo imaging on biological tissues has nonetheless been performed. In vivo imaging is not yet possible due to the low signal-to-noise ratio and speed of the systems.

In rigid endoscopy FF-OCM has been implemented in only one setup that was used for ex vivo imaging of different tissues and in vivo imaging of skin. However the benchtop setup is still bulky and needs to be miniaturized and designed into a handheld probe.

The endoscopic FF-OCM systems give en face images at depths relatively close to the surface: 100 µm, and at high resolution: 1–5 µm. From this point of view, the FF-OCM images are closer to confocal microscopy than other OCT images.

Proof-of-concept of endoscopic FF-OCM has been given; yet the imaging systems need improvements mainly in terms of sensitivity and frame rate to achieve the potential of industrial systems. A breakthrough could be made with the development of novel powerful incoherent light sources, mainly based on light-emitting diodes (LEDs), and of faster cameras with larger CMOS sensors and higher full-well capacities.

Endoscopic FF-OCM systems could be useful in the field of biomedical research, small animal imaging, but should be especially designed for medical imaging.

The flexible system could be inserted inside the working channel of an endoscope during the conventional endoscopy procedure, for example, in gastroenterology for polyp surveillance. While the endoscopist still has access to a large macroscopic field of view of the tissue, the FF-OCM probe can target a suspicious lesion and perform a real-time optical biopsy of it.

The rigid system could have two types of clinical applications. On the one hand, a probe with a diameter of a few millimeters and large field of view would be well suited in dermatology for noninvasive imaging of the skin as other confocal systems do (Caliber Imaging & Diagnostics, Inc., USA). It could also be useful in neurosurgery, where the physician has to open the skull to gain access to the brain and does not dispose of imaging systems at the micrometer scale. On the other hand, a thin probe with a diameter less than one millimeter may be inserted in a biopsy needle to image deep inside organs such as breast and guide the biopsy.

The goal of endoscopic FF-OCM is to provide the physician with real-time in vivo and in situ images allowing a first identification of pathological areas before surgery or biopsy, and in the long term reduce and perhaps even replace biopsy.

References

1. J. T. C. Liu, N. O. Loewke, M. J. Mandella, R. M. Levenson, J. M. Crawford, and C. H. Contag, (2011). Point-of-care pathology with miniature microscopes. *Anal. Cell. Pathol.*, **34**, pp. 81–98.
2. P. H. Tran, D. S. Mukai, M. Brenner, and Z. Chen, (2004). In vivo endoscopic optical coherence tomography by use of a rotational microelectromechanical system probe. *Opt. Lett.*, **29**(11), pp. 1236–1238.
3. G. J. Tearney, M. E. Brezinski, B. E. Bouma, S. A. Boppart, C. Pitris, J. F. Southern, and J. G. Fujimoto, (1997). In vivo endoscopic optical biopsy with optical coherence tomography. *Science*, **276**(5321), pp. 2037–2039.
4. W. Drexler and J. G. Fujimoto, eds. (2008). *Optical Coherence Tomography, Technology and Applications*. Springer.

5. C. Xu, J. M. Schmitt, S. G. Carlier, and R. Virmani, (2008). Characterization of atherosclerosis plaques by measuring both backscattering and attenuation coefficients in optical coherence tomography. *J. Biomed. Opt.*, **13**, p. 034003.
6. J. Yin, X. Li, J. Jing, J. Li, D. Mukai, S. Mahon, A. Edris, K. Hoang, K. K. Shung, M. Brenner, J. Narula, Q. Zhou, and Z. Chen, (2011). Novel combined miniature optical coherence tomography ultrasound probe for in vivo intravscular imaging. *J. Biomed. Opt.*, **16**(6), p. 060505.
7. S. Lam, B. Standish, C. Baldwin, A. McWilliams, J. LeRiche, A. Gazdar, A. I. Vitkin, V. Yang, N. Ikeda, and C. MacAulay, (2008). In vivo optical coherence tomography imaging of preinvasive bronchial lesions. *Clin. cancer Res.*, **14**(7), pp. 2006–2011.
8. A. D. Aguirre, J. Sawinski, S. Huang, C. Zhou, and J. G. Fujimoto, (2010). High speed optical coherence microscopy with autofocus adjustment and a miniaturized endoscopic imaging probe. *Opt. Express*, **18**(5), pp. 6210–6217.
9. X. Li, C. Chudoba, T. Ko, C. Pitris, and J. G. Fujimoto, (2000). Imaging needle for optical coherence tomography. *Opt. Lett.*, **25**(20), pp. 1520–1522.
10. B. Y. Yeo, R. A. Mclaughlin, R. W. Kirk, and D. D. Sampson, (2012). Enabling freehand lateral scanning of optical coherence tomography needle probes with a magnetic tracking system. *Biomed. Opt. Express*, **3**(7), pp. 3894–3896.
11. B. C. Quirk, R. A. Mclaughlin, A. Curatolo, R. W. Kirk, P. B. Noble, and D. D. Sampson, (2011). In situ imaging of lung alveoli with an optical coherence tomography needle probe. *J. Biomed. Opt.*, **16**(3), p. 036009.
12. D. Lorenser, X. Yang, R. W. Kirk, B. C. Quirk, R. A. Mclaughlin, and D. D. Sampson, (2011). Ultrathin side-viewing needle probe for optical coherence tomography. **36**(19), pp. 3894–3896.
13. K. M. Kennedy, B. F. Kennedy, R. A. Mclaughlin, and D. D. Sampson, (2012). Needle optical coherence elastography for tissue boundary detection. *Opt. Lett.*, **37**(12), pp. 2310–2312.
14. V. R. Korde, E. Liebmann, and J. K. Barton, (2011). Design of a handheld optical coherence microscopy endoscope. *J. Biomed. Opt.*, **16**(6), p. 066018.
15. C. Liang, J. Wierwille, T. Moreira, G. Schwartzbauer, M. S. Jafri, C. Tang, and Y. Chen, (2011). A forward-imaging needle-type OCT probe for image guided stereotactic procedures. *Opt. Express*, **19**(27), pp. 26283–26294.

16. H. D. Ford, R. Beddows, P. Casaubieilh, and R. P. Tatam, (2005). Comparative signal-to-noise analysis of fibre-optic based optical coherence tomography systems. *J. Mod. Opt.*, **52**(14), pp. 1965–1979.
17. A. Latrive, (2012). Tomographie de Cohérence Optique Plein Champ pour l'endoscopie: Microscopie in situ et in vivo des tissus biologiques. Université Pierre et Marie Curie.
18. K. L. Reichenbach and C. Xu, (2007). Numerical analysis of light propagation in image fibers or coherent fiber bundles. *Opt. Express*, **15**(5), pp. 2151–2165.
19. T. Vercauteren, A. Perchant, G. Malandain, X. Pennec, and N. Ayache, (2006). Robust mosaicing with correction of motion distortions and tissue deformations for in vivo fibered microscopy. *Med. Image Anal.*, **10**(5), pp. 673–692.
20. J. Han, J. Lee, and J. U. Kang, (2010). Pixelation effect removal from fiber bundle probe based optical coherence tomography imaging. *Opt. Express*, **18**(7), pp. 1168–1172.
21. D. Ruprecht and H. Müller, (1992). Image warping with scattered data interpolation methods. In *Forschungsberichte des Fachbereichs Informatik der Universität Dortmund*, no. 443, Dekanat Informatik, Univ.
22. T. Vercauteren, (2008). Image Registration and Mosaicing for Dynamic In Vivo Fibered Confocal Microscopy. École des Mines de Paris.
23. S. Lee, G. Wolberg, and S. Y. Shin, (1997). Scattered data interpolation with multilevel B-splines. *IEEE Trans. Vis. Comput. Graph.*, **3**(3), pp. 228–244.
24. H. D. Ford and R. P. Tatam, (2011). Characterization of optical fiber imaging bundles for swept-source optical coherence tomography. *Appl. Opt.*, **50**(5), pp. 627–640.
25. J. Bures, (2009). *Optique guidée: fibres optiques et composants passifs tout-fibre*. Presses Internationales Polytechnique.
26. H. D. Ford and R. P. Tatam, (2007). Fibre imaging bundles for full-field optical coherence tomography. *Meas. Sci. Technol.*, **18**, pp. 2949–2957.
27. A. Ghatak and K. Thyagarajan, (1998). *An Introduction to Fiber Optics*. Cambridge University Press.
28. W.-Y. Oh, B. E. Bouma, N. Iftimia, R. Yelin, and G. J. Tearney, (2006). Spectrally-modulated full-field optical coherence microscopy for ultrahigh-resolution endoscopic imaging. *Opt. Express*, **14**(19), pp. 8675–8684.

29. C. Boccara, A. Latrive, F. Harms, and B. L. C. de Poly, Full-field optical coherence tomography system for imaging an object. WO 2012035148 A12012.
30. A. Latrive and A. C. Boccara, (2011). In vivo and in situ cellular imaging full-field optical coherence tomography with a rigid endoscopic probe. *Biomed. Opt. Express*, **2**(10). p. 2897.
31. H. Zhong, Z. Guo, H. Wei, C. Zeng, H. Xiong, Y. He, and S. Liu, (2010). In vitro study of ultrasound and different-concentration glycerol–induced changes in human skin optical attenuation assessed with optical coherence tomography. *J. Biomed. Opt.*, **15**(3), p. 036012.
32. T. Xie, S. Guo, Z. Chen, D. Mukai, and M. Brenner, (2006). GRIN lens rod based probe for endoscopic spectral domain optical coherence tomography with fast dynamic focus tracking. *Opt. Express*, **14**(8), pp. 3238–3246.

Chapter 10

Full-Field Optical Coherence Tomography and Microscopy Using Spatially Incoherent Monochromatic Light

Dalip Singh Mehta,[a] Vishal Srivastava,[b] Sreyankar Nandy,[a] Azeem Ahmad,[a] and Vishesh Dubey[a]

[a]*Department of Physics, Indian Institute of Technology Delhi, Hauz Khas, New Delhi 110016, India*
[b]*Instrument Design Development Centre, Indian Institute of Technology Delhi, Hauz Khas, New Delhi 110016, India*
mehtads@physics.iitd.ac.in, vishalsrivastava17@gmail.com

10.1 Introduction

Optical coherence tomography (OCT) is a noncontact, noninvasive, and high-resolution cross-sectional imaging technique for the visualization of subsurface structures of biological and nonbiological objects [1–6]. Over the last two decades OCT has been widely used in biomedical imaging, diagnosis and research, as well as in industrial applications, the details about which can be found in numerous research papers, review articles and many books [1–16]. The basic principle of OCT imaging is low coherence

---

Handbook of Full-Field Optical Coherence Microscopy: Technology and Applications
Edited by Arnaud Dubois
Copyright © 2016 Pan Stanford Publishing Pte. Ltd.
ISBN 978-981-4669-16-0 (Hardcover), 978-981-4669-17-7 (eBook)
www.panstanford.com

interferometry (LCI) using single mode fiber optic Michelson interferometer [1–16]. In most of the OCT systems, a temporally low coherent (spectrally broadband) and spatially highly coherent light, generated from sources, such as superluminescent diodes (SLDs), white light halogen lamps, and Ti:sapphire lasers are used. The very first demonstration of OCT is based on time domain detection [5] in which the echo-time delay of backscattered light from the multilayer object is measured by mechanically sweeping a mirror in the reference arm. The backscattered light from the different depths of the object is interfered with the light reflected from the reference mirror and the interference pattern is detected by a point photodetector. The interference takes place only when the optical path difference (OPD) between the reference and object arm is within the coherence length of light which is based on the temporal frequency spectrum of the light source. The principles of time gating, optical sectioning, and optical heterodyning are combined to reconstruct the cross-sectional images of multilayer objects [1–10]. A single-mode fiber works as a confocal gate aperture to achieve better signal-to-noise ratio (SNR). To reconstruct the 3D image both the depth scan (also called A-scan) which is achieved by scanning the reference mirror in the axial direction and lateral scanning (also called B-scan) which is implemented either by scanning objective lens or by scanning sample itself in the (lateral) *XY* direction or by scanning sample beam using the combination of two mirrors in *XY* directions [1, 2, 10–16]. Depths upto 1–2 mm can be imaged in turbid media or tissues, such as skin or arteries; greater depths are possible in transparent tissues, such as the eye. Temporal coherence of the light source determines the depth resolution, whereas spatial coherence plays a role in both lateral resolution and depth resolution of LCI and OCT [1–10].

Later on frequency domain OCT (FD-OCT) systems were developed which have higher speed, and sensitivity [17–20]. There are mainly two types of FD-OCT systems, spectral domain OCT (SD-OCT) [21–25] and swept-source OCT (SS-OCT) [26–28]. Both SD-OCT and SS-OCT systems are based on single-mode fiber optic interferometers in which a light beam is scanned point by point on the sample in the lateral directions and the backscattered light is interfered with the light reflected back from the reference mirror. In contrast to time domain OCT (TD-OCT) the reference mirror is kept fixed in FD-OCT, that is, no axial scanning of reference mirror is required and hence has higher acquisition speed, improved sensitivity, and better SNR compared to TD-OCT [17–28]. But, still the lateral (XY) mechanical scanning mirrors are required in the FD-OCT systems to build two-dimensional OCT image. To avoid the mechanical scanning of reference mirror and sample arm the OCT systems were extended to full-field detection mode to get faster data acquisition [28–34]. In full-field OCT (FF-OCT) solid-state detector arrays, such as charge-coupled device/complementary metal-oxide semiconductor (CCD/CMOS) detectors are used, that record 2D – interferometric images and thus making the need of lateral scan redundant [28–34]. The FF-OCT techniques have recently become more popular, in which the depth scan (A-scan) and lateral scan (B-scan) can simultaneously be obtained. In FF-OCT systems all the implementations of OCT, that is, TD, SD, and SS versions are being used [1–3, 28–34].

The FF-OCT systems that utilize high-magnification and high-numerical-aperture (NA) microscope objective lenses to focus light tightly on the object surfaces are known as optical coherence microscopes, which give en face high resolution images [1–3, 35–46]. To perform en face OCT, Izatt et al. [35] first developed an OCM technique in which a tightly focused probe beam is scanned across the sample at a fixed depth. Since an en face image can be detected without transverse scanning (XY scanning) by 2D detector arrays (CCD/CMOS), a 3D data set was acquired in a single longitudinal scan [35–46]. Therefore, the OCT system is extended to parallel, full-field/wide-field detection to get even faster data acquisition and area detection in which both A-scan and B-scan are simultaneously obtained [35–46]. The transverse resolution of FF-OCT is similar to conventional microscopy whereas the axial resolution is determined by the spectral properties of the illumination source [1, 2].

Most of the conventional OCT systems based on temporal coherence uses a broad spectral bandwidth light source which suffers from dispersion problem in many practical applications, such as biological applications, and requires dispersion compensation optics and suitable index matching materials [1–16]. Also spectral response of many sample media is inhomogeneous when broadband light propagates inside the medium; either the light's phase (dispersion) or its amplitude (inhomogeneous absorption) might be changed. Thus the statistical properties of the light field changes which may reduce the performance of system. Therefore, the use of a single-wavelength or narrow-band, that is, temporally highly coherent and spatially low coherent (extended), light source can be advantageous in such cases and the system becomes independent from spectral dependence, light source stability is increased which reduces intensity noise, and preferred wavelength can be chosen since one can fit the frequency source to the low-absorbing spectral window of the medium. In this case dispersion compensation is not required and true color spectroscopy can be performed. True color OCT are performed with light-emitting diodes (LED) but LED have higher bandwidth as compared to laser, hence poor-resolution images as compared to monochromatic light source.

In this chapter we describe the full-field spatial coherence gated tomography and topography using temporally highly coherent and spatially incoherent light source. First, briefly the basic principle of LCI, Weinner–Khintchine theorem and resolution of conventional OCT systems is described using spectrally broadband light source. Then the spatial (both transverse and longitudinal) coherence of a spatially extended monochromatic light source is described. The applications of longitudinal spatial coherence (LSC) spatially extended monochromatic light for tomography and topography is detailed. The main purpose in this chapter is first to highlight the potential of performing high-resolution full-field spatial coherence tomography based on the LSC rather than the temporal coherence length of the source. In addition, a simplified model is developed to obtain a high optical sectioning of biological samples using a spatially incoherent monochromatic light source.

10.2 Principle of Low Coherence Interferometry Based on Temporal Coherence of the Light Source

Figure 10.1 shows the schematic diagram of a free space version of an OCT system. Light from a low coherence (broadband) light source is passed through a beam expander (BE) and a spatial filter (SF) unit and is collimated by a lens. The collimated beam is further passed through a beam splitter (BS) and the input beam is then divided into two beams, one propagates towards the reference mirror and the other propagates towards a multilayer sample [1–3]. A low-NA microscope objective lens is then used to focus the light onto the multilayer sample. Light backscattered from the multilayer sample and backreflected from reference mirror is collected by the lens and both the beams are superposed at the BS. The interference between the two beams takes place only when the OPD between the light reflected from the reference mirror and backscattered from the object is less than the coherence length of the light source. In TD-OCT path length of the reference arm is translated longitudinally in time. Interference takes place between the light reflected from the reference mirror and reflections from the different layers of the multilayer sample, as shown in the inset of Fig. 10.1 [1–3].

Due to low coherence of the light source a series of interferograms are recorded as shown in the inset of the Fig. 10.1 by means of sweeping the reference mirror in the axial direction. Each

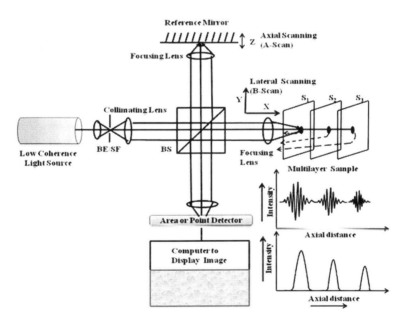

Figure 10.1 Schematic diagram illustrating the experimental setup based on time domain OCT.

interferogram is related to different depths in the sample. If a point detector is used then the envelope of this modulation is recorded at each point. A 2D and 3D image is built by scanning the beam in XYZ directions. On the other hand, if an area detector is used then 2D inteferometric image is recorded at each depth which does not require 2D XY scanning of the beam on the sample. The intensity $I(x, y, \tau)$ of the interference of two partially coherent light beams can be expressed as [1, 3].

$$I(x, y, \tau) = I_R(x, y) + I_S(x, y) + 2\sqrt{I_R(x, y)I_S(x, y)}\mathrm{Re}\gamma(\tau), \quad (10.1)$$

where $I_R(x, y)$ and $I_S(x, y)$ are the intensities of the reference and sample beams, respectively, and $\gamma(\tau)$ is the complex degree of temporal coherence, that is, the interference envelope and τ is the time delay. Due to the coherence gating effect of OCT the complex degree of coherence is represented as a Gaussian function expressed as [1, 2]

$$\gamma(\tau) = \exp\left[-\left(\frac{\pi \Delta \nu \tau}{2\sqrt{\ln 2}}\right)^2\right] \cdot \exp(-j 2\pi \nu_0 \tau), \quad (10.2)$$

where $\Delta \nu$ is the width of the temporal frequency spectrum and ν_0 is the centre frequency of the source. The relationship between the coherence function $\gamma(\tau)$ and the spectral density of the broadband light source is described in the following section.

10.2.1 The Wiener–Khintchine Theorem of Temporal Coherence

This theorem gives the relation between the spectral density of the light source and the correlations of light vibrations (waves) $V(r, t)$ and $V(r, t + \tau)$ at two instance of times t and $t + \tau$ (τ being the time delay between them) and at a fixed space point in the longitudinal direction [47, 48]. According to this theorem the autocorrelation function $\Gamma(\tau)$ of a statistically stationary random process and the source spectral density $S(\omega)$ of the process form a Fourier Transform pair [47, 48]

$$\Gamma(\tau) = \int_{-\infty}^{\infty} S(\omega) e^{-i\omega\tau} d\omega \quad (10.3)$$

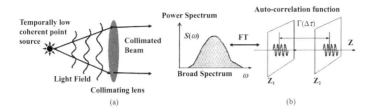

Figure 10.2 (a) Illustrating the light field generated by a broadband point source (b) illustrating the Wiener–Khintchin theorem, that is, the relationship between autocorrelation function and power spectrum of the temporally low-coherent (broadband) light source.

and

$$S(\omega) = \pm \frac{1}{2\pi} \int_{-\infty}^{\infty} \Gamma(\tau) e^{i\omega\tau} d\tau, \quad (10.4)$$

where $\omega = 2\pi\,{}^{\nu}\!/_{c}$ is the angular frequency of light vibrations and c is the speed of light in vacuum. The autocorrelation function is defined as [47, 48]

$$\Gamma(\tau) = \langle V^*(r,t) V(r, t+\tau)\rangle = \lim_{T\to\infty} \frac{1}{2\pi} \int_{-T}^{T} V^*(r,t) V(r, t+\tau) dt, \quad (10.5)$$

where $V^*(r,t)$ is the complex conjugate. Figure 10.2a shows the light field generated by temporally a low coherent (broad spectral width) point source. This kind of source has a very narrow angular frequency spectrum. According to the Wiener–Khintchin theorem, the autocorrelation function and the power spectrum of the temporally low-coherent (broadband) light source form a Fourier transform pair, as shown in Fig. 10.2b [47]. In a conventional OCT system this principle is fully utilized for optical sectioning using low temporal coherence of light.

A measure of the coherence which is sensitive to the optical intensity detected by the detector is provided by the normalized autocorrelation function which is called the complex degree of temporal coherence $\gamma(\tau)$ and is expressed by the following expression [47, 48]:

$$\gamma(\tau) = \frac{\Gamma(\tau)}{\Gamma(0)} = \frac{\langle V^*(r,t) V(r, t+\tau)\rangle}{\langle V^*(r,t) V(r, t)\rangle}; \quad 0 \le |\gamma(\tau)| \le 1. \quad (10.6)$$

The degree of temporal coherence can be determined from the fringe visibility measurement in a Michelson interferometer. For a Gaussian spectral distribution function the shape of $\gamma(\tau)$ is also Gaussian and is given by Eq. 10.2. Coherence time and source spectral bandwidth are inversely proportional to each other. Coherent time is the measure of the time interval during which the appreciable amplitude and phase correlations of the light vibrations at a particular point Q in a fluctuating optical field persists [47]:

$$\Delta\tau = \frac{1}{\Delta\nu}. \tag{10.7}$$

A measure of these correlations of the light vibrations at point Q in a stationary optical field during the time interval τ is the self-coherence or the autocorrelation function and given by Eqs. 10.3 and 10.5 [47, 48]

$$(\Delta\tau)^2 = \frac{\int_{-\infty}^{\infty} \tau^2 |\Gamma(\nu)|^2 S^2(\nu) d\nu}{\int_{-\infty}^{\infty} S^2 |\Gamma(\tau)|^2 d\tau}, \tag{10.8}$$

where $(\Delta\tau)^2$ is coherence time as the normalized root-mean-square (rms) width of the squared modulus of $\Delta\tau$. Now the spectral density is given by [48]

$$S(\nu) = \int_{-\infty}^{\infty} \Gamma(\tau) e^{i2\pi\nu\tau} d\tau. \tag{10.9}$$

The effective spectral width $\Delta\nu$ is

$$(\Delta\nu)^2 = \frac{\int_0^{\infty} (\nu - \bar{\nu})^2 S^2(\nu) d\nu}{\int_0^{\infty} S^2(\nu) d\nu}, \tag{10.10}$$

where

$$\bar{\nu} = \frac{\int_0^{\infty} \nu S^2(\nu) d\nu}{\int_0^{\infty} S^2(\nu) d\nu}. \tag{10.11}$$

In a Michelson interferometer the interference fringes are formed only when the following condition is satisfied [47, 48]:

$$\Delta v \times \Delta \tau \leq 1; \quad \text{or} \quad \Delta \tau = \frac{1}{\Delta v} = \frac{\lambda^2}{c\Delta\lambda}. \tag{10.12}$$

Since $v = c/\lambda$, where λ is the wavelength of light, $\Delta v \approx c\Delta\lambda/\lambda^2$. Equation 10.12 shows that the interference fringes with appreciable fringe visibility can be detected by the detector in Fig. 10.1 only when the condition shown in Eq. 10.12 is satisfied, that is, when the OPD between the two beams arriving at the BS after the round trip travel is [1–3]

$$\Delta l_c = l_2 - l_1 = c\Delta\tau = \frac{c}{\Delta v} = \frac{\lambda^2}{c*\Delta\lambda}, \tag{10.13}$$

where c is the speed of light in vacuum and Δl_c is known as the longitudinal coherence length of the light of vibrations which is related to temporal frequency spectrum of light. Equation 10.13 is the basis for calculating the axial resolution of conventional OCT systems.

10.2.2 Resolution of OCT Systems Based on Temporally Low Coherent Light

The axial resolution of an OCT system which is governed by the longitudinal coherence length which is inversely proportional to the spectral bandwidth of the light source, whereas the transverse resolution depends on the NA of the objective lens and focusing a beam to a small spot size [1–3]. In conventional OCT systems low-NA objective lenses are used to generate cross-sectional images resulting in low transverse resolution for the biological cell imaging. In Eq. 10.2 the Gaussian envelope is amplitude modulated by an optical carrier. The peak of this envelope represents the location of sample under test with amplitude dependent on the reflectivity of the surface (please see inset of Fig. 10.1). The axial and lateral resolutions of OCT are decoupled from one another, the former being an equivalent to the coherence length of the light source and the later being a function of the focusing optics. The coherence length (l_c) based on the temporal frequency spectrum of a light source and hence the axial resolution δz of OCT is given by Eq. 10.14 below

[1–10]. The coherence length (l_c) is the spatial width of the field autocorrelation envelope produced by the interferometer which is purely based on the temporal coherence of the light source. Thus the axial resolution is given by the width of the field autocorrelation function which is inversely proportional to the spectral bandwidth of the light source [1, 3]

$$\delta z = \frac{l_c}{2} = \frac{2 \ln 2}{\pi} \frac{\lambda_0^2}{\Delta \lambda} \qquad (10.14)$$

or

$$\delta z = \frac{l_c}{2 n(\lambda)} = \frac{2 \ln 2}{\pi \, n(\lambda)} \frac{\lambda_0^2}{\Delta \lambda}, \qquad (10.15)$$

where l_c and $\Delta \lambda$ are the full-width at half-maximum (FWHM) of the autocorrelation function and power spectrum of the light source, respectively, λ_0 is the source center wavelength, and $n(\lambda)$ is the refractive index of the medium which is to be imaged. From Eq. 10.15 it is clear that the larger is the spectral bandwidth of the light source, the higher is the axial resolution. Therefore, for obtaining very high axial resolution a large spectral bandwidth of the light can be used. But the use of a large spectral bandwidth of the light source leads to dispersion effect and the refractive index of the sample is wavelength dependent and is given by the following expression [1–3]:

$$n(\lambda) = c \frac{dk}{d\omega} = n - \frac{dn}{d\lambda}. \qquad (10.16)$$

From Eqs. 10.15 and 10.16, it is clear that dispersion also affects the axial resolution of OCT. If the dispersion is zero then $dn/d\lambda$ is zero, which simply gives $n(\lambda) = n$, where n is the refractive index of the sample. Then we get the axial resolution as [1, 2]

$$\delta z = \frac{l_c}{2} = \frac{2 \ln 2}{\pi \, n} \frac{\lambda_0^2}{\Delta \lambda}. \qquad (10.17)$$

This equation is valid only for nondispersive media [1–3] and hence the larger is the bandwidth of the light source, the smaller is the l_c, and the better is the axial resolution. But most of the biological samples are dispersive in nature and hence the axial resolution is calculated using Eq. 10.15 only. Further, the dispersion degrades the image quality and hence all the OCT systems based on broadband light require dispersion compensation optics.

Lateral or transverse resolution depends upon the focusing condition and pixel size of the CCD. For a focused Gaussian beam by a lens of focal length f and diameter D, transverse resolution is given by [1, 2]

$$\Delta X = \frac{4}{\pi}\lambda F\#, \quad \text{where} \quad F\# = \frac{f}{D}. \quad (10.18)$$

In conventional OCT a low-NA objective lens is used; therefore, the transverse resolution of the system is limited by above equation. The transverse resolution for the OCT imaging is the same as for the conventional optical microscopy and is determined by the focusing properties of an optical beam. The minimum spot size to which an optical beam can be focused is inversely proportional to the NA of the angle of focus of the beam. The transverse resolution is [1, 2]

$$\Delta x = \frac{4\lambda}{\pi n}\frac{f}{d}, \quad (10.19)$$

where d is the spot size of the beam on the objective lens and f is the focal length. From the above expression it is clear that transverse resolution is directly proportional to the wavelength λ. If the λ of light is large, then Δx is large, leading to lower transverse resolution of the OCT image. This is the reason for lower transverse resolution in conventional methods like ultrasound where sound waves (having larger wavelength) are used as compared to the OCT technique. In terms of NA of the objective lens the transverse resolution of conventional OCT can be expressed as [1–3]

$$\Delta x = 1.22\frac{\lambda}{2\text{NA}} = 0.61\frac{\lambda}{\text{NA}}. \quad (10.20)$$

Equation 10.20 shows that to improve the transverse resolution high-NA objectives are used in OCT which is known as optical coherence microscopy (OCM) [2, 4]. The use of high-NA objectives in OCM leads to better transverse resolution and in this case en face imaging is more effective than cross-sectional imaging. In OCM the unwanted scattered light coming from the sample is rejected by coherence gating and it become more useful for imaging tissues because of high contrast and improved imaging depth.

In this chapter we describe the full-field spatial coherence gated tomography using temporally highly coherent and spatially incoherent light source. Spatial coherence of the light source was reduced by generating multiple point sources with a combination of static diffuser and vibrating multimode fiber bundle (MMFB). Due to low spatial coherence of the light source the resolution of the system was achieved similar to that of conventional OCT systems using a broadband light source, that is, temporally low coherent. The main advantage of the present technique is that it is nearly free of chromatic dispersion [49–57]. The use of a high-NA objective lens gives both high axial and transverse resolutions. The speckles which arise due to the highly coherent laser source can be removed by using a vibrating MMFB [58]. The details are described below.

10.3 Principle of Spatial Coherence Gated Tomography with a Monochromatic Light Source

Attempts have been made by various research groups to utilize the property of LSC in the field of surface profilometry [49–70]. The effect of spatial coherence has been studied for the layer thickness determination and optical tomography by Safrani et al. [57] using a narrow spectral bandwidth light source obtained by passing the white light through a band pass filter of 10 nm bandwidth. However, the effect of the spatial coherence on optical sectioning of biological samples has not been demonstrated using a spatially incoherent, highly monochromatic light source like a laser so far. In this case instead of using a temporally low coherent point source one can use a spatially incoherent monochromatic extended source. In this case the van Cittert–Zernike theorem of spatial coherence and source intensity distribution is utilized. Figure 10.3a shows the light field generated by temporally highly coherent (monochromatic) and spatially incoherent extended source. This kind of source has a very large angular frequency spectrum. According to the van Cittert–Zernike theorem, the mutual coherence function in the far zone and source intensity distribution form a Fourier transform pair, as shown in Fig. 10.3b [47, 48]. Figure 10.3b gives the concept of both lateral (transverse) and LSC function of light vibrations generated by a spatially incoherent monochromatic extended source. In this case although the temporal coherence length of light source is large but the correlations between the light vibrations, at two spatial points r_1

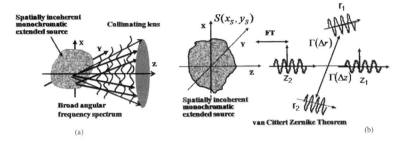

Figure 10.3 (a) Illustrating the light field generated by a spatially incoherent monochromatic extended light source (b) illustrating the van Cittert–Zernike theorem, that is, the relationship between mutual coherence function and intensity distribution across spatially extended light source.

and r_2 in the transverse direction (perpendicular to the direction of propagation z) and in the longitudinal direction at two spatial points z_1 and z_2, are dependent of the spatial coherence but not on the temporal coherence of light source. More details are illustrated in Fig. 10.4a.

Figure 10.4a shows the schematic diagram of monochromatic extended light source illuminating the two pinholes in the lateral as well as in the longitudinal directions. Generally in the practical situations both the size and temporal frequency spectrum of light source is finite. That is if the source is having extended size, the angular spectrum is finite and the temporal frequency spectrum is also broad (broadband source). In this case the two beam interference law can be written in the generalized form [47, 48, 50]

$$I(x, y) = I_1(x, y) + I_2(x, y) + 2\sqrt{I_1(x, y) I_2(x, y)} \, |\gamma_{12}(\Delta r, \Delta z, \Delta t)| \cos$$
$$\times \left\{ \frac{2\pi}{\lambda_0} 2\Delta z + \Lambda \phi_{12}(r, \Lambda z) \right\}, \quad (10.21)$$

where $\gamma_{12}(\Delta r, \Delta z, \Delta t)$ is the normalized degree of coherence defined as

$$\gamma_{12}(\Delta r, \Delta z, \Delta t) = \frac{\Gamma_{12}(\Delta r, \Delta z, \Delta t)}{\Gamma_{11}(\Delta r, \Delta z, \Delta t)\Gamma_{22}(\Delta r, \Delta z, \Delta t)}, \quad (10.22)$$

where $\Delta r = r_1 - r_2$, $\Delta z = z_1 - z_2$, and $\Delta t = t_1 - t_2$. Equations 10.21 and 10.22 are generalized and from these equations we can explain the lateral (transverse) and LSC separately under the assumption that light source is monochromatic and spatially incoherent extended source using Fig. 10.4a,b [52].

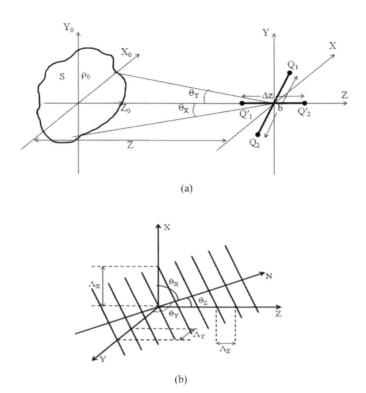

Figure 10.4 (a) Monochromatic extended light source illuminating the two pinholes in the lateral as well as in longitudinal directions; (b) illustrating the angular frequency spectrum of the light vibrations [50, 52].

10.3.1 Transverse (Lateral) Spatial Coherence of Light Vibrations

If the two spatial points $Q_1(r_1, \tau)$ and $Q_2(r_2, \tau)$ are located in a plane perpendicular to the direction of propagation z, as shown in Fig. 10.4a, then the intensity interference law Eq. 10.21 can be written as [50, 52]

$$I(x, y) = I_1(x, y) + I_2(x, y) + 2\,|\Gamma(\Delta r, \Delta z = 0, \Delta t = 0)|\cos$$
$$\times \left(\frac{2\pi}{\lambda}\Delta R + \beta_{12}(0)\right), \qquad (10.23)$$

where $\Gamma(\Delta r, \Delta z = 0, \Delta t = 0)$ is the lateral spatial coherence function, λ is the wavelength of light, and $\beta_{12}(0)$ is the phase of the degree of spatial coherence $\Delta R = R_1 - R_2$ difference between geometrical distances traveled by light vibrations coming from two pinholes and meeting on the observation point. This is also called the equal-time correlation function which is defined as correlation of two light vibrations arriving simultaneously at points $Q_1(r_1)$ and $Q_2(r_2)$ and is given by $\Gamma(\Delta r, \Delta t = 0) = \langle V(\overline{r_1}, t) V^*(\overline{r_2}, t) \rangle$ and the normalized coherence function is given by [47, 48]

$$\gamma(r_1, r_2, 0) = \frac{\Gamma(r_1, r_2, 0)}{\Gamma(r_1, r_1, 0)\Gamma(r_2, r_2, 0)}. \quad (10.24)$$

The van Cittert–Zernike theorem [47, 48, 52] is used for the determination of lateral or transverse spatial coherence properties of an optical field (light vibrations). This can be achieved by setting up Young's two-pinhole experiment in which a quasi-monochromatic ($\Delta \nu \ll \nu_0$) light generated from an extended source is used to illuminate a screen containing two pinholes Q_1 and Q_2 in the far zone. In this case the two pinholes are located in the lateral direction perpendicular to the direction of propagation (longitudinal direction) as shown in Fig. 10.4a. In this situation the time of arrival of light vibrations at two pinholes Q_1 and Q_2 is the same. If these two pinholes are within the coherence area of the extended light source, projecting on the screen containing two pinholes, the interference fringes will be formed on the observation screen. From the visibility measurements as a function of separation of two pinholes b the degree of lateral spatial coherence of light vibrations illuminating two pinholes can be determined from the following formula [47, 48]:

$$V_{12}(0) = |\gamma_{12}(0)| = \frac{I_{\max} - I_{\min}}{I_{\max} + I_{\min}} = \left| \frac{2J_1\left(\frac{ab\omega}{2\varepsilon r}\right)}{\left(\frac{ab\omega}{2\varepsilon r}\right)} \right|, \quad (10.25)$$

where $I_{\max}$ is the maximum value of intensity measured on a bright fringe and $I_{\min}$ is the intensity measured in the dark fringe, a is the source size, $\omega = 2\pi \nu$ is the angular frequency, r is the distance between the sources and plane containing two pinholes, and J_1 is a Bessel function of the first order and first kind. This is also called equal-time degree of spatial coherence. According to the van Cittert–Zernike theorem the Fourier transform of the degree of spatial coherence can be used to reconstruct the source distribution and angular size of the source [47, 48], as shown in Fig. 10.3b.

For lateral spatial coherence the temporal delay $\Delta t = 0$ is assumed. This is also called the mutual coherence function which is the correlation of light vibrations at two different spatial points $Q_1(r_1, \tau)$ and $Q_2(r_2, \tau)$. For a statistically stationary quasi-homogeneous optical field in which the statistical parameters are slowly varying in space compared to the transverse correlation length l_c, which is varying fast. It is possible to use the Weiner–Khintchin theorem for the determination of correlation of light vibrations. In accordance with this theorem the lateral spatial coherence function of statistically stationary quasi-homogeneous optical field is [50, 52]

$$\Gamma(r_1, r_2, \Delta t = 0) = \Gamma(\Delta r, \Delta t = 0) = \iint_{-\infty}^{\infty} W(\vec{k_{xy}}) e^{[i(\vec{k_{xy}} \Delta \vec{r})]} d\vec{k_{xy}}$$

$$= \iint_{-\infty}^{\infty} W(k_x, k_y) e^{[i(k_x \Delta x + k_y \Delta y)]} dk_x dk_y, \quad (10.26)$$

where $\Delta \vec{r}(\Delta x, \Delta y)$ is the difference of transverse coordinates of points $Q_1(r_1)$ and $Q_2(r_2)$, $W(k_x, k_y)$ is the spatial power spectral density of the field in the plane (x, y) perpendicular to the main propagation direction z of the field, and k_x and k_y are the circular spatial frequency components of the field in the directions x and y [50, 52]

$$k_x = \frac{2\pi}{\Lambda_x} = \frac{\omega}{c} \cos \theta_x, \quad (10.27a)$$

$$k_y = \frac{2\pi}{\Lambda_y} = \frac{\omega}{c} \cos \theta_y, \quad (10.27b)$$

where Λ_x and Λ_y are the spatial periods of the wave in the x and y directions and θ_x and θ_y are the angles between wave propagation direction, as shown in Fig. 10.2b. For a spatially incoherent quasi-monochromatic planer light source, the relationship between the spatial power spectral density $W(k_x, k_y)$ in the observation plane and the intensity distribution $I(x_0, y_0)$ in the source plane can be obtained by the van Cittert–Zernike theorem [50–55]:

$$\Gamma(\Delta x, \Delta y) = \int \int_{-\infty}^{\infty} I(x_0, y_0) e^{[i \frac{2\pi}{\lambda_0 z}(x_0 \Delta x + y_0 \Delta y)]} dx_0 dy_0, \quad (10.28)$$

where we have assumed

$$\cos\theta_x = \frac{x_0}{z}, \cos\theta_y = \frac{y_0}{z}, k_x = \frac{2\pi x_0}{\lambda_0 z}, k_y = \frac{2\pi y_0}{\lambda_0 z}; \quad (10.29)$$

$$W(k_x, k_y)dk_x dk_y \approx I(x_0, y_0)dx_0 dy_0. \quad (10.30)$$

And the lateral coherence lengths can be given by [52]

$$\rho_x = \frac{2\pi}{\Delta k_x}, \quad (10.31a)$$

$$\rho_y = \frac{2\pi}{\Delta k_y}. \quad (10.31b)$$

It can be seen from Eqs. 10.31a and 10.31b that the lateral spatial coherence lengths are inversely proportional to the bandwidth of the angular frequency spectrum. Therefore, the larger is the bandwidth of the angular frequency spectrum, the smaller is the lateral spatial coherence length.

10.3.2 Longitudinal Spatial Coherence of Light Vibrations

As shown in Fig. 10.4a if the two pinholes Q_1' and Q_2' are located in the direction of propagation (i.e., in the longitudinal direction) separated by a distance Δz and the light vibrations reaching at points Q_1' and Q_2' are in phase then the correlation between these two light vibrations can be called LSC [50–55]. Under these circumstances a generalized form of the van Cittert–Zernike theorem can be used to determine the LSC of the optical field of having sufficiently narrow frequency spectrum (quasi-monochromatic light). A manifestation of LSC of light vibrations generated by a quasi-monochromatic extended source having sufficiently wide angular spectrum have been observed using a Michelson-type interferometer [50–55]. Based on these observations one can conclude that the temporal coherence is determined by the temporal frequency spectrum with large spectral bandwidth and spatial coherence can be determined by the angular spectrum (having a narrow temporal frequency spectrum). If the size of the light source is small (i.e., a point source) then the coherence length or the coherence time or simply the longitudinal coherence properties of the light vibrations can be purely determined by the temporal frequency spectrum (spectral distribution) of the light source using a Michelson interferometer [1–10]. Therefore, the larger is the bandwidth of the light source, the smaller is the coherence length or the smaller is the coherence time $\Delta\tau$. In the case of a point source and a collimated light field the angular spectrum of the light source is extremely narrow or negligible and hence does not contribute to the longitudinal coherence of the light source.

On the other hand, if the size of the light source is large (extended) and the temporal frequency spectrum is narrow (quasi-monochromatic, $\Delta\nu \ll \nu_0$) then the angular frequency spectrum of the light source is wide and hence the longitudinal coherence properties can be purely determined by the angular frequency spectrum of the light source. In this case the longitudinal coherence length is small, whereas the coherence length due to temporal frequency spectrum is very large ($l_{lsc} \ll l_c$). Such a light source can be realized by passing the laser light through a rotating ground glass, which is also known as a pseudothermal light source [50–58]. If the light from a such as a pseudothermal light source passes through a Michelson interferometer then the longitudinal coherence properties of the light vibrations can be purely determined by the angular frequency spectrum of the light source, even though the coherence length due to temporal frequency spectrum of the light source is very large ($l_{lsc} \ll l_c$). Another method for generating such a light source is by means of passing the broadband light through a narrow spectral transmission filter, such as a holographic notch filter and an interference filter, and making the size of the source large [56–57]. But in this case the spectral bandwidth of the light source is still large ($\sim$10 nm) and hence the effect of temporal coherence due to the temporal frequency spectral bandwidth will also be contributing to the axial resolution of the OCT system. According to Fig. 10.4a, if the two spatial points Q_1' and Q_2' are located in the direction of propagation and if the optical field is statistically stationary and quasi-homogeneous (if its statistical properties are changed slowly over the longitudinal coherence length) the longitudinal coherence function can be written using Wiener–Khintchin theorem [50, 52]

$$\Gamma(r_1, r_2, \Delta t = 0) = \int_{-\infty}^{\infty} W(k_z) e^{[i(k_z \Delta z)]} dk_z, \qquad (10.32)$$

where k_z is angular spatial frequency in the longitudinal direction. $W(k_z)$ is the spatial power spectral density in the longitudinal direction:

$$k_z = \frac{2\pi}{\Lambda_z} = \frac{2\pi}{\lambda}\cos\theta_z = \frac{\omega}{c}\cos\theta_z. \tag{10.33}$$

The longitudinal coherence length is [50, 52]

$$l_{\text{lsc}} = \frac{2\pi}{\Delta k_z}, \tag{10.34}$$

where Δk_z is the longitudinal spatial frequency range that is the width of the spatial frequency spectrum [50, 52]:

$$I(x,y) = I_1(x,y) + I_2(x,y) + 2\sqrt{I_1(x,y)I_2(x,y)}$$
$$|\Gamma(\Delta r = 0, \Delta z, \Delta t = 0)|\cos\left(\frac{2\pi}{\lambda_0}2\Delta z + \Delta\phi_{12}(r,\Delta z)\right). \tag{10.35}$$

For the observation of LSC, the length of LSC l_{lsc} should be smaller than the temporal coherence length, $l_{\text{lsc}} \ll l_c$ where $l_{\text{lsc}} \approx \frac{2\lambda}{\theta^2}$. To achieve this, the angular frequency spectrum of the light field should be sufficiently broad and temporal coherence length should be sufficiently large ($\Delta\lambda \ll \lambda_0$), that is, the temporal frequency bandwidth of light source should be very small. Then the longitudinal coherence of light will be purely determined by the LSC, that is, angular frequency spectrum of the extended monochromatic light source.

$$\Gamma(\Delta r = 0, \Delta z, \Delta t = 0) = \Gamma(\Delta z). \tag{10.36}$$

The interference fringes will appear only when $l_{\text{lsc}} \leq \Delta z$ and the fringes will disappear when $l_{\text{lsc}} \geq \Delta z$ and even though the OPD, $2\Delta z < l_c$. Therefore, a spatially extended quasi-monochromatic light source can be used to optically section the multilayer structure of the sample even though the temporal coherence length is large enough. Such a light source have been realized by many authors by passing the laser light through a rotating ground glass, which is also known as a pseudothermal light source and have been used for optical coherence tomography and profilometry [56–71].

10.4 Principle of Spatial Coherence Gated Microscopy with Monochromatic Light

The FF-OCT systems that utilize high-magnification and high-NA microscope objective lenses to focus light tightly on the object surfaces are known as optical coherence microscopes, which give en face high-resolution images [56, 57]. It has been shown that the interference signal recorded by a CCD camera for a spatially extended incoherent light source is given by [56, 57, 71]

$$I_D(x', y', \Delta z) \propto |E_{image}(x', y', \Delta z)|^2 \propto I_0(x, y)\{1 + \gamma(x, y) A(\Delta z - l(x, y)) \cos 2\pi f_z [\Delta z - l(x, y)]\}, \quad (10.37)$$

(x', y') and (x, y) are the lateral coordinates in image and object space, respectively. I_D is total intensity projected on each pixel of the camera and E_{image} is total electric field projected on each pixel (x', y'), I_0 is the direct current component, γ is the fringe contrast, f_z is the spatial carrier frequency, A is the axial response, Δz is the scanning distance, and l is the sample distance from the top surface of the sample [56, 57, 71].

$$\gamma(x, y) = 2\sqrt{R_R R_S(x, y, l)} / [R_{scat}(x, y) + R_R R_{Stotal}(x, y)], \quad (10.38)$$

$$A(\Delta z - l(x, y)) = \text{sinc}\left\{\frac{n_0 NA^2 \pi}{\lambda}\{\Delta z - l(x, y)\}\right\}, \quad (10.39)$$

$$f_z = \frac{n_0}{\lambda/2}\left(1 - \frac{NA^2}{4}\right). \quad (10.40)$$

In Eq. 10.38, R_R is the reference reflectivity, R_S is the sample reflectivity, and R_{Stotal} and R_{scat} are the total reflectivity coefficients of all resolvable and irresolvable microstructures, respectively. In Eq. 10.39, n_0 is the index of matching material and NA is the numerical aperture in air [56, 57, 71]:

$$\Delta z(x, y)_{peak_N} = \sum_{i=0}^{N} l_i(x, y) \frac{n_0}{n_i}. \quad (10.41)$$

By moving the sample in the upward and downward direction or by changing Δz, the output interference signal can be recorded from the different layers of the multilayered objects. In Eq. (10.41), the Nth peak corresponds to the $N+1$ interface and $l_i(x, y)$ is the thickness of the ith layer width $l_0(x, y) = 0$. By defining the resolution of the system according to the first zero of the sinc function, one obtains the following relation [56, 57, 71]:

$$\Delta Z^i_{\text{axial}} = \frac{\lambda n_i}{\text{NA}^2}. \qquad (10.42)$$

Hence, by using a high-NA one can obtain the high longitudinal spatial axial resolution. It has been shown that when a narrow-band light source is combined with a high-NA microscope objective, the longitudinal coherence does not depend on the spectral bandwidth $\left(\Delta Z \sim \frac{\lambda^2}{\Delta \lambda}\right)$ but on the NA of the objective lens $\left(\Delta Z \approx \frac{\lambda}{\text{NA}^2}\right)$ [56, 57, 71], where λ is the central wavelength, $\Delta \lambda$ is the spectral bandwidth of the source, NA is the numerical aperture of the objective lens, and ΔZ is the axial resolution. Thus even with a monochromatic light source whose temporal coherence is very large, one can perform optical sectioning and obtain high-resolution tomographic images by utilizing the short LSC property.

10.5 Experimental Details for Spatial Coherence Gated Tomography with Monochromatic Light

The wide-field spatial coherence gated tomographic system was constructed by using a laser, an MMFB, a diffuser plate, a Mirau interferometer, and a CCD camera [58, 71].

A schematic diagram of the experimental setup is shown in Fig. 10.5 [71]. A He Ne laser (wavelength 632.8 nm, and power 15 mW) was made incident onto the beam expander (BE) and was collimated by a lens. The collimated beam was made incident on the stationary diffuser. The diffused light is collected by the MMFB of core diameter of 0.1 mm each fiber. The scattered light at the output of the MMFB is collected by the lens. The light was then collimated and made incident onto the Mirau objective (Model No. 503210, 50X/0.55 DI, WD 3.4 Nikon, Japan) installed on the reflection mode vertical microscope (NIKON ECLIPSE 50i, made in Japan). This objective lens is attached with a piezoelectric

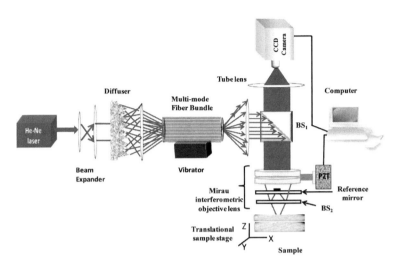

Figure 10.5 Schematic diagram of a spatial coherence gated tomographic system [71].

transducer (PZT) (Peizo, Jena, MIPOS 3). The PZT is driven by an amplifier which is controlled by a personal computer and it moves the objective lens with the in-built reference mirror in the vertical direction for introducing the phase shifts between the reference and object beams. A relay lens is used to image the object onto the CCD camera (Q-Imaging, 01-QIClick-F-M-12, Canada, 1392 × 1040 pixels with pixel size 6.25 × 6.25 μm^2). This is nearly common path and compact interferometric objective lens with high magnification and hence very well suitable for simultaneous imaging of biological cells and generating interferograms quickly because it is simple to align. The phase-shifted interferograms are recorded by a CCD camera. The biological sample was placed on the 3D (XYZ) translational object stage. The expanded laser beam illuminates the static diffuser plate, which generates the multiple scattering points (as shown in Fig. 10.5). The scattered rays are collected by the lens and focus toward the MMFB which consists of hundreds of fibers of equal length. The MMFB supports a multitude of propagating modes with different phase velocities as various rays propagate down the fiber with varying distances depending on their angle of propagation, and hence suffer different phase delays. The MMFB was vibrated to further reduce the speckles. Thus the output of the MMFB acts as mutually spatially incoherent point sources generating M independent speckle patterns, which add on an intensity basis therefore, the phase is completely randomized and hence the light is completely spatially incoherent.

10.5.1 Determination of Longitudinal Spatial Coherence Length

To show the influence of NA on the LSC length or on axial resolution, the spatial coherence tomographic technique was applied on variety of samples with different NA objective lenses. To obtain the LSC of the present system a number of interferograms were grabbed through the scanning process with different NA Mirau objective lenses in the axial direction in air, taking an aluminum-coated reflecting mirror as a sample. The data shown in Fig. 10.6a was taken by scanning the sample mirror vertically for determining the visibility of the interference fringes. These interferograms were stacked and the peak visibility is detected for each pixel. Figure 10.6b shows the visibility profile of these stacked interferograms for 50X (NA = 0.55) Mirau objectives in the axial direction at a particular xy pixel.

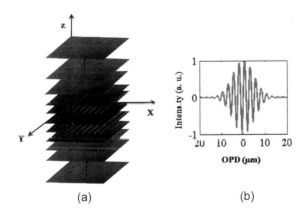

Figure 10.6 (a) Stacked interferograms for 50X Mirau objective in the axial direction, and (b) line profile of stacked interferograms as a function of OPD [71].

Different NA Mirau objective lenses were used to measure the LSC length. The obtained signal is very similar to the 1D TD-OCT signal. A similar procedure has been repeated for the 10X (NA = 0.3) and 20X (NA = 0.4), respectively. The data was obtained by translating the mirror in the vertical direction with 2 μm steps (for 10X) for several micrometers till the visibility of the fringes drops nearly zero. The recorded 2D interferograms were analyzed by fast Fourier transform (FFT) and the visibility and complex degree of longitudinal spatial coherence was calculated by taking the ratio of the first-order peak to that of the zero-order peak in the Fourier spectrum. Figure 10.7a–c shows the variation of the degree of longitudinal spatial coherence with the OPD, using Mirau interferometric objectives of varying magnifications. It is

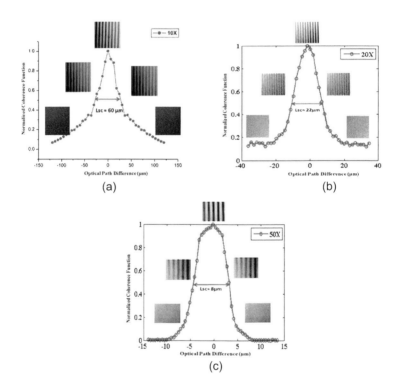

Figure 10.7 LSC length for (a) 10X (NA = 0.30), (b) 20X (NA = 0.40), and (c) 50X (NA = 0.55) NA objective lens.

observed that the fringe contrast and the corresponding degree of longitudinal spatial coherence is maximum for zero path difference and decreases as the OPD increases. The FWHM of the longitudinal spatial coherence curve is calculated to 60 μm and this gives an estimate of the LSC length of the system for 10X (NA = 0.3) magnification. As the present system is based on reflection mode, hence the axial resolution is half that of the round-trip coherence length ($\Delta z = \frac{LSC}{2}$), and calculated to be 30 μm. A similar procedure was followed to calculate the visibility and degree of spatial coherence by using 20X (NA = 0.4) and 50X (NA = 0.55) Mirau objectives, as shown in Figs. 10.7b and 10.7c, respectively. The recorded interferograms with varying OPDs are shown in the inset of Fig. 10.7a–c. The calculated values of LSC and ΔZ are 22 μm and 11 μm for the 20X objective, while for 50X, the corresponding values are 8 μm and 4 μm in air, respectively. In Fig. 10.7a–c the experimental results are presented for different NA objective lenses, that is, 10X, 20X, and 50X for which LSC length comes out to be 60 μm, 22 μm, and 8 μm, respectively. It is observed from Fig. 10.7a–c that the higher is the NA of objective lens, the higher is the axial resolution which satisfies $\Delta Z \approx \frac{\lambda}{NA^2}$ [58, 71].

10.5.2 Application of Longitudinal Spatial Coherence in Profilometry

The feasibility of the proposed method is verified by measuring the shape variation of a discontinuous three-dimensional object. First we placed an integrated circuit (IC) on the sample stage. Step size between the IC and sample stage was obtained using Mirau interferometric objectives of 10X magnification combined with spatially incoherent monochromatic illumination ($\lambda = 632.8$ nm). A step size of 160 μm has been analyzed. Figure 10.8a shows the image of the sample and Fig. 10.8b the interference pattern recorded using a 10X Mirau objective lens at different depths which can be clearly seen. Figure 10.8c shows the surface profile of the object. The main purpose was to study the effect of spatial coherence on axial resolution of the system. Another sample, one rupee Indian coin was clamped to the sample stage. Cross-sectional images of features on

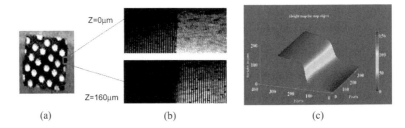

Figure 10.8 (a) Image of the step object, (b) interference pattern recorded at 0 μm and 30 μm along the axial direction, and (c) reconstructed 3D height map of the step object corresponding to (b).

a 1 rupee Indian coin were obtained using Mirau interferometric objectives of varying magnification (10X and 20X), combined with spatially incoherent monochromatic illumination ($\lambda = 632.8$ nm). The main purpose was to study the effect of increase in the NA of the objective lens on the LSC length. A feature size of 25 μm has been analyzed as shown in Fig. 10.9a. Figure 10.9b shows the interference pattern recorded using a 10X Mirau objective lens. It is clearly seen that while using a lower magnification objective lens (10X, NA 0.3) one cannot distinguish the height difference between the two layers as the feature height lies less than the axial resolution for 10X NA (30 μm, in air) of the system. As we move toward higher magnification (high NA) the axial resolution improves. As the higher magnification objective lens (20X, NA 0.4) is used, the coherence length decreases and sectional images of the two layers can be distinguished, as shown in Fig. 10.9b [71]. Five phase-shifted interferograms were recorded and analyzed. The height difference of 25 μm was analyzed and height variation reconstructed and is shown in Fig. 10.9c.

10.5.3 High-Resolution Longitudinal Spatial Coherence Gated Tomography

Finally, in Fig. 10.10 we present the application of the system to image a transparent multilayer onion sample and red blood cells. First, we placed two overlapping onion skin layers on the reflecting mirror. The thickness of the sample is not uniform. Interferograms were recorded from each onion layer. It is clearly seen from

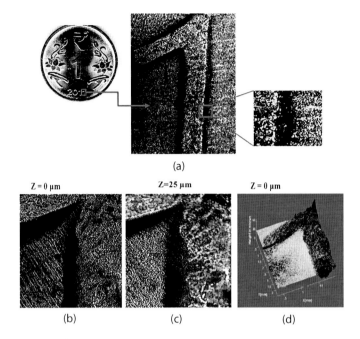

Figure 10.9 (a) Image of the coin and interference pattern recorded using 10X (NA = 0.3), (b) interference pattern recorded at 0 μm and 25 μm along the axial direction, and (c) reconstructed 3D height map of the coin.

Fig. 10.10a,b that when the interference fringes are formed from reflected light from the first layer of the onion skin then there is no interference between the light reflected from the second layer of the onion skin. As we move the translational stage upward in the axial direction (Z direction) the interference of light reflected from the second layer takes place and no interference takes place between the light reflected from first layer of the onion skin. This way we have recorded en face interferometric images of the multilayer object. The high contrast en face interferometric images of multilayer object are first recorded by translating the object stage (with a step of 0.5 μm for 50X) in the axial direction. At every step after locating the high contrast en face interferometric image of a particular object layer, five phase-shifted interferograms for each layer were recorded and the local variation of the phase map of the object surface is then reconstructed [71]. The interferograms were processed using the

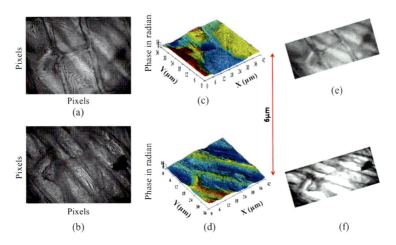

Figure 10.10 (a and b) Interferogram of upper and bottom layers, (c and d) corresponding unwrapped phase maps, and (g and h) amplitude images of the two onion skin layers, respectively [71].

five-step phase-shifted algorithm to reconstruct the phase maps. Figure 10.10c,d shows the corresponding unwrapped phase maps of Fig. 10.10a,b of the two different layers of the onion skin, which are 6 μm apart. $\Delta\phi(x, y)$ is calculated using phase map of reference plane and that of the object surface. We also reconstructed the amplitude images which are shown in Fig. 10.10e,f separated by 6 μm. From these unwrapped phase maps the refractive index of the sample can be calculated of the known sample thickness which will be helpful for quantitative phase imaging.

In the second sample we sandwiched prepared red blood cells (RBCs) between the cover slip and reflecting mirror. Figure 10.11a–c shows the recorded interferograms of the light reflected from the top layer, the bottom layer of the cover slip, and from the RBCs kept on the reflecting mirror, respectively. Figure 10.11d–f shows the corresponding unwrapped phase maps. The two layers of the cover slips are 160 μm apart. The RBC surface is 8 μm from the bottom layer of the cover slip. We also reconstructed the amplitude images which are shown in Fig. 10.11g–i, in which the two layer of cover slip are separated by 160 μm apart and the RBC surface is 8 μm from the bottom layer of the cover slip. This analysis shows the depth of penetration as well as axial sectioning power of the present setup.

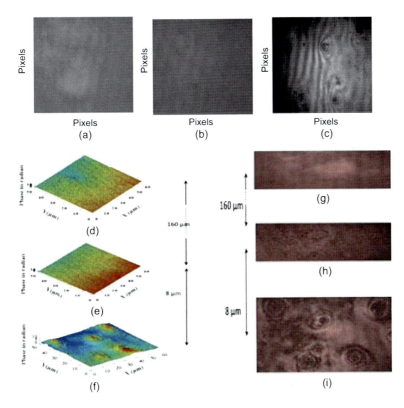

Figure 10.11 (a–c) Interferogram of upper, middle, and bottom layers of the sample, (d–f unwrapped phase maps, and (g–i) amplitudes images of top layer cover slip, middle layer cover slip, and human RBCs, respectively [71].

The obtained phase maps are useful in finding the refractive index of the samples. Measurement of the refractive index of biological cells is important for extracting vital information, such as hemoglobin concentration, average cell mass, thickness, etc.

10.6 Coherence Holography

Recently, the application of LSC has been demonstrated for coherence holography [72–75]. In this method an object recorded in a hologram is reconstructed as the 3D distribution of a complex spatial coherence function, rather than as the complex amplitude distribution of the optical field itself [72]. In this method the LSC of the monochromatic light source is engineered by passing the light through a rotating ground glass. With this method one can create an optical field distribution with a desired spatial coherence function, and visualize the coherence function in real time as the contrast and phase variations in an interference fringe pattern [72–75]. The reconstructed image of the complex coherence function can be quantified with the Fourier transform method of fringe pattern analysis. Because of its unique capability of controlling and synthesizing spatial coherence of quasi-monochromatic optical fields in 3D space [60], coherence holography has been applied for dispersion-free spatial coherence tomography [62] and profilometry [59, 60]. This opens up a new era in the area profilometry, tomography, and 3D imaging by holography using a temporally highly coherent and spatially low coherent light source.

10.7 Conclusion

This chapter gives detailed analysis of full-field spatial coherence tomography using a narrowband illumination light source (laser), which is based on the short LSC length rather than on the temporal coherence length of the source. The chapter also outlined the problem of temporal-coherence-based OCT systems and gives the solution. It is also shown that as the NA increases the axial resolution improves. Some examples of full-field spatial coherence tomography applications are added to underline the potential of spatial coherence gated detection. Additionally, the method was applied to for profilometry and for quantitative imaging of 3D biological and multilayered samples. It is also shown that a narrowband light source and a high-NA objective may be used for obtaining en face imaging of a highly scattering media as well.

Acknowledgments

Financial assistance from the Department of Science and Technology Delhi, Government of India, for Project No. SR/S2/LOP-0021/2008 is gratefully acknowledged.

References

1. Bouma, B. E., and Jearney, G. J. (2002). *Handbook of Optical Coherence Tomography*, Marcel Dekker, Inc.
2. Drexler, W., and Fujimoto, J. G. (2008). *Optical Coherence Tomography Technology and Applications* 1st Ed., Springer, USA.
3. Schuman, J. S., Puliafito, C. A., Fujimoto, J. G., and Duker, J. S. (2007). *Optical Coherence Tomography of Ocular Diseases* 3rd Ed., Springer.
4. Fercher, A. F., Drexler, W., Hitzenberger, C. K., and Lasser, T. (2003). Optical coherence tomography- principles and applications. *Rep. Prog. Phys.*, **66**, pp. 239–303.
5. Huang, D., Swanson, E. A., Lin, C. P., Schuman, J. S., Stinson, W. G., Chang, W., Hee, M. R., Flotte, T., Gregory, K., Puliafito, C. A., and Fujimoto, J. G. (1991). Optical coherence tomography. *Science*, **254**, pp. 1178–1181.
6. Fercher, A. F. (1996). Optical coherence tomography. *J. Biomed. Opt.*, **11**, pp. 157–173.
7. Stifter, D. (2007). Beyond biomedicine: a review of alternative applications and developments for optical coherence tomography. *Appl. Phys. B*, **88**, pp. 337–357.
8. Fujimoto, J. G., Brezinski, M. E., Tearney, G. J., Boppart, S. A., Bouma, B. E., Hee, M. R., Southern, J. F., and Swanson, E. A. (1995). Optical biopsy and imaging using optical coherence tomography. *Nat. Med.*, **1**, pp. 970–972.
9. Tearney, G. J., Bouma, B. E., Boppart, S. A., Golubovic, B., Swanson, E. A., and Fujimoto, J. G. (1996). Rapid acquisition of in-vivo biological images by use of optical coherence tomography. *Opt. Lett.*, **21**, pp. 1408–1410.
10. Fercher, A. F., Mengedoht, K., and Werner, W. (1998). Eye-length measurement by interferometry with partially coherent light. *Opt. Lett.*, **13**, pp. 186–188.
11. Schmitt, J. M., Knuttel, A., and Bonner, R. F. (1993). Measurement of optical properties of biological tissues by low-coherence reflectometry, *Appl. Opt.*, **32**, pp. 6032–6042.
12. Boppart, S. A., Bouma, B. E., Pitris, C., Southern, J. F., Brezinski, M. E., and Fujimoto, J. G. (1998). *In vivo* optical coherence tomography cellular imaging. *Nat. Med.*, **4**, pp. 861–865.
13. Podoleanu, A. G., Rogers, J. A., Jackson, D. A., and Dunne, S. (2000). Three dimensional OCT images from retina and skin. *Opt. Express*, **7**, pp. 292–298.

14. Hitzenberger, C. K., Drexler, W., and Fercher, A. F. (1992). Measurement of corneal thickness by laser Doppler interferometry. *Invest. Ophthalmol. Vis. Sci.*, **33**, pp. 98–103.
15. Fercher, A. F., Mengedoht, K., and Werner, W. (1998). Eye-length measurement by interferometry with partially coherent light. *Opt. Lett.*, **13**, pp. 186–188.
16. Schmitt, J. M., Knüttel, A., and Bonner, R. F. (1993). Measurement of optical properties of biological tissues by low-coherence reflectometry. *Appl. Opt.*, **32**, pp. 6032–6042.
17. Chinn, S. R., Swanson, E. A., and Fujimoto, J. G. (1997) Optical coherence tomography using a frequency-tunable optical source. *Opt. Lett.*, **22**, pp. 340–342.
18. Leitgeb, R., Hitzenberger, C. K., and Fercher, A. F. (2003). Performance of Fourier domain versus time domain optical coherence tomography. *Opt. Express*, **11**, pp. 889–894.
19. Leitgeb, R. A., Drexler, W., Unterhuber, A., Hermann, B., Bajraszewski, T., Le, T., Stingl, A., and Fercher, A. F. (2004). Ultrahigh resolution Fourier domain optical coherence tomography. *Opt. Express*, **12**, pp. 2156–2165.
20. Choma, M. A., Sarunic, M. V., Yang, C., and Izatt, J. A. (2003). Sensitivity advantage of swept source and Fourier domain optical coherence tomography. *Opt. Express*, **11**, pp. 2183–2189.
21. Deboer, J. F., Cense, B., Park, B. H., Pierce, M. C., Tearney, G. J., and Bouma, B. E. (2003). Improved signal-to-noise ratio in spectral-domain compared with time-domain optical coherence tomography. *Opt. Lett.*, **28**, pp. 2067–2069.
22. Wojtkowski, M., Bajraszewski, T., Targowski, P., and Kowalczyk, A. (2003). Real time in vivo imaging by high-speed spectral optical coherence tomography. *Opt. Lett.*, **28**, pp. 1745–1747.
23. Leitgeb, R. A., Drexler, W., Unterhuber, A., Hermann, B., Bajraszewski, T., Le, T., Stingl, A., and Fercher, A. F. (2004). Ultrahigh resolution Fourier domain optical coherence tomography. *Opt. Express*, **12**, pp. 2156–2165.
24. Choi, K. J. C., Soh, K. S., Ho, D. S., and Kim, B. M. (2005) Real spectral domain optical coherence tomography using a superluminescent diode. *J. Korean Phys. Soc.*, **47**, pp. 375–379.
25. Lim, H., de Boer, J. F., Park, B. H., Lee, E. C. W., Yelin, R., and Yun, S. H. (2006). Optical frequency domain imaging with a rapidly swept laser in the 815–870 nm range. *Opt. Express*, **14**, pp. 5937–5944.

26. Srinivasan, V. J., Huber, R., Gorczynska, I., and Fujimoto, J. G. (2007). High-speed, high-resolution optical coherence tomography retinal imaging with a frequency-swept laser at 850 nm. *Opt. Lett.*, **32**, pp. 361–363.
27. Sarunic, M. V., Choma, M. A., Yang, C., and Izatt, J. A. (2005). Instantaneous complex conjugate resolved spectral and swept-source OCT using 3×3 fiber couplers. *Opt. Express*, **13**, pp. 957–967.
28. Dubois, A., Vabre, L., Boccara, A. C., Beaurepaire, E. (2002). High resolution full-field optical coherence tomography with a Linnik microscope. *Appl. Opt.*, **41**, pp. 805–812.
29. Vabre, L., Dubois, A., and Boccara, A. C. (2002). Thermal-light full-field optical coherence tomography. *Opt. Lett.*, **27**, pp. 530–532.
30. Dubois, A., Grieve, K., Moneron, G., Lecaque, R., Vabre, L., and Boccara, C. (2004). Ultrahigh-resolution full-field optical coherence tomography. *Appl. Opt.*, **43**, pp. 2874–2883.
31. Dubois, A., Moneron, G., Grieve, K., and Boccara, A. C. (2004). Three-dimensional cellular-level imaging using full-field optical coherence tomography. *Phys. Med. Bio.*, **49**, pp. 1227–1234.
32. Yu, L., and Kim, M. K. (2004). Full-color three-dimensional microscopy by wide-field optical coherence tomography. *Opt. Express*, **12**, pp. 6632–6641.
33. Grieve, K., Dubois, A., Simonutti, M., Paques, M., Sahel, J., Le Gargasson, J. F., and Boccara, C. (2005). In vivo anterior segment imaging in the rat eye with high speed white light full-field optical coherence tomography. *Opt. Express*, **13**, pp. 6286–6295.
34. Moneron, G., Boccara, A. C., and Dubois, A. (2005). Stroboscopic ultrahigh-resolution full-field optical coherence tomography. *Opt. Lett.*, **30**, pp. 1351–1353.
35. Wang, H. W., Rollins, A. M., and Izatt, J. A. (1999). High-speed full-field optical coherence microscopy. *Proc. SPIE*, **3598**, pp. 204–212.
36. Sarunic, M. V., Weinberg, S., and Izatt, J. A. (2006). Full-field swept-source phase microscopy. *Opt. Lett.*, **31**, pp. 1462–1464.
37. Akiba, M., Chan, K. P., and Tanno, N. (2003). Full-field optical coherence tomography by two-dimensional heterodyne detection with a pair of CCD cameras. *Opt. Lett.*, **28**, pp. 816–818.
38. Oh, W. Y., Bouma, B. E., Iftimia, N., Yun, S. H., Yelin, R., and Tearney, G. J. (2006). Ultrahigh resolution full-field optical coherence microscopy using InGaAs camera. *Opt. Express*, **14**, pp. 726–735.

39. Grieve, K., Dubois, A., Moneron, G., Guyot, E., and Boccara, A. C. (2005). Full-field OCT: ex vivo and in vivo biological imaging applications. *Proc. SPIE,* **5690**, pp. 31–38.
40. Watanabe, Y., Hayasaka, Y., Sato, M., and Tanno, N. (2005). Full-field optical coherence tomography by achromatic phase shifting with a rotating polarizer. *Appl. Opt.,* **44**, pp. 1387–1392.
41. Dubey, S. K., Anna, T., Shakher, C., and Mehta, D. S. (2007). Fingerprint detection using full-field swept-source optical coherence tomography. *Appl. Phys. Lett.,* **91**, pp. 181106–181108.
42. Dubey, S. K., Mehta, D. S., Anand, A., and, Shakher, C. (2008). Simultaneous topography and tomography of latent fingerprints using full-field swept-source optical coherence tomography. *J. Opt. A: Pure Appl. Opt.,* **10**, pp. 015307–015315
43. Anna, T., Shakher, C., and Mehta, D. S. (2009). Simultaneous tomography and topography of silicon integrated circuits using full-field swept-source optical coherence tomography. *J. Opt. A: Pure Appl. Opt.,* **11**, pp. 045501–045509.
44. Anna, T., Shakher, C., and Mehta, D. S. (2010). Three-dimensional shape measurement of microlens arrays using full-field swept-source optical coherence tomography. *Opt. Lasers Eng.,* **48**, pp. 1145–1151.
45. Anna, T., Srivastava, V., Shakher, C., and Mehta, D. S. (2011). Transmission mode full-field swept-source optical coherence tomography for simultaneous amplitude and quantitative phase imaging of transparent objects. *IEEE Photon. Lett.,* **23**, pp. 899–901.
46. Anna, T., Srivastava, V., Mehta, D. S., and Shakher, C. (2011). High resolution full-field optical coherence microscopy using Mirau interferometer for the quantitative imaging of biological cells. *Appl. Opt.,* **50**, pp. 6343–6351.
47. Mandel, L., and Wolf, E. (1995). *Optical Coherence and Quantum Optics* 1st Ed., Cambridge University Press, UK.
48. Born, M., and Wolf, E. (2002). *Principle of Optics* 7th Ed., Cambridge University Press, New York.
49. Abdulhalim, I. (2006). Competence between spatial and temporal coherence in full field optical coherence tomography and interference microscopy. *J. Opt. A,* **8**, pp. 952–958.
50. Ryabukho, V., Lyakin, D., and Lobachev, M. (2004). Influence of longitudinal spatial coherence on the signal of a scanning interferometer. *Opt. Lett.,* **29**, pp. 667–669.

51. Ryabukho, V., Lyakin, D., and Lobachev, M. (2005). Longitudinal pure spatial coherence of a light field with wide frequency and angular spectra. *Opt. Lett.*, **30**(3), pp. 224–226.
52. Ryabukho, V. P., Lyakin, D. V., Grebenyuk, A. A., and Klykov, S. S. (2013). Wiener–Khintchin theorem for spatial coherence of optical wave field. *J. Opt.*, **15**, pp. 025405– 025415.
53. Ryabukho, V. P., Lyakin, D. V., and Lobachev, M. I. (2004). Manifestation of longitudinal correlations in scattered coherent fields in an interference experiment. *Opt. Spectrosc.*, **97**, pp. 299–304.
54. Ryabukho, V. P., and Lyakin, D. V. (2005). The effects of longitudinal spatial coherence of light in interference experiments. *Opt. Spectrosc.*, **98**, pp. 273–283.
55. Ryabukho, V., Lyakin, D., and Lobachev, M. (2005). Longitudinal pure spatial coherence of a light field with wide frequency and angular spectra. *Opt. Lett.*, **30**, pp. 224–226.
56. Safrani, A., and Abdulhalim, I. (2011). Spatial coherence effect on layers thickness determination in narrowband full field optical coherence tomography. *Appl. Opt.*, **50**, pp. 3021–3027.
57. Safrani, A., and Abdulhalim, I. (2012). Ultra high resolution full field optical coherence tomography using spatial coherence gating and quasi monochromatic illumination. *Opt. Lett.*, **37**, pp. 457–460.
58. Mehta, D. S., Naik, D. N., Singh, R. K., and Takeda, M. (2012). Laser speckle reduction by multimode optical fiber bundle with combined temporal, spatial, and angular diversity. *Appl. Opt.*, **51**, pp. 1894–1904.
59. Rosen, J., and Takeda, M. (2000). Longitudinal spatial coherence applied for surface profilometry. *Appl. Opt.*, **39**, pp. 4107–4111.
60. Gokhler, M., Duan, Z., Rosen, J., and Takeda, M. (2003). Spatial coherence radar applied for tilted surface profilometry. *Opt. Eng.*, **42**(3), pp. 830–836.
61. Wang, W., Kozaki, H., Rosen, J., and Takeda, M. (2002). Synthesis of longitudinal coherence functions by spatial modulation of an extended light source: a new interpretation and experimental verifications. *Appl. Opt.*, **41**(10), pp. 1962–1971.
62. Duan, Z., Miyamoto, Y., and Takeda, M. (2006). Dispersion-free optical coherence depth sensing with a spatial frequency comb generated by an angular spectrum modulator. *Opt. Express*, **14**(25), pp. 12109–12121.
63. Baleine, E., and Dogariu, A. (2004). Variable coherence tomography. *Opt. Lett.*, **29**(11), pp. 1233–1235.

64. Pavliček, P., Halouzka, M., Duan, Z., and Takeda, M. (2009). Spatial coherence profilometry on tilted surfaces. *Appl. Opt.*, **48**, pp. H40–H47.
65. Duan, Z., Miyamoto, Y., and Takeda, M. (2006). Coherence holography by achromatic 3-D field correlation of generic thermal light with an imaging Sagnac shearing interferometer. *Opt. Express*, **14**, pp. 12109–12121.
66. Liu, Z., Gemma, T., Rosen, J., and Takeda, M. (2010). Improved illumination system for spatial coherence control. *Appl. Opt.*, **49**, pp. D12–D16.
67. Abdulhalim, I., and Dadon, R. (2008). Multiple interference and spatial frequencies effect on the application of frequency-domain optical coherence tomography to thin films metrology. *Meas. Sci. Technol.*, **20**, pp. 015108–015118.
68. Abdulhalim, I., Friedman, R., Liraz, L., and Dadon, R. (2007). Full field frequency domain common path optical coherence tomography with annular aperture. *Proc. SPIE*, **6627**, p. 662719.
69. Abdulhalim, I. (2012). Spatial and temporal coherence effects in interference microscopy and full-field optical coherence tomography. *Ann. Phys.*, **524**, pp. 787–804.
70. Dhalla, Al-H., Migacz, J. V., and Izatt, J. A. (2010). Crosstalk rejection in parallel optical coherence tomography using spatially incoherent illumination with partially coherent source. *Opt. Lett.*, **35**, pp. 2305–2307.
71. Srivastava, V., Nandy, S., and Mehta, D. S. (2013). High-resolution full-field optical coherence tomography using a spatially incoherent monochromatic light source. *Appl. Phys. Lett.*, **103**, pp. 103702–103706.
72. Takeda, M., Wang, W., Duan, Z., and Miyamoto, Y. (2005). Coherence holography. *Opt. Express*, **13**(23), pp. 9629–9635.
73. Naik, D. N., Ezawa, T., Miyamoto, Y., and Takeda, M. (2010). Real-time coherence holography. *Opt. Express*, **18**(13), pp. 13782–13787.
74. Naik, D. N., Ezawa, T., Miyamoto, Y., and Takeda, M. (2009). 3-D coherence holography using a modified Sagnac radial shearing interferometer with geometric phase shift. *Opt. Express*, **17**(13), pp. 10633–10641.
75. Naik, D. N., Ezawa, T., Miyamoto, Y., and Takeda, M. (2010). Phase-shift coherence holography. *Opt. Lett.*, **35**(10), pp. 1728–1730.

Chapter 11

Real-Time and High-Quality Online 4D FF-OCT Using Continuous Fringe Scanning with a High-Speed Camera and FPGA Image Processing

P. C. Montgomery, F. Anstotz, D. Montaner, and F. Salzenstein

ICube (Laboratoire des Sciences de l'Ingénieur, de l'Informatique et de l'Imagerie), University of Strasbourg-CNRS, 23 rue du Loess, 67037 Strasbourg, France
paul.montgomery@unistra.fr

While progress has been considerable in instrumentation using scanning interferometry for the measurement of microscopic surface roughness and buried structures in transparent and biological layers, achieving real-time measurement, while maintaining high image quality remains a challenge. The main reasons for this are the requirements of a high bandwidth for the data acquisition by the detector, the transfer of data to the processor, and the processing itself. One solution is to use continuous fringe scanning over the depth of the sample together with a high-speed camera and cabled logic for the processing. We present here a summary of the 4D

Handbook of Full-Field Optical Coherence Microscopy: Technology and Applications
Edited by Arnaud Dubois
Copyright © 2016 Pan Stanford Publishing Pte. Ltd.
ISBN 978-981-4669-16-0 (Hardcover), 978-981-4669-17-7 (eBook)
www.panstanford.com

microscopy system we have developed for inline measurement of 3D surface roughness using a high-speed CMOS camera and parallel processing with an FPGA that enables data processing rates of up to 1.28 Gb/s. Details of the system are presented together with the two fringe detection algorithms that have been implemented based on the detection of the maximum fringe intensity and on the maximum of the fringe modulation function. The practical performance is demonstrated on the measurement of changing samples, with 3D image rates of up to 22 i/s being achieved for an image size of 256 × 320 pixels, and 3 i/s for an image size of 640 × 1024 pixels over a depth of 5 µm. Depths of up to 20 µm can be measured. While online 4D microscopy opens up new applications for characterizing soft and surfaces moving in a nonperiodic way, a discussion is then presented of how such a system could be adapted to the measurement of volumic images for tomography.

11.1 Introduction

Progress has been considerable over the last decade in instrumentation using scanning interferometry for the measurement of microscopic surfaces, layers and biological specimens. Surface roughness can be measured with nm axial resolution over large depths of several mm [1], buried structures can be imaged with an axial resolution of 0.5 µm in medium-thickness transparent layers of several micrometers [2] and cell details can be imaged with an axial resolution of 5 µm to depths of several hundred µm in biological layers [3]. While most applications have concerned measurements of static samples, new applications of great interest have appeared concerning dynamic phenomena such as surface chemical reactions, transients in microelectromechanical systems (MEMS), changes in soft materials, and living matter in biological specimens. To reduce 3D image blurring, the temporal resolution needs to be increased and this is often at the expense of the signal-to-noise ratio, which tends to lead to a reduction in the quality of the results [4]. Achieving real-time measurement while maintaining high image quality therefore remains a challenge. The main reasons for this are the requirements of a high bandwidth for the data acquisition by the detector, the transfer of data to the processor and the processing itself.

In the field of imaging biological specimens in vivo, there is a tremendous motivation to develop new instrumentation capable of helping in the accurate differentiation between cancerous and noncancerous cells for example [5]. While full-field optical coherence tomography (FF-OCT) is capable of providing the high resolution required as well as having the advantage of not requiring markers (fluorescent dyes or nanomarkers), the low measurement speeds of present systems tends to degrade the quality of volumic images in living, soft material due to blurring [6]. This same problem has occurred in materials science in the analysis of microscopic surface roughness of materials and components that change over time in a nonperiodic way [7]. For example, in order to measure changing surface morphology due to the movement in soft materials or a surface changing due to a chemical reaction, or transients in microsystems, a temporal resolution well below one second is required. Such a temporal resolution is generally considered as being in real time.

As mentioned, the main obstacles to achieving real-time measurement in both 3D surface measurement and tomographic imaging are due to limits in the probe scanning speed and the bandwidth of the data acquisition, transmission, and processing. For example, in atomic force microscopy (AFM) by using resonant scanning tips [8], the 3D image acquisition rate can be increased to several tens of images per second (i/s), but with severe limitations in terms of the field size covered (<10 µm $\times$ 10 µm or $<128 \times 128$ pixels), the difficulty in calibrating the axial measurements and the dynamic range.

In confocal microscopy, while a rate of several tens of image slices per second has been achieved for studying biological samples [9], the image volume rate is still several seconds or more for one volumic image to give sufficient depth resolution. Using laser beam galvanometer scanning, a typical data acquisition rate of 20 Mb/s can be achieved. For an image size of 256 $\times$ 256 pixels (x8 bits depth), this results is an acquisition rate of 38 image slices per second and a volumic image rate of 1 i/s for a 38-slice volumic image. To achieve a larger image slice size (512 $\times$ 512 pixels or more) and greater depth sampling (100 slices or more per volumic image) would require a higher scanning speed and much higher data rates.

Digital holographic microscopy (DHM) overcomes the limitation in scanning speed by not requiring any mechanical scanning, leading to a higher 3D image rate of up to 15 i/s when measuring surface shape [10]. In practice, though, these higher speeds are limited to the case of smooth surfaces, since the processing requires phase unwrapping due to the use of monochromatic illumination. The dynamic range without axial scanning is also limited by the depth of field of the objective. In the improved resolution tomographic mode of DHM, tomographic diffractive microscopy can give volumic images at a rate of 6 s per image using CPU processing and 1 s per image with assistance from a GPU [11].

White light scanning interferometry (WLSI), also known as coherence probe microscopy (CPM) or coherence scanning interferometry (CSI), as in FF-OCT, also uses white light interferometry but mainly for microscopic surface roughness and shape measurement [1]. The technique is of particular interest in quantitative microscopy since no physical scanning is necessary in the XY plane as is the case with confocal microscopy and scanning stylus and tip techniques. High roughness from tens to hundreds of micrometers and even millimeters can be measured with an axial sensitivity of between 1 nm and 15 nm, depending on the type of surface [12].

In WLSI, the scanning of white light fringes over the depth of the roughness to be measured together with the associated image processing in the PC processor or a dedicated DSP requires several seconds for shallow depths (1 μm to 2 μm) and up to several minutes for large depths (200 m) [12] due to the fine step sampling (down to 20 nm) required for the high axial resolution achieved for full field imaging over 1024 × 640 pixels. One way of achieving real-time measurement of surface roughness is to use an integrated analogue processing circuit associated with a dedicated complementary metal-oxide semiconductor (CMOS) camera as in the parallel optical coherence tomography (pOCT). While demonstrating a 3D image rate of 14 i/s over a depth of 1.4 mm, the axial resolution is limited to 4.5 μm [13]. Even though the technique is of interest in certain types of specific applications, the system has the disadvantages of a low image resolution (90 × 144 pixels), photodiodes with a very low fill factor (10%) and a nonconfigurable processing system.

Another way of performing real-time measurement with WLSI is to use continuous fringe scanning together with a high-speed camera and high-speed image processing. To achieve online measurements, highly parallel processing is required to carry out processing on the fly. While recent microprocessors could in theory achieve the necessary processing rates, the architectures are not adapted to the input rate from a high-speed camera. Using a home-built high-speed charge-coupled device (CCD) camera and cabled logic processing (Virtex XCV1000 from Xilinx field-programmable gate array [FPGA]), we have already demonstrated a data acquisition and processing rate of 520 Mb/s in a first prototype system [4, 14]. An online 3D image rate of 5 images/s was demonstrated for measurements over a depth of 5 µm.

We then developed a second prototype of the 4D microscopy system based on a commercial CMOS camera (500–20,000 i/s) connected via a CameraLink connection to a second generation Virtex-II Pro FPGA (Xilinx) processing board [7]. Two algorithms to perform the signal processing were developed specifically for implementation in cabled logic. The first algorithm is known as peak fringe scanning microscopy (PFSM), in which the maximum fringe intensity along the optical axis is detected [15]. It has a variable step size of 20 nm to 100 nm and is very compact to implement. The second one is the five-step adaptive (FSA) algorithm [16] that uses a fixed step size of around 80 nm (or $\pi/2$ phase step) and measures the fringe modulation function, or visibility. This algorithm gives more robust measurements, but is more difficult to implement. Data processing rates of 800 Mb/s–1.28 Gb/s are achieved with this system, resulting in real online measurements and recordable 3D results at a rate of 22 i/s.

As mentioned previously, one of the main limitations of FF-OCT is the low temporal resolution. The aim of this paper is therefore to present a summary of the 4D microscopy system as applied to surface roughness measurement and to see how it could be used in FF-OCT.

Firstly, the theory of interference microscopy is presented, together with some 1D simulation results and a description of the two algorithms (PFSM and FSA) implemented. Then the experimental system is described, followed by a summary of the system performance. The following section presents some applications of 4D microscopy for the measurement of moving surfaces. Finally, a discussion of the future developments of the technique is given and in particular in relation to its possible use in FF-OCT.

11.2 Theory

11.2.1 *Interference Microscopy*

Under monochromatic illumination, an interference microscope separates the illumination beam into two, each of which is at the same wavelength and so capable of interfering, giving place to interference fringes on the detector. Each beam takes a different optical path, in the two arms of the interference system (Michelson, Mirau, or Linnik-type objectives), the first beam reflecting on the reference mirror and the second on the sample. The difference in the optical path created by the roughness of the sample surface with respect to the flat surface of the reference mirror leads to constructive or destructive interference at each point in the image, giving a luminous intensity linked to the path difference and therefore the height of the surface relief at the point considered. By changing the path length between the objective and the sample, by moving either the sample or the objective along the optical axis, an interferogram is created at each point from which the height of the roughness can be calculated at each point in the image. The intensity at a point with coordinates (x, y) in the interferogram can be expressed as an equation of the following type:

$$I(x, y) = I_0(x, y)[1 + \gamma_0 \cos(\Phi(x, y))], \quad (11.1)$$

where $I_0(x, y)$ is the intensity of the incident light, γ_0 the fringe visibility, and $\Phi(x, y)$ the phase of the signal.

Using polychromatic light (white light), interferograms are also created from two interfering beams from the same source in the plane of the detector but localized near to the region of zero path difference due to the limited coherence length resulting from the wide spectral band. Under these conditions, each wavelength interferes with itself to give a superposition of interference fringes that are de-correlated. The sum of these interference fringes at the

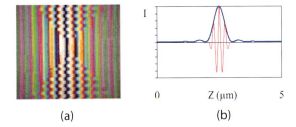

Figure 11.1 The optical probe used in white light interference microscopy. (a) Polychromatic (white light) interference fringes viewed in color superimposed on a reference surface (MRS3). (b) A single packet of white light fringes measured along the optical axis at a given pixel with a monochrome detector, forming the optical probe.

detector gives a particular intensity profile along the optical axis with a main central fringe (of zero order) either side of which the higher-order fringes decrease in amplitude further from the peak. Viewed in color (Fig. 11.1a), the central fringe is black and white and the higher-order fringes are colored. This is due to the degree of coherence that decreases further from the central fringe which corresponds to a zero difference in the optical path. Using a monochrome detector to measure the fringe intensity along the optical axis, the higher-order fringes rapidly fall off in intensity (Fig. 11.1b).

To make a measurement, the fringes are scanned over the entire depth of the surface roughness. The height of the sample at a given point corresponds to the position of the peak of the fringe envelope at that point with respect to the initial position. The positions along the optical axis are known from the calibrated stepper used to move the sample or objective. By using specialized algorithms to detect the peak of the fringe envelope at each pixel in the image, the complete 3D shape of the sample surface can be measured. A useful way of considering the measurement principal is to imagine a virtual, thin probe plane formed by the peak of the fringe envelope at each point in the image that is scanned over the depth of the surface roughness. In reality, this probe consists of the superposition of the fringes on an image of the sample surface in the plane of the detector during the scanning of the fringes along the optical axis.

11.2.2 Modeling of the Optical Probe in Interference Microscopy

In polychromatic light, it is possible to extend Eq. 11.1 to the many wavelengths making up the illumination beam. Thus the intensity $I_\lambda(x, y)$ at the detector (a CCD camera, for example) for a given wavelength λ, can be written as

$$I_\lambda(x, y) = I_0(\lambda)[1 + \gamma_0(\lambda) \cdot \cos(\Phi(x, y))], \quad (11.2)$$

where $I_0(\lambda)$ is the intensity of the wavelength λ of the incident light, $\gamma_0(\lambda)$ is the coherence function, and $\Phi(x, y)$ is the phase of the signal. Since the illumination spectrum is wide but limited (from λ_1 to λ_2), the intensity $I_{WL}(x, y)$ received at a point on the detector can be written according to the following relationship:

$$I_{WL}(x, y) = \int_{\lambda_1}^{\lambda_2} I_0(\lambda)[1 + \gamma_0(\lambda) \cdot \cos(\Phi(x, y))] \cdot d\lambda. \quad (11.3)$$

The shape of the intensity profile of the interference fringes in white light depends on several parameters such as the nature of the sample (the refractive index, the coefficient of reflection, the phase change on reflection, and the spectral absorption), the illumination source (the spectrum), and the response of the optical system as a function of the wavelength. The measurement of surface roughness therefore depends a great deal on the interaction between the optical probe and the rest of the system.

Software for simulating the interference phenomenon has been developed with the aim of helping to better understand the measurement process [17]. As a first step, to simplify the process, the calculations only take into account the central illumination ray and the different optical parameters that can influence the formation of the interference fringes. These parameters are modelled and taken into account in the terms $I_0(\lambda)$ and $\gamma_0(\lambda)$.

The principal of calculation of the interferograms, based on a simplified version of Eq. 11.3, is given in the following relationship, where the integral is replaced with a discrete sum:

$$I_b(x, y) = \sum_{\lambda_1}^{\lambda_2} I_0(\lambda_i)[1 + \gamma_0(\lambda_i) \cdot \cos(\Phi(x, y))]. \quad (11.4)$$

This 1D model has proven quite satisfactory to account for the principal interactions between the beam and the system. For

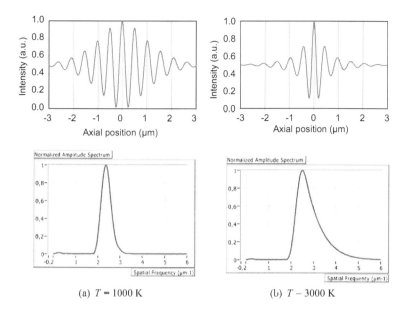

Figure 11.2 Simulation of the effects of the temperature of the light source: reduction in the width of the fringe envelope with increasing temperature due to the change in the spectral bandwidth of the illumination source (objective NA = 0.60).

example, using this model we have shown that the overall shape of the fringe envelope depends on the several parameters already cited above, namely the black-body emission of the illumination source and therefore its temperature (Fig. 11.2), the spectral response of the optical system (objective, mirror, camera, sample, filters), and the numerical aperture (NA) of the objective (Fig. 11.3).

The fringe visibility factor depends on the different phenomena contributing to the reduction in the spatial coherence of the fringes, such as the NA of the objective, the aperture diaphragm of the illumination and the type of illumination (Köhler illumination, for example). We have previously confirmed how an increase in lateral resolution could be achieved using high-intensity UV sources by means of a high-NA objective to reduce the envelope width, even in the case of a coherent source, due to the reduction in the depth of field [18].

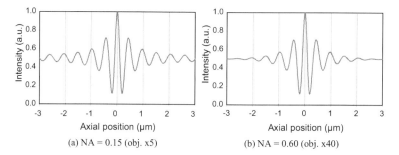

Figure 11.3 Simulation of the effects of the numerical aperture: reduction in the envelope width for an increasing NA due to a reduction in the spatial coherence of the light transmitted by the objective.

In addition, we have confirmed that the shape of the fringe profile within the envelope is linked to the nature of the material to be analyzed as well as that of the reference mirror. The difference between the two materials (equal to the difference in the refractive indices) leads to a phase difference between the light coming from the reference mirror and that from the sample, leading to an asymmetry in the intensity profile [17]. This asymmetry can lead to a measurement error according to the algorithm used and an apparent step height measured between different materials at the same height. A comparison between the results of the simulation and real measurements confirms the validity of the approach.

This first simulation model has nonetheless certain limits in its ability to interpret certain phenomena when measuring structures such as those that are high and narrow (2 μm wide comb structures in MOEMS) [19] or multilayer materials. Taking into account the 3D form of the illumination could explain certain measurement artifacts observed such as the widening of narrow structures and straight edges due to phantom fringes and shadowing or the masking of parts of the illumination beam. The study of these phenomena could be used to refine the model.

11.3 Envelope Detection Algorithms

Many algorithms to perform envelope detection of the fringe signal along the optical axis have been published in the literature, varying greatly in terms of mathematical complexity, calculation cost, processing time, robustness, and axial measurement uncertainty. Most of the methods are based on an AM-FM signal model, which is compatible to the equation in light intensity measured along the optical axis of an interference microscope. Envelope detection can be performed using a demodulation procedure [20, 21], or measurement of the fringe visibility [22] at a given sample, along the optical axis z Other methods based on Fourier transform [20, 23] or wavelets [24] have also been proposed but require more involved computing means. The techniques that are more costly in terms of calculation complexity are better adapted to measuring static surfaces [12] or measurements involving image acquisition with a high-speed camera followed by postprocessing in which the time for processing a single 3D image of more than a second is not a limitation [4].

On the contrary, to perform real-time measurements online in which the processing needs to be carried out on the fly within a time of less than a second per 3D image, the use of complex mathematical functions in a PC processor, a DSP or a GPU is not possible at present above an image acquisition rate of 90 i/s. The processing of images coming from a high-speed camera with a rate above 90 i/s requires the use of dedicated components or cabled logic such as an FPGA. The choice of the algorithm is therefore determined by the possibilities of implementing it in an FPGA that limits the mathematical functions available to simple operators such as adders, multipliers, and comparators. Two of the simplest algorithms proposed in the literature, based on the detection of the peak intensity of the fringe signal and on the detection of the peak of the fringe modulation function, are now described.

11.3.1 Peak Fringe Detection Algorithm

The PFSM algorithm [15] is based on the determination of the peak of the fringe signal intensity, $Z(I_{max})$, along the optical axis (Figs. 11.4 and 11.5a) that corresponds to the central zero-order fringe. This is true as long as the background intensity remains smaller than this value along the whole of the scanning length. The interest of this algorithm is its simplicity, requiring only one

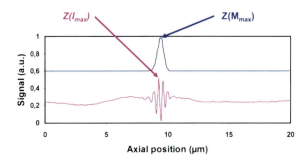

Figure 11.4 Determination of the peak of the fringe intensity using the PFSM algorithm (pink curve) or the peak of the fringe modulation using the FSA algorithm (blue curve) of the fringe signal in white light interferometry.

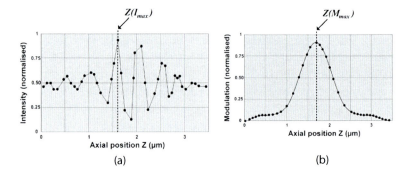

Figure 11.5 Two algorithms adapted to cabled logic processing for detecting the surface position. (a) PFSM algorithm: detection of the peak intensity (zero-order fringe). (b) FSA algorithm: detection of the peak of the modulation function of the fringes.

comparator and a reduced number of images stored in memory, which makes it well adapted to implementation in cabled logic. The axial resolution is dependent on the displacement step between each image, with typical values being between 20 nm to 80 nm.

11.3.2 The Fringe Modulation Algorithm

The FSA algorithm [16, 28] is an efficient method for measuring the modulation or the visibility of the fringe signal from a series of five consecutive intensity measurements along the optical axis:

$$M = C\sqrt{((I_2 - I_4)^2 - (I_1 - I_3)(I_3 - I_5))}, \quad (11.5)$$

where C is a standardization constant. This formula is exact when the phase change between interferograms is $\pi/2$. The position of the surface along the Z axis at a given pixel can be found at the peak of the modulation of the fringes, $Z(M_{max})$ (Figs. 11.4 and 11.5b). Because of its relative simplicity, the FSA algorithm is also well adapted to implementation in an FPGA. The axial resolution of the FSA algorithm in its simplest form (without interpolation) is limited to a phase step of 90°, typically around a value of 80 nm in white light.

The FSA method corresponds to a discrete version of the Teager–Kaiser energy operator (TKEO) [24], applied to a derivative of the signal which is used to eliminate additional lowfrequency components:

$$\Psi[x(n)] = x^2(n) - x(n+1)x(n-1). \quad (11.6)$$

The TKEO is an efficient nonlinear and local algorithm for envelope detection and phase retrieval from AM-FM signals. The TKEO has found applications in speech processing [24], image processing [26], and pattern recognition [27]. This operator is an energy-tracking operator which instead of measuring the signal energy measures the energy of the source or the system that produces the signal. After the envelope detection step, to provide a more flexible detection, it is also possible to use spline smoothing and interpolation methods, instead of using a Gaussian shape as many authors do. Spline smoothing and interpolation following the FSA algorithm enable the axial resolution of static surface roughness measurement to be improved to between 10 and 15 nm [28] Maragos and Potamianos [29] have proposed an extension of the TKEO, called the k-order differential energy operator (DEO) k, given by

$$\Psi_k[x(t)] = x^{(1)}(t)x^{(k-1)}(t) - x^{(0)}(t)x^{(k)}(t). \quad (11.7)$$

For $k = 2$ the TKEO corresponds to the second-order DEO. We have also proposed a general extension of these operators in [30], namely

$$\Psi_{p,q,m,l}[x(t)] = x^{(p)}(t)x^{(q)}(t) - x^{(m)}(t)x^{(l)}(t). \quad (11.8)$$

Two discrete versions of the higher-order methods have been derived. Maragos et al. have proposed the following one:

$$\Phi_{km}[x(n)] = x(n)x(n+k) - x(n-m)x(n+m+k). \quad (11.9)$$

Thus for $k = 0$ and $m = 1$, this corresponds to the discrete TKEO. Let us call x_1 the derivative signal: it is possible to extract the local frequency and envelope information in the following way:

$$\hat{\Omega}^2 = \frac{\Phi_{km}[x_1(n)]}{\Phi_{km}[x(n)]},$$

$$\hat{A} = \sqrt{\frac{\Phi_{km}[x(n)]}{\sin(m\Omega)\sin((m+k)\Omega)}}.$$

We have proved in Ref. [31] that these algorithms tend to be more robust than the TKEO, due to their asymmetric properties and thus their ability to de-correlate the additive noise. We have also proposed a discrete version of Eq. 11.8 by

k even:

$$\Psi_{H^d_{k,p,m}}[x(n)] = \frac{1}{2}[x(n+p)x(n+q) + x(n-p)x(n-q) \\ -(x(n+m)x(n+l) + x(n-m)x(n-l))]$$

k odd:

$$\Psi_{H^d_{k,p,m}}[x(n)] = \frac{1}{2}[x(n+m)x(n+l) + x(n-m)x(n-l) \\ -(x(n+p)x(n+q) + x(n-p)x(n-q))].$$

The demodulation process is thus given by

$$\hat{\Omega}^2 = \frac{\Psi_{H^d_{k,p,m}}[x_1(n)]}{\Psi_{H^d_{k,p,m}}[x(n)]},$$

$$\hat{A} = \sqrt{\frac{\Psi_{H^d_{k,p,m}}[x(n)]}{\sin[(m-p)\Omega]\sin[(m+p-k)\Omega]}}.$$

Finally it is possible to extract the information of interest by several methods, which has been studied in Ref. [30] but which needs more appropriate further studies.

11.4 Experimental

The main aim of the design of the 4D microscopy system is the ability to make 3D measurements online using continuous fringe scanning [4]. To achieve such real-time quantified measurement requires high-speed fringe scanning, image acquisition, and processing together with the storage of the measurement data and the generation of 3D results for visualization. With all this in mind, the system was built around an adapted Leica DMR-X microscope equipped with an interference objective mounted on a piezo scanner together with a high-speed camera for image acquisition [7, 32]. Details of the system can be seen in the schematic layout in Fig. 11.6 and the photo in Fig. 11.7 [33, 34].

A choice of Michelson (x5) and Mirau (x10 and x40) interference objectives (Leica) are available (Fig. 11.8a).

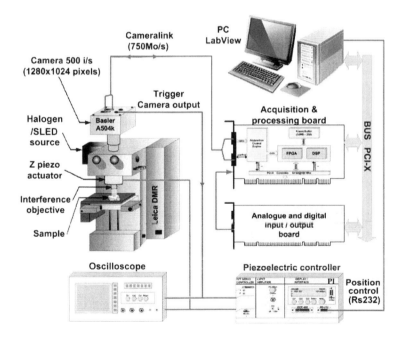

Figure 11.6 Schematic layout of the 4D microscopy system.

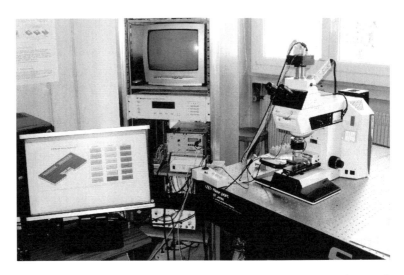

Figure 11.7 A general view of the 4D microscopy system and control electronics.

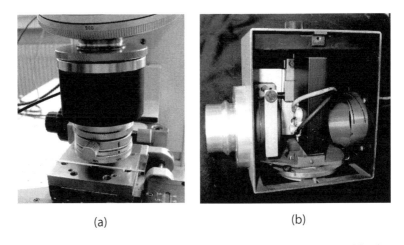

Figure 11.8 Details of the adapted scanner and illumination. (a) The interference objective mounted in the high-speed piezoscanner. (b) The high power SLED illuminator with cooling radiator.

The piezo actuator (model P726 PIFOC from PI) was chosen for its high stiffness and linearity and high scanning speed so as to be to operate with a saw tooth function for Z-scanning over a range of 100 μm. Capacitive position detection provides linear feedback for closed loop control, giving an axial sensitivity of 0.4 nm. Illumination is provided by either a halogen lamp or a pulsed SLED (Fig. 11.8b). The high-speed camera used is a CMOS camera (Basler A504k, 1280 × 1024 pixels, 500 images per second in full-field mode) connected by a Camera Link connection to a Dalsa Coreco "Anaconda" acquisition board equipped with a Virtex II FPGA programmable chip.

System synchronization is provided by a clock signal generated by a digital input/output board function generator controlled with LabView. This timing signal is used to generate the ramp signal for the piezo actuator, to operate the fringe image acquisition with the camera, to control the acquisition and processing rates on the Coreco board, and to synchronize the 3D image acquisition and processing by the PC (Fig. 11.9).

The different signals are monitored on a digital oscilloscope. Surface roughness measurements are made during both the rising and falling slopes of the piezo scanning signal.

Details of the image processing unit (IPU) designed around the Virtex 2Pro FPGA with the associated memory units are shown in Fig. 11.10. The memory consists of four high-speed SRAM memories of 2 Mb and two DDR memories of 64 Mb.

Complex software of high and low levels was developed with LabView and C^{++} respectively to initialize and configure the different electronic components of the system (Fig. 11.11). The

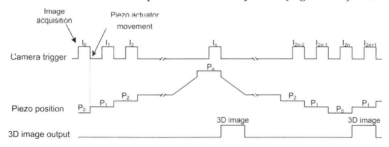

Figure 11.9 Details of the synchronization of the different timing signals.

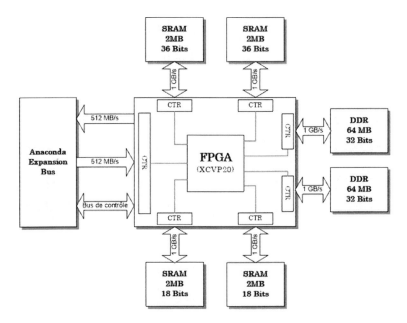

Figure 11.10 Details of the IPU consisting of the Virtex-II Pro FPGA and associated memory units. © Dalsa Anaconda.

software can be used to control the choice of the fringe image acquisition resolution and rate, the frequency and depth for Z-scanning, the choice of PFSM or FSA algorithms, and the rate of production of 3D images. The different types of results to be stored on the hard disc of the computer can also be chosen, consisting of the direct fringe images, the maximum intensity image (fringe visibility), the measured altitude, or the 3D views.

The two algorithms used for envelope detection, PFSM and FSA, were implemented in cabled logic by first simulating them in very high-speed integrated circuit hardware description language (VHDL) [32] and then synthesizing and optimizing them for the "Anaconda" FPGA board. The architecture of the PFSM algorithm is shown in Fig. 11.12. The 64-bit camera signal is first unpacked to 32 bits for processing in the FPGA. Different portions of the incoming images are dispatched to internal memory and a comparator is used to compare the intensities of the incoming image with the previous one. When the intensity increases, the maximum intensity is stored

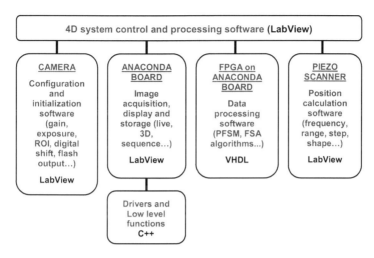

Figure 11.11 Details of the general architecture of the 4D microscopy control and processing software.

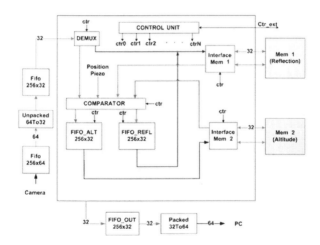

Figure 11.12 Architecture of the implementation of the PFSM algorithm in the FPGA.

in the first memory (Mem 1, Reflection) and the piezo position is noted in the second memory (Mem 2, Altitude).

A simplified logic diagram of the second algorithm based on the calculation of the fringe modulation is shown in Fig. 11.13, requiring

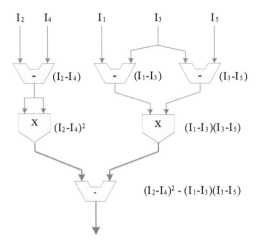

Figure 11.13 Use of subtraction and multiplication functions to implement envelope modulation algorithm (FSA) in cabled logic in the FPGA.

the use of 4 independent memories. Despite the relative simplicity of the algorithms, requiring only subtraction, multiplication and comparator functions, much work was necessary to develop the specific architecture for the cabled logic. Many difficulties had to be overcome for the interfacing between the camera and the acquisition and processing board using a Camera Link connection as well as between the FPGA and PC through the PCI-X bus.

The control of the different memory units around the FPGA was also challenging, requiring simultaneous access of several images at the same time as well as the continuous updating of the measured modulation and altitude images. In addition to online measurement, the system can also be used for off-line processing by storing a series of fringe images for subsequent processing on the PC.

11.5 System Performance

In this section we present the theoretical performance of the acquisition and processing system in terms of 3D image rate as a function of scanning depth over a range of 1–10 μm for different axial resolutions (Fig. 11.14) [7]. A step of 76 nm was used between

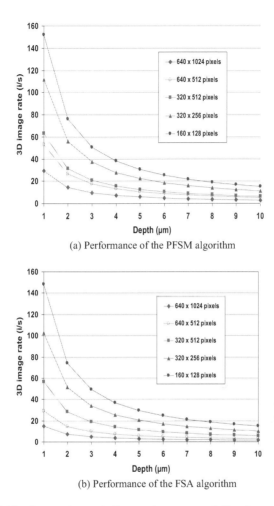

Figure 11.14 Comparison of the performance of the two algorithms implemented in the FPGA, in terms of the 3D image rate as a function of the scanning depth and the image resolution for an axial scanning step of 76 nm.

fringe images. The calculation of the performance is complex, depending on many parameters, amongst which the main limitation to the image rate comes from the FPGA and its local environment [33]. The theoretical 3D image rates for the PFSM algorithm (Fig. 11.14a) are 2.9 i/s to 152 i/s for image sizes of 640 × 1024 pixels to 160 × 128 pixels and scanning depths of 1 μm to 10 μm, respectively. For the same conditions, the FSA algorithm shows slightly lower 3D image rates of 1.5 i/s to 148 i/s (Fig. 11.14b). The maximum axial dynamic range of the system at present is 20 μm.

A comparison of the performance of the two algorithms implemented shows that the 3D image rates are slightly higher for the PFSM algorithm than for the FSA algorithm for the same image resolutions and scanning depths. This improvement is due to the greater simplicity of the PFSM algorithm compared with that of the FSA algorithm. For example, for a depth of 3 μm and a high image resolution of 640 × 1024 pixels, the 3D image rate is 9.8 i/s with the PFSM algorithm, compared with 4.9 i/s for the FSA algorithm. For a lower image resolution of 160 × 128 pixels, the 3D image rate increases to 50 i/s for the PFSM algorithm compared to 49 i/s for the FSA algorithm. A second advantage of the PFSM algorithm is its greater flexibility compared with that of the FSA algorithm, which has a fixed scanning step, with a value of 76 nm in the conditions used for the tests. For the PFSM algorithm, the step can be varied over a practical range of 20–100 nm, which enables higher axial resolutions to be attained directly. If the step is reduced though, there is a corresponding increase in processing time due to the greater number of images to be processed, and a decrease in the 3D image rate.

A comparison of the upper limits in scanning or processing rates is presented in Fig. 11.15. With the present design, the main

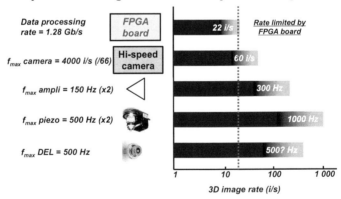

Figure 11.15 Theoretical limits of the different elements in the 4D microscopy system.

limitation is due to the FPGA processing rate of 1.28 Gb/s which limits the 3D measurement rate to 22 i/s for an average image size of 320 × 256 pixels [33].

11.6 Applications

In this section we present some results of the 4D microscopy system for different samples, image sizes, and measurement rates.

11.6.1 Laterally Moving Microfluxgate Surface Measured with the PFSM Algorithm

An efficient way of demonstrating the functioning of the 4D microscopy system is to measure a sample that is undergoing lateral translation. The first results are of a CMOS magnetic probe, the microfluxgate, developed at ICube using a process derived from a standard Bipolar technique for Si [35]. The surface structure consists of metallic contact pads and thin wire connections to the central probe, with a covering of a transparent passivation layer that varies in thickness from 0.8 μm to 1.6 μm. Both the thin wires and the presence of the transparent layer make the measurement more difficult. The results in Fig. 11.16 show the interface between the passivation layer and the metallic contacts and connections to the probe [34]. The sample was translated laterally at a constant speed of 20 μm/s during the measurement with the PFSM algorithm, resulting in a 3D image rate of 1.9 i/s. The 3D results shown are those produced in real time by the 4D system using the Imaq Vision module of LabView as visualized on the computer screen.

The objective used was a x5 Michelson with an image size of 1024 × 640 pixels (2.4 mm × 1.5 mm). The fringe image acquisition rate was 240 i/s, and the scanning step was 40 nm over a depth of 5 μm. The axial resolution of these results is limited to the sampling step of 40 nm.

Experience from the development of the first real-time prototype showed that when moving from static measurements to real-time measurements one of the biggest problems is a deterioration in the quality of the results [4]. The 3D results from the first

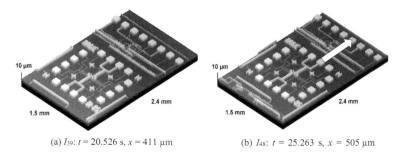

(a) I_{39}: $t = 20.526$ s, $x = 411$ μm   (b) I_{48}: $t = 25.263$ s, $x = 505$ μm

Figure 11.16 Two 3D real-time images of a CMOS magnetic probe being translated sideways at 20 μm/s, measured at a 3D image rate of 1.9 i/s with the x5 Michelson objective and the PFSM algorithm (image size = 1024 × 640 pixels).

prototype showed a high level of noise and aberrations, mainly from the camera, but also from the signal processing as well as blurring from the sample movement. In the present case with the second prototype, the results were much improved due to the better image quality from the CMOS camera and the more robust FSA algorithm. The main defect that remained was the blurring on the leading and trailing edges of the square step structures perpendicular to the direction of motion which could only be reduced by increasing the temporal resolution.

The same sample was then measured at the same lateral translation speed of 20 μm/s with a x10 Mirau objective to give finer details and with a smaller image resolution of 320 × 256 pixels (375 μm × 300 μm) to give a higher 3D image rate of 10 i/s. The 3D results shown in Fig. 11.17 and hereafter, while being first processed on the FPGA to produce the measurement data, were then converted to multicolor 3D images using MountainsMap® 6 (Digital Surf) image topography software. The image acquisition rate was 1010 i/s and the scanning step was 40 nm over a scan range of 5 μm. While the image quality was generally very good, with the axial resolution being limited to the axial sampling step of 40 nm, some artifacts of $\lambda/2$ in height were visible on measurements made during the downward movement of the scan. These artifacts were due to the detection of secondary fringes instead of the central zero-

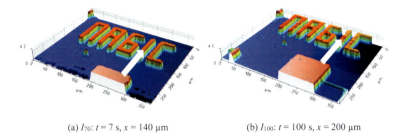

(a) I_{70}: $t = 7$ s, $x = 140$ μm (b) I_{100}: $t = 100$ s, $x = 200$ μm

Figure 11.17 Two 3D real-time images of a CMOS magnetic probe being translated sideways at 20 μm/s, measured at a 3D image rate of 10 i/s with the x10 Mirau objective and the PFSM algorithm (image size = 320 × 256 pixels).

order fringe by the PFSM algorithm when measuring through the transparent layer.

11.6.2 Laterally Moving Microfluxgate Surface Measured with the FSA Algorithm

The same sample was then measured with the FSA algorithm, the x5 Michelson objective, and an axial scan step of 76 nm over a dynamic range of 5 μm [34]. For the large image size (1024 × 640 pixels, 2.4 mm × 1.5 mm), the lateral translation speed was 20 μm/s and the fringe image acquisition was made at a rate of 190 i/s, giving a 3D image rate of 2.88 i/s (Fig. 11.18).

For the smaller image size (320 × 256 pixels, 750 μm × 600 μm), the lateral translation speed was 200 μm/s and the fringe image acquisition rate was 1320 i/s, resulting in a 3D image rate of 20 i/s (Fig. 11.19).

11.6.3 Laterally Moving GaN Surface Measured with the FSA Algorithm

A second example is given in Fig. 11.20 of a GaN surface after chemical etching that was translated laterally at a speed of 200 μm/s and measured with the PFSM algorithm, resulting in a 3D image rate of 6 i/s. The objective used was a x10 Mirau, the image size was 320 × 256 pixels (375 μm × 300 μm), the fringe image acquisition rate

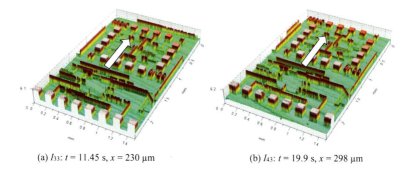

(a) I_{33}: $t = 11.45$ s, $x = 230$ μm (b) I_{43}: $t = 19.9$ s, $x = 298$ μm

Figure 11.18 Two large-area 3D real-time images of a CMOS magnetic probe being translated sideways at 20 μm/s, measured at a 3D image rate of 2.88 i/s with the x5 Michelson objective and the FSA algorithm (image size = 1024 × 600 pixels, 2.4 mm × 1.5 mm).

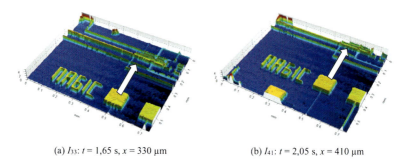

(a) I_{33}: $t = 1{,}65$ s, $x = 330$ μm (b) I_{41}: $t = 2{,}05$ s, $x = 410$ μm

Figure 11.19 Two small-area 3D real-time images of a CMOS magnetic probe being translated sideways at 200 μm/s, measured at a 3D image rate of 20 i/s with the x5 Michelson objective and the FSA algorithm (image size = 256 × 320 pixels, 600 μm × 750 μm).

was 400 i/s, and the scanning step was 76 nm over a scan range of 5 μm. The image quality was very good, with the axial resolution being limited to the axial sampling step of 76 nm [9].

A profile analysis performed off-line using MountainsMap software (Fig. 11.21) shows the good quality of the real-time measurements.

The results of the FSA algorithm showed an axial measurement uncertainty equivalent to the sampling step, in this case 76 nm. There are less artifacts present than the results with the PFSM

Applications | 419

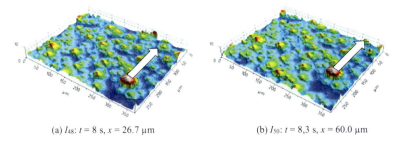

(a) I_{48}: $t = 8$ s, $x = 26.7$ μm

(b) I_{50}: $t = 8,3$ s, $x = 60.0$ μm

Figure 11.20 Two 3D real-time images of a GaN surface after chemical etching moving laterally at 10 μm/s, measured at a 3D image rate of 3 i/s with the x40 Mirau objective and the PFSM algorithm (image size = 320 × 256 pixels).

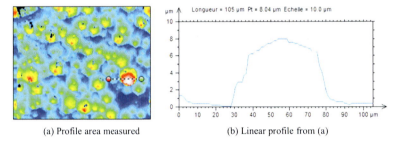

(a) Profile area measured

(b) Linear profile from (a)

Figure 11.21 Profile analysis of results in Fig. 11.16 using the MountainsMap software.

algorithm. These improved results of the FSA algorithm are due to the removal of the low-frequency variation of the background intensity, giving an improved robustness compared with the PFSM algorithm.

11.6.4 Drying Drop of Liquid Correction Whitener Measured with the FSA Algorithm

In the following test, the change in absolute shape of a drop of liquid correction whitener (Tipp-Ex) was measured during the evaporation of the solvent (Fig. 11.22). The measurement was made over a period of more than 8 minutes using the FSA algorithm with a sampling step of 76 nm, over a scan depth of 19 μm [33].

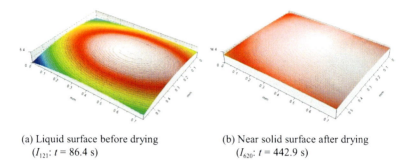

(a) Liquid surface before drying
(I_{121}: t = 86.4 s)

(b) Near solid surface after drying
(I_{620}: t = 442.9 s)

Figure 11.22 Two 3D real-time images of a drop of drying correction whitener (Tipp-Ex) measured at a 3D image rate of 1.4 i/s over a period of more than 8 minutes with the FSA algorithm (image size = 640 × 512 pixels).

The x10 Mirau objective was used for an image size of 640 × 512 pixels (750 μm × 600 μm) and a fringe image acquisition rate of 350 i/s, giving a 3D image rate of 1.4 i/s. The point of interest of this particular measurement was the quantity of data recorded; 722 3D images were stored on the hard disk, making full use of the real online capability of the technique developed. The two results taken from the 3D image series show the smooth liquid drop near the beginning (Figs. 11.22a and 11.23a) and some surface structure appearing after drying (Figs. 11.22b and 11.23b), at a time of 442.9 s.

A summary of the results of the 4D measurements and the associated parameters is given in Table 11.1.

11.7 Future Developments and Potential Applications in FF-OCT

The 4D system presented has been designed specifically for the real-time measurement of surface roughness and 3D shape. One of the applications targeted is the quantitative analysis of biomaterial layer growth (hydroxyapatite), necessitating the development of an immersion head for in situ measurement [33]. With the existence of signals coming from buried interfaces and structures, the next obvious step for the 4D system is to adapt it to tomographic analysis. In the case of the results on the microfluxgate, measurements

Table 11.1 Summary of measurement parameters for samples measured

Figure	Sample	Lat. speed (μm/s)	Image size (pix)	Image size (μm)	Scan depth/step	Algo.	Camera rate (i/s)	3D rate (i/s)
11.16	CMOS x5	20	1024 × 640	2400 × 1500	5 μm 40 nm	PFSM	240	1.9
11.17	CMOS x10	20	320 × 256	375 × 300	5 μm 40 nm	PFSM	1010	10
11.18	CMOS x5	20	1024 × 640	2400 × 1500	5 μm 76 nm	FSA	190	3
11.19	CMOS x5	200	320 × 256	750 × 600	5 μm 76 nm	FSA	1320	20.5
11.20, 11.21	GaN x10	200	320 × 256	750 × 600	5 μm 76 nm	PFSM	400	6
11.22, 11.23	Tipp-Ex x_0	–	640 × 512	750 × 600	19 μm 76 nm	FSA	350	1.4

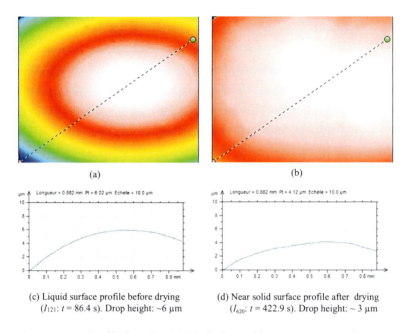

Figure 11.23 Profile from Fig. 11.18 of a drop of drying correction whitener.

were made on the buried interface underneath the transparent passivation layer due to the signal from the interface being larger than that from the air/passivation layer surface (Figs. 11.12–11.15). The interest of the 4D system is in the reduction in the quantity of data, enabling much higher sampling rates than the 3D image rates. With the data processing and transfer rate of the system being limited to 1.28 Gb/s, the quantity of data is reduced by a factor equal to the number of sampling steps. Thus for an image size of 320 × 256 pixels over a depth of 5 μm, the transfer of processed images is limited to 22 i/s for an image acquisition of 1320 i/s and the data reduction corresponds to a factor of about 63. It is the use of the FPGA to carry out preprocessing to find the pertinent information of the height at each pixel that allows the reduction in the data volume.

In FF-OCT for the tomography of transparent layers what is required is either a full image volume x, y, z or single image slices across the image stack either *en face* or over the depth [3]. If the 4D system was to be used for FF-OCT, it would not be possible to produce a full image volume in real time since the data produced would be at the same rate as that measured. In the previous case cited, this would slow down the measurement to a rate of 3 s per image stack over 5 µm. For a more practical depth of 200 µm, the image stack would take 120 s for a sampling step of 76 nm.

On the other hand, a similar architecture using the FSA algorithm to calculate the fringe visibility could be used to produce image slices consisting of the calculated values of the visibility at each pixel instead of storing the height associated with the peak value. A single image slice would then be produced for every complete scan. Depending on the camera architecture, it would be possible to choose an *xy* image slice or *xz* slice for a given value of *y* or a *yz* slice for a given *x*. For the example given (320 × 256 pixels) an *xz* or *yz* image slice could be produced at a rate of 22 i/s over a depth of 5 µm and 0.55 i/s over a depth of 200 µm.

With the present camera used, it would actually be possible to increase the image rate by two different means. The first would be by reducing the sampling depth to about 1 µm for visualizing an *xy* image, and the second would be to reduce the width of the *xy* image to 1024 × 8 pixels for visualizing an *xz* or *yz* image. The higher measurement rates could be used to integrate to reduce noise and still be within the real-time regime.

11.8 Conclusions

In this work we have summarized the development of second prototype 4D microscopy system for measuring 3D surface roughness in real-time, online measurements. Continuous scanning white light fringe interferometry is used in combination with a high-speed CMOS camera and FPGA cabled logic processing. The performance results of two algorithms for processing the fringes implemented in cabled logic have been presented. The first uses a maximum fringe intensity detection algorithm (PFSM) and the second, the maximum of the fringe modulation function (FSA). The algorithm that gives the highest 3D image measurement rates is the PFSM algorithm due to its greater simplicity and highest processing speed on the FPGA.

Online results have been demonstrated of the real-time measurement of a CMOS chip and a rough GaN surface after chemical etching being translated laterally at 20 μm/s and 200 μm/s and measured over a depth of 5 μm. For large images (1024 × 640 pixels), a 3D measurement rate of up to 3 i/s has been demonstrated and up to 21 i/s for a smaller image size of 320 × 256 pixels). While the results with the PFSM algorithm are slightly faster than those for the FSA algorithm and can be used with a smaller sampling step, the latter is more robust than the PFSM algorithm due to the removal of the slowly varying background intensity. The 3D image rates could be increased proportionally for a decrease in the scan depth.

The capacity of such an online measurement system has also been demonstrated with the measurement in the change in absolute shape of a drop of liquid correction whitener drying over a period of more than 8 minutes, the results consisting of over 700 3D images.

Such a real-time, online measurement system with great flexibility in terms of image size, 3D rate and scan depth opens up new applications in surface metrology for the quantitative characterization of a large variety of surfaces that change over time in a nonperiodic way, such as with transients in MEMS, soft materials, layer growth, and surface chemical reactions. The same system could be used for processing in FF-OCT for tomographic analysis by extracting fringe visibility image slices in real time.

Acknowledgments

The authors would like to acknowledge the financial support of this work from Oséo with the CAM4D Connectus project and the practical assistance from colleagues and technical staff at ICube (D-ESSP).

References

1. Schmit, J., Reed, J., Novak, E., and Gimzewski, J. K. (2008). Performance advances in interferometric optical profilers for imaging and testing. *J. Opt. A: Pure Appl. Opt.*, **10**, pp. 64001–64007.

2. Montgomery, P., Montaner, D., and Salzenstein, F. (2012). Tomographic analysis of medium thickness transparent layers using white light scanning interferometry and XZ fringe image processing. *Proc. SPIE*, **8430**, p. 843014.
3. Dubois, A., Moreau, J., and Boccara, A. C. (2008). Spectroscopic ultrahigh-resolution full-field optical coherence microscopy. *Opt. Express*, **16**, pp. 17082–17091.
4. Montgomery, P. C., Draman, C., Uhring, W., and Tomasini, F. (2004). Real time measurement of microscopic surface shape using high speed cameras with continuously scanning interference microscopy. *Proc. SPIE*, **5458**, pp. 101–108.
5. Jain, M., Narula, N., Salamoon, B., Shevchuk, M. M., Aggarwal, A., Altorki, N., Stiles, B., Boccara, C., and Mukherjee, S. (2013). Full-field optical coherence tomography for the analysis of fresh unstained lobectomy specimens. *J. Pathol. Inform.*, **4**(1), pp. 26–33.
6. Sacchet, D., Brzezinski, M., Moreau, J., Georges, P., and Dubois, A. (2010). Motion artifact suppression in full-field optical coherence tomography. *Appl. Opt.*, **45**, pp. 1480–1488.
7. Montgomery, P. C., Anstotz, F., Johnson, G., and Kiefer, R. (2008). Real time surface morphology analysis of semiconductor materials and devices using 4D interference microscopy. *J. Mater. Sci. Mater. Electron.*, **19**(1), pp. 194–198.
8. Schick, A., and Breitmeier, U. (2004). Fast scanning confocal sensor provides high quality height profiles in the microscopic range. *Proc. SPIE*, **5457**, pp. 115–125.
9. Horber, J. K. H., and Miles, M. J. (2003). Scanning probe evolution in biology. *Science*, **302**, pp. 1002–1005.
10. Charrière, F., Kühn, J., Colomb, T., Montfort, F., Cuche, E., Emery, Y., Weible, K., Marquet, P., and Depeursinge, C. (2006). Characterization of microlenses by digital holographic microscopy. *Appl. Opt.*, **45**(5), pp. 829–835.
11. Bailleul, J., Simon, B., Debailleul, M., Liu, H., and Haeberlé, O. (2012). GPU acceleration towards real-time image reconstruction in 3D tomographic diffractive microscopy. *Proc. SPIE*, **8437**, p. 843707.
12. Petitgrand, S. (2006). Méthodes de microscopie interférométrique 3D statiques et dynamiques, pour la caractérisation de la technologie et du comportement des microsystèmes. Thèse d'université, IEF, Université Paris XI, France.

13. Beer, S., and Seitz, P. (2005). Real-time tomographic imaging without x-rays: a smart pixel array with massively parallel signal processing for real-time optical coherence tomography performing close to the physical limits. Proc. IEEE of Int. Conf. on PhD Research in Microelectronics and Electronics (PRIME 2005), Lausanne, Switzerland.
14. Dubois, A., Vabre, L., Boccara, A. C., Montgomery, P. C., Cunin, B., Reibel, Y., and Draman, C. (2002). Real-time high-resolution topographic imagery using interference microscopy. *Eur. Phys. J. Appl. Phys.,* **20**, pp. 169–175.
15. Montgomery, P. C., and Fillard, J. P. (1992). Peak fringe scanning microscopy (PFSM): submicron 3D measurement of semiconductor components. *Proc. SPIE,* **1755**, pp. 12–23.
16. Larkin, K. G. (1996). Efficient nonlinear algorithm for envelope detection in white light interferometry. *J. Opt. Soc. Am.,* **13**, pp. 832–843.
17. Montaner, D., Montgomery, P. C., Pramatarova, L., and Pecheva, E. (2008). Analyses locales de couches épaisses complexes par modélisation de la sonde optique en microscopie interférométrique. *Actes du neuvième colloque francophone, Méthodes et techniques optiques pour l'industrie, Club CMOI,* Nantes, France, 17–21 novembre 2008, Société Française d'Optique.
18. Montgomery, P. C., and Montaner, D. (1999). Deep submicron 3D surface metrology for 300 mm wafer characterization using UV coherence microscopy. *Microelectron. Eng.,* **45**, pp. 291–297.
19. Montgomery, P. C., Montaner, D., Manzardo, O., Flury, M., and Herzig, H. P. (2004). The metrology of a miniature FT spectrometer MOEMS device using white light scanning interference microscopy. *Thin Solid Films,* **450**, pp. 79–83.
20. Chim, S. S. C., and Kino, G. S. (1990). Correlationmicroscope. *Opt. Lett.,* **15**(10), pp. 579–581.
21. Caber, P. J. (1993). Interferometric profiler for rough surfaces. *Appl. Opt.,* **32**(19), pp. 3438–3441.
22. Davidson, M., Kaufman, K., Mazor, I., and Cohen, F. (1987). An application of interference microscopy to integrated circuit inspection and metrology. *Proc. SPIE,* **775**, pp. 233–247.
23. de Groot, P., and Deck, L. (1995). Surface profiling by analysis of white-light interferograms in the spatial frequency domain. *J. Mod. Opt.,* **42**(2), pp. 389–401.
24. Sandoz, P. (1997). Wavelet transform as a processing tool in whitelight interferometry. *Opt. Lett.,* **22**(14), pp. 1065–1067.

25. Maragos, P., Quatieri, T. F., and Kaiser, J. F. (1991). Speech nonlinearities, modulations, and energy operators. Proc. IEEE Int. Conf. Acoustics, Speech, Signal Processing (ICASSP '91), vol. 1, pp. 421–424, Toronto, Ontario, Canada.
26. Havlicek, J. P., and Bovik, A. C. (2000). Image modulation models. In *Handbook for Image and Video Processing*, pp. 313–324, Academic Press, New York, NY, USA.
27. Cexus, J. C., and Boudraa, A. O. (2004). Teager-Huang analysis applied to Sonar Target recognition. *Int. J. Signal Proc.*, **1**(1), pp. 1–5.
28. Montgomery, P. C., Salzenstein, F., Montaner, D., Serio, B., and Pfeiffer, P. (2013). Implementation of a fringe visibility based algorithm in coherence scanning interferometry for surface roughness measurement. *Proc. SPIE*, **8788**, p. 87883G.
29. Maragos, P., and Potamianos, A. (1995). Higher order differential energy operators. *IEEE Signal Proc. Lett.*, **2**(8), pp. 152–154.
30. Salzenstein, F., Boudraa, A. O., and Cexus, J. C. (2007). Generalized higher-order nonlinear energy operators. *J. Opt. Soc. Am. A*, **24**(12), pp. 3717–3727.
31. Salzenstein, F., Montgomery, P. C., Montaner, D., and Boudraa, A. O. (2005). Teager-Kaiser energy and higher-order operators in white-light interference microscopy for surface shape measurement. *EURASIP J. Adv. Signal Proc.*, **17**, pp. 2804–2815.
32. Johnson, G., Montgomery, P., Anstotz, F., and Kieffer, R. (2007). Development of a 3D dynamic measurement system using a high speed camera with white light scanning interference microscopy associated with real-time FPGA image processing. *Proc. SPIE*, **6616**, p. 661633.
33. Montgomery, P. C., Anstotz, F., Montaner, D., Pramatarova, L., and Pecheva, E. (2010). Towards real time 3D quantitative characterisation of in situ layer growth using white light interference microscopy. *J. Phys. Conf. Ser.*, **253**, p. 012017.
34. Montgomery, P. C., Anstotz, F., Montagna, J. (2011). La mesure de changement de forme de surfaces microscopiques par la microscopie 4D. edited by Société Française d'Optique. *12ème Colloque International Francophone sur les Méthodes et Techniques Optiques pour l'Industrie (CMOI 2011)*, Nov 2011, Lille, France.
35. Kurban, U., Kammerer, J. B., Hehn, M., Montaigne, F., Hébrard, L., and French, P. (2006). Design and modeling of micro-fluxgate sensor without pick-up coil. In *Proceedings of the 20th Eurosensors Conference*, Göteborg (Sweden).

Chapter 12

Digital Interference Holography for Tomographic Imaging

Lingfeng Yu, Mariana C. Potcoava, and Myung K. Kim
Department of Physics, University of South Florida, Tampa, FL 33620, USA
lingfengyu@gmail.com

12.1 Introduction

One of the important challenges for biomedical optics is noninvasive 3D imaging, and various techniques have been proposed and available. For example, confocal scanning microscopy provides high-resolution sectioning and in-focus images of a specimen. However, it is intrinsically limited in frame rate due to serial acquisition of the image pixels. Ophthalmic imaging applications of laser scanning in vivo confocal microscopy have been recently reviewed [1]. Another technique, optical coherence tomography (OCT), is a scanning imaging technique with micrometer scale axial and lateral resolution, based on low coherence or white light interferometry to coherently gate backscattered signal from different depths in the

---

Handbook of Full-Field Optical Coherence Microscopy: Technology and Applications
Edited by Arnaud Dubois
Copyright © 2016 Pan Stanford Publishing Pte. Ltd.
ISBN 978-981-4669-16-0 (Hardcover), 978-981-4669-17-7 (eBook)
www.panstanford.com

object [2, 3]. Fourier domain optical coherence tomography (FD-OCT) is a significant improvement over the time domain optical coherence tomography (TD-OCT) [4–6], in terms of the acquisition speed and sensitivity. A related technique of wavelength scanning interferometry uses the phase of the interference signal, between the reference light and the object light which varies in the time while the wavelength of a source is swept over a range. A height resolution of about 3 μm has been reported using a Ti:sapphire laser with wavelength scanning range of about 100 nm [7, 8]. The technique of structured illumination microscopy provides wide-field depth-resolved imaging with no requirement for time-of-flight gated detection [9].

The principle of holography was introduced by Dennis Gabor [10] in 1948, as a technique where wavefronts from an object were recorded and reconstructed in such a way that not only the amplitude but also the phase of the wave field were recovered. In 1967, J. Goodman demonstrated the feasibility of numerical reconstruction of holographic image using a densitometer-scanned holographic plate [11]. Schnars and Jueptner, in 1994, were the first to use a charge-coupled device (CCD) camera connected to a computer as the input, completely eliminating the photochemical process, in what is now referred to as digital holography [12–14]. Various useful and special techniques have been developed to enhance the capabilities and to extend the range of applications. Phase-shifting digital holography allows elimination of zero-order and twin-image components even in on-axis arrangement [15–17]. Optical scanning holography can generate holographic images of fluorescence [18]. Three-channel color digital holography has been demonstrated [19]. Application of digital holography in microscopy is especially important, because of the extremely narrow depth of focus of high-magnification systems [20, 21]. Numerical focusing of holographic images can be accomplished from a single exposed hologram. Direct accessibility of phase information can be utilized for numerical correction of various aberrations of the optical system, such as field curvature and anamorphism [22]. Digital holography has been particularly useful in metrology, deformation measurement, and vibrational analysis [23–25]. Microscopic imaging by digital holography has been applied to imaging of microstructures and biological systems [23, 26, 27]. We have developed digital interference holography (DIH) for optical tomographic imaging [28–33] as well as multiwavelength quantitative phase contrast digital holography for high-resolution microscopy [34–37].

In the last few years, the scanning wavelength technique in various setups has been adopted by researchers for 3D imaging of microscopic samples. When digital holography is combined with optical coherence tomography, a series of holograms are obtained by varying the reference path length [38]. A new tomographic method that combines the principle of DIH with spectral interferometry has been developed using a broadband source and a line-scan camera in a fiber-based setup [39]. Subwavelength-resolution phase microscopy has been demonstrated [40] using a full-field swept source for surface profiling. Nanoscale cell dynamics were reported using cross-sectional spectral domain phase microscopy (SDPM) with lateral resolution better than 2.2 µm and axial resolution of about 3 µm [41]. A spectral shaping technique for DIH is seen to suppress the side lobes of the amplitude modulation function and to improve the performance of the tomographic system [42]. Submicrometer resolution of DIH has been demonstrated [43].

Another optical tomographic technique, applied widely for determination of the refractive index [44–49], is based on acquiring multiple interferograms while the sample is rotating. The reconstruction of the phase distribution is performed using filtered backprojection algorithm. Then the phase distribution is scaled to refractive index values. Refractive index distribution reveals information about the cellular internal structure of a transparent or semitransparent specimen.

In this book chapter, we use computer and holographic techniques with DIH to accurately and consistently identify and quantify different objects structure with micrometer resolution. This technique is based on an original numerical method [28], where a 3D microscopic structure of a specimen can be reconstructed by a succession of holograms recorded using an extended group of scanned wavelengths.

12.2 Principle of Digital Interference Holography

12.2.1 Basic Description of DIH [28–33]

Suppose an object is illuminated by a laser beam of wavelength λ. A point r_0 on the object scatters the light into a Huygens wavelet, $A(r_0)\exp(ik|r-r_0|)$, where the object function $A(r_0)$ is proportional to the amplitude and phase of the wavelet scattered or emitted by object points (Fig. 12.1a). For an extended object, the field at r is $E(r) \approx \int A(r_0)\exp(ik|r-r_0|)d^3r_0$, where the integral is over the object volume. The amplitude and phase of this field at the hologram plane $z = 0$ is recorded by the hologram, as $H(x_h, y_h; \lambda)$. The holographic process is repeated using N different wavelengths, generating the holograms $H(x_h, y_h; \lambda_1)$, $H(x_h, y_h; \lambda_2), \ldots, H(x_h, y_h; \lambda_N)$. From each of the holograms, the field $E(x, y, z; \lambda)$ is calculated as a complex 3D array over the volume in the vicinity of the object (Fig. 12.1b). Superposition of these N 3D arrays results in

$$\sum_k \int A(r_0)\exp(ik|r-r_0|)d^3r_0 \sim \int A(r_0)\delta(r-r_0)d^3r_0 \sim A(r).$$

(12.1)

That is, for a large enough number of wavelengths, the resultant field is proportional to the field at the object and is nonzero only at the object points. In practice, if one uses a finite number N of wavelengths, with uniform increment $\Delta(1/\lambda)$ of the inverse wavelengths, then the object image $A(r)$ repeats itself (other than the diffraction/defocusing effect of propagation) at a beat wavelength $\Lambda = [\Delta(1/\lambda)]^{-1}$, with axial resolution $\delta = \Lambda/N$. By use of appropriate values of $\Delta(1/\lambda)$ and N, the beat wavelength Λ can be matched to the axial range of the object, and δ to the desired level of axial resolution.

Addition of a series of N cosines or imaginary exponentials yields $S = \frac{1}{N}\sum_{n=1}^{N}\exp(in\delta kz)$, so $|S|^2 = \frac{\sin^2(N\pi z/\Lambda)}{N^2\sin^2(\pi z/\Lambda)}$, where $\delta k = 2\pi/\Lambda$. The signal-to-noise ratio (SNR) of the peaks at $z = 0, \Lambda, \ldots$ grows proportional to N^2, while the width of the peak narrows as $\delta \approx \Lambda/N$.

This behavior of the SNR and resolution is achieved only if all the amplitudes and phases of cosines are identical. Each hologram captured by the camera is normalized by the 2D average of each

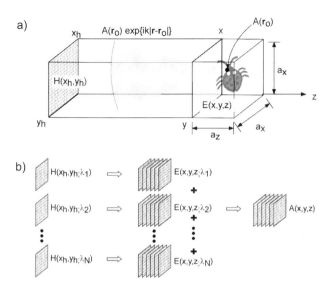

Figure 12.1 (a) Geometry and (b) process of DIH. H, hologram; E, optical field in the object volume; A, object function. See text for more details.

hologram to compensate for the laser power variation across the tuning range.

$$H_{\text{norm}}(x_h, y_h; \lambda_n) = \frac{H(x_h, y_h; \lambda_n)}{\sum_{x_h, y_h} H(x_h, y_h; \lambda_n)/N_x N_y} \quad (12.2)$$

12.2.2 Phase Correction in DIH

The phase calculated by digital holography is given by $\varphi = kZ = 2\pi Z/\lambda$, where Z is the distance of an object point relative to the position of the reference mirror and λ is the wavelength of the laser. Uncertainty in k or Z leads to phase error, which needs to be corrected for. For example, from each hologram $H(\lambda_n)$, a 2D phase profile is calculated $\varphi_n(x, y) = \text{phase}\{E(x, y, z; \lambda_n)\}$ at a suitable value of z that corresponds to a location near the object. Then calculate the difference profiles $\delta\varphi_n(x, y) = \varphi_n(x, y) - \varphi_{n-1}(x, y) = \delta k \cdot Z(x, y)$, where $Z(x, y)$ is the Z-profile of the object being imaged. If δk's are perfectly equally spaced between holograms, then all $\delta\varphi_n$'s for various n's should be identical. Otherwise, the uncertainties introduce phase error ε_n, so that the measured phase differences are

$\delta\varphi'_n(x, y) = (\delta k + \Delta k) \cdot Z = \delta\varphi_n(x, y) + \varepsilon_n$, where Δk is the deviation from the nominally constant δk. The trick is to determine the series $\varepsilon_2, \varepsilon_3, \ldots, \varepsilon_n$ that makes $\delta\varphi_2, \delta\varphi_3, \ldots, \delta\varphi_n$ as identical as possible. To find ε_2, for example, take the difference $\delta\varphi'_3 - \delta\varphi'_2$ plus a value of ε_2, modulo 2π, and take the average over (x, y). Do this for a number of ε_2 values in the range of 0 to 2π until the one is found that minimizes $\delta\varphi'_3 - \delta\varphi'_2$. This procedure works best if $Z(x, y)$ is a well-defined 2D function. With diffuse or multilayered objects, one assumes that the phase is mostly determined by the top surface of a tissue, for example. With the retinal sample objects presented later in this chapter, this seems to be the case, whereas imaging experiments with more diffuse texture, such as mouse skin, the resultant images are noisier and of less contrast. It is also noted that the error $\varepsilon_n = \Delta k \cdot Z$ is larger if Z is larger, so that it is advantageous to keep Z as small as possible, that is, to keep the reference mirror position to match the object position as closely as possible.

12.2.3 Spectral Shaping in DIH

In Eq. 12.2, the reconstructed wave fields from each wavelength were numerically normalized by the average power so that the superposition of all the holographic fields has virtually resulted in a synthetic rectangular spectrum of the light source, which will cause big side lobes in the amplitude modulation function. These side lobes will actually generate severe spurious structures in tomographic imaging and will greatly increase the noise level of tomographic reconstruction [42].

With finite wavenumbers, Eq. 12.1 can be rewritten as:

$$E(\mathbf{r}) \approx \sum_k \int A(\mathbf{r}_0) \exp(ik|\mathbf{r} - \mathbf{r}_0|) d^3\mathbf{r}_0$$
$$\approx \int A(\mathbf{r}_0) \sum_k \exp(ik|\mathbf{r} - \mathbf{r}_0|) d^3\mathbf{r}_0$$
$$\approx \int A(\mathbf{r}_0) M(|\mathbf{r} - \mathbf{r}_0|) d^3\mathbf{r}_0$$
$$\approx A(\mathbf{r}), \quad (12.3)$$

where $M(|\mathbf{r} - \mathbf{r}_0|) = \sum_k \exp(ik|\mathbf{r} - \mathbf{r}_0|)$ is defined as an amplitude modulation function (AMF). As the number of wavelengths goes to infinite, the AMF actually become a delta function; thus, the resultant field is proportional to the field at the object and is nonzero only at the object points.

However, any physically existing light sources have a limited spectrum range of $[k_{min}, k_{max}]$, with a bandwidth of $\Delta k = k_{max} - k_{min}$. Practically, if one uses a finite number N of wavenumbers at regular intervals of $dk = \frac{k_{max} - k_{min}}{N-1}$ from $[k_{min}, k_{max}]$, the above equation can be written as

$$E(\mathbf{r}) \approx \int A(\mathbf{r}_0) \sum_{k=k_{min}}^{k_{max}} \exp(ik|\mathbf{r} - \mathbf{r}_0|) d^3\mathbf{r}_0$$

$$\approx \int A(\mathbf{r}_0) \exp(i\bar{k}|\mathbf{r} - \mathbf{r}_0|) \frac{\sin(Ndk|\mathbf{r} - \mathbf{r}_0|/2)}{\sin(dk|\mathbf{r} - \mathbf{r}_0|/2)} d^3\mathbf{r}_0, \quad (12.4)$$

where $\bar{k} = \frac{k_{max} + k_{min}}{2}$. Thus, except for an exponential term, the amplitude modulation function becomes

$$M(|\mathbf{r} - \mathbf{r}_0|) = \frac{\sin(Ndk|\mathbf{r} - \mathbf{r}_0|/2)}{\sin(dk|\mathbf{r} - \mathbf{r}_0|/2)}. \quad (12.5)$$

Clearly, the above process to obtain the amplitude modulation function is equivalent to the interference of a large number of monochromatic waves with equal intensities and equally spaced frequencies, which results in the generation of a narrow pulse of light. The amplitude modulation function has a periodic sequence of pulse-like peaks with a period (or beat wavelength) of $\Lambda = 2\pi [dk]^{-1}$ and the axial resolution $\delta = \Lambda/N = 2\pi/\Delta k$. If all the wavenumbers within the bandwidth $[k_{min}, k_{max}]$ are continuously scanned for illumination and reconstruction, it can be easily shown that the normalized amplitude modulation function finally become a sinc function as

$$\bar{M}(|\mathbf{r} - \mathbf{r}_0|) = \frac{\sin(\Delta k|\mathbf{r} - \mathbf{r}_0|/2)}{\Delta k|\mathbf{r} - \mathbf{r}_0|/2}. \quad (12.6)$$

Thus, the beat wavelength Λ will become infinitely large, and the axial resolution remains the same as in Eq. 12.5.

From the above equations, it is noticed that the amplitude modulation function actually forms a Fourier transform pair with the spectral shape of the light source as in optical coherence tomography [2]. Because $\mathbf{r}$ and k space form a Fourier transform pair, equally k-spaced wavelengths are always preferred for scanning. A tunable laser is normally used as a light source which is sequentially scanned to obtain the equally k-spaced wavelengths. Each wavelength corresponds to a quasi-Dirac spectrum (very narrow compared to the tuning range) of the source.

When the laser powers of all the scanning illumination wavelengths were normalized, from Eq. 12.5 or Eq. 12.6 we know that the AMF will cause big side lobes which are hard to suppress. These side lobes will generate severe spurious structures in tomographic imaging and increase the average noise level of the reconstruction, as demonstrated in Ref. [42].

To solve these problems, the concept of spectral shaping was introduced to the DIH system. Spectral filtering [50, 51] and shaping [52] were previously reported to obtain Gaussian spectra from non-Gaussian sources in order to improve the point-spread function in OCT. Similarly, in DIH, because of the Fourier transform relationship between the AMF and light source spectra, it follows the Fourier uncertainty relation that Fourier transform of a Gaussian function is another Gaussian, and the product of variances of Fourier transform pairs reach minimum for Gaussian functions [53]. Thus, digitally correcting the non-Gaussian spectra could result in the reduction of side lobes in the amplitude modulation function and eliminate spurious structures in tomographic imaging. Assume that the light source power spectrum is finally shaped to a Gaussian spectral form:

$$S(k - \bar{k}) = \frac{1}{\sqrt{2\pi}\sigma_k} \exp\left[-\frac{(k-\bar{k})^2}{2\sigma_k^2}\right], \qquad (12.7)$$

which has been normalized to unit power,

$$\int_{-\infty}^{\infty} S(k - \bar{k}) dk = 1, \qquad (12.8)$$

where $\bar{k}$ is the center wavenumber and $2\sigma_k$ is the standard deviation power spectral bandwidth.

The resultant wave field in Eq. 12.1 can now be written as

$$\begin{aligned} E(\mathbf{r}) &\approx \int_k S(k - \bar{k}) \int A(\mathbf{r}_0) \exp\left(ik|\mathbf{r} - \mathbf{r}_0|\right) d^3\mathbf{r}_0 \cdot dk \\ &\approx \int A(\mathbf{r}_0) \int_k S(k - \bar{k}) \exp\left(ik|\mathbf{r} - \mathbf{r}_0|\right) dk \cdot d^3\mathbf{r}_0. \end{aligned} \qquad (12.9)$$

Obviously, the amplitude modulation function $M(|\mathbf{r}-\mathbf{r}_0|)$ now becomes the Fourier transform of the Gaussian spectra $S(k-\bar{k})$. Thus, the amplitude modulation function contains a Gaussian envelope as well with a characteristic standard deviation spatial width $2\sigma_x$ that is inversely proportional to the power spectral bandwidth, which means that $\sigma_x \sigma_k = 1$. This is the limiting case of a general inequality on the product of variances of Fourier transform pairs. In general, if S is an arbitrary distribution and M is its Fourier transform, then the product of the variations is greater than one. This confirms the Fourier uncertainty relation that the product of variances of a Fourier transform pair reaches its minimum for Gaussian functions. If the above product is not minimized, then the AMF must not be a Gaussian. In numerical implementation, an N-point Gaussian spectra covering the spectrum range of $[k_1, k_N]$ is obtained by digitizing Eq. 12.7 as

$$S(n+1) = \exp\left[-\frac{1}{2}\left(\alpha\frac{k_1 + ndk - \bar{k}}{(k_N - k_1)/2}\right)^2\right], \quad (12.10)$$

where $0 \leq n \leq N-1$, and α is a parameter introduced to adjust the width of the Gaussian spectra. We have ignored the constant before the exponential term in Eq. 12.7.

12.3 Techniques of DIH

12.3.1 Detailed Description of DIH Experiments

One basic configuration of the apparatus used in our study [28–33] is a Michelson interferometer (Fig. 12.2). The light source is a Coherent 699 ring dye laser, pumped by Millenia V diode-pumped solid-state laser, tunable over a range of 565 nm to 615 nm with an output power of up to 500 mW. The laser output is spatial-filtered and collimated. The focusing lens L2 focuses the laser on the back focus of the objective lens L3, so that the object is illuminated by a collimated beam. The lenses L3 and L5 form a microscope pair, so that the CCD acquires a magnified image of a plane H in the vicinity of the object plane. The reference mirror is an optical conjugate of the plane H through the matching objective lens L4.

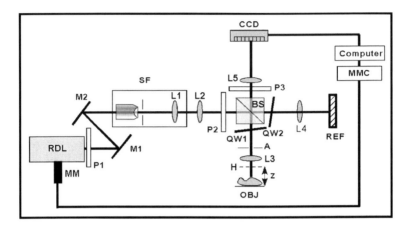

Figure 12.2 DIH optical apparatus. RDL, ring dye laser; M#, mirrors; SF, spatial filter and expander; L#, lenses; P#, polarizers; BS, polarizing beam splitter; QW#, quarter wave plates; A, aperture; H, hologram plane; OBJ, object; REF, reference mirror; MM, motorized micrometer; MMC, controller for MM [54].

Then the image acquired by the CCD is equivalent to a holographic interference between a plane reference wave and the object wave that has propagated (diffracted) over a distance z from the object plane. In general the object plane may be at an arbitrary distance z from the hologram plane H, and the object can be numerically brought back in focus by the digital holography process. But in practice, it is advantageous to keep the object plane in focus to simplify the optical alignment and to help identify the object portion being imaged, as well as minimizing potential secondary aberration effects. The polarization optics—polarizer P2, analyzer P3, quarter wave plates, and polarizing beam splitter—is used to continuously adjust the relative partition of optical power between the object and reference fields and to maximize the interference contrast. The polarizer P1 at the output of laser is used to continuously adjust the overall power input to the interferometer. The CCD camera (Sony XC-ST50) has 780 × 640 pixels with a 9 µm pitch and is digitized with an 8-bit monochrome image acquisition board (NI IMAQ PCI-1407). Slight rotations of the reference mirror and object planes enable the acquisition of off-axis hologram. A variable aperture placed at the back-focal (Fourier) plane of the objective lens L3 can be useful in controlling the angular spectrum of the object field.

As discussed in the previous section, the tunable range of wavelengths determines the axial resolution of the image, while the tuning resolution or the wavelength increment, determines the axial range, or the axial size of an object. The ability to distinguish axial distances of various layers of a tissue is called the axial resolution, δz. This tuning parameter is obtained in the following way:

$$k = \frac{2\pi}{\lambda}, \quad \delta k = \frac{2\pi \delta \lambda}{\lambda^2}, \quad \Lambda = \frac{2\pi}{\delta k} = \frac{\lambda^2}{\delta \lambda}, \quad \delta z = \frac{\Lambda}{N}, \quad (12.11)$$

where λ is the center wavelength, $\delta\lambda$ is the wavelength increment, k is the wavevector, δk is the wavenumber increment, δz is the axial resolution, and Λ is the object axial size. For example, a dye laser tunable over 34 nm range, for 50 wavelengths, centered at 585 nm results in an axial resolution of about $\delta z = 10$ μm and an axial range of $\Lambda = 500$ μm. The scanning process is controlled using a stepper motor that changes the birefringent filter of the laser in small increments, changing the laser wavelength when it rotates.

12.3.2 DIH by the Angular Spectrum Algorithm

We have found that the use of the angular spectrum algorithm has a number of advantages over Fresnel transform or Huygens convolution methods [32] for wave field reconstruction. Suppose $E_0(x_0, y_0)$ represents the 2D optical field at the hologram plane, then its angular spectrum is the Fourier transform

$$F(k_x, k_y; 0) = \iint E_0(x_0, y_0) \exp[-i(k_x x_0 + k_y y_0)] dx_0 dy_0, \quad (12.12)$$

where k_x and k_y are the spatial frequencies. The angular spectrum of an off-axis hologram contains a zero-order and a pair of first-order components, the latter corresponding to the twin holographic images. One of the first-order components in $F(k_x, k_y; 0)$ can be separated from the others with a numerical bandpass filter if the off-axis angle of the reference beam is properly adjusted. The object field can then be rewritten as the inverse Fourier transform of the angular spectrum, properly filtered:

$$E(x_0, y_0) = \iint F(k_x, k_y; 0) \exp[i(k_x x + k_y y)] dk_x dk_y. \quad (12.13)$$

The field distribution after propagation over a distance z is then

$$E(x, y; z) = \iint F(k_x, k_y; 0) \exp[i(k_x x + k_y y + k_z z)] dk_x dk_y$$
$$= \mathcal{F}^{-1}\{\mathcal{F}\{E_0\} \exp[ik_z z]\}, \quad (12.14)$$

where $k_z = [k^2 - k_x^2 - k_y^2]^{1/2}$, $k = 2\pi/\lambda$, and the symbol $\mathcal{F}$ denotes Fourier transform. Distinct advantages of the angular spectrum method include consistent pixel resolution, no minimum reconstruction distance, and easy filtering of noise and background components. Once the angular spectrum at $z = 0$ is calculated by a Fourier transform and filtered, the field at any other z plane can be calculated with just one more Fourier transform, whereas the Fresnel or convolution methods require two Fourier transforms for each value of z. Analytically, the angular spectrum method is equivalent to Huygens convolution method, because

$$E(x, y; z) = E_0 \otimes h = \mathcal{F}^{-1}\{\mathcal{F}\{E_0\} \cdot \mathcal{F}\{h\}\}, \quad (12.15)$$

where $\otimes$ represents convolution, $h \propto \exp(ikr) = \exp\left[ik\sqrt{x^2 + y^2 + z^2}\right]$ is the point spread function, and $F\{h\} \propto \exp(ik_z z)$ is the coherence transfer function. But the numerical calculations over a discretized finite plane H behave very differently between these methods. The Huygens method, based on the convolution of the spherical wavelets, has a minimum distance requirement to avoid aliasing of spherical wavefront of high curvature as well as a maximum distance to be able to account for spherical wavefront of a high enough fringe frequency across the hologram plane. On the other hand, the angular spectrum method accounts for all the recorded spatial frequencies of the hologram and this spectral content of plane waves is independent of the propagation distance. More discussion on various reconstruction methods used in digital holography were reported in Ref. [55].

Figure 12.3 shows an example of a US Air Force resolution target. The 1040×1040 μm², 256×256 pixel area selected in Fig. 12.3a shows group II elements 3 and 4 as well as group IV elements 2 through 6. The smallest visible element, group IV element 6 has a 17.54 μm width. The reconstruction distance z, representing the distance from the object to the hologram plane is 643.56 μm. The

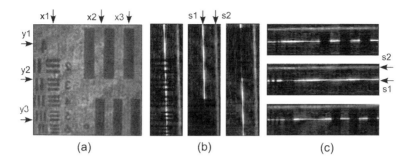

Figure 12.3 The reconstructed volume of the resolution target: (a) x–y cross section, 1040×1040 μm². (b) y–z cross sections at various x values, 500×1040 μm², from left to right, $x1$, $x2$, and $x3$. (c) x–z cross sections at various y values, 1040×500 μm², from top to bottom, $y1$, $y2$, and $y3$. s1 is the chromium-coated glass surface and s2 is the surface of the clear tape [54].

complex field of the resolution target is computed separately for 50 wavelengths by numerical diffraction using the angular spectrum method, which gives an axial range of $\Lambda = 500$ μm and axial resolution of $\delta z = 10$ μm. All 3D electric fields are added together to obtain a 3D volume structure of the object being imaged. Cross sections of the volume can be taken in the x, y, and z planes. Cross-sectional images in the y–z planes (Fig. 12.3b) and x–z planes (Fig. 12.3c) are shown above.

The resolution target is an object without internal structure and the reflection of the laser beam takes place at the surface of the resolution target. A piece of clear tape is placed on top of the resolution target to provide a second surface for demonstration of tomographic imaging. The first layer s1 in Fig. 12.3b,c is the reflection that comes from the chromium-coated glass surface. The second layer s2 is the reflection from the attached tape surface. We have tested the improvement of SNR with increasing number of holograms, N. As described above, the SNR is expected to grow as N^2. As seen in Fig. 12.4, the fourfold increase in N from 100 to 400 lowers the noise from -30 dB to about -45 dB, which is consistent with $10 \times \log 16 = 12$ dB. This data set is without the clear tape attachment.

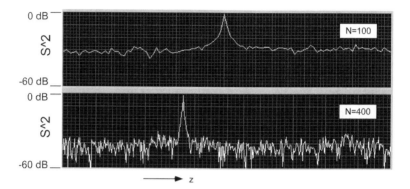

Figure 12.4 Improvement of SNR with number of holograms, N. The peak in each semilog graph represents the surface of resolution target. As N increases fourfold the SNR increases by 12 dB, as expected [54].

12.3.3 *Variable Tomographic Scanning [31, 56]*

In most of the 3D microscopy systems including OCT and DIH reported above, the 3D volume is reconstructed as a set of scanning planes with the scanning direction along the optical axis of the system. If a tomographic image on a plane not parallel to the original reference mirror is required, it can be reconstructed by combining or interpolating points from different tomographic layers, however the quality will be compromised especially when the lateral resolution does not match well with the axial resolution. To get better results, the whole process needs to be physically repeated with the reference mirror tilted or the object rotated to a desired orientation.

Since the advantage of digital holography is that a single hologram records the entire 3D information of the object, the object wave distributions and the synthesized tomographic image at an arbitrarily tilted plane can be reconstructed rigorously, and selective tomographic scanning with different orientation is possible. This tilted reconstruction plane then functions as the scanning plane in DIH.

Diffraction from a tilted plane based on the Rayleigh–Sommerfeld diffraction formula [57] was previously studied for computer-generated holograms [58] by D. Leseberg et al. and was later applied for numerical reconstruction of digital holography with

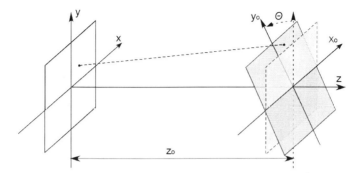

Figure 12.5 Reconstruction of the wave field on a tilted x_o–y_o plane for a given wave distribution on the x–y plane (hologram plane).

changed viewing angles [59] and on tilted planes [31]. Suppose the object wave distribution $o(x, y)$ on the hologram (at the $z = 0$ plane) is already known. For simplicity, reconstruction of the wave distribution on a tilted plane, $x_o - y_o$, with its normal in the y-z plane is considered, as shown in Fig. 12.5. The Rayleigh–Sommerfeld diffraction integral gives:

$$E(x_o, y_o, z_o) = \frac{iC_0}{\lambda} \iint o(x, y) \frac{\exp[ikr(x, y, x_o, y_o)]}{r(x, y, x_o, y_o)} \times \chi \\ \times (x, y, x_o, y_o) dx dy, \qquad (12.16)$$

where k is the wavenumber given by $k = 2\pi/\lambda$. C_0 is a constant and $\chi(x, y, x_o, y_o)$ is the inclination factor, which is approximately unitary under the Fresnel approximation and is omitted from the following equations. The inverse length $1/r$ can be replaced by $1/r_o$ and the $r(x, y, x_u, y_u)$ in the argument of the exponential can be expressed as

$$r = \sqrt{(z_o - y_o \sin\theta)^2 + (x - x_o)^2 + (y - y_o \cos\theta)^2}, \qquad (12.17)$$

which can be expanded as a power series of $r_o = (z_o^2 + x_o^2 + y_o^2)^{1/2}$ If only the first two lowest order terms in the expanded series are considered, and a further approximation is introduced, $ik(x^2 + y^2)/2r_o \approx ik(x^2 + y^2)/2z_o$, which holds almost the same restriction as the Fresnel condition, Eq. 12.16 can be finally expressed as

$$E(\xi, \eta, z_o) = \frac{iC_0}{\lambda r_o} \exp\left[ik\left(r_o - \frac{z_o y_o \sin\theta}{r_o}\right)\right]$$
$$\times \iint o(x, y) \exp\left[\frac{ik}{2z_o}(x^2 + y^2)\right] \exp$$
$$\times [-i2\pi(\xi x + \eta y)]\, dx\, dy, \qquad (12.18)$$

with:

$$\xi = \frac{x_o}{\lambda r_o}, \quad \text{and} \quad \eta = \frac{y_o \cos\theta}{\lambda r_o}. \qquad (12.19)$$

Equation 12.18 can be implemented with the fast Fourier transform (FFT) algorithm and a coordinate transform is made to get the wave distribution in the (x_o, y_o) coordinate, as indicated in Eq. 12.19.

In the discrete implementation of Eq. 12.18, the resolution of the reconstructed plane is determined according to Shannon theory, and is given approximately as

$$\Delta x_o = \frac{\lambda z}{N \Delta x}, \quad \Delta y_o = \frac{\lambda z}{N \Delta y \cos\theta}, \qquad (12.20)$$

where Δx_o and Δy_o are the resolutions of the tilted plane, Δx and Δy, are the resolutions of the hologram plane and $N \times N$ is the array size of a square area on the CCD. Note that if the tilted angle θ is equal to zero, then Eq. 12.18 simplifies to the well-known Fresnel diffraction formula [13]. Although the above algorithm only considers the situation that the angle θ lies in the y-z plane, it can easily be extended to any tilted angle θ in space [56]. From the above, the wave fields at an arbitrarily tilted plane can be reconstructed from the holograms, and the tomographic images in DIH, synthesized from multiple wave distributions, can be flexibly adjusted to variable orientations.

Figure 12.6 shows an example of a prepared slide of a chick embryo, of area 2.26×2.26 mm^2. The embryonic blood vessel is located in a tilted plane with $\theta = 2.5°$. Figure 12.6a shows one tomographic image with normal scanning. If the scanning plane is intentionally tilted with a proper angle that aligns with the vessel, now the whole blood vessel shows up at a single tomographic reconstruction in Fig. 12.6b. Since the holograms have recorded all the information of the object, the intention of variable tomographic

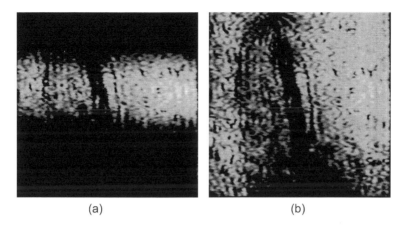

(a) (b)

Figure 12.6 Contour images of a chick embryo in a 2.26 × 2.26 mm^2 area: (a) normal tomographic reconstruction with the Fresnel diffraction formula; (b) tilted tomographic reconstruction with the proposed algorithm. Adapted from Ref. [31].

scanning is to calculate rigorous wave field distributions on titled planes directly from the recorded holograms and to better observe interesting structures or features on planes that are not parallel to the original scanning plane.

Instead of using the Rayleigh–Sommerfeld diffraction formula in Eq. 12.16, a few other methods were also reported to calculate wave field diffraction to tilted planes. The angular spectrum method was extended to reconstruct diffraction between tilted planes in Chapter 9 of Ref. [55]. Another way to reconstruct tilted wave field for variable tomographic imaging is to calculate the inclination phase factors with corresponding to the tilt of the desired observation plane and then compensate that to the complex object images of a so-called illumination-angle-scanning DIH [60]. The computational load is relatively low, and the pixel size automatically remains the same as the angular spectrum method.

12.3.4 DIH Based on Spectral Interferometry

Recently the principle of DIH was also implemented with a fiber-based spectral interferometry [39]. Instead of using tunable

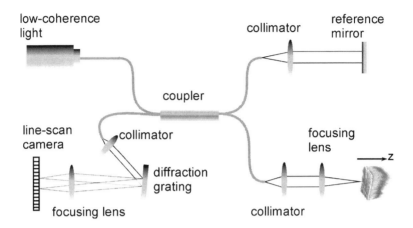

Figure 12.7 Apparatus for DIH based on spectral interferometry [39].

lasers, a relatively inexpensive broadband source was used as the light source. A line-scan camera was used for detection. Hundreds of 2D synthesized holograms (or object wave fields) were obtained by transversal scanning of a probe beam within a few seconds. Holographic images of an object volume were numerically reconstructed from each synthesized hologram and tomographic images were obtained by superposition of all the image volumes.

The design of the spectral-interferometer-based DIH system is illustrated in Fig. 12.7. A fiber-based Michelson interferometer is illuminated by a superluminescent diode with a spectrum centered at 1315 nm and a total delivered power of 8 mW. About half of the power goes to the sample arm. The full-width at half-maximum of the source is ~95 nm, and its coherence length is ~8 µm. Backreflected lights from the reference and sample arms are guided into a spectrometer and sampled by a 1×1024 line-scan camera (SU1024-1.7T, Sensors Unlimited) at 7.7 kHz. The captured spectrogram in the camera is linearly interpolated as 1024 evenly k-spaced wavelengths extending from 1250.623 nm to 1373.190 nm, corresponding to a spectral resolution of 0.12 nm and an imaging depth of 3.6 mm in vacuum. The interference between the two arms is recorded in the frequency domain as [61, 62]

$$I(k) = S(k)R^2 + S(k) \int_{-\infty}^{\infty}\int_{-\infty}^{\infty} O(\Delta z)O(\Delta z')$$
$$\times \exp\left[ik(\Delta z - \Delta z') + \varphi(\Delta z) - \varphi(\Delta z')\right]d\Delta z d\Delta z'$$
$$+ 2S(k)R \int_{-\infty}^{\infty} O(\Delta z)\cos(k\Delta z + \varphi(\Delta z))d\Delta z, \quad (12.21)$$

where R and O represent the amplitude of the reference and object signals, respectively. Δz and $\Delta z'$ denote the double-pass path length difference between the sample and the reference mirror, and φ is the phase term of object signals. The first two terms yield DC and low-frequency noises and the last term contains the depth information of the object and can be obtained by an inverse FFT from k to z space. Note that inverse FFT of the last term gives symmetric positive and negative images in space. According to Eq. 12.21, the nonoverlapping positive image can be extracted by properly adjusting the relative position of the reference mirror, and its corresponding spectral information can be Fourier transformed back to k space as

$$I'(k) = S(k)R \int_{-\infty}^{\infty} O(\Delta z)\exp[ik\Delta z + \varphi(\Delta z)]d\Delta z. \quad (12.22)$$

Note that $I'(k)$ is the spectral information of a single line (A-line) along the z axis when the probe beam is illuminating a specific (x, y) position on the sample. Because phase fluctuations exist during 2D scanning of the system, a microscope cover glass (or other surface) is intentionally used as a reference to eliminate the phase fluctuation of different transversal scans. For each A-line scanning, the phase of the reference pixel in the corresponding A-line (z axis) profile, which represents the front surface of the cover slip, is calculated and subtracted along all pixels of the A-line, then the spectral information $I'(k)$ of each A-line in Eq. 12.22 is calculated by taking FFT of the filtered and phase-corrected positive complex image. Thus, by 2D scanning of the galvo-system, object wave fields $I'(x, y, k_n)$ with ($n = 1, 2, \ldots 1024$) different wavenumbers are readily available. We noticed that the information $I'(x, y, k)$ at a designated k was equivalent to the extracted object wave field from a 2D hologram recorded with a wavenumber k. According to the principle of DIH, a holographic 3D object volume was numerically reconstructed from each $I'(x, y, k)$ by use of a diffraction algorithm, and all the 3D arrays are numerically superposed together, resulting in an accumulated field distribution that represents the 3D object structure.

Also note that the system shown in Fig. 12.7 is similar to FD-OCT [61–63], a powerful imaging technique that has attracted a great deal of attention recently. The axial resolution of the proposed arrangement is determined by the coherence length of the light source, and high axial resolution can be achieved independently of the beam-focusing conditions. The lateral resolution is determined by the diameter d of a probe beam and the focal length f of the objective as $\Delta x = 4\lambda f/(\pi d)$ [64]. However, only a very small range around the depth of field (DOF) exhibits the desired lateral resolution; the probe beam outside DOF will be largely expanded and the obtained image will be blurred.

Compared to FD-OCT, DIH reconstructs the object wave fields at different depth compensating the out-of-focus blur, high-resolution details can be recovered from outside of the depth-of-focus region with little loss of lateral resolution. Another simplified way to improve lateral resolution in FD-OCT imaging is to use monochromatic numerical diffraction method [65] to simulate the wave propagation process from out-of-focus scatterers and find back the focus.

Figure 12.8 shows one example of a ∼400 μm thick onion slice placed ∼100 μm away from the DOF region of the probe beam. The use of an objective (20 ×, Nachet) in this example improved the lateral resolution to ∼3.5 μm but decreased the DOF to 14.4 μm. Figure 12.8a shows the spectral interferogram $I(k)$ of one A-line scan and Fig. 12.8b shows its inverse FFT. The peak F in the figure represents the front surface of an added microscope cover glass and peak B shows its back surface. The phase of the peak F was used as the reference for all pixels of the A-line. Then, the phase-corrected positive image was filtered and Fourier transformed back to obtain a complex $I'(k)$, whose amplitude and phase information are plotted in Figs. 12.8c and 12.8d, respectively. After 2D scanning, the total acquisition time of 1024 synthesized holograms takes ∼10 seconds, which could be further improved by using a line-scan camera of higher speed. Figure 12.8e shows two of these holograms.

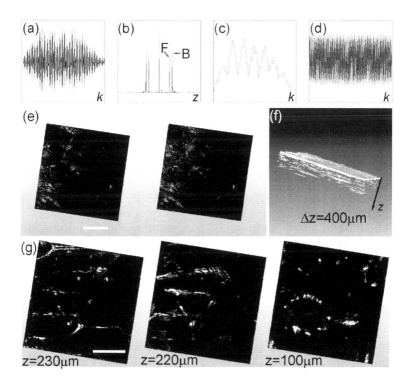

Figure 12.8 (a) Spectral interferogram and (b) its inverse FFT for one A-line scanning; the phase-corrected positive image is Fourier-transformed to get the (c) amplitude and (d) phase of a complex spectral counterpart $I'(k)$ of the positive image; (e) absolute object wave fields $I'(x, y, k)$ for k_{256} and k_{512}, respectively; (f) 3D reconstruction of the onion slice; (g) tomographic images of different depths. The vertical scales in (a–d) are normalized and the scale bar represents 100 μm [39].

Based on all (or part of) these holograms, the algorithm of DIH is then utilized to reconstruct the tomographic images of the sample. Figure 12.8f shows a 3D reconstruction of the sample and Fig. 12.8g shows several reconstructed tomographic images located at different depths. Note that artifacts may exist in the tomographic reconstruction if the phase information is not correctly obtained because of multiple factors, such as the phase instability of the system or existence of multiple scatterings, etc.

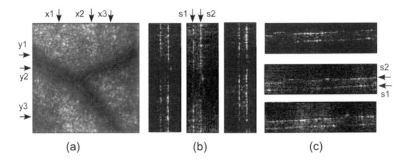

Figure 12.9 The reconstructed volume of the retina with blood vessels: (a) $x-y$ en face image, 670×670 μm^2; (b) from left to right, $z-y$ cross sections along $x1$, $x2$, and $x3$, 500×670 μm^2; (c) from top to bottom, $x-z$ cross sections along $y1$, $y2$, and $y3$, 670×500 μm^2. s1 is the choroidal surface and s2 is the retinal surface [54].

12.4 Applications of DIH

12.4.1 *Animal Tissue*

In the following, we present a few examples of tomographic imaging of biological specimens using DIH. Figures 12.9 and 12.10 are images of a porcine eye tissue provided by the Ophthalmology Department at the USF. It was preserved in formaldehyde, refrigerated and a piece of the sclera, with retinal tissue attached, was cut out for imaging. The holographic image acquisition and computation of the optical field are carried out for each of 50 wavelengths in the range from 565 nm to 602 nm. Superposition of images, in DIH processes described above, reveals the principal features of the retinal anatomy. The imaged surface areas are 0.67×0.67 mm^2 for Fig. 12.9 and 1.04×1.04 mm^2 for Fig. 12.10. The axial range $\Lambda = 500$ μm and axial resolution $\delta z = 10$ μm for both image sets. The measured SNR for these images was about 45–55 dB.

In Fig. 12.9, the images reveal convex surfaces of blood vessels, as well as about 150 μm thick layer of retina, s2, on top of the choroidal surface, s1. The blood vessels in Fig. 12.9 were apparently fixed with blood in them, while in Fig. 12.10, the preparation and handling of the tissue sample resulted in tear of some of the retinal tissue. Thus the upper right half of Fig. 12.10 has intact retinal

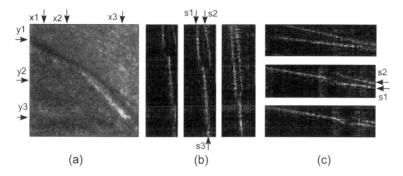

Figure 12.10 The reconstructed volume of torn retina on choroid: (a) x–y en face image, 1040 × 1040 μm^2; (b) from left to right, y–z cross sections along $x1$, $x2$, and $x3$, 500 × 1040 μm^2; (c) from top to bottom, x–z cross sections along $y1$, $y2$, and $y3$, 1040 × 500 μm^2. s1 and s3 are choroidal surfaces and s2 is the retinal surface [54].

tissue, while the lower left half is missing the retinal layer and the choroidal surface is exposed. In Fig. 12.10b, the boundary marked s3 is the bare choroidal surface, while the surface marked s1 is the choroidal surface seen through the retinal surface s2. The index of refraction of the retinal layer causes the choroidal surface to appear at a different depth compared to the bare surface, causing the break in the outline of the choroidal surface in Fig. 12.10c. In fact, it is possible to estimate the index of the retinal layer from the change in apparent depth of choroidal surface, and is consistent with the expected value in the range of 1.35–1.40. Figure 12.11 is the tomographic image data on a mouse skin tissue. One can discern the uppermost layer of epidermis (stratum corneum) as distinct high reflectivity layer, while the rest of the dermis appears as a region of diffuse scattering. The SNR is lower than the retina images because of the lower reflectivity of the tissue sample and the more diffuse nature of the tissue.

12.4.2 Human Retina

Figure 12.12 shows amplitude images obtained from a healthy excised human eye supplied to us by the Lions Eye Institute for Transplant & Research of Tampa, FL. The experimental setup

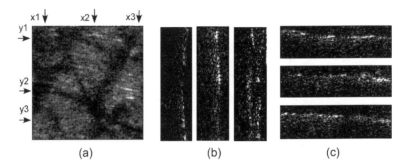

Figure 12.11 The reconstructed volume of mouse skin: (a) x–y en face image, 670×670 μm^2; (b) from left to right, y–z cross sections along $x1$, $x2$, and $x3$, 500×670 μm^2; (c) from top to bottom, x–z cross sections along $y1$, $y2$, and $y3$, 670×500 μm^2 [54].

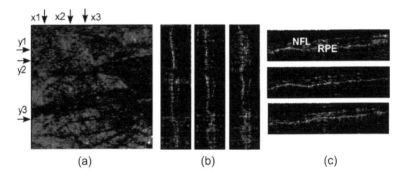

Figure 12.12 The reconstructed volume of the human macula sample: (a) x–y cross section, 5020×5020 μm^2; (b) y–z cross sections at various x values, 5020×209.75 μm^2, from left to right, $x1$, $x2$, and $x3$; (c) x–z cross sections at various y values, 209.75×5020 μm^2, from top to bottom, $y1$, $y2$, and $y3$; $\Delta\lambda = 0.560$–0.600 μm; $\delta z = 4.19$ μm; $\delta x = \delta y = 19.6$ μm; NX = NY = 256 pixels; NZ = 50 pixels [66].

was shown in Fig. 12.2. The holographic image acquisition and computation of the optical field of the macula sample are carried out for about 50 wavelengths in the $\Delta\lambda$ range of 560–600 nm. Superposition of images, in the DIH processes described above, reveals the topographic mapping within the macular tissue, Fig. 12.12a, and clearly delineates borders of blood vessel segments. The imaged surface area is 5020×5020 μm^2 with a physical axial

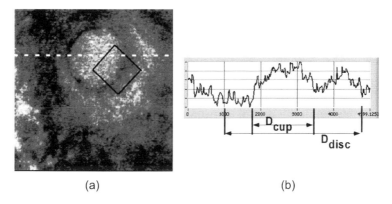

Figure 12.13 The reconstructed volume of the human optic nerve sample: (a) $x-y$ cross section, FOV = 5020×5020 µm^2; $Z_0 = 446$ µm; $\Delta\lambda = 0.563$–0.605 µm; $\delta z = 4.06$ µm; $\delta x = \delta y = 19.6$ µm; NX = NY = 256 pixels; NZ = 50 pixels; (b) cross section along the dashed line in Fig. 12.5a; D_{cup} = cup diameter; D_{disc} = disc diameter [66].

range $\Lambda = 209.75$ µm and physical axial resolution $\delta z = 4.19$ µm. Different layers are distinguishable in the cross-sectional images of the human macula (Fig. 12.12b,c). The optical thickness between the retinal nerve fiber layer (NFL) and the retinal pigment epithelial (RPE) layer is about 84 µm.

Figure 12.13a, represents the en face reconstructed 3D structure of the optic nerve region with an area of 5020×5020 µm^2. We can identify the scleral ring (disc) diameter $D_{disc} = 1750$ µm and the cup diameter $D_{cup} = 660$ µm (Fig. 12.13b). Our measurements cannot be clinically correlated with normal anatomic values as the tissue was postmortem and edematous, falsely enlarging the disc diameter with swelling of the surrounding nerve fiber layer tissue. Vitreous papillary adhesions also had to be removed with forceps from the optic nerve tissue, and any remnants lying atop the nerve tissue rim could have falsely enlarged the measurements. The optic nerve sample was slightly tilted when it was imaged, resulting in the left side of the scleral ring appearing to be darker than the right. The optic cup has a well distinguished shape with high reflectivity, depicted by brighter colors on 3D imaging.

Figure 12.14a, represents the en face reconstructed optical field of the rhombus region from Fig. 12.13a. The cross-sectional images

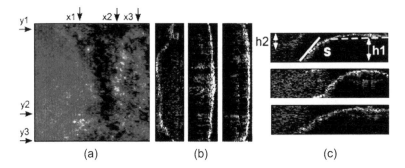

Figure 12.14 The reconstructed volume of the human optic nerve sample (the rhombus-shaped volume from Fig. 12.5a): (a) x–y cross section, FOV = 1100×1100 μm^2; (b) y–z cross sections at various x values, 1100×280.35 μm^2, from left to right, $x1$, $x2$, and $x3$; (c) x–z cross sections at various y values, 280.35×1100 μm^2, from top to bottom, $y1$, $y2$, and $y3$; $Z = 29.7$ μm; $\Delta\lambda : 0.565 - 0.595$ μm; $\delta z = 5.61$ μm; $\delta x = \delta y = 4.32$ μm; NX = NY = 256 pixels; NZ = 50 pixels; s = slope; $h1$, $h2$ = heights [66].

in the y–z planes (Fig. 12.14b) and x–z planes (Fig. 12.14c), are also shown. In practice, if one uses a finite number of wavelengths, N with a uniform increment δk of inverse wavelengths, then the object image repeats itself at a beat wavelength $\Lambda = \frac{2\pi}{\delta k}$, with axial resolution $\delta z = \Lambda/N$. This occurred in the x–z cross sections (Fig. 12.14c, top), where the bottom of the cup depth, $h2$, is found at the top. Therefore, to find the cup depth, $h1$ is added to $h2$, where $h1$ (relative to the baseline height) and $h2$ are the upper and the bottom part of the cup, respectively. By using appropriate values of δk and N, the physical axial size (beat wavelength), Λ, can be matched to the axial range of the object, and δz to the desired level of physical axial resolution. The height of the x–z cross sections (Fig. 12.14c, top) represents the physical axial size $\Lambda = 280.35$ μm made of 50 pixels (the number of wavelengths being scanned) with a physical axial resolution $\delta z = 5.61$ μm. Using this information one can quantify the cup depth as being $h = 355.11$ μm and a cup slope, s, of about $47°$.

12.4.3 DIH for Biometric Application

Biometric identification technologies make use of various intrinsic physical or behavioral traits, such as face, iris, voice, signature, gait,

DNA, and fingerprints. The fingerprinting has the longest history, with evidence of its use since ancient Babylonian and Chinese civilizations. In 1880, Henry Faulds considered skin furrows of the hand as an identifying mark [67]. After a few years of experimental work, the Galton–Henry system of fingerprint classification was published and quickly introduced in the USA in 1901 for criminal identification records [68]. For biometry, three types of fingerprints are considered: latent, patent, and plastic [69]. Latent prints are invisible (or not easily visible) traces left on hard surfaces. Patent prints are easily visible, typically produced by applying ink on the finger and pressing onto paper, during the process of enrolling. Plastic prints are formed when the finger presses onto a soft surface such as wax, soap, or putty.

Though it has a long history, fingerprinting continues to develop in many aspects. In particular, 3D imaging is believed to offer additional information and security against spoofing.

DIH was used to image 3D surface of plastic prints. For example, in Fig. 12.15, a 5.02×5.02 mm^2 field of view (FOV) is imaged while the laser wavelength is scanned over the range 559–599 nm at 50 intervals. This results in a 3D tomographic image volume of $5.02 \times 5.02 \times 0.210$ mm^3 with axial resolution of 4.19 μm after accounting for the reflection geometry. The resultant image volume can be represented in any convenient forms. An en face cross-sectional view of the plastic print is shown in Fig. 12.15a and several x or y cross-sectional views are shown in Figs. 12.15b and 12.15c. The cross-sectional views clearly depict the crests and valleys of the friction ridges, with about a 750 μm ridge periodicity and about a 50 μm depth of grooves. A surface-rendering software can be used to render the tomography data into a more user-friendly format. Note that the crest of this negative print corresponds to the valley of the grooves on the actual finger and vice versa. These are plastic prints of clay material, so the subsurface structure is of no interest, but there is clear evidence of signals from subsurface scattering points. This indicates feasibility of tomography of subsurface tissue and vein structures in live fingers, which can also be a basis of biometry. Another example of DIH topography is shown in Fig. 12.15d–f, with a wider FOV but a similar depth $10.4 \times 10.4 \times 0.213$ mm^3. Figure 12.15d is presented as an en face flat view, that is, the entire 3D array

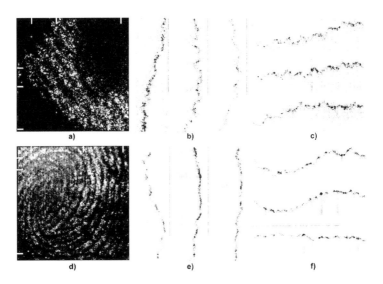

Figure 12.15 Tomographic images of plastic fingerprint by digital interference holography (DIH): (a) z section of the 3D volume image; (b) x sections along three vertical lines indicated with ticks in (a); (c) y sections along three horizontal lines indicated with ticks in (a). The image volume for (a–c) is $5.02 \times 5.02 \times 0.211$ mm^3. Another example of a 3D volume image is shown in (d) en face flat view, (e) x sections, and (f) y sections. The image volume for (d–f) is $10.4 \times 10.4 \times 0.213$ mm^3 [70].

viewed from the top along the z direction. This sample appears to have a larger overall undulation of the surface, whereas the friction ridges are shallower than the other example.

In Fig. 12.16, another example of 3D volume image, $4.86 \times 4.86 \times 0.213$ mm^3, is rendered in perspective views at a few different viewing angles. Notice here an important level 2 (minutiae) feature called crossover, which is a short ridge that runs between two parallel ridges. The length of the ridge between the end points is 1700 µm.

Biometry in general and fingerprinting in particular are becoming increasingly important in diverse areas from electronic commerce to homeland security, and there are heightened interests in developing new methods and devices for faster and more reliable acquisition of biometric information. Three-dimensional, high-sensitivity, nondestructive imaging by digital holography and

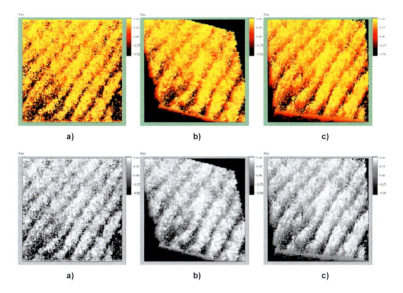

Figure 12.16 Perspective rendering of 3D volume image data by DIH from a few different viewing angles. The image volume is 4.86 × 4.86 × 0.210 mm^3 [70].

interferography appear to have potential to make significant contributions in these areas.

12.4.4 Submicron Tomography of Cells with DIH

Instead of using DIH to image thick scattering tissue, Jonas Kühn and his colleagues used DIH to image fixed red blood cells (RBCs) with high magnification and retrieve the reflection phase signal of those RBCs. By sequentially acquiring reflection holograms and summing 20 wavefronts equally spaced in spatial frequency in the 485–670 nm range, slice-by-slice tomographic reconstruction of the RBCs with submicron axial resolution of 0.6–1 μm was reported [71].

12.5 Discussions

In conclusion, we have presented the recent developments of DIH techniques for tomographic imaging. The sensitivity of our DIH system is less than 60 dB [32] without image accumulation or lock-in, which is now lower than the performance of the standard OCT, (~100 dB) or full-field optical coherence tomography (FF-OCT) (~90 dB) [72] where 200 images are locked in to obtain one tomographic image.

Tissue absorption and scattering are wavelength dependent and are critical for system optimization. For ophthalmic imaging, the light source has to be chosen as a function of the characteristics of the reflected spectrum. This spectrum varies with three histological parameters, RPE melanin, hemoglobin, and choroidal melanin. The intensity decreases in the green region once the melanin increases and the RPE response becomes weaker. Also, the absorption of the ophthalmic tissue depends on hemoglobin in visible and water in infrared [73, 74]. Current OCT systems for ophthalmic applications use light sources centered at 800 nm. The wavelength band centered on 830 nm has been employed in clinical OCT instruments. An ultrahigh-resolution (UHR) OCT technology using a broadband Ti:Al2O3 laser (centered on 800 nm) at an axial resolution of 2–3 μm has been demonstrated for in vivo imaging of retinal and corneal morphology [75, 76]. A center wavelength at 1 μm provides an optimal wavelength for imaging choroidal morphology and microvasculature below the retinal pigment epithelium, because increasing center wavelengths longer than 800 nm increases imaging depth due to reduced scattering at longer wavelengths; also, there is a local minimum of water absorption at 1 μm. In addition, eye-safe exposure at 1 μm is a few times larger than that at 800 nm, which permits the use of a higher incident power for ocular imaging. Choroidal imaging with a spectral domain OCT system at 1064 nm was reported in Ref. [77].

Considering all the above factors, to optimize the DIH system for ophthalmic applications, we can replace the dye laser (560–600 nm) with a Ti:sapphire laser (of longer wavelengths). Also, the sensitivity of DIH system can be improved to more than 90 dB by introducing a high-speed camera and increasing the number of holograms from 50 to 500. One challenge in ophthalmic imaging applications is to ensure that the level of laser radiation on the eye is not damaging to the vision. The sensitivity of the system needs to be sufficient so that a radiation level weak enough to safe can still generate good-quality images.

In vivo imaging of human eye needs to be fast enough to avoid blurring due to eye movement (tremors, drifts, saccades). At this stage of development, the acquisition time is limited by the need to manually scan the laser wavelength. In the near future it can be accomplished by a motorized micrometer under computer control. Then the limiting factor of scanning will be the camera frame rate. In the above examples, the scanning time is a few tens of seconds long. A Ti:sapphire laser with an appropriate actuator and a sweep function parameterized by time is a good option that works very well with the swept-source optical coherence tomography (SS-OCT) systems. Also, a retinal tracker system for 3D retinal morphology and function can be developed and integrated in the DIH setup.

With increased speed, sensitivity and resolution, DIH has the potential to provide a significant improvement in terms of information captured, both for the diagnosis of disease and for the understanding of normal histopathology and physiology. With better axial resolution and greater axial range, we expect to be able to extract more information about retinal thickness and structure. With full development of its capabilities, DIH may provide another option in ocular imaging, providing high-resolution 3D information, which could potentially aid in guiding the diagnosis and treatment of many ocular diseases.

References

1. Patel, D. V., and McGhee, C. N. J. (2007). Contemporary in vivo confocal microscopy of the living human cornea using white light and laser scanning techniques: a major review. *Clin. Experiment. Ophthalmol.*, **35**, pp. 71–88.
2. Huang, D., Swanson, E. A., Lin, C. P., Schuman, J. S., Stinson, W. G., Chang, W., Hee, M. R., Flotte, T., Gregory, K., Puliafito, C. A., et al. (1991). Optical coherence tomography. *Science*, **254**, pp. 1178–1181.
3. Fujimoto, J. G., Brezinski, M. E., Tearney, G. J., Boppart, S. A., Bouma, B. E., Hee, M. R., Southern, J. F., and Swanson, E. A. (1995). Optical biopsy and imaging using optical coherence tomography. *Nat. Med.*, **1**, pp. 970–972.

4. Leitgeb, R., Hitzenberger, C. K., and Fercher, A. F. (2003). Performance of fourier domain vs. time domain optical coherence tomography. *Opt. Express*, **11**, pp. 889–894.
5. De Boer, J. F., Cense, B., Park, B. H., Pierce, M. C., Tearney, G. J., and Bouma, B. E. (2003). Improved signal-to-noise ratio in spectral-domain compared with time-domain optical coherence tomography. *Opt. Lett.*, **28**, pp. 2067–2069.
6. Choma, M. A., Sarunic, M. V., Yang, C., and Izatt, J. A. (2003). Sensitivity advantage of swept source and Fourier domain optical coherence tomography. *Opt. Express*, **11**, pp. 2183–2189.
7. Yamamoto, A., Kuo, C. C., Sunouchi, K., Wada, S., Yamaguchi, I., and Tashiro, H. (2001). Surface shape measurement by wavelength scanning interferometry using an electronically tuned Ti: sapphire laser. *Opt. Rev.*, **8**, pp. 59–63.
8. Yamaguchi, I., Yamamoto, A., and Yano, M. (2000). Surface topography by wavelength scanning interferometry. *Opt. Eng.*, **39**, pp. 40–46.
9. Ansari, Z., Gu, Y., Siegel, J., Jones, R., French, P. M. W., Nolte, D. D., and Melloch, M. R. (2002). Wide-field, real-time depth-resolved imaging using structured illumination with photorefractive holography. *Appl. Phys. Lett.*, **81**, pp. 2148–2150.
10. Gabor, D. (1971). Holography Nobel Lecture http://nobelprize.org/nobel_prizes/physics/laureates/1971/gabor-lecture.pdf.
11. Goodman, J. W., and Lawrence, R. W. (1967). Digital image formation from electronically detected holograms. *Appl. Phys. Lett.*, **11**, pp. 77–79.
12. Schnars, U. (1994). Direct phase determination in hologram interferometry with use of digitally recorded holograms. *J. Opt. Soc. Am. A*, **11**, pp. 2011–2015.
13. Schnars, U., and Juptner, W. (1994). Direct recording of holograms by a Ccd target and numerical reconstruction. *Appl. Opt.*, **33**, pp. 179–181.
14. Schnars, U., and Juptner, W. P. O. (2002). Digital recording and numerical reconstruction of holograms. *Meas. Sci. Technol.*, **13**, pp. R85–R101.
15. Zhang, T., and Yamaguchi, I. (1998). Three-dimensional microscopy with phase-shifting digital holography. *Opt. Lett.*, **23**, pp. 1221–1223.
16. Yamaguchi, I., Kato, J., Ohta, S., and Mizuno, J. (2001). Image formation in phase-shifting digital holography and applications to microscopy. *Appl. Opt.*, **40**, pp. 6177–6186.
17. Yamaguchi, I., and Zhang, T. (1997). Phase-shifting digital holography. *Opt. Lett.*, **22**, pp. 1268–1270.

18. Poon, T.-C. (2003). Three-dimensional image processing and optical scanning holography. *Adv. Imag. Electron Phys.*, **126**, pp. 329–350.
19. Yamaguchi, I., Matsumura, T., and Kato, J. (2002). Phase-shifting color digital holography. *Opt. Lett.*, **27**, pp. 1108–1110.
20. Barty, A., Nugent, K. A., Paganin, D., and Roberts, A. (1998). Quantitative optical phase microscopy. *Opt. Lett.*, **23**, pp. 817–819.
21. Cuche, E., Bevilacqua, F., and Depeursinge, C. (1999). Digital holography for quantitative phase-contrast imaging. *Opt. Lett.*, **24**, pp. 291–293.
22. Ferraro, P., De Nicola, S., Finizio, A., Coppola, G., Grilli, S., Magro, C., and Pierattini, G. (2003). Compensation of the inherent wave front curvature in digital holographic coherent microscopy for quantitative phase-contrast imaging. *Appl. Opt.*, **42**, pp. 1938–1946.
23. Xu, L., Peng, X., Miao, J., and Asundi, A. K. (2001). Studies of digital microscopic holography with applications to microstructure testing. *Appl. Opt.*, **40**, pp. 5046–5051.
24. Pedrini, G., and Tiziani, H. J. (1997). Quantitative evaluation of two-dimensional dynamic deformations using digital holography. *Opt. Laser Technol.*, **29**, pp. 249–256.
25. Picart, P., Leval, J., Mounier, D., and Gougeon, S. (2005). Some opportunities for vibration analysis with time averaging in digital Fresnel holography. *Appl. Opt.*, **44**, pp. 337–343.
26. Haddad, W. S., Cullen, D., Solem, J. C., Longworth, J. W., McPherson, A., Boyer, K., and Rhodes, C. K. (1992). Fourier-transform holographic microscope. *Appl. Opt.*, **31**, pp. 4973–4978.
27. Xu, W. B., Jericho, M. H., Meinertzhagen, I. A., and Kreuzer, H. J. (2001). Digital in-line holography for biological applications. *Proc. Natl. Acad. Sci. U S A*, **98**, pp. 11301–11305.
28. Kim, M. K. (1999). Wavelength scanning digital interference holography for optical section imaging. *Opt. Lett.*, **24**, p. 1693.
29. Kim, M. K. (2000). Tomographic three-dimensional imaging of a biological specimen using wavelength-scanning digital interference holography. *Opt. Express*, **7**, pp. 305–310.
30. Dakoff, A., Gass, J., and Kim, M. K. (2003). Microscopic three-dimensional imaging by digital interference holography. *J. Electron. Imag.*, **12**, pp. 643–647.
31. Yu, L. F., and Kim, M. K. (2005). Wavelength scanning digital interference holography for variable tomographic scanning. *Opt. Express*, **13**, pp. 5621–5627.

32. Yu, L. F., and Kim, M. K. (2005). Wavelength-scanning digital interference holography for tomographic three-dimensional imaging by use of the angular spectrum method. *Opt. Lett.*, **30**, pp. 2092–2094.
33. Kim, M. K., Yu, L. F., and Mann, C. J. (2006). Interference techniques in digital holography. *J. Opt. A: Pure Appl. Opt.*, **8**, pp. S518–S523.
34. Gass, J., Dakoff, A., and Kim, M. K. (2003). Phase imaging without 2-pi ambiguity by multiwavelength digital holography. *Opt. Lett.*, **28**, pp. 1141–1143.
35. Mann, C. J., Yu, L. F., Lo, C. M., and Kim, M. K. (2005). High-resolution quantitative phase-contrast microscopy by digital holography. *Opt. Express*, **13**, pp. 8693–8698.
36. Parshall, D., and Kim, M. K. (2006). Digital holographic microscopy with dual-wavelength phase unwrapping. *Appl. Opt.*, **45**, pp. 451–459.
37. Mann, C. J., Yu, L. F., and Kim, M. K. (2006). Movies of cellular and subcellular motion by digital holographic microscopy. *Biomed. Eng. Online*, **5**, p. 21.
38. Massatsch, P., Charriere, F., Cuche, E., Marquet, P., and Depeursinge, C. D. (2005). Time-domain optical coherence tomography with digital holographic microscopy. *Appl. Opt.*, **44**, pp. 1806–1812.
39. Yu, L., and Chen, Z. (2007). Digital holographic tomography based on spectral interferometry. *Opt. Lett.*, **32**, pp. 3005–3007.
40. Sarunic, M. V., Weinberg, S., and Izatt, J. A. (2006). Full-field swept-source phase microscopy. *Opt. Lett.*, **31**, pp. 1462–1464.
41. Ellerbee, A. K., Creazzo, T. L., and Izatt, J. A. (2007). Investigating nanoscale cellular dynamics with cross-sectional spectral domain phase microscopy. *Opt. Express*, **15**, pp. 8115–8124.
42. Yu, L. F., and Chen, Z. P. (2007). Improved tomographic imaging of wavelength scanning digital holographic microscopy by use of digital spectral shaping. *Opt. Express*, **15**, pp. 878–886.
43. Montfort, F., Colomb, T., Charriere, F., Kuhn, J., Marquet, P., Cuche, E., Herminjard, S., and Depeursinge, C. (2006). Submicrometer optical tomography by multiple-wavelength digital holographic microscopy. *Appl. Opt.*, **45**, pp. 8209–8217.
44. Charriere, F., Marian, A., Montfort, F., Kuehn, J., Colomb, T., Cuche, E., Marquet, P., and Depeursinge, C. (2006). Cell refractive index tomography by digital holographic microscopy. *Opt. Lett.*, **31**, pp. 178–180.

45. Meneses-Fabian, C., Rodriguez-Zurita, G., and Arrizon, V. (2006). Optical tomography of transparent objects with phase-shifting interferometry and stepwise-shifted Ronchi ruling. *J. Opt. Soc. Am. A*, **23**, pp. 298–305.
46. Gorski, W. (2006). Tomographic microinterferometry of optical fibers. *Opt. Eng.*, **45**, pp. 125002–125012.
47. Charriere, F., Pavillon, N., Colomb, T., Depeursinge, C., Heger, T. J., Mitchell, E. A. D., Marquet, P., and Rappaz, B. (2006). Living specimen tomography by digital holographic microscopy: morphometry of testate amoeba. *Opt. Express*, **14**, pp. 7005–7013.
48. Vishnyakov, G. N., Levin, G. G., Minaev, V. L., Pickalov, V. V., and Likhachev, A. V. (2004). Tomographic interference microscopy of living cells. *Microsc. Anal.*, **18**, pp. 15–17.
49. Choi, W., Fang-Yen, C., Badizadegan, K., Oh, S., Lue, N., Dasari, R. R., and Feld, M. S. (2007). Tomographic phase microscopy. *Nat. Methods*, **4**, pp. 717–719.
50. Drexler, W., Morgner, U., Kartner, F. X., Pitris, C., Boppart, S. A., Li, X. D., Ippen, E. P., and Fujimoto, J. G. (1999). In vivo ultrahigh-resolution optical coherence tomography. *Opt. Lett.*, **24**, pp. 1221–1223.
51. Hartl, I., Li, X. D., Chudoba, C., Ghanta, R. K., Ko, T. H., Fujimoto, J. G., Ranka, J. K., and Windeler, R. S. (2001). Ultrahigh-resolution optical coherence tomography using continuum generation in an air-silica microstructure optical fiber. *Opt. Lett.*, **26**, pp. 608–610.
52. Zhang, Y., Sato, M., and Tanno, N. (2001). Resolution improvement in optical coherence tomography by optimal synthesis of light-emitting diodes. *Opt. Lett.*, **26**, pp. 205–207.
53. Fercher, A. F., Drexler, W., Hitzenberger, C. K., and Lasser, T. (2003). Optical coherence tomography - principles and applications. *Rep. Prog. Phys.*, **66**, pp. 239–303.
54. Potcoava, M. C., and Kim, M. K. (2008). Optical tomography for biomedical applications by digital interference holography. *Meas. Sci. Technol.*, **19**, p. 074010.
55. Kim, M. K. (2011). *Digital Holographic Microscopy: Principles, Techniques, and Applications*. Springer, New York.
56. Yu, L. F., and Kim, M. K. (2006). Variable tomographic scanning with wavelength scanning digital interference holography. *Opt. Commun.*, **260**, pp. 462–468.
57. Goodman, J. W. (1996). *Introduction to Fourier Optics*, 2nd Ed. McGraw Hill, Boston.

58. Leseberg, D., and Frere, C. (1988). Computer-generated holograms of 3-D objects composed of tilted planar segments. *Appl. Opt.*, **27**, pp. 3020–3024.
59. Yu, L. F., An, Y. F., and Cai, L. L. (2002). Numerical reconstruction of digital holograms with variable viewing angles. *Opt. Express*, **10**, pp. 1250–1257.
60. Jeon, Y., and Hong, C. K. (2010). Optical section imaging of the tilted planes by illumination-angle-scanning digital interference holography. *Appl. Opt.*, **49**, pp. 5110–5116.
61. Fercher, A. F., Hitzenberger, C. K., Kamp, G., and El-Zaiat, S. Y. (1995). Measurement of intraocular distances by backscattering spectral interferometry. *Opt. Commun.*, **117**, pp. 43–48.
62. Häusler, G., and Lindner, M. W. (1998). "Coherence radar" and "spectral radar" - new tools for dermatological diagnosis. *J. Biomed. Opt.*, **3**, pp. 21–31.
63. Wojtkowski, M., Srinivasan, V. J., Ko, T. H., Fujimoto, J. G., Kowalczyk, A., and Duker, J. S. (2004). Ultrahigh-resolution, high-speed, Fourier domain optical coherence tomography and methods for dispersion compensation. *Opt. Express*, **12**, pp. 2404–2422.
64. Bouma, B. E., and Tearney, G. J. (2002). *Handbook of Optical Coherence Tomography*. Marcel Dekker New York.
65. Yu, L., Rao, B., Zhang, J., Su, J., Wang, Q., Guo, S., and Chen, Z. (2007). Improved lateral resolution in optical coherence tomography by digital focusing using two-dimensional numerical diffraction method. *Opt. Express*, **15**, pp. 7634–7641.
66. Potcoava, M. C., Kay, C. N., Kim, M. K., and Richards, D. W. (2010). In vitro imaging of ophthalmic tissue by digital interference holography. *J. Mod. Opt.*, **57**, pp. 115–123.
67. Faulds, H. (1880). On the skin-furrows of the hand. *Nature*, **22**, p. 605.
68. Galton, F. (1889). Personal identification and description. *J. Roy. Anthropol. Inst.*, **18**, pp. 177–191.
69. O'Gorman, L. (1998). An overview of fingerprint verification technologies. *Inform. Security Tech. Rep.*, **3**, pp. 21–32.
70. Potcoava, M. C., and Kim, M. K. (2009). Fingerprint biometry applications of digital holography and low-coherence interferography. *Appl. Opt.*, **48**, pp. H9–H15.
71. Kühn, J., Montfort, F., Colomb, T., Rappaz, B., Moratal, C., Pavillon, N., Marquet, P., and Depeursinge, C. (2009). Submicrometer tomography

of cells by multiple-wavelength digital holographic microscopy in reflection. *Opt. Lett.*, **34**, pp. 653–655.

72. Dubois, A., Grieve, K., Moneron, G., Lecaque, R., Vabre, L., and Boccara, C. (2004). Ultrahigh-resolution full-field optical coherence tomography. *Appl. Opt.*, **43**, pp. 2874–2883.

73. Srinivasan, V. J., Huber, R., Gorczynska, I., Fujimoto, J. G., Jiang, J. Y., Reisen, P., and Cable, A. E. (2007). High-speed, high-resolution optical coherence tomography retinal imaging with a frequency-swept laser at 850 nm. *Opt. Lett.*, **32**, pp. 361–363.

74. Srinivasan, V. J., Ko, T. H., Wojtkowski, M., Carvalho, M., Clermont, A., Bursell, S.-E., Song, Q. H., Lem, J., Duker, J. S., and Schuman, J. S. (2006). Noninvasive volumetric imaging and morphometry of the rodent retina with high-speed, ultrahigh-resolution optical coherence tomography. *Invest. Ophthalmol. Vis. Sci.*, **47**, pp. 5522–5528.

75. Drexler, W., Morgner, U., Ghanta, R. K., Kärtner, F. X., Schuman, J. S., and Fujimoto, J. G. (2001). Ultrahigh-resolution ophthalmic optical coherence tomography. *Nat. Med.*, **7**, pp. 502–507.

76. Wollstein, G., Paunescu, L. A., Ko, T. H., Fujimoto, J. G., Kowalevicz, A., Hartl, I., Beaton, S., Ishikawa, H., Mattox, C., and Singh, O. (2005). Ultrahigh-resolution optical coherence tomography in glaucoma. *Ophthalmology*, **112**, pp. 229–237.

77. Yasuno, Y., Hong, Y., Makita, S., Yamanari, M., Akiba, M., Miura, M., and Yatagai, T. (2007). In vivo high-contrast imaging of deep posterior eye by 1-um swept source optical coherence tomography and scattering optical coherence angiography. *Opt. Express*, **15**, pp. 6121–6139.

PART III
ADDITIONAL IMAGING MODALITIES

Chapter 13

Technological Extensions of Full-Field Optical Coherence Microscopy for Multicontrast Imaging

Arnaud Dubois

Laboratoire Charles Fabry, CNRS, Institut d'Optique Graduate School,
Univ. Paris-Saclay, 2 av. Augustin Fresnel, 91127 Palaiseau, France
arnaud.dubois@institutoptique.fr

13.1 Introduction

Optical coherence tomography (OCT) detects the intensity of light reflected or backscattered from structures within the sample to provide cross-sectional images with micrometer resolution. Several technological extensions of OCT have been developed for revealing features of biological or material samples not readily visible in conventional OCT and/or for noninvasive functional imaging (see Fig. 13.1). These extensions provide for example measurements of the polarization [1–3] or spectroscopic [4–7] response of the sample. OCT has been combined with other imaging modalities, including in particular fluorescence microscopy, to obtain microstructure and molecular information simultaneously [8–14]. Elastic

---

Handbook of Full-Field Optical Coherence Microscopy: Technology and Applications
Edited by Arnaud Dubois
Copyright © 2016 Pan Stanford Publishing Pte. Ltd.
ISBN 978-981-4669-16-0 (Hardcover), 978-981-4669-17-5 (eBook)
www.panstanford.com

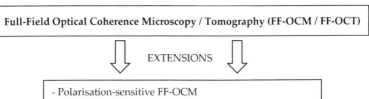

Figure 13.1 Technological extensions of FF-OCM (also termed "full-field optical coherence tomography", FF-OCT) for multicontrast imaging.

properties of biological tissues can be used as a contrast mechanism to form images in optical coherence elastography, which is based on OCT as an underlying imaging modality [15–18]. Work has also been carried out on the development of OCT systems operating at multiple wavelengths to maximize either the imaging penetration depth or the spatial resolution and/or to enable better visualization and differentiation of specific structures within the sample [19–22].

Similar technological developments of the full-field optical coherence microscopy (FF-OCM) technique have been carried out (see Fig. 13.1). Polarization-sensitive FF-OCM takes advantage of the additional information carried by the polarization state of light backscattered from the sample. Spectroscopic FF-OCM is another technical extension that provides information on the spectral content of backscattered light by detection and processing of the whole interferometric signal. As alternative methods to take advantage of the spectroscopic response of the sample and/or optimize the spatial resolution or the accessible depth, multiband FF-OCM systems have been proposed. FF-OCM has been combined with fluorescence microscopy to provide biochemical and metabolism information on biological samples using targeted fluorophores. At last, FF-OCM was recently combined with elastography to create a virtual palpation map at the micrometer scale [23].

Most of these technological extensions of FF-OCM are presented in this chapter. Some of them are described in more detail in other chapters of this book.

13.2 Polarization-Sensitive FF-OCM

The polarization state of light can be modified by various light–matter interactions. This phenomenon can be used to generate additional imaging contrasts in OCT. Mechanisms responsible for the modification of the polarization include birefringence [1, 24], diatenuation [25, 26], and depolarization [27]. In most biological tissues, diatenuation is very weak [26]. Birefringence is found in fibrous tissues, whereas depolarization can be caused by multiple light scattering at large particles or scattering at nonspherical particles. Birefringence measurement with OCT is especially useful in ophthalmology for the diagnosis of glaucoma [28, 29] and for cornea characterization [30]. Changes in collagen structure and distribution in skin can be detected by polarization-sensitive optical coherence tomography (PS-OCT) for the evaluation of burn depth [31] and for the diagnosis of photo-aged skin [32]. PS-OCT can also be used in dentistry for monitoring the onset and progression of caries [33, 34]. Depolarization has been suggested for diagnosing of the retinal pigment epithelium in age-related macular degeneration [35]. The distribution of the birefringence within a material, either caused by internal stress fields or by structural sample anisotropies, can serve as additional source of information for material characterization and evaluation [3]. Initially implemented in time-domain optical coherence tomography (TD-OCT) [1, 36], PS-OCT techniques were later adapted to the faster and more sensitive spectral-domain version of OCT [37, 38].

Several full-field versions of PS-OCT have been developed. The initial version required manual rotation of a polarizer between image acquisitions for calculation of two-dimensional birefringence maps [39]. The first polarization-sensitive FF-OCM system capable of real-time imaging [40] is presented in the next section of this chapter. This full-field version of PS-OCT was based on the principle of early scanning PS-OCT systems [1, 36]; it used two synchronized charge-coupled device (CCD) cameras for simultaneous acquisition of images in orthogonal linear polarization states. Contrast-enhanced imaging of biological tissues was demonstrated with this system [40, 41]. It was also applied to the characterization of various semitransparent materials [41–43], as will be shown in this chapter and in Chapter 21. Similar polarization-sensitive FF-OCM systems were developed later, using a single camera with polarization modulation [44], or including a spectroscopic imaging modality [41].

13.2.1 Experimental Setup and Principle

The experimental setup of conventional polarization-sensitive FF-OCM is shown schematically in Fig. 13.2. It is based on a Linnik interferometer with two identical microscope objectives and two synchronized image sensors (CCD or complementary metal-oxide semiconductor [CMOS]) for simultaneous acquisition of images in orthogonal linear polarization states. In the system reported in [40], low coherence illumination is achieved by using a 100 W halogen lamp. The incident light is linearly polarized by a polarizer oriented at a 0° angle (direction orthogonal to the plane of Fig. 13.2) and split into the reference and sample arms by a broadband polarization-insensitive beam splitter (BS). Light in the reference arm passes through an achromatic quarter-wave plate (AQWP) oriented at a 22.5° angle to the incident polarization. After reflection on the reference mirror (~2% reflectivity), light passes through the AQWP again. Reference light is thus linearly polarized at a 45° angle to the incident polarization. Light in the sample arm passes through an identical AQWP oriented at a 45° angle so that the sample is illuminated with circularly polarized light. Returning from the sample and passing through the AQWP, light has a modified polarization state due to the optical anisotropy properties of the sample. After recombination at the BS, the sample and reference optical fields interfere. The interferometric signal is split into two orthogonal polarization components by a polarizing beam splitter (PBS). These polarization directions are denoted by symbols ∥ (parallel to the polarization at the entrance of the interferometer) and ⊥ (orthogonal to the polarization at the entrance of the interferometer). A doublet lens (achromatic, 300 mm focal length) forms interferometric images onto two identical CCD image sensors (Dalsa 1M15, 1024 × 1024 pixels, 12 bits). Because of the spectral response of the silicon-based CCDs and the use of a long-wave pass filter, the effective power spectrum of illumination has a width of $\Delta \lambda$

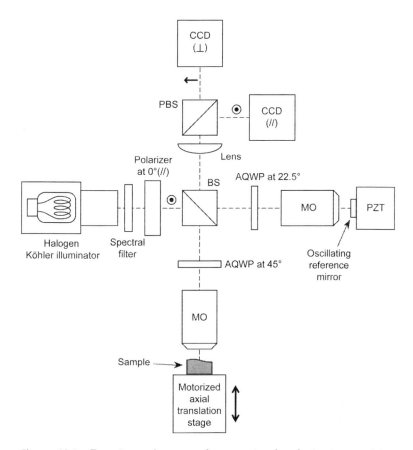

Figure 13.2 Experimental setup of conventional polarization-sensitive FF-OCM. AQWP, achromatic quarter-wave plate; BS, beam splitter; PBS, polarization beam splitter; MO, microscope objective; PZT, piezoelectric transducer.

= 300 nm (full-width at half-maximum [FWHM]) and is centered at $\lambda = 750$ nm, which fits the bandwidth acceptance of the achromatic wave plates. The reference mirror is mounted on a piezoelectric translation stage to make it oscillate at frequency $f = 3.5$ Hz. This oscillation generates a sinusoidal phase modulation in the interferometer. The CCD cameras are triggered at frequency $4f = 14$ Hz and are synchronized with the piezoelectric transducer (PZT) oscillation. The CCD cameras capture four images per modulation period. N series of four images are accumulated ($N = 5\text{--}10$ typically). An algebraic combination of the acquired images gives the intensity and phase retardation due to birefringence as will be explained in the next section. Images are produced at a frame rate of several hertz, depending on the number N of accumulated images (3.5 Hz for $N = 1$).

13.2.2 Image Calculation

The Jones matrix formalism can be used to establish the expression of the light intensity detected by each camera. For simplicity, the light source is first supposed to be monochromatic with wavenumber k. After the polarizer, light is linearly polarized at an angle of 0° with respect to the normal to the plane of the interferometer. This direction is denoted by symbol $||$. This polarized light is described by the Jones vector

$$\mathbf{E} = E_0 \begin{pmatrix} 1 \\ 0 \end{pmatrix}. \tag{13.1}$$

The BS divides the incident light by amplitude evenly between the sample and reference arms of the interferometer, such that the Jones vectors describing the light entering the reference and sample arms—$\mathbf{E}_{r,i}$ and $\mathbf{E}_{s,i}$, respectively—are given by

$$\mathbf{E}_{r,i} = \mathbf{E}_{s,i} = \frac{1}{\sqrt{2}} E_0 \begin{pmatrix} 1 \\ 0 \end{pmatrix}. \tag{13.2}$$

The polarization state of light reflected from the reference mirror, after the BS, can be calculated by multiplying by the Jones matrices of the optical elements in the reference optical path:

$$\mathbf{E}_r = \frac{1}{\sqrt{2}} \mathbf{J}_{\lambda/4,\,22.5°} \sqrt{R_r} \mathbf{J}_{\lambda/4,\,22.5°} \mathbf{E}_{r,i} e^{-2ikl_r}. \tag{13.3}$$

Here, R_r represents the reflectivity of the reference mirror, $\mathbf{J}_{\lambda/4,\,22.5°}$ the Jones matrix of the quarter-wave plate oriented at 22.5° and l_r the optical length of the reference arm. After substitutions, the following expression is obtained:

$$\mathbf{E}_r = \frac{i}{2\sqrt{2}} E_0 \sqrt{R_r} \begin{pmatrix} 1 \\ 1 \end{pmatrix} e^{-2ikl_r}. \tag{13.4}$$

The polarization state of light returning from the sample arm can be calculated similarly. The sample is assumed to be a linearly birefringent material that introduces a phase retardation δ between the slow and fast axes (orthogonal directions). The fast axis of the birefringent sample is at an angle α to the $\parallel$ (0°) direction. The Jones matrix of the sample is

$$\mathbf{J}_s = \begin{pmatrix} e^{i\frac{\delta}{2}}\cos^2\alpha + e^{-i\frac{\delta}{2}}\sin^2\alpha & 2i\sin\frac{\delta}{2}\cos\alpha\sin\alpha \\ 2i\sin\frac{\delta}{2}\cos\alpha\sin\alpha & e^{-i\frac{\delta}{2}}\cos^2\alpha + e^{i\frac{\delta}{2}}\sin^2\alpha \end{pmatrix}. \quad (13.5)$$

The light returning from the sample arm, after the BS, is described by the vector

$$\mathbf{E}_s = \frac{1}{\sqrt{2}} \mathbf{J}_{\lambda/4,45°} \mathbf{J}_s \sqrt{R_s} \mathbf{J}_s \mathbf{J}_{\lambda/4,45°} \mathbf{E}_{s,i} e^{-2ikl_s}, \quad (13.6)$$

where R_s is the scalar reflectivity of the sample at a given depth, $\mathbf{J}_{\lambda/4,45°}$ the Jones matrix of the quarter-wave plate oriented at 45°, and l_s the optical length of the sample arm. Since

$$\mathbf{J}_{\lambda/4,45°} = \frac{1}{\sqrt{2}} \begin{pmatrix} 1 & i \\ i & 1 \end{pmatrix}, \quad (13.7)$$

Eq. 13.6 simplifies to:

$$\mathbf{E}_s = \frac{i}{2} E_0 \sqrt{R_s} \begin{pmatrix} \sin\delta e^{2i\alpha} \\ \cos\delta \end{pmatrix} e^{-2ikl_s}. \quad (13.8)$$

The light intensity upon each camera corresponds to the intensity detected in the $\parallel$ and $\perp$ polarization directions. It can be described by the two-dimensional vector

$$\begin{pmatrix} I^\parallel \\ I^\perp \end{pmatrix} \propto |\mathbf{E}_s + \mathbf{E}_r|^2 = |\mathbf{E}_s|^2 + |\mathbf{E}_r|^2 + 2\text{Re}\{\mathbf{E}_s \mathbf{E}_r^*\}, \quad (13.9)$$

where Re{} represents the real part of a complex number. Developing the previous equation yields:

$$I^\parallel = I_0 \left\{ R_s \sin^2\delta + \frac{R_r}{2} + 2\sqrt{\frac{R_r R_s}{2}} \sin\delta \cos(2k\Delta l + 2\alpha) \right\}, \quad (13.10)$$

and

$$I^\perp = I_0 \left\{ R_s \cos^2\delta + \frac{R_r}{2} + 2\sqrt{\frac{R_r R_s}{2}} \cos\delta \cos[2k\Delta l] \right\}, \quad (13.11)$$

where $\Delta l = l_s - l_r$ and I_0 is a proportionality factor.

If broadband light is considered now, the two above equations have to be integrated over the harmonic content of the light source. As seen in Chapter 1, this generates a coherence envelope that is related to the Fourier transform of the light spectral power distribution. Here, instead of introducing a coherence envelope, we consider for simplicity that light returning from the sample and collected by the microscope objective can be decomposed into two parts depending on whether it can interfere or not (see Chapter 1). Light that can interfere is regarded as if it originates from a plane en face surface with reflectivity distribution R_s, located at a given depth within the sample. This light interferes with the light reflected by the plane reference surface. The light backscattered from the sample and collected by the microscope objective that does not interfere is represented by an equivalent reflectivity coefficient denoted as R_{inc}. This quantity, made up from all reflected and backscattered components throughout the sample except those in the coherence volume, can be considered as constant. The quantity R_{inc} is decomposed as $R_{inc} = R_{inc}^{\perp} + R_{inc}^{\parallel}$, where $R_{inc}^{\perp}$ and $R_{inc}^{\parallel}$ denote the proportions detected in both polarization states. Under these conditions, Eqs. 13.10 and 13.11 can be rewritten as

$$I^{\parallel} = I_0 \left\{ R_{inc}^{\parallel} + R_s \sin^2 \delta + \frac{R_r}{2} + 2\sqrt{\frac{R_r R_s}{2}} \sin \delta \cos(2k\Delta l + 2\alpha) \right\}, \quad (13.12)$$

and

$$I^{\perp} = I_0 \left\{ R_{inc}^{\perp} + R_s \cos^2 \delta + \frac{R_r}{2} + 2\sqrt{\frac{R_r R_s}{2}} \cos \delta \cos[2k\Delta l] \right\}. \quad (13.13)$$

Due to the very short coherence length (broadband light is used), the amount of light returning form the sample that interfere and that is detected is small compared to all the light incident upon the image sensors. This implies that $R_s \ll R_{inc}^{\perp}$, $R_{inc}^{\parallel}$ and $R_s \ll R_r$. Taking into account the sinusoidal phase modulation induced by the oscillation of the reference surface (as described in Chapter 1), one can finally write simplified expressions for the intensity upon both cameras:

$$I^{\parallel}(t) = I_0 \left\{ R_{inc}^{\parallel} + \frac{R_r}{2} + 2\sqrt{\frac{R_r R_s}{2}} \sin \delta \cos[2k\Delta l + 2\alpha + \psi \sin(2\pi ft + \theta)] \right\}, \quad (13.14)$$

$$I^\perp(t) = I_0 \left\{ R_{inc}^\perp + \frac{R_r}{2} + 2\sqrt{\frac{R_r R_s}{2}} \cos\delta \cos[2k\Delta l + \psi \sin(2\pi ft + \theta)] \right\}.$$
(13.15)

The cameras integrate the intensity over the four quarters of the modulation period T. Each camera delivers series of four images that can be expressed in terms of photo-electrons as

$$\varepsilon_{p=1,2,3,4}^{\|} \propto N \int_{(p-1)T/4}^{pT/4} I^{\|}(t)\, dt, \tag{13.16}$$

$$\varepsilon_{p=1,2,3,4}^{\perp} \propto N \int_{(p-1)T/4}^{pT/4} I^{\perp}(t)\, dt, \tag{13.17}$$

where N is the number of images that are accumulated. Using the calculations with Bessel functions presented in Chapter 1, the following expressions can be established:

$$\left(\varepsilon_1^{\|} - \varepsilon_2^{\|}\right)^2 + \left(\varepsilon_3^{\|} - \varepsilon_4^{\|}\right)^2 = 256\, (\Gamma/\pi)^2\, (N\xi_{sat})^2 \frac{R_r}{\left(R_r/2 + R_{inc}^{\|}\right)^2} R_s \sin^2\delta, \tag{13.18}$$

$$\left(\varepsilon_1^{\perp} - \varepsilon_2^{\perp}\right)^2 + \left(\varepsilon_3^{\perp} - \varepsilon_4^{\perp}\right)^2 = 256\, (\Gamma/\pi)^2\, (N\xi_{sat})^2 \frac{R_r}{\left(R_r/2 + R_{inc}^{\perp}\right)^2} R_s \cos^2\delta, \tag{13.19}$$

and

$$\frac{\left(\varepsilon_3^{\|} - \varepsilon_4^{\|}\right)^2}{\left(\varepsilon_1^{\|} - \varepsilon_2^{\|}\right)^2} = \tan(2k\Delta l + 2\alpha + \pi/4), \tag{13.20}$$

$$\frac{\left(\varepsilon_3^{\perp} - \varepsilon_4^{\perp}\right)^2}{\left(\varepsilon_1^{\perp} - \varepsilon_2^{\perp}\right)^2} = \tan(2k\Delta l + \pi/4). \tag{13.21}$$

One can see that the scalar reflectivity of the sample and the phase retardation due to the sample birefringence, at a given depth in the sample, can be obtained respectively from the following image combinations:

$$\left[\left(\varepsilon_1^{\|} - \varepsilon_2^{\|}\right)^2 + \left(\varepsilon_3^{\|} - \varepsilon_4^{\|}\right)^2\right] + \left[\left(\varepsilon_1^{\perp} - \varepsilon_2^{\perp}\right)^2 + \left(\varepsilon_3^{\perp} - \varepsilon_4^{\perp}\right)^2\right] \propto R_s, \tag{13.22}$$

$$\sqrt{\frac{\left(\varepsilon_1^{\|} - \varepsilon_2^{\|}\right)^2 + \left(\varepsilon_3^{\|} - \varepsilon_4^{\|}\right)^2}{\left(\varepsilon_1^{\perp} - \varepsilon_2^{\perp}\right)^2 + \left(\varepsilon_3^{\perp} - \varepsilon_4^{\perp}\right)^2}} = |\tan\delta|. \qquad (13.23)$$

The orientation of the fast axis of the birefringent sample can be obtained from

$$\frac{1}{2}\left\{\tan^{-1}\left[\frac{\left(\varepsilon_3^{\|} - \varepsilon_4^{\|}\right)^2}{\left(\varepsilon_1^{\|} - \varepsilon_2^{\|}\right)^2}\right] - \tan^{-1}\left[\frac{\left(\varepsilon_3^{\perp} - \varepsilon_4^{\perp}\right)^2}{\left(\varepsilon_1^{\perp} - \varepsilon_2^{\perp}\right)^2}\right]\right\} = \alpha. \qquad (13.24)$$

13.2.3 Validation

Moderate-numerical-aperture (NA = 0.3 typically) microscope objectives are typically used, which results in a theoretical transverse resolution of ~1 μm without introducing significant polarization artifacts that may be detected with higher NA. The axial imaging resolution is then determined by the effective coherence length of the illumination source. In a medium of refractive index $n = 1.33$ (water), the axial resolution was measured to be $\Delta z \sim 1$ μm.

The instrument accuracy for linear retardance measurements was tested by measuring the phase retardation introduced by a Babinet compensator placed in the sample arm between the quarter-wave plate and the microscope objective. A glass plate with appropriate thickness was placed in the other arm of the interferometer to compensate for the dispersion mismatch. A phase retardation varying between 0° and 225° was generated by the calibrated Babinet compensator. Figure 13.3 compares the measured phase retardation with the true phase retardation introduced by the compensator. The root mean square (RMS) error was 3° (7% average error over the whole scale). The error may result from the calculations, which assumed monochromatic illumination. Moreover, when imaging a birefringent sample with a microscope objective, the angle between the optical rays and the birefringence optic axis is not as well defined. It was shown that a substantial discrepancy between measured retardance and true retardance can be induced by variations in this angle [45]. A rotation of the Babinet compensator in the plane orthogonal to the optical axis had no effect on the measured value of the phase retardation, which

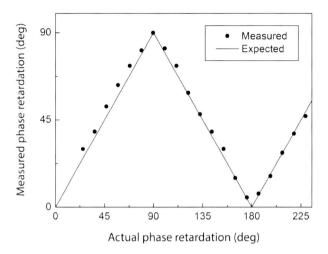

Figure 13.3 Measured versus actual phase retardation introduced by a calibrated Babinet compensator.

was a confirmation of the insensitivity of the setup to the sample orientation.

To illustrate the imaging capabilities of the polarization-sensitive FF-OCM system, fresh muscle tissues extracted from a shrimp tail were imaged. The tissues were immersed in phosphate-buffered saline (PBS, pH 7). En face images are shown in Fig. 13.4. Linear retardance varies from 0° (black) to 90° (white). Intensity, on a logarithmic scale, varies from −90 dB (black) to −45 dB (white). Intensity images and phase images exhibit different and uncorrelated contrasts. Optical anisotropy unhomogeneities in the tissue are revealed. It is worth noting that the phase retardation at a depth z under the sample surface does not represent a strictly depth-resolved measurement [46], but the accumulated linear retardance, that is, the linear retardance integrated from the surface to the depth z and over the optical ray angle distribution.

As another illustration, images of tendon from a rat tail obtained with a similar polarization-sensitive FF-OCM system [44] are shown in Fig. 13.5. Several strands of tendon with crimp patterns adjoin each other, as shown in Fig. 13.5a. Secondary fiber bundles (fascicles) and even subfascicles (digits 1 to 4) of the tendon, which are invisible with conventional PS-OCT in general, are clearly visible

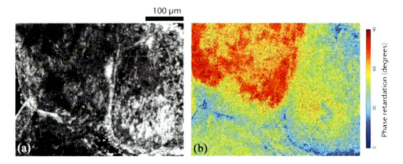

Figure 13.4 En face polarization-sensitive FF-OCM images of shrimp tail muscle (in vitro) at ~50 μm depth. (a) The intensity-based image and (b) the corresponding linear-birefringence-induced phase retardation image.

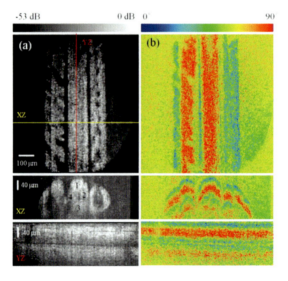

Figure 13.5 Ex vivo images of rat tail tendon by polarization-sensitive FF-OCM: (a) En face intensity images at a depth of 43 μm; XZ and YZ cross-sectional images. (b) En face retardation image at the same depth; XZ and YZ cross-sectional retardation images.

in the XY and YZ images, respectively. Figure 13.5b shows the linear retardance image taken at the same depth. The phase retardation from 0° to 90° is expressed with pseudo-color mapping. The cross-sectional image shows the variation of phase retardation along and across the tendon.

Compared with conventional FF-OCM, polarization-sensitive FF-OCM provides additional information related to the sample polarization properties. This technological extension has been successfully demonstrated for imaging of biological tissues exhibiting form birefringence. It has also been applied to the characterization and evaluation of various materials through measurement of their birefringence caused by either internal stress or structural anisotropies (see Chapter 21). With a moderate increase in experimental setup, complexity and price, polarization-sensitive FF-OCM offers larger potentialities than conventional FF-OCM with similar performance in terms of spatial resolution and image acquisition speed. Further work has now to be carried out at the application and validation levels for polarization-sensitive FF-OCM to become a useful and widespread imaging technique in the fields of material characterization, biological studies and medical diagnosis.

13.3 Spectroscopic FF-OCM

Spectroscopic OCT is an extension of OCT that is used for performing morphological imaging and spectroscopic imaging simultaneously [4–6]. Information on the spectral content of light backscattered from the sample is obtained by detection and processing of the OCT interferometric signal. This method allows the spectral power distribution of backscattered light to be measured over the entire available optical bandwidth. Using ultrabroadband Ti:sapphire laser technology, spectroscopic measurements was demonstrated over a spectral range from 650 nm to 1000 nm [4].

The first spectroscopic extension of FF-OCM, based on Fourier transform spectroscopy with thermal light, was introduced in 2008 [17]. Similar systems were developed later, still using a halogen lamp as illumination source to take advantage of the broad and smooth spectral power distribution of thermal light [41, 48].

13.3.1 Experimental Setup

The experimental setup of the initial spectroscopic FF-OCM system is identical to the one of conventional FF-OCM [49, 50], that is, an interference microscope in the Linnik configuration, except that the reference mirror is fixed. The light source is a halogen lamp. Identical glass plates are placed in both arms of the interferometer at an optimized angle to compensate for residual dispersion mismatch induced by the microscope objectives and the beam splitter. The numerical aperture of the water immersion microscope objectives is 0.3. The image sensor is a CCD camera with the following characteristics: 1024 × 1024 pixels, frame rate 15 Hz, 12 bits digitization. The sample to image is displaced along the optical axis (z) by a computer-controlled PZT in steps Δz. At each step p, an adjustable number of identical interferometric images is captured by the CCD and summed up. The resulting image E_p is then written to the computer hard disk. A stack of N interferometric images $E_{p=1,\ldots,N}$ is recorded by scanning the sample in the z direction. The full interferometric signal as a function of depth (z) in the sample is thus acquired simultaneously by each pixel of the CCD. Due to the spectral response of the CCD and the optical elements of the experimental setup, the detected spectral range stretches from 600 nm to 900 nm. The PZT step is adjusted to $\delta z = 35$ nm, which ensures the sampling of the interferometric signal to be several times above the Nyquist limit. The PZT is equipped with absolute-measuring, direct-metrology capacitive sensors to minimize errors and fluctuations in the nominal step value. After recording the full interferometric signal, intensity and spectral information is obtained by data postprocessing.

13.3.2 Intensity-Based Tomographic Imaging

The method for obtaining intensity-based tomographic images differs from that used in conventional FF-OCM. In the spectroscopic FF-OCM system, the length of the interferometer reference arm is fixed (the reference mirror does not oscillate). The phase shift is generated here by scanning the sample in the axial direction. The fringe envelope is calculated using a five-frame nonlinear phase-shifting algorithm [51]. This algorithm requires a phase step of $\pi/2$. With a mean wavelength of $\lambda = 750$ nm and a PZT step of $\delta z = 35$ nm, the phase step between successive images in the acquired stack of interferometric images is $\sim\pi/4$ (assuming the refractive index of the sample to be ~ 1.4). The algorithm is applied by taking one image in two in the stack. Each intensity-based image I_k is thus calculated according to:

$$I_k = (E_{k+2} - E_{k+6})^2 - (E_k - E_{k+4})(E_{k+4} - E_{k+8}), k = 1, \ldots, N-8. \quad (13.25)$$

$I_{k=1,\ldots,N-8}$ represents a stack of en face intensity-based tomographic images. Finally, a compressed stack I^* is calculated by summing 10 adjacent elements of I. Adjacent intensity-based tomographic images from I^* are thus separated by a distance of $10 \times \delta z = 350$ nm, which is less than the axial resolution.

Detection sensitivity and dynamic range in FF-OCM rely on several experimental parameters including the full well capacity of the camera. This parameter can be virtually increased by pixel binning and image accumulation. In order to keep the acquisition time relatively short, only 5 identical interferometric images are acquired and accumulated at each PZT step and 2×2 pixel binning is applied. The intensity detection sensitivity was then measured to be ~ 85 dB.

13.3.3 Spectroscopic Measurements

Spectroscopic information is obtained by Fourier analysis of the full interferometric signal, as done in the first spectroscopic extension of conventional OCT [4]. Spatially resolved power density spectra are calculated using a short-time Fourier transform. Fast Fourier transforms (FFTs) of the interferometric data acquired by each CCD pixel (vectors of N points) are calculated over a moving Gauss window. The window is moved in steps of 10 points so that a power density spectrum is associated with each element of the three-dimensional intensity data set I^*. At any location in the sample, both the intensity and the power density spectrum of light backscattered from that location can thus be measured.

The choice of the width of the Gauss window is the result of a compromise between spectral resolution and spatial resolution of the spectral data. Spectra are measured here with low spectral resolution to favor the spatial resolution. A narrow Gauss window is chosen, of width equal to the imaging axial resolution, that is, an FWHM of 0.8 µm. High spatial resolution of the spectral information is thus guaranteed. The spectral resolution, defined as the width (FWHM) of the spectral impulse response (Fourier transform of the Gauss window), is equal to 240 nm. Due to the low spectral resolution, the measured spectra are always peak functions. Only spectral shifts can be measured. Despite a low spectral resolution, high spectral sensitivity is achieved, which is required to measure weak spectral shifts.

The size of the Gauss window is fixed to 4096 points. Zero padding is applied if necessary. The spectrum sampling step is then $\delta\sigma = (2 \times 4096 \times \delta z)^{-1} = 35$ cm^{-1} ($\sim$2 nm at $\lambda = 750$ nm). The FFTs are then calculated to get a local spectrum associated to each point of the three-dimensional intensity image I^*. The center of mass of each spectrum is chosen as metric for spectroscopic information representation. For simultaneously displaying intensity and spectroscopic information, a hue-saturation-luminance (HSL) color map is used. Hue encodes the spectroscopic information (center of mass of the spectra), whereas luminance and saturation together encode the intensity information, as done in [6]. The hue variation is limited from 0 to 0.5 (the full available range being from 0 to 1) so that the color varies from red to light blue. Luminance ranges from 0 to 1 and saturation from 0.5 to 1.

13.3.4 Interpretation of Spectroscopic Measurements

The spectral shifts detected by spectroscopic FF-OCM result from elastic scattering and absorption of light during its propagation in the sample. Absorption occurs over the entire optical path, while scattering takes place at the local interfaces of refractive index unhomogeneities. Absorption measurements are therefore not really spatially resolved. In contrast, scattering measurements are more localized indicators. The contributions of both effects, however, are difficult to distinguish [52–54].

The spectral changes due to single-particle backscattering are related to the size of that particle [55]. In theory, spectroscopic FF-

OCM could be used to determine the size of scattering particles. In practice, this is particularly difficult for several reasons:

- The wavelength-dependent attenuation of light due to absorption and scattering during the propagation from the sample surface to the imaged particle and back from the imaged particle to the sample surface.
- The presence of multiple scattering particles in the coherence volume generates interference, causing spectral modulations [6, 52, 54]. The number of scatterers in the coherence volume affects the spectral modulation. If the number of scatterers is large, the spectral modulation tends to be averaged out spatially. If the number of scatterers is small, modulation effects on the spectrum occur. This situation is likely to happen in FF-OCM because the coherence volume is small (~ 1 μm^3). Even if the spectral modulations are not resolved, they may cause a spectral shift. Since it is often difficult to know whether one or several particles are present in the coherence volume, it is difficult to be certain whether the measured spectral shift results from single-particle backscattering or multi-particle backscattering. Consequently, the determination of particle size is considerably hampered.
- The sample is illuminated with a distribution of incidence angles authorized by the numerical aperture of the objective. Compared to plane wave illumination, this geometrical effect modifies the fringe period of the interferometric signal causing an apparent shift of the calculated spectra [56, 57].

Work has been published on wavelength-dependent scattering measurements achieved by spectroscopic OCT [53, 54, 58]. For the reasons cited above and due to the inhomogeneous properties of tissue scatterers, such as variations in size, refractive index, and density, the quantitative interpretation of scattering effects detected by spectroscopic FF-OCM is a real challenge.

In contrast, a larger number of studies have focused on measuring the wavelength-dependent absorption from either endogenous or exogenous absorbers [52, 53, 59]. According to Beer's law, for an absorber to significantly shift the spectrum of backscattered light, the absorber either needs to have a large absorption cross section, be present in high concentration, or the length of the absorber region needs to be large. In order to maximize the imaging penetration depth, near-infrared light is used to benefit from a minimum of absorption by the biological tissues (so-called therapeutic window). Except for a few biological components such as melanin and blood (absorption peak of deoxyhemoglobin at 760 nm and hemoglobin isosbestic point at around 800 nm), endogeneous biological molecules have a very weak absorption cross section in the near-infrared spectral region. Finally, since the FF-OCM imaging penetration depth does not exceed $\sim$1 mm, the effect of absorption is integrated over a small optical path. Consequently, the spectral modifications induced by absorption are generally weak.

13.3.5 Validation of the Spectroscopic Measurements

Experiments were carried out to validate the spectroscopic measurements achieved with the spectroscopic FF-OCM system. The surface of a glass microscope slide covered by a layer of colored agarose gel was imaged (see Fig. 13.6). A glass microscope coverslip was placed over the gel at an angle such that the gel acts as a wedge-shaped absorbing medium. The coloration was obtained by introducing blue ink during the preparation of the gel. The optical dispersion introduced by the microscope coverslip was compensated for by placing an identical coverslip in the other arm of the interferometer between the microscope objective and the beam splitter. A typical transmission spectrum of the ink was measured using a conventional grating-based spectrometer (see Fig. 13.7a). Since the sensitivity of FF-OCM spreads from 600 nm to 900 nm, the absorption of the blue ink is expected to induce detectable spectral changes.

FF-OCM spectroscopic measurements are reported in Fig. 13.7b. As expected, the spectrum of light reflected by the microscope slide is shifted toward longer wavelengths at positions where the wedge is thicker. The theoretical shift was calculated using the measured spectral transmission of the ink, and assuming a

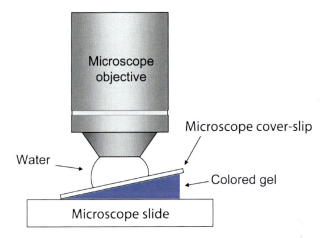

Figure 13.6 Experimental arrangement for absorption measurement validation using spectroscopic FF-OCM. An agarose gel colored with blue ink was placed between a microscope slide and a coverslip. The wedge-shaped gel acts as an absorbing medium with variable thickness.

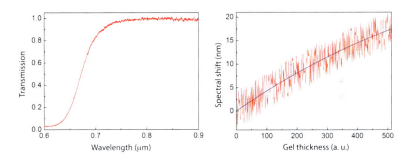

Figure 13.7 Left: Typical transmission spectrum of the blue ink measured by conventional spectrometry. Right: Shift of the power spectrum center of mass of light reflected by a glass plate after passing twice through an agarose gel colored with blue ink (experiment depicted in Fig. 13.6); spectroscopic FF-OCM measurement (in red) and calculation (in blue).

Gaussian illumination spectrum centered at 750 nm with an FWHM of 240 nm. In order to fit the experiment, the ink concentration was the only adjustable numerical parameter. The RMS noise in the plot shown in Fig. 13.7b is an indicator of the sensitivity of the spectral shift measurements. We estimate the sensitivity to be

~5 nm, which corresponds to the PZT nominal step fluctuations. This estimation is based on the reflection from a glass plate covered with a gel (reflectivity ≈0.35%), which is higher than the reflectivity of most biological structures. The spectral measurement sensitivity when deep imaging inside scattering samples is certainly degraded because of much lower signal-to-noise ratio.

13.3.6 Demonstration of Imaging Contrast Enhancement

The capability of the spectroscopic FF-OCM to provide contrast-enhanced images is illustrated by taking as a sample a *Xenopus Laevis* (African frog) tadpole. The sample (ex vivo) was placed in a tank filled with phosphate-buffered saline (PBS). The physical dimensions of the imaged volume was 530 µm × 530 µm × 175 µm ($x \times y \times z$). A vertical (xz) tomographic image extracted from the three-dimensional data set is shown in Fig. 13.8 with combined spectroscopic/intensity-based contrasts.

The image reveals the internal tissue morphology. Pleomorphic mesenchymal cells can be visualized. Cell membranes, cell nuclei and melanocytes appear highly backscattering as compared to cytoplasm. Melanocytes (indicated by two white arrows) can be

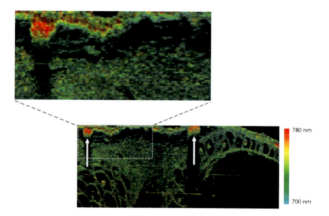

Figure 13.8 Combined intensity/spectroscopy FF-OCM image of an ex vivo *Xenopus Laevis* tadpole. The center of mass of the optical spectra is used as the spectroscopic metric. The white arrows indicate the presence of two melanocytes.

clearly distinguished thanks to the spectroscopic-based contrast. They appear red in this color representation indicating a spectral shift of ~35 nm as compared to the average of the rest of the image. This spectral shift is probably mainly due to the absorption of melanin, which decreases monotonically with the wavelength in the visible and near infrared [4, 60]. A slight red shift of the measured spectra with increased imaging depth can be observed, as reported by other groups [4, 6]. This tendency is consistent with the fact that longer wavelengths penetrate deeper than shorter wavelengths due to reduced scattering. However, the imaging depth is not large enough (175 μm) to make this phenomenon clearly visible.

The acquisition time of the three-dimensional data set ($\sim 10^9$ "points," 5 Gbits) was ~110 min. This relatively long acquisition time was limited by the CCD frame rate and the time spent accessing and writing the data on the computer hard disk. Because of the long acquisition time, the technique was not appropriate for in vivo imaging. Nevertheless, as demonstrated here, it was applicable on ex vivo or motionless samples.

13.3.7 Spectroscopic Polarization-Sensitive FF-OCM

An extended FF-OCM system was developed recently, capable of measuring simultaneously the power spectrum, the intensity and the birefringence-induced phase retardation of light backscattered by the structures of a sample [41]. This multimodal imaging system combines the spectroscopic FF-OCM and polarization-sensitive FF-OCM techniques presented previously in one single imaging system. The experimental setup is identical to the one of polarization-sensitive FF-OCM depicted in Fig. 13.2, except that a nanometer-resolution piezoelectric-actuated linear translation stage is used instead of motorized stage to translate the sample in the axial direction in steps of a few nanometers. The full interferometric signal can thus be acquired and then processed to provide the spectroscopic information, as described previously. The oscillation of the reference mirror is stopped for that measurement. If spectroscopic information is not desired, high-speed en face imaging with intensity and phase retardation contrasts only is also possible, by making the reference surface oscillate, as done in the conventional polarization-sensitive FF-OCM system described in Section 13.2 [40].

To demonstrate the capability of this multimodal imaging system to measure simultaneously the intensity, the power spectrum, and the phase retardation, a sample presenting both birefringence and absorption variations in the detection spectral region, that is, between 500 nm and 1000 nm, was imaged. The sample was an infrared viewing card (Thorlabs, model VRC5), characterized by a spectral sensitivity in the range between 700 nm and 1400 nm, with a maximum of sensitivity around 1000 nm. The absorption of the phosphore-based photosensitive material, strictly increasing with wavelength in the imaging band, is therefore expected to modify the power spectrum of light having travelled a certain distance through the photosensitive material. This photosensitive material is encapsulated between two plastic layers. Since many plastic materials are birefringent, phase retardation changes of backscattered light emanating from that region of the sample are expected to be detected.

Figure 13.9 shows an *xz*-oriented section extracted from a three-dimensional image acquisition, with intensity-based contrast (a), spectroscopic contrast (b) and phase retardation contrast (c).

The intensity-based image (Fig. 13.9a) reveals part of the internal structure of the viewing card. Three distinct layers can be distinguished. The upper layer corresponds to plastic material having a thickness of 170 µm. A random distribution of scattering structures is revealed inside this polymer material. The inhomoge-

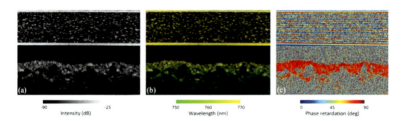

Figure 13.9 Cross-sectional multimodal FF-OCM images of an infrared viewing card. Intensity-based image (a), spectroscopic image (b), and phase retardation image (c). The size of each image is 650 µm (horizontal) × 450 µm (vertical).

neous bottom layer corresponds to the photosensitive material. The middle layer, in contrast, is homogeneous since no scattered light is detected from this region. This layer is assumed to be the glue used for maintaining the photosensitive material attached to the plastic layers.

The spectroscopic image (Fig. 13.9b) is a combination of spectroscopic and intensity measurements displayed using an HSL color map. Hue encodes the spectroscopic information, that is, the center of mass of the spatially resolved spectra, whereas luminance encodes the intensity information. Hue variation is restricted in the range from 56° to 117°, whereas luminance ranges from 0 to 1. The saturation parameter is set to 1. A shift of the spectrum center of mass of light emanating from the photosensitive material toward shorter wavelengths is observed as depth increases. The maximal value of the detected spectral shifts reaches 18 nm. In order to verify the validity of this result, the viewing card was placed in front of a halogen lamp and the spectrum of light transmitted by the card was measured using a silicon-based spectrometer. Figure 13.10 compares this measured spectrum with the spectrum of the lamp itself. One can see the effect of the photosensitive material which

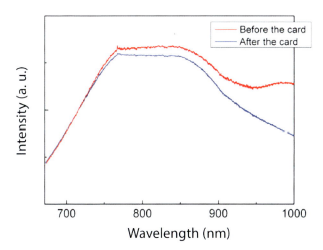

Figure 13.10 Normalized spectral power distribution of light emitted by the halogen lamp before and after passing through the infrared viewing card.

attenuates more the long wavelengths than the short wavelengths in the spectrometer detection band. The centers of mass of the two spectra are calculated over the wavelength range 670–1000 nm. They are located at 865 nm and 888 nm for the transmitted and incident light, respectively. The corresponding spectral shift of 23 nm is consistent with the measurements provided by the FF-OCM spectroscopic modality.

Figure 13.9c shows an image reconstructed from FF-OCM polarization-sensitive measurements. The phase retardation is coded with a color map varying from blue (0°) to red (90°). Fringes parallel to the surface are observed in the superficial layer with a constant period of $p = 25 \pm 2$ μm. This reveals a homogeneous birefringence of this layer made of plastic. Birefringence is common in plastic materials because their molecules are frozen in a stretched conformation when the plastic is molded. The value of the birefringence can be deduced from $\delta n = \lambda/(4p) = 0.007 \pm 0.001$, λ being the mean wavelength (750 nm). This value is consistent with measurements done later, using another method [41]. The birefringence of the intermediate layer cannot be measured since no backscattered light is detected from this region. The lower layer, corresponding to the photosensitive material, does not exhibit any measurable birefringence.

13.4 Multispectral FF-OCM

The technological approach of spectroscopic FF-OCM presented previously involves detection of the whole interferometric signal and its analysis using Fourier mathematics [47]. This method requires intensive computation and high mechanical stability to avoid degradation of phase-sensitive information during the acquisition time. Another technique was developed to exploit the spectroscopic response of the imaged sample. The method consists of imaging the sample in different bands. It has been applied in conventional OCT [21, 22, 61]. The implementation of the method in FF-OCM, referred to as multispectral FF-OCM, is presented in this section. The first multispectral FF-OCM system was capable of imaging in two bands simultaneously [62]. Other systems were developed later for three-band imaging [63, 64]. The dual-band and three-band systems reported in Refs. [62] and [64], respectively, are presented in the following sections.

13.4.1 Dual-Band FF-OCM

13.4.1.1 Method

Silicon-based cameras are used in FF-OCM for detection of optical wavelengths between 500 nm and 1000 nm typically [49, 50], whereas Indium Gallium Arsenide (InGaAs) cameras are used for imaging at wavelengths longer than ∼1000 nm [65, 66]. A dual-band FF-OCM system capable of imaging simultaneously in two distinct broad infrared bands centered at ∼800 nm and ∼1200 nm was developed by using two different cameras [62]. The experimental setup, represented schematically in Fig. 13.11, is based on a conventional FF-OCM system with a halogen lamp as the low coherence light source [49]. Water immersion microscope objectives (10x, NA = 0.3) are used. Light is split into the reference and sample arms of the Linnik-type interferometer by a broadband beam splitter, with a variation of the ratio transmission/reflection less than 10% over the entire wavelength region of interest (700–1600 nm). The surface of an $Y_3Al_5O_{12}$ (YAG) crystal is used to provide a reference surface of ∼2% reflectivity for both wavelength regions. A motorized axial translation stage allows axial scanning of the sample for acquiring an image stack. A dichroic mirror with high reflectivity (>98%) in the 650–950 nm band and high transmission (>85%) in the 1100–1500 nm band separates both spectral bands before detection. Two achromatic doublet lenses (L1, L2) are used to project simultaneously the interferometric images onto the sensor of the two array cameras, a silicon CCD camera (Model CA-D1 from Dalsa, 256 × 256 pixels, 8 bits, 180 Hz) with a maximal sensitivity at ∼800 nm (650–950 nm wavelength region) and an InGaAs area scan camera (Model SU320MS from Sensors Unlimited, 320 × 256 pixels, 12 bits, 20 Hz) with a maximal sensitivity at ∼1300 nm (900–1700 nm wavelength region). An optical density is set ahead the InGaAs camera to work close to saturation simultaneously for both cameras.

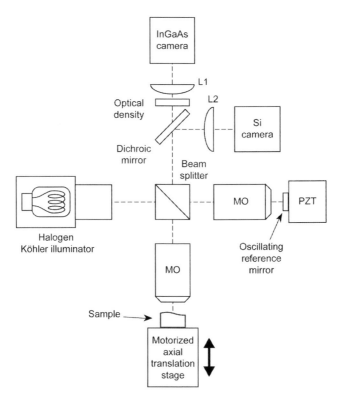

Figure 13.11 Experimental setup of dual-band FF-OCM. MO, microscope objective; L1, L2, achromatic doublet lenses; PZT, piezoelectric transducer.

En face tomographic images in both bands are obtained by subtraction of two phase-shifted interferometric images acquired by each camera [67]. A sinusoidal phase modulation is generated by the oscillation of the reference surface using a PZT actuator. The amplitude of the PZT oscillation is adjusted so that the phase shift between the two interferometric images used for the calculation of the tomographic image is close to π in both bands. In order to maximize the operation speed, the frequency of the PZT oscillation is set to 100 Hz, which corresponds to half the frame rate of the fastest camera (Silicon camera, 200 Hz). Both cameras have the same exposure time (maximal acquisition time of the fastest camera). The frame rate of the slowest camera (InGaAs camera) is adjusted to

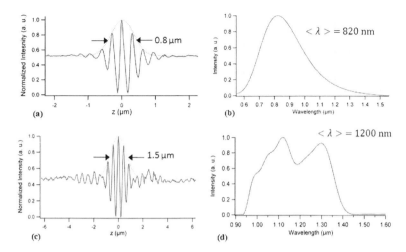

Figure 13.12 Experimental interferograms acquired with the Si camera (a) and with the InGaAs camera (c). The Fourier transform of the interferograms yields the shape of two imaging bands centered at 820 nm (c) and 1200 nm (d).

an odd fraction of the one of the Silicon camera. Equalization of detection sensitivities in both bands is achieved by adjusting the frame rate of the slowest camera to 40 Hz. A detection sensitivity of ∼90 dB requires the accumulation of 250 images with the Silicon camera and 50 images with the InGaAs camera.

The interferogram response was measured in both bands, using a mirror as a sample (see Fig. 13.12). The Fourier transform of the interferograms was calculated to characterize the bands. The center wavelengths were found at 820 nm and 1200 nm. The FWHM of the fringe envelope was found to be 0.8 µm and 1.5 µm at 820 nm and 1200 nm center wavelength respectively. Although the spectral response of the InGaAs camera extends to 1700 nm, absorption of light by the immersion medium (water) limits the effective spectral range above 1400 nm. In air, the theoretical axial resolution would be ∼1.0 µm at 1300 nm center wavelength. However, even if the absorption by water significantly degrades the axial resolution, a configuration with water immersion objectives was preferred to reduce dispersion mismatch (and consequently reduce the degradation of the axial resolution as the imaging depth increases) and avoid the need for dynamic focusing adjustments.

The theoretical transverse imaging resolution was calculated to be 1.7 µm and 2.0 µm for the Si and InGaAs cameras respectively. The experimental values were 1.8 µm and 2.5 µm, respectively. The degradation between experimental and theoretical transverse resolutions with the InGaAs camera was attributed to optical aberrations of the microscope objectives, which are not optimized in the 1200 nm wavelength region.

13.4.1.2 Simultaneous dual-band imaging

Scattering and absorption of light are the dominant factors that limit the imaging penetration depth in most biological tissues. The intensity of light scattered by particles depends strongly on their size and nature. Generally, this intensity varies with the optical wavelength λ as $1/\lambda^k$ where k is a parameter which depends on the size, geometry and refractive index of the scattering particles ($k > 1$). This law indicates that the intensity of scattered light in biological tissues decreases when wavelength increases [19, 68].

Absorption of light in most biological tissues comes essentially from the presence of water. The absorption of light by water is minimal at the wavelength of 500 nm and globally increases with the wavelength [69].

The optimal wavelength to maximize the imaging penetration results from the best trade-off between absorption and scattering. Previous work has shown better penetration depth in highly scattering tissues for OCT operating in the 1300 nm wavelength region compared to 830 nm [19].

The effect of scattering on imaging penetration depth with FF-OCM was studied by imaging a highly contrasted sample through a scattering medium. This sample was a mask for optical lithography. It was made of metallic (chromium) structures on a polished glass substrate. The scattering medium (3 mm thick) was of 4% Intralipid solution. This medium is commonly used to simulate scattering by biological tissues [70, 71]. It was used as immersion medium for the microscope objective in the sample arm. The reflection coefficient of the metallic structures was almost independent of wavelength

Figure 13.13 Dual-band FF-OCM imaging of metallic structures deposited on a glass substrate, through a 3 mm thick Intralipid solution, acquired simultaneously in the 0.82 μm (a) and 1.20 μm (b) wavelength regions.

in the infrared region. Without Intralipid solution, images at both wavelengths had similar contrasts. With the Intralipid solution as immersion medium, the image contrast was considerably degraded at 820 nm whereas it was almost unaltered at 1200 nm due to the wavelength dependence of light scattering (see Fig. 13.13). In biological media, the situation is however expected to be quite different because the intensity of light backscattered by the structures inside the imaged sample is wavelength dependent. The signal itself will be reduced at longer wavelengths.

Simultaneous dual-band FF-OCM imaging of an African frog tadpole *Xenopus laevis* was performed. A 250 μm × 1200 μm cross-sectional (*XZ*) image was extracted from a three-dimensional data set acquired in the tadpole head. Images obtained at 820 nm and 1200 nm are displayed in Fig. 13.14a and 13.14b, respectively. Epidermis, different stages of mesenchymal cells and structures from the digestive system are revealed. As expected, the axial and transverse resolutions are better with the Silicon camera, allowing for example better visualization of the morphology of the mesenchymal cell membranes and nuclei. As seen before, the degradation of the axial and transverse resolutions at 1200 nm is due to wings in the interferogram response, whose contribution is enhanced by the logarithmic scale, and optical aberrations of the microscope objectives. This degradation leads to blurred images with the InGaAs camera compared to the silicon camera, which explains the slightly apparent difference between both images. On the other hand, a higher penetration depth can be achieved with the

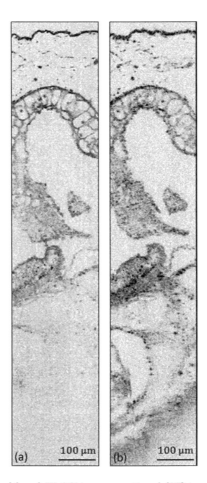

Figure 13.14 Dual-band FF-OCM cross-sectional (XZ) images of an African tadpole *Xenopus laevis*, ex vivo, at 820 nm (a) and 1200 nm (b) center wavelengths.

InGaAs camera, making possible the imaging of scattering structures at depth larger than ∼1 mm.

As a second example, simultaneous dual-band FF-OCM imaging of the rabbit trachea, ex vivo, is presented in Fig. 13.15. When comparing the fine structures in the single-band images, like the cartilaginous ring above the epithelium, it is clear that the spatial resolution is higher at 820 nm compared to 1200 nm, as expected.

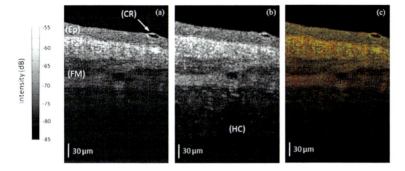

Figure 13.15 Dual-band FF-OCM cross-sectional images of rabbit trachea, ex vivo, at 820 nm (a) and 1200 nm (b). (c) The differential color image. Ep, epithelium; CR, cartilaginous ring; FM, fibrous membrane; HC, hyaline cartilage.

In the other hand, there is an evident benefit in penetration depth at 1200 nm. For better visualization of the differences between both single-band images, a color image representation was applied similar to the one presented in [61]. An HSL value mapping was used to represent the intensity and enhance the spectroscopic information content of the dual-band OCT images. To obtain this color image, both OCT images were normalized and the hue attribute was fixed (in green and red hues, respectively, for the 820 nm and the 1200 nm images). Then, the saturation and luminance attributes of the differential color map were set as the average OCT signal amplitude of both images. Thus, the structures with higher intensity at 1200 nm appear red in the color image, while structures which are dominant at 820 nm are colored in green. Due to the degradation of the axial and transverse resolutions at 1200 nm, great care was taken for perfect image superposition. The color image is presented in Fig. 13.15c. In the superficial layer, corresponding to the trachea epithelium, the backscattered light intensity is similar for both wavelength bands. In contrast, in the medium layer, the fibrous membrane structures have various wavelength dependent properties. At larger depth, the image color turns to red, due to a better penetration at 1200 nm compared to 820 nm.

13.4.2 Three-Band FF-OCM

The dual-band FF-OCM system presented above can image simultaneously in two distinct wavelength regions. Compared to Fourier-transform-based spectroscopic FF-OCM [47], which requires the acquisition and processing of the entire interferometric signal, this system dramatically reduces the time to obtain images. Unfortunately, the spectroscopic information provided by dual-band FF-OCM is limited to measurements in only two bands. Imaging in more than two bands would provide further spectroscopic information. However, the approach implemented in dual-band FF-OCM is not technologically convenient for multiband imaging since each band requires a dedicated detector. Using a single detector and illuminating sequentially at different wavelengths seems more appropriate for multiband imaging. This approach is, however, challenging in OCT due to the relationship between the imaging axial resolution and the spectral width of the illumination light. Because of the limited spectral range covered by the detector, a trade-off has to be made between axial resolution and spectral resolution. Three-band imaging can be considered as a good trade-off, allowing the production of color images using the conventional RGB three color channels. Three-band FF-OCM, using a single camera and three distinct light-emitting diodes (LEDs) in the red (R), green (G), and blue (B) bands, was reported to produce true color images with an axial resolution of ~ 5 µm [63]. This system is described in Chapter 15. In this section, a higher-resolution, high-sensitivity three-band FF-OCM system using a single camera and a single light source is reported. This system covers a spectral range from 530 nm to 1700 nm [64].

13.4.2.1 Materials and methods

The layout of the three-band FF-OCM imaging system reported in [64] is depicted in Fig. 13.16. A halogen lamp is used as the low-coherence light source. The interferometric images are acquired by an InGaAs-based area camera (OWL SW1.7-CL-HS VIS-SWIR manufactured by Raptor Photonics). The camera has an extended sensitivity in the visible to offer a spectral sensitivity ranging from

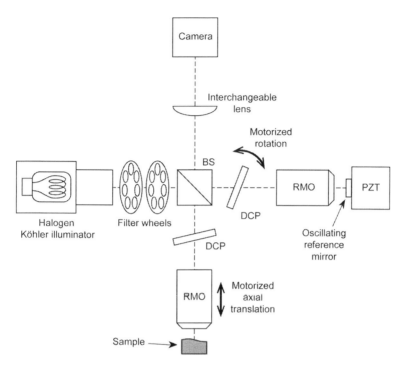

Figure 13.16 Achromatic three-band FF-OCM setup. RMO, reflective microscope objective; BS, beam splitter; DCP, dispersion compensation plate; PZT, piezoelectric transducer.

400 nm to 1700 nm (see Fig. 13.17a). The number of pixels is 320×256. The frame rate can reach 346 Hz. En face tomographic images are obtained by arithmetic combination of four phase-shifted interferometric images [50]. The phase shift is generated by making the reference surface oscillate using a piezoelectric actuator synchronized with the camera trigger signal. Reflective microscope objectives are used ($15\times$, 0.28 NA) to avoid chromatic aberrations. They are based on a Schwarzschild two-mirror design consisting of a small convex primary mirror and a larger concave secondary mirror. The mirrors are broadband coated with gold for optimal reflectivity from visible to infrared. The objectives are corrected for the primary spherical, coma and astigmatic aberrations.

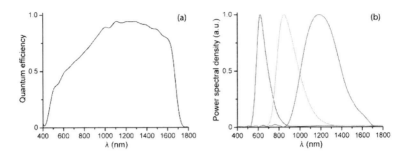

Figure 13.17 (a) Variation of the quantum efficiency (in electron/photons) of the extended InGaAs camera as a function of wavelength. (b) Normalized spectral detection sensitivity associated to the three imaging bands.

A plane reference surface of ~4% reflectance is obtained with a BK7 glass plate. The optical path length of the reference arm is adjusted by rotating a 12 mm thick fused silica glass plate placed in the reference arm, using a motorized rotation stage. Another identical glass plate is placed at a fixed angle in the sample arm to compensate for the dispersion mismatch introduced by the other plate. The zero-delay position can thus be varied to get en face tomographic images at different depths in the sample, while compensating the dispersion mismatch. The microscope objective in the sample arm is mounted on a linear motorized translation stage and is moved appropriately to maintain the focus. Dynamic focusing is thus achieved as the imaging depth is varied [66, 72].

Three illumination bands (see Fig. 13.17b) within the spectral range of sensitivity of the camera are selected by using three filter sets. The three bands, centered at 635 nm, 870 nm, and 1170 nm, spread out over the spectral range 530–1700 nm.

Two different 400 mm focal length doublet lenses, one optimized for the 600–1000 nm wavelength range and the other one for 1000–1600 nm, are used depending on the filter combination, to project the interferometric images onto the camera sensor. Kinematic bases are employed for interchanging the lenses quickly and ensuring that still exactly the same part of the sample is imaged. Changing the filter set and the tube lens takes only a few seconds. The sample must be stationary during that time.

Due to the broad spectra considered here, dispersion mismatch between the two arms of the interferometer has to be reduced much as possible to avoid severe loss of performance in terms of signal contrast and axial resolution. Dispersion mismatch is partially avoided by the rotating glass plate placed in the reference arm, whose role is to adjust the zero-delay position as well as to keep the dispersion mismatch minimized as the imaging depth in the sample is changed. Theoretically, a perfect compensation is achieved if the group velocity dispersion of the glass plate is equal to the one of the sample. By assuming that biological samples are mainly made of water, fused silica is well suited with regard to its dispersion properties [73]. Both dispersion compensation plates (DCPs) are initially oriented at a same angle, θ_0 (see Fig. 13.16). According to numerical simulations, $\theta_0 = 30°$ appears to be a good trade-off for minimizing both the deviation of the light beam and the losses by reflection on the DCP surfaces when the plate is rotated by an additional angle $\delta\theta$. This value is then chosen as fixed angle for the DCP in the sample arm and as starting angle for the reference DCP rotation. For the three bands, the variation of the optical path length difference d, with respect to $\delta\theta$ was fitted with a polynomial function of order 3 in order to set the rotating movement parameters of the reference DCP during the acquisition. A rotation angle $\delta\theta$ of $10°$ changes d to ~ 1 mm. Due to optical dispersion in the DCPs, the value of d varies by 2% for $\delta\theta \approx 10°$ depending on the band. When imaging deeply into a sample, a shift of the imaging depth can be observed depending on the band because of dispersion in the sample. These phenomena are corrected by appropriate axial linear dilatation of the image in the three bands. This procedure, based on image correlation analysis, is required for superposition of the three acquired images resulting in a single RGB image.

The rotation of the reference DCP also induces a progressive shift of the light beam, which leads to a distortion of the image. This artifact is corrected by digital image processing.

13.4.2.2 System characteristics and image results

The width of each of the three bands was adjusted so as to have a quasi-constant axial resolution Δz in the three imaging bands. The

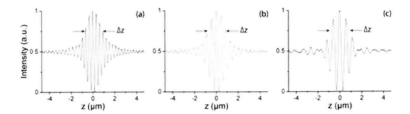

Figure 13.18 Interferograms measured in the three imaging bands centered at 635 nm (a), 870 nm (b), and 1170 nm (c).

interferometric signal, measured in the three bands using a mirror as a sample, is shown together with its envelope in Fig. 13.18. The axial resolution, defined as the FWHM of the fringe envelope, was then measured to be ∼1.9 μm.

The transverse resolution is essentially determined by the numerical aperture of the microscope objectives and the mean wavelength λ_0. The transverse resolution was measured in the three bands by imaging 0.1–μm diameter gold beads embedded in an agarose gel. Measurements are similar to theoretical values and confirm the expected differences between the transverse resolutions in the three bands. Like the axial resolution, it is desirable to have the same transverse resolution in the three imaging bands for superposition of the images. For that purpose every en face image is convolved with a suited two-dimensional Gaussian function so that the transverse resolution is constant in the three bands.

The exposure time of the camera was adjusted depending on the band to equalize the detection sensitivity in the three bands. The number N of accumulated images was set depending on the band to compensate for variation of the fringe contrast. This provided a detection of reflectivity variations with a precision of ∼10%. By using a BK7 glass plate whose reflectivity varies by less than 5% between the bands, N was chosen equals to 30 (band 1), 26 (band 2), and 24 (band 3) to obtain a quasi-constant FF-OCM signal in the three bands given the precision. The acquisition frame rate of en face tomographic images varied then from 1 Hz to 4 Hz, depending on the band, while the detection sensitivity defined as the lowest-detectable sample reflectivity [74] was close to 85 dB for the three bands. Thus, for each band, the performance of the system in terms

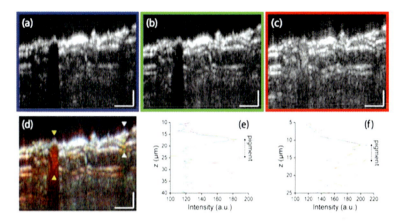

Figure 13.19 Three-band FF-OCM images (*xz* sections) of a wood sample partially covered with Prussian blue and Cerulean blue oil paintings at 635 nm (a), 870 nm (b), 1170 nm (c), and with RGB representation (d–f) Cuts along the *z* axis of the RGB image intensity at the pigment locations. The positions of the cuts are indicated by yellow (e) and white (f) arrows in the RGB image (d). The scale bar is 50 μm along the *x* axis (horizontal) and 10 μm along the *z* axis (vertical).

of spatial resolution and detection sensitivity is similar to the one of conventional FF-OCM [50, 66].

Three-band imaging can reveal the presence of details in a sample, owing to their wavelength-dependent scattering and absorption properties, which are imperceptible otherwise. A piece of wood with two drops of different pigment colors deposited at the surface was imaged (see Fig. 13.19). Since the reflectivity of wood structures does not change significantly in the three bands, the surface of the sample appears white. Red stains are visible on the surface in the RGB representation due to side lobes in the interferogram envelop of the band at 1170 nm (Fig. 13.18c). A slight red hue can also be observed on the wood layers. This tendency is consistent with the fact that longer wavelengths penetrate deeper than shorter wavelengths due to reduced optical scattering [4, 47].

One can notice that while it is almost impossible to distinguish the two painted regions with only one or two bands, three-band imaging with RGB representation clearly reveals them. Intensity profiles of the RGB image are shown to highlight the presence of

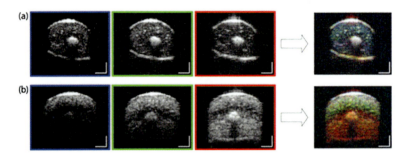

Figure 13.20 Three-band FF-OCM images (xz sections) of a light (a) and a dark (b) human hair. From left to right at 635 nm, 870 nm, 1170 nm, and with RGB representation. The scale bar is 20 µm in the two directions.

the two pigments and provide a measurement of their size. One can deduce from Fig. 13.19 that one of the pigment absorbs light between 600 nm and 1000 nm (bands 1 and 2) and appears red in the RGB image, whereas the other pigment absorbs the visible (band 1) and infrared light beyond ~1.3 µm (part of band 3) and then appears greenish in the RGB representation. This is consistent with the spectroscopic properties of the Prussian blue and Cerulean blue oil paintings [75].

An example of how three-band FF-OCM can be used for the characterization of biological samples is shown in Fig. 13.20. Two distinct human hairs, a light one and a dark one, were imaged. Hair darkness is relative to the quantity of eumelanin pigments. Since eumelanin absorption decreases as wavelength increases [76], dark hair absorbs visible light and less infrared light, much more readily than light hair. This is why a difference of imaging penetration depth can be observed between the hairs in band 1 but not in band 3. Scattering reinforces this difference since it is weaker in the infrared and thus the imaging penetration depth is less limited. However, the penetration here is dominated by absorption since the difference between absorption coefficients is much more important than between scattering coefficients for distinct eumelanin concentrations [76]. Three-band imaging can thus reveal spectral absorption properties. One can notice that the sections of the hairs do not appear circular in the images. This artifact comes from a distortion of the optical wavefront due to the curved shape of the upper interface between the sample and air, and to the unhomogeneity of the refractive index distribution within the sample.

Compared to conventional one-band imaging and even dual-band imaging, three-band FF-OCM provides image contrast enhancement and improves the sample differentiation capability without significant loss of spatial resolution. Although three successive image acquisitions are required to produce a RGB image, unlike Refs. [62] and [63], one single camera and one single light source are employed, which makes the system more workable. Besides, this technological approach could be easily extended for imaging in more than three bands. This would provide real spectroscopic analysis of the imaged sample at the price of longer acquisition times and tradeoffs between spectral resolution and spatial resolution.

13.5 Combination of FF-OCM with Fluorescence Microscopy

Fluorescence microscopy has become an essential tool in life sciences, as well as in material sciences. Various samples exhibit fluorescence when they are illuminated. This phenomenon has been exploited in the fields of botany, petrology, and in the semiconductor industry. In contrast, the study of biological objects by fluorescence microscopy is often complicated by either extremely faint or bright, nonspecific autofluorescence. Exogenous fluorophores, excited by specific optical wavelengths, can then be used. They are often highly specific in their attachment targeting and have a significant fluorescence efficiency. Because OCT is an interferometric technique, it cannot sense the incoherent light emitted from fluorescent agents. Therefore, fluorescence cannot be used directly as an image contrast enhancement mechanism in OCT. However, fluorescence imaging and OCT can be coupled for dual-modality imaging. The scanning nature of standard OCT makes this technology particularly suitable for being coupled with scanning fluorescence microscopy techniques based on linear excitation [8, 9, 11, 12, 77, 78], or nonlinear excitation [79, 80]. In contrast, the combination of scanning fluorescence microscopy techniques with

FF-OCM is not so straightforward, since FF-OCM is not a point-by-point imaging technique. Fluorescence imaging, with wide-field illumination instead of scanning illumination, is more appropriate to be coupled with FF-OCM. Epifluorescence microscopy is the most popular wide-field fluorescence microscopy technique. However, it suffers from a poor optical sectioning ability, which does not allow three-dimensional image reconstructions as can be accomplished with FF-OCM. Structured illumination microscopy (SIM) is a wide-field imaging technique with an optical sectioning capacity similar to the one of confocal microscopy. This technique relies on the projection of a grid pattern onto the image focal plane to reject out-of-focus blur [81]. The combination of fluorescence SIM and FF-OCM was proposed by three research groups almost simultaneously [82–84]. One of these combinations is described in the following section. The one reported in [84] is presented in Chapter 16.

13.5.1 Materials and Methods

The layout of the dual-modality imaging system reported in [82] is shown in Fig. 13.21. The FF-OCM modality is based on a conventional setup as described in [50]. A halogen lamp is used as the low-coherence light source in a Köhler illumination system. Water immersion microscope objectives ($20\times$, 0.5 NA) are employed. The length of the reference arm can be varied by a motorized translation stage. The microscope objective in the sample arm is mounted on another motorized translation stage to maintain the focus while the depth in the sample is scanned. Infrared light ($\lambda > 650$ nm) at the interferometer output is transmitted by a dichroic beam splitter and focused by an achromatic doublet lens ($f = 300$ mm) onto the sensor of a CCD camera (Dalsa, Dalstar 1M15, 1024×1024 pixels, 15 fps). An oscillation of the reference mirror is generated by a PZT. Multiple phase-shifted interferometric images are acquired for various positions of the reference mirror. A mathematical combination of the acquired images allows reconstructing a FF-OCM image, that is, an en face tomographic image of the sample. An algorithm based on the acquisition of only two phase-shifted interferometric images is used [67].

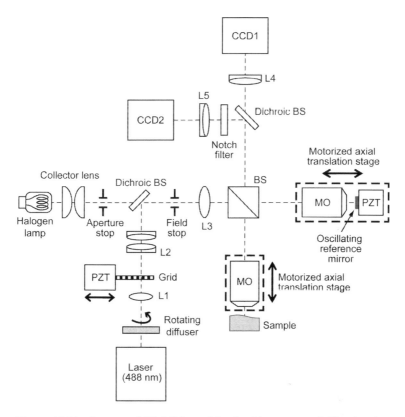

Figure 13.21 Layout of FF-OCM combined with structured illumination fluorescence microscopy. BS, beam splitter; MO, microscope objective; PZT, piezoelectric transducer; L1/L3, singlet lenses; L2, pair of achromatic doublets; L4/L5, achromatic doublets; CCD, charge-coupled device.

The fluorescence imaging modality uses a continuous wave diode-pumped Titanium Sapphire laser (Coherent, 488-200 CRDM) as a source of coherent light at the wavelength of 488 nm. The laser beam is coupled into the system by a dichroic beam splitter that reflects visible light ($\lambda < 700$ nm) and transmits infrared light ($\lambda > 700$ nm). A rotating ground glass diffuser is placed at the output of the laser to suppress the effect of speckle. A lens (L1, $f = 100$ mm) expands the size of the illumination field on the sample. The optical subsystem consisting of lenses L1, L2, and L3 images the ground surface of the diffuser onto the back focal plane of the microscope objectives. Visible fluorescence light ($\lambda < 650$ nm) is reflected by the dichroic beam splitter and focused onto another CCD detector (CCD2, Photometrics, CoolSnap HQ, 1392 × 1040 pixels, 10 fps) by achromatic doublet lens L5 ($f = 300$ mm). A notch filter (Semrock, NF-488) eliminates the excitation light. SIM is implemented using a 20-line-pair/mm grid placed in the laser beam path after L1 and imaged by L2 on the field stop plane that is conjugate to the focal planes of the microscope objectives. L2 consists of a pair of achromatic doublets ($f = 60$ mm for each) achieving 1:1 imaging. The grid is mounted on a PZT that controls its transverse positioning. Images of the sample are acquired for three positions of the grid. These positions are shifted by one third of the grid period with respect to each other. The collected images are processed to get rid of the grid pattern and compute the optically sectioned image of the sample [81]. The conventional wide field fluorescence image is obtained from a mathematical combination of the three raw acquired images.

The images obtained from the FF-OCM imaging modality and the fluorescence imaging modality can be displayed separately or overlaid to show different information. The detectors used in the two imaging modalities have different format in terms of pixel number and pixel size. To overlay the fluorescence and FF-OCM images, it is necessary to crop and then interpolate the initial FF-OCM image so that it shows the same field of view and has the same number of data points as the fluorescence image.

13.5.2 System Performance and Image Results

The axial resolution of the FF-OCM imaging modality was measured to be 1.5 µm in air (1.1 µm in tissue assuming a refractive index of 1.4). The theoretical lateral resolution was calculated to be 0.9 µm at 800 nm center wavelength. The detection sensitivity was measured to be 89 dB by accumulating 10 images.

The fluorescence point-spread function was measured by imaging fluorescent beads. A lateral resolution of 0.7 µm was determined, which is comparable to the theoretical value of 0.6 µm at a wavelength of 520 nm. The axial resolution was measured to be 6.0 µm. A 4.5 µm theoretical axial resolution was computed. The

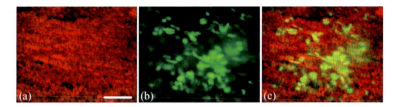

Figure 13.22 FF-OCM image (a) and optically sectioned fluorescence image (b) of mouse colon tissue stained with acridine orange. Both images are merged in (c). Scale bar is 50 μm.

discrepancy between measured and theoretical values probably results from the presence of optical aberrations in the imaging system, including the aberrations associated with the imaging of the grid into the sample.

Samples of excised mouse tissues were topically stained with acridine orange and set between a glass slide and a coverslip. Acridine orange is a nucleic acid stain with dual peak emission wavelengths around 520 nm and 650 nm when bound to DNA and RNA, respectively. A drop of distilled water was deposited on the coverslip as the immersion medium for the microscope objective. Images of mouse colon tissue are displayed in Fig. 13.22. Figure 13.22a shows an en face tomographic image acquired at a depth of 40 μm. Figure 13.22b shows the co-registered optically sectioned fluorescence images. The FF-OCM image and the optically sectioned fluorescence image are colored in red and green, respectively, and overlaid to form Fig. 13.22c for direct comparison. Most out-of-focus light is rejected in the optically sectioned image. The fluorescence signal reveals individual cell nuclei, while the FF-OCM data shows a fibrous pattern that is most probably connective tissue. The fluorescence modality dictates the maximum imaging depth while being able to co-register a fluorescence image and a FF-OCM image at the same plane location in a specimen. This is because visible light is more scattered than infrared light in biological tissues [85] and intensity-based detection techniques are less sensitive than interferometric detection techniques [79].

The imaging system presented here is flexible in that other target-specific fluorophores can be used, provided that they emit in the range of wavelengths below 650 nm and that the appropriate excitation light source is used. Microscope objectives with higher NA could also be used to improve the lateral resolution in both modalities and the axial resolution of the fluorescence SIM technique. Simultaneous acquisition enables capturing an image of the same tissue location regardless of sample motion. However, this motion needs to be negligible during the acquisition time to avoid image blurring. Overlaying the images obtained from each imaging modality highlights their differences.

The fluorescence signal could be associated with biochemical processes at the cellular level. An integrated imaging system capable of optically sectioned fluorescence imaging and FF-OCM imaging could potentially be valuable in biomedical applications by providing complementary information during biological tissue screening. This has the potential for better understanding cellular functions, and improving the ability to detect pathologies by monitoring morphological and biochemical changes in tissues. The reported imaging system could be used in conjunction with a biopsy procedure. Excised tissue samples could be explored with minimal preparation as opposed to standard histopathology process.

13.6 Conclusion

The imaging contrast of FF-OCM results from variations of the intensity of light scattered by the sample microstructures. This source of contrast, due to unhomogeneities of the refractive index within the sample, provides morphological information. However, in many instances the change in scattering properties between structures of different types is small and difficult to measure. Among the most significant advances in the development of FF-OCM, since the introduction of the initial technology more than fifteen years ago, is the introduction of several technological extensions for multicontrast and multimodality imaging. Polarization-sensitive FF-OCM takes advantage of the additional information carried by the polarization state of light scattered from the sample. Spectroscopic FF-OCM provides information on the spectral content of scattered light by detection and processing of the whole interferometric signal. As alternative methods to take advantage of the spectroscopic response of the sample, dual-band and three-band FF-OCM systems have been proposed. Though having been improved by these extensions, FF-OCM itself is still not sensitive enough to provide absorption contrast or biochemical and molecular information. Therefore, multimodal optical imaging systems with FF-OCM combined to other imaging modalities, have been developed. FF-OCM has been combined with fluorescence microscopy to potentially provide biochemical and metabolism information in addition to morphological structure. FF-OCM was recently combined with elastography for adding to morphological images a map of the mechanical properties of the sample.

Most technical extensions of FF-OCM are currently in their early stages of development. They have only been demonstrated and validated on test samples. Further work has to be done for their optimization and more complete validation. Other technical concepts already applied to conventional OCT have to be developed. Intensive studies have now also to be carried out at the application level for these various contrast-enhanced and/or multimodal FF-OCM techniques to become widely used in the fields of material characterization, biological studies, and medical diagnosis.

References

1. De Boer, J. F., Milner, T. E., Van Gemert, M. J. C., and Nelson, J. Stuart. (1997). Two-dimensional birefringence imaging in biological tissue by polarization-sensitive optical coherence tomography. *Opt. Lett.*, **22**, pp. 934–936.
2. Hitzenberger, C. K., Götzinger, E., Sticker, M., and Fercher, A. F. (2001). Measurement and imaging of birefringence and optic axis orientation by phase resolved polarization sensitive optical coherence tomography. *Opt. Express*, **9**, pp. 780–790.
3. Wiesauer, K., Pircher, M., Goetzinger, E., Hitzenberger, C. K., Engelke, R., Ahrens, G, Gruetzner, G., and Stifter, D. (2006). Transversal ultrahigh-resolution polarization sensitive optical coherence tomography for strain mapping in materials. *Opt. Express*, **14**, pp. 5945–5953.
4. Morgner, U., Drexler, W., Kärtner, F. X., Li, X. D., Pitris, C., Ippen, E. P., and Fujimoto, F. J. (2000). Spectroscopic optical coherence tomography. *Opt. Lett.*, **25**, pp. 111–113.

5. Leitgeb, R., Wojtkowski, M., Kowalczyk, A., Hitzenberger, C. K., Sticker, M., and Fercher, A. F. (2000). Spectral measurement of absorption by spectroscopic frequency-domain optical coherence tomography. *Opt. Lett.*, **25**, pp. 820–822.
6. Adler, D., Ko, T., Herz, P., and Fujimoto, J. G. (2004). Optical coherence tomography contrast enhancement using spectroscopic analysis with spectral autocorrelation. *Opt. Express*, **12**, pp. 5487–5501.
7. Faber, D. J., Mik, E. G., Aalders, M. C. G., and Van Leeuwen, T. G. (2005). Toward assessments of blood oxygen saturation by spectroscopic optical coherence tomography. *Opt. Lett.*, **30**, pp. 1015–1017.
8. Yuan, S., Roney, C. A., Wierwille, J., Chen, C. W., Xu, B., Griffiths, G., Jiang, J., Ma, H., Cable, A., Summer, R. M., and Chen, Y. (2012). Co-registered optical coherence tomography and fluorescence molecular imaging for simultaneous morphological and molecular imaging. *Phys. Med. Biol.*, **55**, pp. 191–206.
9. Dai, C., Liu, X., and Jiao, S. (2012). Simultaneous optical coherence tomography and autofluorescence microscopy with a single light source. *J. Biomed. Opt.*, **17**, 080502-1–080502-3.
10. Kuranov, R. V., Sapozhnikova, V. V., Shakhova, N. M., Gelikonov, V. M., Zagainova, E. V., and Petrova, S. A. (2002). Combined application of optical methods to increase the information content of optical coherent tomography in diagnostics of neoplastic processes. *Quantum Electron.*, **32**, pp. 993–998.
11. Bradu, A., Ma, L., Bloor, J. W., and Podoleanu, A. (2009). Dual optical coherence tomography/fluorescence microscopy for monitoring of Drosophila melanogaster larval heart. *J. Biophotonics*, **2**, pp. 380–388.
12. Park, J., Jo, J. A., Shrestha, S., Pande, P., Wan, Q., and Applegate, B. E. (2010). A dual-modality optical coherence tomography and fluorescence lifetime imaging microscopy system for simultaneous morphological and biochemical tissue characterization. *Biomed. Opt. Express*, **1**, pp. 186–200.
13. Iftimia, N., Iyer, A. K., Hammer, D. X., Lue, N., Mujat M., Pitman, M., Ferguson, R. D., and Amiji, M. (2012). Fluorescence-guided optical coherence tomography imaging for colon cancer screening: a preliminary mouse study. *Biomed. Opt. Express*, **3**, pp. 178–191.
14. Singh, A. D., Belfort, R. N., Sayanagi, K., and Kaiser, P. K. (2010). Fourier domain optical coherence tomographic and auto-fluorescence findings in indeterminate choroidal melanocytic lesions. *Br. J. Ophthalmol.*, **94**, pp. 474–478.

15. Schmitt, J. (1998). OCT elastography: imaging microscopic deformation and strain of tissue *Opt. Express*, **3**, pp. 199–211.
16. Wang, R. K., Kirkpatrick, S., and Hinds, M. (2007). Phase-sensitive optical coherence elastography for mapping tissue microstrains in real time *Appl. Phys. Lett.*, **90**, p. 164105.
17. Liang, X., Crecea, V., and Boppart, S. A. (2010). Dynamic optical coherence elastography: a review *J. Innov. Opt. Health Sci.*, **3**, pp. 221–233.
18. Kennedy, B. F., Liang, X., Adie, S. G., Gerstmann, D. K., Quirk, B. C., Boppart, S. A., and Sampson, D. D. (2011). In vivo three-dimensional optical coherence elastography. *Opt. Express*, **19**, pp. 6623–6634.
19. Schmitt, J. M., Knuttel, A., Yadlowsky, M., and Eckhaus, M. A. (1994). Optical coherence tomography of a dense tissue: statistics of attenuation and backscattering. *Phys. Med. Biol.*, **39**, pp. 1705–1720.
20. Wang, Y., Nelson, J. S., Chen, Z., Reiser, B. J., Chuck, R. S., and Windeler R. S. (2003) Optimal wavelength for ultrahigh-resolution optical coherence tomography *Opt. Express* **11**, pp. 1411–1417.
21. Aguirre, A. D., Nishizawa, N., Seitz, W., Ledere, M., Kopf, D., and Fujimoto, J. G. (2006). Continuum generation in a novel photonic crystal fiber for ultrahigh resolution optical coherence tomography at 800 nm and 1300 nm. *Opt. Express*, **14**, pp. 1145–1160.
22. Alex, A., Považay, B., Hofer, B., Popov, S., Glittenberg, C., et al. (2010). Multispectral in vivo three-dimensional optical coherence tomography of human skin *J. Biomed. Opt.*, **15**, p. 026025.
23. Nahas, A., Bauer, M., Roux, S., and Boccara, A. C. (2013). 3D static elastography at the micrometer scale using full-field OCT *Biomed. Opt. Express* **4**, pp. 2138–2149.
24. Everett, M., Schoenenberger, J. K., Colston, B. W. Jr., and Da Silva, L. B. (1998). Birefringence characterization of biological tissue by use of optical coherence tomography. *Opt. Lett.*, **23**, pp. 228–230.
25. Todorović, M., Jiao, S, Wang, L. V., and Stoica, G. (2004). Determination of local polarization properties of biological samples in the presence of diattenuation by use of Mueller optical coherence tomography.*Opt. Lett.*, **29**, pp. 2402–2404.
26. Kemp, N., Zaatari, H., Park, J, H., Rylander, G., and Milner, T. (2005). Form-biattenuance in fibrous tissues measured with polarization-sensitive optical coherence tomography. *Opt. Express*, **13**, pp. 4611–4628.
27. Pircher, M., Goetzinger, E., Leitgeb, R., and Hitzenberger, C. (2004). Three dimensional polarization sensitive OCT of human skin in vivo. *Opt. Express*, **12**, pp. 3236–3244.

28. Cense, B., Chen, T. C., Park, B. H., Pierce, M. C., and De Boer, J. F. (2002). In vivo depth-resolved birefringence measurements of the human retinal nerve fiber layer by polarization-sensitive optical coherence tomography. *Opt. Lett.*, **27**, pp. 1610–1612.
29. Cense, B., Chen, T. C., Park, B. H., Pierce, M. C., and De Boer, J. F. (2004). Thickness and birefringence of healthy retinal nerve fiber layer tissue measured with polarization-sensitive optical coherence tomography. *Invest. Ophthalmol. Vis. Sci.*, **45**, pp. 2606–2612.
30. Götzinger, E., Pircher, M., Sticker, M., Fercher, A. F., and Hitzenberger, C. K. (2004). Measurement and imaging of birefringent properties of the human cornea with phase-resolved, polarization-sensitive optical coherence tomography. *J. Biomed. Opt.*, **9**, pp. 94–102.
31. Park, B. H., Saxer, C., Srinivas, S. H., Nelson, J. S., and De Boer, J. F. (2001). In vivo burn depth determination by high-speed fiber-based polarization sensitive optical coherence tomography. *J. Biomed. Opt.*, **6**, pp. 474–479.
32. Pierce, M. C., Strasswimmer, J., Park, B. H., Cense, B., and De Boer, J. F. (2004). Birefringence measurements in human skin using polarization sensitive optical coherence tomography *J. Biomed. Opt.*, **9**, pp. 287–291.
33. Baumgartner, A., Dichtl, S., Hitzenberger, C. K., Sattmann, H., Robl, B., Moritz, A., Fercher, A. F., and Sperr, W. (2000). Polarization-sensitive optical coherence tomography of dental structures. *Caries Res*, **34**, pp. 59–69.
34. Fried, D., Xie J., Shafi, S., Featherstone J. D. B., Breunig T. M., and Le C. (2002). Imaging caries lesions and lesion progression with polarization sensitive optical coherence tomography *J. Biomed. Opt.*, **7**, pp. 618–627.
35. Pircher, M., Goetzinger, E., Findl, O., and Hitzenberger, C. K. (2005). Polarization preserving and depolarizing ocular tissues studied with polarization sensitive optical coherence tomography *Invest. Ophthalmol. Vis. Sci.*, **46**, E-Abstract 4267.
36. Hee, M. R., Huang, D., Swanson, E. A., and Fujimoto, J. G. (1992) Polarization-sensitive low-coherence reflectometer for birefringence characterization and ranging. *J. Opt. Soc. Am. B*, **9**, pp. 903–908.
37. Yasuno, Y., Makita, S., Sutoh, Y., Itoh, M., and Yatagai, T. (2002). Birefringence imaging of human skin by polarization-sensitive spectral interferometric optical coherence tomography. *Opt. Lett.*, **27**, pp. 1803–1805.
38. Götzinger, E., Pircher, M., and Hitzenberger, C. (2005). High speed spectral domain polarization sensitive optical coherence tomography of the human retina. *Opt. Express*, **13**, pp. 10217–10229.

39. Moreau, J., Loriette, V., and Boccara, A. C. (2003). Full-field birefringence imaging by thermal-light polarization-sensitive optical coherence tomography II, Instrument and results *Appl. Opt.*, **42**, pp. 3811–3818.
40. Moneron, G., Boccara, A. C., and Dubois, A. (2007). Polarization-sensitive full-field optical coherence tomography. *Opt. Lett.*, **32**, pp. 2058–2060.
41. Dubois, A. (2012). Spectroscopic polarization-sensitive full-field optical coherence microscopy. *Opt. Express*, **20**, pp. 9962–9977.
42. Heise, B., Schausberger, S. E., Buchroithner, B., Aigner, M., Milosavljevic, I., Hierzenberger, P., Bernstein, S., Häuser, S., Jesacher, A., Bernet, S., Ritsch-Marte, M., and Stifter, D. (2013). Full field optical coherence microscopy for material testing: contrast enhancement and dynamic process monitoring. *Proc. SPIE*, **8792**, pp. 87921A (1–6).
43. Heise, B., Buchroithner, B., Schausberger, S. E., Hierzenberger, P., Eder, G., and Stifter, D. (2014). Simultaneous detection of optical retardation and axis orientation by polarization-sensitive full-field optical coherence microscopy for material testing *Laser Phys. Lett.,* **11,** p. 055602.
44. Park, K. S., Choi W. J., Eom T. J., and Lee B. H. (2013). Single-camera polarization-sensitive full-field optical coherence tomography with polarization switch. *J. Biomed. Opt.*, **18**, p. 100504.
45. Ugryumova, N., Gangnus, S. V., and Matcher, S. J. (2006). Three-dimensional optic axis determination using variable-incidence-angle polarization-optical coherence tomography. *Opt. Lett.*, **31**, pp. 2305–2307.
46. Guo, S., Zhang, J., Wang, L., Nelson, J. S., and Chen, Z. (2004). Depth-resolved birefringence and differential optical axis orientation measurements with fiber-based polarization-sensitive optical coherence tomography. *Opt. Lett.*, **29**, pp. 2025–2027.
47. Dubois, A., Moreau, J., and Boccara, A. C. (2008). Spectroscopic ultrahigh-resolution full-field optical coherence microscopy. *Opt. Express*, **16**, pp. 17082–17091.
48. Latour, G., Moreau, J., Elias, M., and Frigerio, J. M. (2010). Microspectrometry in the visible range with full-field optical coherence tomography for single absorbing layers *Opt. Commun.*, **283**, pp. 4810–4815.
49. Vabre, L., Dubois, A., and Boccara, A. C. (2002). Thermal-light full-field optical coherence tomography. *Opt. Lett.*, **27**, pp. 530–532.
50. Dubois, A., Grieve, K., Moneron, G., Lecaque, R., Vabre, L., and Boccara, A. C. (2004). Ultrahigh-resolution full-field optical coherence tomography. *Appl. Opt.*, **43**, pp. 2874–2882.

51. Larkin, K. G. (1996). Efficient nonlinear algorithm for envelope detection in white light interferometry. *J. Opt. Soc. Am. A*, **13**, pp. 832–843.
52. Hermann, B., Bizheva, K., Unterhuber, A., Považay, B., Sattmann, H., Schmetterer, L., Fercher, A., and Drexler, W. (2004). Precision of extracting absorption profiles from weakly scattering media with spectroscopic time-domain optical coherence tomography. *Opt. Express*, **12**, pp. 1677–1688.
53. Xu, C., Ye, J., Marks, D. L., and Boppart, S. A. (2004). Near-infrared dyes as contrast-enhancing agents for spectroscopic optical coherence tomography. *Opt. Lett.*, **29**, pp. 1647–1649.
54. Xu, C., Carney, P., and Boppart, S. (2005). Wavelength-dependent scattering in spectroscopic optical coherence tomography. *Opt. Express*, **13**, pp. 5450–5462.
55. Van de Hulst, H. C. (1957). *Light Scattering by Small Particles*, Wiley, New York.
56. Dubois, A., Selb, J., Vabre, L., and Boccara, A. C. (2000). Phase measurements with wide-aperture interferometers. *Appl. Opt.*, **39**, pp. 2326–2331.
57. Dubois, A. (2004). Effects of phase change on reflection in phase-measuring interference microscopy. *Appl. Opt.*, **43**, pp. 1503–1507.
58. Xu, C., Vinegoni C., Ralston, T. S., Luo, W., Tan, W., and Boppart, S. A. (2006). Spectroscopic spectral-domain optical coherence microscopy. *Opt. Lett.*, **31**, pp. 1079–1081.
59. Faber, D. J., Mik, E.G., Alders, M. C. G., and Van Leeuwen, T. G. (2003). Light absorption of (oxy-)hemoglobin assessed by spectroscopic optical coherence tomography. *Opt. Lett.*, **28**, 1437–1439.
60. Delacretaz, G. P., Steiner, R. W., Svaasand, L. O., Albrecht, H. J., and Meier, T. H. (1996). Epidermal melanin absorption in human skin, in laser-tissue interaction and tissue optics. *Proc. SPIE*, **2624**, pp. 143–154.
61. Spöler, F., Kray, S., Grychtol, P., Hermes, B., Bornemann, J., Först, M., and Kurz, H. (2007). Simultaneous dual-band ultra-high resolution optical coherence tomography. *Opt. Express*, **15**, pp. 10832–10842.
62. Sacchet, D., Moreau, J., Georges, P., and Dubois, A. (2008). Simultaneous dual-band ultrahigh-resolution full-field optical coherence tomography. *Opt. Express*, **16**, pp. 19434–19446.
63. Yu, L. F., and Kim, M. K. (2004). Full-color three-dimensional microscopy by wide-field optical coherence tomography. *Opt. Express*, **12**, pp. 6632–6641.

64. Federici, A., and Dubois, A. (2014). Three-band 1.9 micron-axial resolution full-field optical coherence microscopy over a 580–1600 nm wavelength range using a single camera. *Opt. Lett.*, **39**, pp. 1374–1377.
65. Oh, W. Y., Bouma, B. E., Iftimia, N., Yun, S. H., Yelin, R., and Tearney, G. J. (2006). Ultrahigh-resolution full-field optical coherence microscopy using InGaAs camera. *Opt. Express*, **14**, pp. 726–735.
66. Dubois, A., Moneron, G., and Boccara, A. C. (2006). Thermal-light full-field optical coherence tomography in the 1.2 micron wavelength region. *Opt. Commun.*, **266**, pp. 738–743.
67. Dubois, A., Moneron, G., Grieve, K., and Boccara, A. C. (2004). Three-dimensional cellular-level imaging using full-field optical coherence tomography. *Phys. Med. Biol.*, **49**, pp. 1227–1234.
68. Parsa, P., Jacques, S. L., and Nishioka, N. S. (1989). Optical properties of rat liver between 350 and 2200 nm. *Appl. Opt.*, **28**, pp. 2325–2330.
69. Hale, G. M., and Querry, M. R. (1973). Optical constants of water in the 200 nm – 200 µm wavelength region. *Appl. Opt.*, **12**, pp. 555–563.
70. Flock, S. T., Jacques, S. L., Wilson, B. C., Star, W. M., and Van Gemert M. J. C. (1992). Optical properties of Intralipid: a phantom medium for light propagation studies. *Lasers Surg. Med.*, **12**, pp. 510–519.
71. Van Staveren, H. G., Moes, C. J. M., Van Marle, J., Prahl, S. A., and Van Gemert, M. J. C. (1991). Light scattering in Intralipid-10% in the wavelength range of 400–1100 nanometers. *Appl. Opt.*, **30**, pp. 4507–4515.
72. Labiau, S., David, G., Gigan, S., and Boccara, A. C. (2009). Defocus test and defocus correction in full-field optical coherence tomography. *Opt. Lett.*, **34**, pp. 1576–1578.
73. Malitson, H. (1965). Interspecimen comparison of the refractive index of fused silica. *J. Opt. Soc. Am.*, **55**, pp. 1205–1209
74. Dubois, A., Vabre, L., Boccara, A. C., and Beaurepaire, E. (2002). High-resolution full-field optical coherence tomography with a Linnik microscope. *Appl. Opt.*, **11**, pp. 805–812.
75. Liang, H., Lange, R., Peric, B., and Spring, M. (2013). Optimum spectral window for imaging of art with optical coherence tomography. *Appl. Phys. B*, **111**, pp. 589–602.
76. Simpson, C. R., Kohl, M., Essenpreis, M., and Cope, M. (1998). Near-infrared optical properties of ex vivo human skin and subcutaneous tissues measured using the Monte Carlo inversion technique. *Phys. Med. Biol.*, **43**, pp. 2465–2478.

77. Dunkers, J., Cicerone, M., and Washburn, N. (2003). Collinear optical coherence and confocal fluorescence microscopies for tissue engineering. *Opt. Express*, **11**, pp. 3074–3079.
78. Makhlouf, H. Rouse, A. R., and Gmitro, A. F. (2010). A dual modality fluorescence confocal and optical coherence tomography microendoscope. *Proc. SPIE*, **7558**, Endoscopic Microscopy V.
79. Beaurepaire, E., Moreaux, L., Amblard, F., and Mertz, J. (1999). Combined scanning optical coherence and two-photon-excited fluorescence microscopy. *Opt. Lett.*, **24**, pp. 969–971.
80. Tang S., Krasieva, T. B., Chen, Z., and Tromberg, B. J (2006). Combined multiphoton microscopy and optical coherence tomography using a 12-fs broadband source. *J. Biomed. Opt.* **11**, p. 020502.
81. Neil, M. A. A., Juskaitis, R., and Wilson, T. (1997). Method of obtaining optical sectioning by using structured light in a conventional microscope. *Opt. Lett.*, **22**, pp. 1905–1907.
82. Makhlouf, H., Perronet, K., Dupuis, G., Levêque-Fort, S., and Dubois, A. (2012). Simultaneous optically sectioned fluorescence and optical coherence microscopy with full-field illumination. *Opt. Lett.*, **37**, pp. 1613–1615.
83. Auksorius, E., Bromberg, Y., Motiejūnaitė, R., Pieretti, A., Liu, L., Coron, E, Aranda, J., Goldstein, A. M., Bouma, B. E., Kazlauskas, A., and Tearney, G. J. (2012). Dual-modality fluorescence and full-field optical coherence microscopy for biomedical imaging applications. *Biomed. Opt. Express*, **3**, pp. 661–666.
84. Harms, F., Dalimier, E., Vermeulen, P., Fragola, A., and Boccara, A. C. (2012). Multimodal full-field optical coherence tomography on biological tissue: toward all optical digital pathology. *Proc. SPIE*, **8216**, Multimodal Biomedical Imaging VII, 821609.
85. Taroni, P., Pifferi, A., Torricelli, A., Comelli, D., and Cubeddu, R. (2003). In vivo absorption and scattering spectroscopy of biological tissues. *Photochem. Photobiol. Sci.*, **2**, pp. 124–129.

Chapter 14

Spectroscopic Full-Field Optical Coherence Tomography

Julien Moreau
Laboratoire Charles Fabry, Institut d'Optique Graduate School, Palaiseau, France
julien.moreau@institutoptique.fr

Full-field optical coherence tomography (FF-OCT) is based on a Michelson interferometer illuminated by a spatially and temporally incoherent source. Acquisition with a digital camera of transverse (en face) tomographic images at the output of the interferometer allows 3D imaging of a sample, including ex vivo biological tissues. Ultrahigh resolution in the three dimensions as low as 1 μm can be achieved by adding microscope objectives in the arms of the interferometer. Spectroscopic FF-OCT is an extension of this technique where cross-sectional tomographic imaging and spectroscopic data are acquired simultaneously. In this chapter, we will show how information on the spectral content of backscattered light can be obtained by processing the OCT interferometric signal as it is classically done in a Fourier transform spectrometer, but with the added benefit of keeping the ultrahigh spatial resolution in the tomographic images. Influence of some key data processing parameters is simulated and discussed. Some experimental spectra reconstruction from tomographic data will be shown.

Handbook of Full-Field Optical Coherence Microscopy: Technology and Applications
Edited by Arnaud Dubois
Copyright © 2016 Pan Stanford Publishing Pte. Ltd.
ISBN 978-981-4669-16-0 (Hardcover), 978-981-4669-17-7 (eBook)
www.panstanford.com

14.1 FF-OCT Principle

In this section, we will give a brief mathematical description of the type of signal obtained with full-field optical coherence tomography (FF-OCT) and the different key parameters of this technique.

In a typical FF-OCT system [1, 2] such as the one shown in Fig. 14.1, the incident light coming from the illumination source is split into two arms and reflected by a reference mirror and by the studied sample. These beams are recombined on the beam splitter and the interference pattern is recorded by a digital camera.

It is straightforward to show that for a Michelson interferometer illuminated by a monochromatic light source at a frequency ν, the intensity (the square modulus of the complex electrical field) in the transverse-oriented (xy) image, corresponding to a given pixel (m, n)

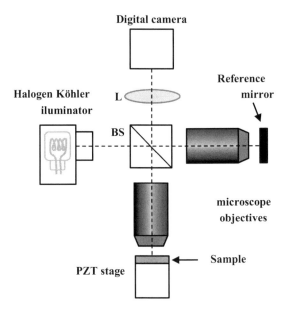

Figure 14.1 Experimental arrangement of a full-field optical coherence tomographic system. BS, beam splitter (broadband, nonpolarizing); microscope objectives (water immersion); L, achromatic doublet lens. The object is translated in the axial direction using a piezoelectric transducer (PZT) translation stage. The reference mirror has a reflectivity of $\sim 2\%$.

of the digital camera is given by

$$I_{m,n}(\nu) = I_0(\nu)T(\nu)S(\nu)\left(R_{\text{ref}}(x_0, y_0) + R_s(x_0, y_0)\right.$$
$$\left. +2\sqrt{R_{\text{ref}}(x_0, y_0)\, R_s(x_0, y_0)}.\cos\left(\frac{2\pi}{c}\delta\right)\right), \quad (14.1)$$

where $R_{\text{ref}}(x_0,y_0)$ and $R_s(x_0,y_0)$ are, respectively, the reflectivity of the reference mirror and of the object at the coordinates (x_0,y_0) conjugated with the pixel (m, n), $I_0(\nu)$ is the monochromatic intensity of the source at the entrance of the interferometer, $T(\nu)$ in the transmission supposedly identical in each arm, $S(\nu)$ is the spectral sensitivity of the digital camera, and δ is the optical path difference between both arms. The presence of microscope objectives does not significantly change this result as long as there are similar in the two arms and with low numerical aperture, which is the case in FF-OCT.

FF-OCT, to achieve a high spatial resolution, must use a broadband light source such as a tungsten halogen lamp which shows a continuous spectrum over the whole visible and near infrared (very close in fact to a black-body emission at 3500 K). Such a broadband source can be considered as a superposition of monochromatic sources, incoherent between themselves. In other word, the output intensity for a broadband illumination is simply the sum of the previous monochromatic signal over the whole spectral transfer function of the setup.

$$I_{m,n}(\delta) = \int I_{m,n}(\nu).dV. \quad (14.2)$$

Figure 14.2 shows an example [3] of such a normalized interferogram, at a given point on a homogenous flat sample, for an experimental spectral transfer function with a full width at half maximum (FWHM) of 120 nm, typical of what we have with a halogen source and a charge-coupled device (CCD) detector. As we can expect with such an incoherent source, interference only occurs when the optical path difference between the two arms of the interferometer is very close to zero. In fact, it can be shown that the FWHM of the interferogram is on a first approximation, inversely proportional to the width of the spectral transfer function. A very broad source and

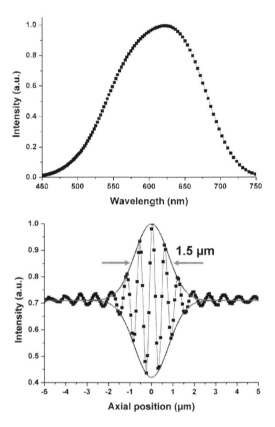

Figure 14.2 (Top) Typical experimental transfers function in an FF-OCT working with a halogen lamp and a CCD detector. (Bottom) Interferogram recorded from a mirror in water and its envelope extracted using the Larkin algorithm. An FWHM of 1.5 μm is measured. From Ref. [3], Copyright © 2010 Elsevier.

detector and therefore a highly localized interferogram are the key to the ultrahigh resolution obtained in FF-OCT.

For now, we will make the hypothesis that the reflectivity of the reference mirror and of the object is independent of wavelength. This leads to the following expression for the interferogram:

$$I_{m,n}(\delta) = I_{m,n}(\infty) + \sqrt{R_{\text{ref}}.R_{\text{s}}}f(\delta). \quad (14.3)$$

The first term $I_{m,n}(\infty)$ is of no interest at is simply the signal without any interference effect ($\delta = \infty$). The second term, however, contains the tomographic information R_s and strongly depends on the optical path difference δ. Many methods have been proposed in the literature to extract the term $\sqrt{R_{ref}.R_s}$. The general idea is, for a given position of the sample, to acquire several images at different δ by moving the reference mirror. Some combination of these images allows removing the constant term and extracting the tomographic information. One of the most robust algorithms is a five-frame nonlinear phase shifting, known as the Larkin algorithm [4], which has the following expression:

$$\sqrt{R_{ref}.R_s} \propto \frac{1}{4}\left[4.(I_2 - I_4)^2 + (-I_1 + 2I_3 - I_5)^2\right]^{1/2}, \quad (14.4)$$

where $I_{1...5}$ are image of the sample taken for different positions of the reference mirror corresponding to the following path difference: $\delta_{1...5} = \{-\frac{\lambda}{2}; -\frac{\lambda}{4}; 0; \frac{\lambda}{4}; \frac{\lambda}{2}\}$. This corresponds to an axial translation step for the piezoelectric transducer behind the reference mirror of around 35 nm between each frame.

The envelope of the interferogram shown in Fig. 14.2 is an example of such a reconstruction using the Larkin algorithm. It is obtained with almost no residual oscillation and its amplitude is proportional to the local reflectivity $\sqrt{R_{ref}(x, y)}$ of the sample. By doing this for different positions z of the sample, a 3D imaging of the sample is realized. The axial resolution (z) in these tomographic images is limited by the FWHM of the interferogram and therefore the spectral transfer function of the system. The lateral resolution depends only on the numerical aperture of the microscope objective and on the sampling of the image by the CCD pixels. It is usually better to choose the numerical aperture of the objective such that the lateral resolution in the images is more or less equal to the axial resolution. Also, for samples consisting mainly of water, it is preferable to use identical water immersion microscope objectives in the two interferometer arms in order to minimize dispersion mismatch introduced by refractive index heterogeneities. The axial resolution degradation is thus minimized when the imaging depth increases.

Figure 14.3 shows an example of the kind of tomographic image that can be obtained with an FF-OCT system. Here, a varnished layer of Viridian green pigments before restoration of a painting. The

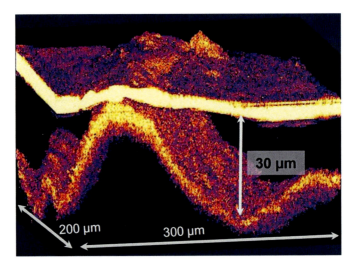

Figure 14.3 Three-dimensional image (false colors) obtained by FF-OCT of a varnished viridian green painting with physical dimensions.

upper interface corresponds to the air/varnish layer and the second one to the interface varnish/paint.

14.2 Spectroscopic FF-OCT

FF-OCT and more generally OCT systems give a 3D image of a sample where the contrast in the image is related to a local change of refractive index. However, in a complex sample, a variation of the optical index can have many causes: a higher density of materials (such as the inside of a cell in a biological tissue), a change in the nature of the material, like in multilayers sample, or even a local modification of the absorption or scattering properties inside the sample. OCT, by itself, does not allow distinguishing between these causes. It only gives a structural image with no direct information to the underlying composition. It was therefore tempting to combine OCT with other imaging modality in order to have access to additional information on the sample composition. In particular, developments in OCT technology have been carried out in order to exploit the spectroscopic response of the imaged sample

[5, 6]. This technique, referred to as spectroscopic OCT, detects and processes the interferometric signal to provide spatially resolved spectroscopic information. It can be used to enhance image contrast, permitting better differentiation through spectroscopic properties.

In particular, adding this spectroscopic modality to FF-OCT seems especially well adapted as this technique relies on the use of a broadband visible source to illuminate the sample. Going back to the intensity at the output of the FF-OCT system, if we no longer consider that the local spectral reflectivity R_s of the sample is flat and independent of wavelength but has a certain spectral signature due to absorption and scattering, then Eq. 14.2 can be written as

$$I_{m,n}(\delta) = \int (R_{\text{ref}} + R_s)\,T.S.I_0 dv + 2\,\text{Re}\left[\int \sqrt{R_{\text{ref}} R_s}.T.S.I_0 e^{\frac{2\pi v}{c}\delta} dv\right] \quad (14.5)$$

or

$$I_{m,n}(\delta) = \int (R_{\text{ref}} + R_s).T.S.I_0 dv + 2\,\text{Re}\left[\text{FT}\left(\sqrt{R_{\text{ref}} R_s}.T.S.I_0\right)\right], \quad (14.6)$$

where FT is the spectral Fourier transformation. The explicit dependency on x, y, and v of the different quantities have been removed for sake of simplicity.

It follows that the spectral reflectivity $R_s(\lambda)$ of the sample can be extracted from the interferometric data by simply doing an inverse FT of the measured interferogram. This type of signal processing is what is classically achieved in any FT spectrometer, albeit here, this operation can be done on every pixel (m, n) of the camera, therefore combining the imaging capability with the spectrometric modality. We then have

$$R_s(\lambda) = \frac{1}{4 R_{\text{ref}}(\lambda) T(\lambda)^2 S(\lambda)^2 I_0(\lambda)^2} \left|\text{FT}^{-1}\left(I_{m,n}(\delta) - I_{m,n}(\infty)\right)\right|^2. \quad (14.7)$$

The normalization term $R_{\text{ref}}(\lambda) S(\lambda)^2 T(\lambda)^2 I(\lambda)^2$ contains all the spectral characteristics of the setup, mainly the spectrum of the source, the transmission of the various optical elements, and the reflectivity of the reference mirror. This term can be easily measured at the beginning of an experiment simply by putting a glass slide in the sample arm and acquiring the interferogram of the air/glass interface (or water/glass interface in the case of a water immersion objective). By doing this inverse FT for each interferogram recorded by all the pixel of the camera, a spectral reflectivity map at a given depth inside the sample can be reconstructed. It is important to note here that $R_s(\lambda)$ will contain all the absorption and scattering processes undergo by the light between the surface of the sample and the imaging depth z. In order to do achieve such a reconstruction of the local spectral characteristics, a sampling of the interferograms well below the Nyquist limit is needed. This is done by moving the sample in very small steps and acquiring an image at each position. In this configuration, the reference mirror can stay fixed as this stack of images can also be used to reconstruct the envelope of the interferogram. Numerically, the signal processing is based on the fast Fourier transform (FFT) algorithm of the interferogram recorded by each pixel of the CCD.

To illustrate the effectiveness of this technique, the spectrum of a colored glass filter (BG36 from Schott) obtained with a spectroscopic full field OCT system and the algorithm based on Eq. 14.7, is shown in Fig. 14.4. In this very simple experiment, the incident light of the halogen lamp passes through the glass filter and is reflected by identical mirrors in both arms of the interferometer. Due to the very high signal-to-noise ratio (SNR) in this measurement, the reconstructed spectrum is in excellent agreement with the transmittance measured independently by a classical spectrometer. Only at longer wavelengths (above 700 nm), can we notice a degradation of the reconstruction, mainly due to the use of an apodization window to limit the set of data on which FFT is applied The different size of Hanning windows (12 µm and 24 µm total widths compared to the 25 µm width interferogram) leads to the different spectra shown in Fig. 14.4.

The partial loss of spectral information at the longest wavelengths and no reconstruction of the very fine variations in the sample spectra are well-known effects of apodization functions on FT [7]. In practice the choice of the size of the apodization window is always a compromise as taking a wide apodization window or no window at all, introduces noise in FFT and strongly degrades the reconstructed spectrum.

Figure 14.5 shows some numerical examples of the robustness of the spectrum reconstruction toward noise, using an optimized

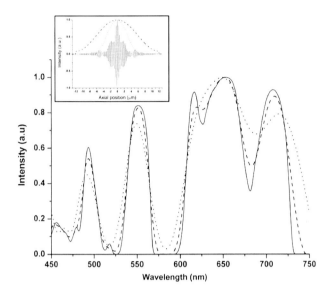

Figure 14.4 Comparison of the transmittance spectrum of the BG36 filter (solid black line) with the spectral information calculated by FFT from OCT data. Reconstructed spectrum is shown for a 24 µm (dashed line) and a 12 µm (dotted line) width Hanning window. (Insert) Interferometric signal, before apodization, with both Hanning windows. From Ref. [3], Copyright © 2010 Elsevier.

apodization window. To give a rough estimate, in order to reconstruct simple spectral features, such as the difference between a red and a green dye for example, an SNR for the acquired interferograms of at least 10 is necessary. For a complex spectral signature with thin spectral line, an SNR of 100 is probably needed. It is also necessary to increase the size of the interferogram to 4096 points by zero padding in order to reduce the spectrum sampling step and obtain a good spectral resolution. With 4096 points typically, the spectrum sampling step in the visible is between 2 and 4.5 nm after FFT.

Finally, to illustrate the principle of microspectrometry coupled with 3D imaging, a liquid crystal display (LCD) pixel made of three different organic dyes was imaged with a spectroscopic full field OCT system. Figure 14.6 shows a classical en face microscopic image of the surface of the pixel where the RGB dyes appear as thin bands, around 100 µm wide, on a glass substrate.

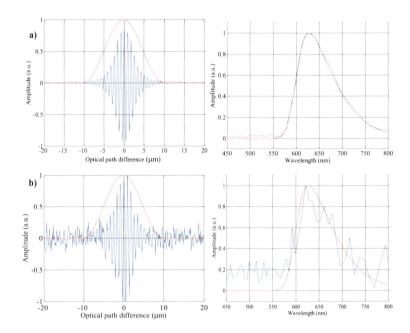

Figure 14.5 Example of numerical interferograms (left) with an optimum apodization window (dashed red line) and the reconstructed spectra (right) obtained by FFT (full line). The theoretical spectrum, corresponding to a red absorbing thick layer, is also shown (dashed red line). A Gaussian noise with a relative intensity of 1% (a) and 10% (b) was added to the ideal interferograms to show the impact on the reconstructed spectrum.

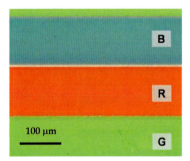

Figure 14.6 Image of an LCD pixel taken under an optical microscope. The blue, red, and green dyes that constitute the pixel are, respectively, label B, R, and G. From Ref. [3], Copyright © 2010 Elsevier.

(a)

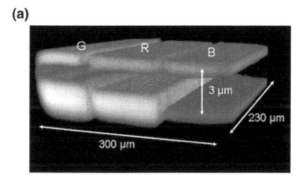

(b)

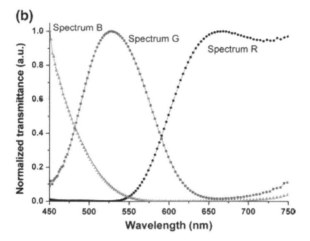

Figure 14.7 (a) A 3D ultrahigh tomographic view (300 × 230 × 10 μm stack) of the three bands of dyes (respectively, G, green; R, red; and B, blue) and (b) normalized transmittance spectra calculated from OCT data for the three different layers (light gray for the B spectrum, gray for the G spectrum, and black for the R spectrum). From Ref. [3], Copyright © 2010 Elsevier.

Tomographic image and spectroscopic information were extracted from the acquired set of data using the previously described method. Figure 14.7 shows both the 3D tomographic view of the LCD pixel using Linnick algorithm and the normalized reflectivity spectrum extracted by FFT. The two interfaces, first the one at the top between air and dyes and then the second interface, between the dyes and the glass, can be clearly dissociated even if the dyes are only 3 μm thick. This is made possible by the excellent axial resolution of FF-OCT. It can be noticed that the second interface appears much broader than the upper one for the green and red dye. This is explained by the fact that the first tomographic signal on the air–dye interface is obtained with the whole white light illumination spectrum of the setup which corresponds to the optimum 1.5 μm axial resolution, whereas for the deepest tomographic signal, light has been partially absorbed by the dye before reaching the interface, thus reducing the effective illumination spectrum and the axial resolution.

The normalized spectra extracted from FFT of the interferograms are presented in Fig. 14.7b. These spectra are exactly equivalent to the transmittance of the dyes. Indeed, the incident light from the setup is transmitted and partially absorbed by the given dye before being reflected by the substrate interface, crossing a second time the dye and finally being collected by the microscope objective. The normalization is realized, as previously described, by taking into account the spectral transfer function of the setup. No significant difference between calculated spectra, for a given dye, across the bottom interface was measured. It can be seen that the spectrum of the dye "R" is typical of a red dye with a total absorption under 550 nm and a maximum of transmission for longer wavelengths. On the opposite, the dye "B" has its maximum of transmission well below 500 nm, giving an effective blue color. Finally, the "G" spectrum presents a maximum around 530 nm, as expected for a green dye. These three spectra, obtained simultaneously with tomographic data, correspond well to the expected transmittance spectra for an RGB-colored display and were obtained on $\sim$100 μm^2 areas, too small to be measured with traditional spectrometers.

14.3 Limitations of Spectroscopic FF-OCT

The first obvious main limitation of spectroscopic FF-OCT is the large number of images that must be acquired in order to do FFT analysis. Hundreds or even thousands of en face images can be needed for a full spectroscopic tomographic analysis of a 3D sample. This almost forbids any in vivo applications as any displacement of the sample during acquisition will degrade the interferometric OCT signal. Also, as shown above, to accurately reconstruct the complete spectrum of the backscattered light, a high SNR in the interferograms is necessary. This can be difficult to obtain in biological tissues where the index contrasts inside the sample and therefore the amount of backscattered light that will be collected by the OCT system, are usually very small. Another limitation which is not specific to the full field technique but common to all spectroscopic OCT systems comes from the origin of the spectroscopic signal. In particular, in term of acquired data, one must distinguish the effect of absorption and scattering on the spectral shape of the light transmitted or reflected by the sample. The simplest case from a theoretical point of view is a purely absorbing sample where, as expected by Beer's law, the intensity of the spectral lines in the backscattered signal will be proportional to the length of the absorber region as well as the concentration and cross section of the absorber. In practice, the quality of the spectroscopic data that can be acquired in absorbing samples will usually be limited by the penetration depth in OCT images, typically 1 mm at most for highly transparent biological tissues and less than a few tens of microns in absorbing inorganic materials. These small optical paths will limit the detection of spectral signatures. On the other hand, scattering which also change the spectrum of the reflected light, is a much more complex problem. There is no simple relationship between the physical properties of an object and its backscattered spectrum, except in some very specific cases. Interpretation of these scattering spectra, measured with spectroscopic OCT as with any other optical technique is therefore very difficult.

References

1. Vabre, L., Dubois, A., and Boccara, A. C. (2002). Thermal-light full-field optical coherence tomography. *Opt. Lett.*, **27**, pp. 530–533.
2. Dubois, A., Grieve, K., Moneron, G., Lecaque, R., Vabre, L., and Boccara, A. C. (2004). Ultrahigh-resolution full-field optical coherence tomography. *Appl. Opt.*, **43**, pp. 2874–2883.

3. Latour, G., Moreau, J., Elias, M., and Frigerio, J. M. (2010). Microspectrometry in the visible range with full-field optical coherence tomography for single absorbing layers. *Opt. Commun.*, **283**, pp. 4810–4815.

4. Larkin, K. G. (1996). Efficient nonlinear algorithm for envelope detection in white light interferometry. *J. Opt. Soc. Am. A*, **13**, pp. 832–843.

5. Morgner, U., Drexler, W., Kärtner, F. X., Li, X. D., Pitris, C., Ippen, E. P., and Fujimoto, J. G. (2000). Spectroscopic optical coherence tomography. *Opt. Lett.*, **25**, pp. 111–113.

6. Dubois, A., Moreau, J., and Boccara, A. C. (2008). Spectroscopic ultrahigh-resolution full-field optical coherence microscopy. *Opt. Express*, **16**(21), pp. 17082–17091.

7. Bracewell, R. N. (2000). *The Fourier Transform and Its Applications*. McGraw Hill Higher Education, New York.

Chapter 15

Multiwavelength Full-Field Optical Coherence Tomography

Mariana C. Potcoava, Nilanthi Warnasooriya, Lingfeng Yu, and Myung K. Kim

Department of Physics, University of South Florida, 4202 East Fowler Ave, Tampa, FL 33620-5700, USA
mkkim@usf.edu, mariana@intelligent-imaging.com

The success of full-field optical coherence tomography (FF-OCT) or full-field optical coherence microscopy (FF-OCM) in imaging biomedical tissue with cellular-resolution and high-sensitivity scores is already recognized in the biomedical applications. Various techniques and applications have been developed to take advantage of the FF-OCT characteristics, based on low-coherence interferometry of a light source along with phase shifting. In this chapter we review the basic principle of FF-OCT and its extension in multiwavelength FF-OCT for phase measurement and full-color imaging.

15.1 Introduction

Full-field optical coherence instruments utilize the principle of single-shot interferometry with low-coherence sources. These kind

Handbook of Full-Field Optical Coherence Microscopy: Technology and Applications
Edited by Arnaud Dubois
Copyright © 2016 Pan Stanford Publishing Pte. Ltd.
ISBN 978-981-4669-16-0 (Hardcover), 978-981-4669-17-7 (eBook)
www.panstanford.com

of sources have been used in interferometry since Michelson and Morley [1] conducted interferometry experiments, for the first time, to determine that electromagnetic waves propagate in vacuum. Interferometers are phase-sensitive instruments and the phase difference between the two interferometer arms determines the object structure. The advantage of using low-coherence sources is that it confers optical sectioning capability, while suppressing the parasitic interferences and imaging parts of the object that are not within the coherent length. Examples of low-coherence sources are tungsten lamp, superluminescent diode (SLD) or light-emitting diodes (LEDs). Optical coherence tomography (OCT) is a noncontact, raster-scanning, noninvasive 3D optical imaging technique that also uses a low-coherence source to record cross-sectional images of a sample under study with high lateral and axial resolution (close to 1 μm). The OCT technique was first introduced and developed by J. Fujimoto's group at the MIT Lincoln Laboratory in the early 1990s [2, 3] and A. F. Fercher [4]. Since then, OCT has already applied in many clinical and diagnosis techniques, especially in studying the human retina and macula [2, 3, 5–7]. The most basic form of OCT, time domain optical coherence tomography (TD-OCT), is based on the interference of low-coherence light in a Michelson interferometer. A mirror in the reference arm is mechanically moved over a distance equal to the sample depth range, in order to scan z axis (A-scan), creating constructive interference when the reference mirror is at the same distance as the object's reflecting surface. The distances need to match within the coherence length of the light, which therefore determines the axial resolution. Two additional scans, xy (B-scan), are needed to record volumetric data. This is achieved by raster scanning of the illumination spot along the sample area. TD-OCT provides the necessary resolution, but the main limitation is the mechanical z-scan of the reference mirror. In order to overcome the problem of z-scanning, new technologies in OCT include Fourier domain optical coherence tomography (FD-OCT) in two forms, spectral domain optical coherence tomography (SD-OCT), where the mechanical z-scanning of the TD-OCT is replaced with spectral analysis, and swept-source optical coherence tomography (SS-OCT), where the spectral analysis is replaced with wavelength scanning of the light source [8–11]. For wavelength scanning, a tunable laser can be used, or an acousto-optical tunable filter on a white light source. An axial resolution of 1–2 μm has been reported using a femtosecond laser [10]. Although both methods are similar in imaging characteristics, SD-OCT has a simple and rapid way of recording the spectroscopic information. Two terminologies are characterizing the z-scanning, the axial range and the axial resolution. In TD-OCT the axial scan range of the reference mirror and the coherence length of the light give the axial scan range and the axial resolution in TD-OCT, respectively. The depth information, in SD-OCT, is encoded in spectral frequency domain rather than time domain by using a fixed reference mirror and a broad-bandwidth light source. The axial range depends on the wavelength resolution of the spectrometer and the axial resolution depends on the full wavelength range of the spectrometer. Similar to conventional microscopy, the lateral resolution is mainly characterized by the numerical aperture (NA) of the objectives. More significantly, it is shown that the signal-to-noise ratio (SNR) of SD-OCT is better than TD-OCT by a factor $\sqrt{N}$, where N is the number of detector elements on the spectrometer [12]. In SS-OCT, a single frequency is recorded at a time by a single photodetector and not a spectrometer.

All OCT techniques have been extended recently to full-field (wide-field or en face) detection to take advantage of using a detector array for faster data acquisition [13–15].

In full-field optical coherence tomography (FF-OCT), the process of coherence detection is carried out in parallel of all the pixels. The optical setup is still a Michelson interferometer using a low-coherence light source and phase-shifting interference technique to suppress the background. One can buid a Linnik interferometer by adding microscope objectives in the object and the reference arms. The phase shifting can be achieved by dithering of the reference mirror.

In the original FF-OCT experiment by Boccara et al., use of Michelson objective precluded piezo dithering: instead they used polarization modulation using photoelastic modulator.

It is well known that interferographic images using broadband sources are significantly less affected by coherent noise. To achieve a height resolution of several nanometers and a range of several microns one needs to decrease the coherent noise. In this matter, a LED-based single-wavelength or multiwavelength phase imaging interference microscopy was developed that combines phase-shifting interferometry with single- [16] or multiwavelength optical phase unwrapping [17, 18].

Another optical method of optical tomography for surface and subsurface imaging, based on the principle of FF-OCT, using a multispectral scheme of phase shifting, and capable of providing full-color 3D views of a tissue structure was demonstrated in Ref. [19]. Contour or tomographic images were obtained with an interferometric imaging system using broadband light sources. The technique can provide critical information for the physiological and pathological applications.

In this book chapter, we review a few phase-related techniques using small variations from the FF-OCT principle, with great potential to become powerful imaging techniques.

15.2 Basic Single-Wavelength, Low-Coherence Interferography

The principle of low-coherence interferometry consists of detecting the maximum of the source coherence envelope or the zero-order fringes diffraction maximum. Usually, a low-coherence source has a coherence length of a few tens of microns. Low-coherence interferography was demonstrated in [16]. The experimental setup is shown in Fig. 15.1, in a Michelson interferometer with a lensless configuration in both the object and the reference arm. The beam from the LED is collimated by a low-numerical-aperture lens and illuminates the sample without passing through a microscope objective as in the Linnik configuration. A polarizer–analyzer pair also allows continuous variation of the relative intensity between the two arms. Inserting an additional quarter-wave plate in only the reference arm modulates the phase between vertical and horizontal polarizations. Reflections from a reference mirror and an object surface illuminated by low-coherence light results in interference fringes over a contour of the object at a distance equivalent to the reference surface, while the rest of the object maintains constant background. The images are acquired by a charge-coupled device

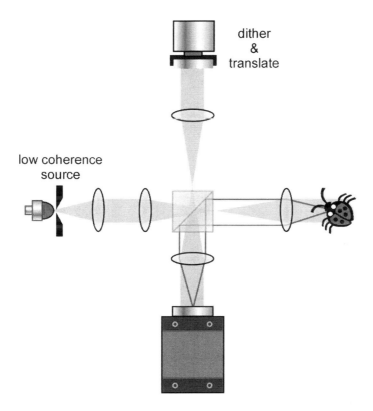

Figure 15.1 Low-coherence interferometry experimental setup.

(CCD). The envelope of the interference fringes can be extracted by a phase-shifting method, where the reference position is dithered at quarter-wavelength intervals [13, 20]. Each interferogram is of the form

$$I(x, y) = I_O(x, y) + I_B(x, y) + I_R(x, y)$$
$$+ 2\sqrt{I_O(x, y) I_R(x, y)} \cos[\varphi_i + \varphi(x, y)], \quad (15.1)$$

where $I_R(x, y)$ is the reference intensity and $I_B(x, y)$ and $I_O(x, y)$ are the incoherent and coherent parts of the object with respect to the reference; $\varphi(x, y)$ is the phase variation over the object surface due to its topography or the relative phase between the object and the reference mirror; and φ_i is the phase shift imposed by moving the mirror by quarter-wavelength intervals.

The intensity and phase map of the object field are given by

$$I_0(x, y) \propto \left[(I_0 - I_\pi)^2 + (I_{\pi/2} - I_{3\pi/2})^2\right], \quad (15.2)$$

$$\varphi(x, y) = \tan^{-1}\left[\frac{I_{3\pi/2} - I_{\pi/2}}{I_0 - I_\pi}\right]. \quad (15.3)$$

Alternatively, the interferograms can be acquired with the illumination turned on for a quarter period while the reference mirror is sinusoidally dithered. Then the interference envelope is given by

$$I_0 \propto \left[(I_0 - I_{\pi/2}) + (I_\pi - I_{3\pi/2})\right]^2 + \left[(I_0 - I_{\pi/2}) - (I_\pi - I_{3\pi/2})\right]^2. \quad (15.4)$$

The thickness of the interference envelope is given by the coherence length of the light source and the whole 3D object surface is imaged by scanning the object's z distance.

The concept of FF-OCT has been applied to 3D topography fingerprinting of plastic prints on Mikrosil and latent prints on crinkled aluminum foil using the LED-based low-coherence interferography described above.

Figure 15.2a is an example of an interferogram with a 15 mm × 12 mm field of view (FOV), where coherent interference speckles are barely visible. The phase-shifting method extracts a contour section of the surface that is within a coherence length with respect to the reference mirror, as shown in Fig. 15.2b. The object's z position is then scanned over a distance of 700 μm to accumulate the surface contours. For these experiments, instead of 3D volume image arrays, the image data consist of a 2D array of z values of maximum interference signals, that is, the z positions of the surface points for the 2D array of pixels. In Fig. 15.2c, therefore, the gray scale represents the z values of the surface points over the 700 μm range, which is rendered as a 3D surface in perspective in Fig. 15.2d. This fingerprint sample displays a level 1 whorl pattern, as well as many of the level 2 and level 3 features, including the bifurcation, ending, crossover, and pores. We also imaged latent prints on pieces of plastic packaging, Fig. 15.3a, and crinkled aluminum foil, Fig. 15.3b.

15.2.1 Digital Focusing in Low-Coherence Interferometry

Once the amplitude and phase information of the object are calculated, the principle of Fourier optics can be used to digitally

Basic Single-Wavelength, Low-Coherence Interferography | 539

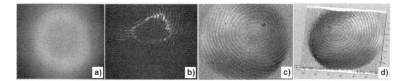

Figure 15.2 Topographic imaging of a plastic fingerprint by low-coherence interferography: (a) an interferogram as captured by camera, (b) a coherent interference envelope extracted by the phase-shifting method, (c) a two-dimensional map of the fingerprint surface topography with the gray scale representing the height of the surface at each pixel, and (d) a three-dimensional perspective rendering of (c). The image volume is 15 mm × 12 mm × 0.70 mm [16].

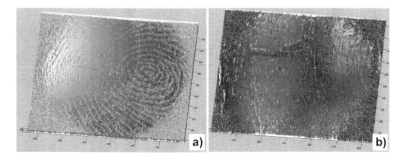

Figure 15.3 Topographic images of latent fingerprints on the surface of (a) a piece of blister-pack plastic packaging and (b) a piece of crinkled aluminum foil. The image volume in both cases is 15 mm × 12 mm × 0.70 mm [16].

focus the object wave field if it is originally out of focus. From Fourier optics [21], if $E(x, y; 0)$ is the en face wave field distribution at plane $z = 0$, the corresponding angular spectrum of the field at this plane can be obtained by taking the Fourier transform:

$$S(k_x, k_y; 0) = \iint E(x, y; 0) \exp[-i(k_x x + k_y y)] dx dy, \quad (15.5)$$

where k_x and k_y are corresponding spatial frequencies of x and y. The field $E(x, y; 0)$ can be rewritten as the inverse Fourier transform of its angular spectrum

$$E(x, y; 0) = \iint S(k_x, k_y; 0) \exp[i(k_x x + k_y y)] dk_x dk_y. \quad (15.6)$$

The complex exponential function $\exp[i(k_x x + k_y y)]$ may be regarded as a projection, onto the plane $z = 0$, of a wave propagating with a wave vector (k_x, k_y, k_z), where $k_z = [k^2 - k_x^2 - k_y^2]^{1/2}$ and $k = 2\pi/\lambda$. Thus, the field $E(x, y; 0)$ can be viewed as a superposition of many wave components propagating in different directions in space and with the complex amplitude of each component equal to $S(k_x, k_y; 0)$. After propagating along the z axis to a new plane, the new angular spectrum, $S(k_x, k_y; z)$, at plane z can be calculated from $S(k_x, k_y; 0)$ as $S(k_x, k_y; z) = S(k_x, k_y; 0) \exp[ik_z z]$. Thus, the complex field distribution of any plane perpendicular to the propagating z axis can be calculated from Fourier theory as

$$E(x, y; z) = \iint S(k_x, k_y; z) \exp[i(k_x x + k_y y)] dk_x dk_y. \quad (15.7)$$

The same principle of digital focusing in low-coherence interferometry is also applicable to fiber-based FD-OCT [22].

Figure 15.4 shows the flow diagram of the whole process for lateral resolution improvement based on an FD-OCT system. After the captured spectral interferogram in a line-scan camera is rescaled as evenly k-spaced, a fast Fourier transform is followed to achieve the complex A-line information along the z axis. The whole 3D complex volume of the sample is obtained by 2D scanning of a galvosystem. Since the sample is located outside the degree-of-freedom (DOF) range of the probe beam, the OCT system suffers greatly from lateral resolution degradation. Note here that the 3D out-of-focus complex volume needs to be resampled in the z direction to achieve a sequence of en face (xy) images $I(x, y; z_i)$. Then the angular spectrum $S(k_x, k_y; z_i)$ of each en face image was calculated by Eq. 15.5, and the new angular spectrum after propagating a distance z was calculated by multiplying $S(k_x, k_y; z_i)$ with a z-dependent exponential term as $\exp[ik_z z]$ with $k_z = [k^2 - k_x^2 - k_y^2]^{1/2}$. Finally, the en face field distribution at plane $(z_i + z)$ was calculated from Eq. 15.9. Thus, by selecting a correct reconstruction distance z, the proposed method can be used to numerically cancel the lateral defocus and improve the lateral resolution.

Basic Single-Wavelength, Low-Coherence Interferography | 541

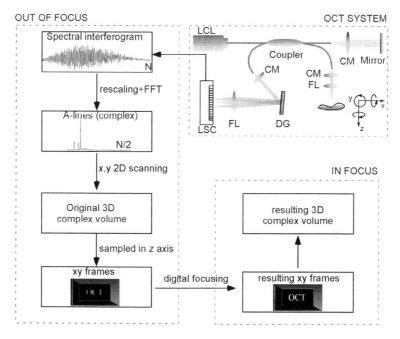

Figure 15.4 Flow diagram of the numerical focusing process (FL, focusing lens; DG, diffraction grating; CM, collimator; LSC, line-scan camera; LCL, low-coherent laser) [22].

Experiments were performed to verify the effectiveness of the proposed idea [22]. The DOF of the system was determined by the diameter (4.8 mm) of the probe beam and the focal length (40 mm) of the objective in the system which turned out to be ~230 μm in air. The lateral resolution was ~13.8 μm. For the purpose of comparison, Figs. 15.5a and 15.5b first show two xz B-scan images and one xy en face image of a focused onion, respectively. Then an onion sample placed about 1.2 mm away from the DOF region was studied. Figure 15.5c shows two xz B-scan images of the defocused onion and Fig. 15.5d shows the en face amplitude images of a layer ~300 μm underneath the onion surface. Since the onion was placed outside the DOF, the probe Gaussian beam was expanded to illuminate a larger area and multiple scatterers in the lateral directions contributed to a single A-line detection. This resulted in a great degradation of the lateral resolution, as is clearly shown in

542 | Multiwavelength Full-Field Optical Coherence Tomography

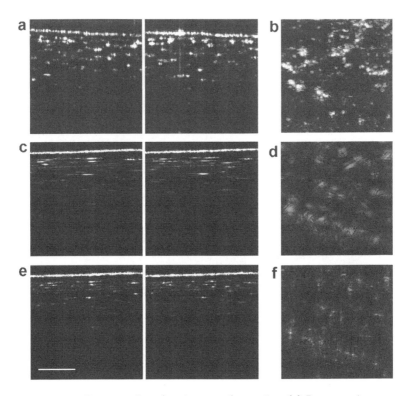

Figure 15.5 B-scan and en face images of an onion: (a) B-scan xz images and (b) en face xy image of a focused onion; (c) B-scan xz images and (d) en face amplitude image of a defocused onion. (e) B-scan images after digital focusing, obtained at the same cross-sectional position as (c). (f) Digitally focused en face image from (d). The scale bar represents 0.5 mm [22].

Fig. 15.5c and 15.5d. Figure 15.5e shows the digitally focused B-scan images at the same cross-sectional positions, as in Fig. 15.5c, and the corrected en face image of Fig. 15.5d is shown in Fig. 15.5f where the scatterers are now greatly sharpened to small bright points. Note that only those scatterers whose centers are located exactly on the observed B-scan image will shrink to bright points, but those scatterers centered on neighboring B-scan images tend to be eliminated or removed.

Figure 15.6 shows another example of onion cells. An objective of 10 mm focal length (20×, Nachet) was used to improve the lateral

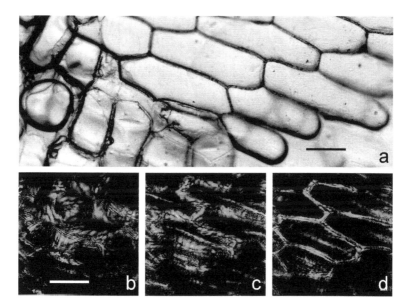

Figure 15.6 En face (xy) images of a defocused onion. (a) Onion cells under the microscope; (b) a defocused onion layer ∼240 μm away from the focal plane. (c and d) Reconstructed images with different focusing distances. The scale bar represents 100 μm [22].

resolution of the system to ∼3.5 μm, but the DOF was narrowed to 14.4 μm. Figure 15.6a shows onion cells under a microscope. Figure 15.6b shows the defocused en face OCT image of a layer ∼240 μm away from the focal plane; Fig. 15.6c shows the reconstructed image when the layer is partially focused, and the digitally focused en face image is finally shown in Fig. 15.6d. The cells are now well in focus and the boundaries in both lateral directions are greatly sharpened.

The above experiment suggests that the wave scattering process from out-of-focus scatterers in OCT can be considered as a 2D scalar diffraction model. The lateral resolution in either x or y direction can be improved by use of a noniterative numerical diffraction algorithm, and high-resolution details can be reconstructed from outside the depth-of-field region without any special hardware in system design.

15.3 Multiwavelength, Optical-Phase-Unwrapping, Low-Coherence Interferography

When obtaining a phase image using a series of interferograms, the use of the mathematical arctangent function, as seen in Eq. 15.3 results 2π ambiguities in the final phase image. Removing these ambiguities is essential to obtain a continuous surface profile of the test object. The process of removing 2π ambiguities in a phase map is called phase unwrapping. Generally, phase unwrapping computationally intensive algorithms can fail in the presence of noise and undersampling [23, 24]. Other drawbacks of these algorithms are that they require long processing times, and some algorithms are specific to certain applications [25]. The principle of multiwavelength phase unwrapping in the context of phase-shifting interferometry was first introduced by Y. Y. Cheng and J. C. Wyant in 1985 [26, 27]. However, known applications have been confined to optical profilers with raster-scanned pointwise interferometry. Recently, multiwave phase unwrapping has been applied to full-frame phase images [17–19, 28]. In this section, we review the technique of multiwavelength optical phase unwrapping using phase-shifting interference microscopy and LEDs at two and three different wavelengths.

The experimental setup is a standard Michelson interferometer as shown in Fig. 15.7. A set of LEDs (Luxeon™ Emitter diodes, Lumileds Lighting LLC) has been used as the light source. The micrometer range coherence wavelengths of LEDs significantly reduce the speckle noise inherent to coherent light sources such as lasers [29, 30]. Interference of coherent waves produces speckle noise, which limits the phase map's information. Ease of use, cost effectiveness, and reduced apparatus dimensions are added advantages of using LEDs.

The light from LEDs is collimated by lens L1 and passed through the polarizer P. The polarizing beam splitter (PBS) splits the incoming beam into two orthogonally polarized beams, S and P. The S-polarized beam is reflected at the beam splitter (BS) to illuminate the sample object, OBJ, and the P-polarized beam is transmitted through the PBS to illuminate the reference mirror, REF. Two quarter-wave plates, Q1 and Q2, in the object and reference

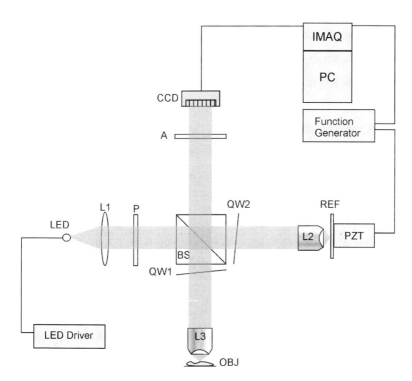

Figure 15.7 Experimental setup for low-coherence phase-shifting interferometry. L1, collimating lens; L2, L3, 20X microscope objectives; P, polarizer; BS, polarizing beam splitter; QW1, QW2, quarter-wave plates; A, analyzer; REF, reference mirror; OBJ, object; PZT, piezoelectric transducer; CCD, charged-coupled device camera [17, 18].

arms change the polarization states of the two reflected beams, thus avoiding light traveling back to the LED and directs all reflected light to the CCD. The analyzer A combines the two light beams into a common polarization state so that interference can occur on the CCD plane. The polarizer–analyzer pair also controls the variation of the relative intensity between the two arms.

The reference mirror is mounted on a piezoelectric transducer (PZT). A function generator supplies a ramp signal to the PZT to dither the reference mirror by a distance of one wavelength. Images are recorded at quarter-wavelength intervals. Two 20X microscope

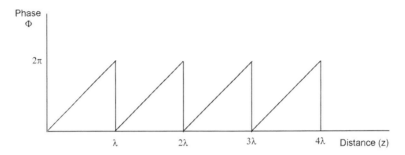

Figure 15.8 Phase vs. distance. 2π ambiguities occur in the phase profile when the distance is a multiple of the wavelength [18].

objectives L2 and L3 are placed in front of the object and the reference mirror to acquire images with high resolution.

Images acquired by the CCD are sent to an image acquisition board (National Instruments IMAQ PCI™-1407) installed in an Intel Pentium® 4 CPU 2.80 GHz computer. The CCD used in the experiment is a Sony XC-ST50 black-and-white camera module, with a 6.4 mm × 4.8 mm sensing area, 768 × 494 pixels, and an 8.4 μm × 9.8 μm pixel size. Image analysis and calculations are done by using LabVIEW programs.

Using intensity $I(x, y)$ of the light captured by CCD at quarter-wavelength intervals at $\phi_i = 0, \pi/2, \pi$, and $3\pi/2$, the final phase profile of the test object can be derived, as described in Section 15.2.

When using a wavelength smaller than the object's height, the phase image contains 2π ambiguities, as shown in Fig. 15.8. It is impossible to obtain an accurate surface profile of the test object without unwrapping the phase image, because there are many distances value for a given phase value.

When an object is imaged with m number of wavelengths, the surface profile Z_m for the m-th wavelength can be expressed as a function of its phase ϕ_m as follows:

$$Z_m = \frac{\lambda_m \varphi_m}{2\pi}. \qquad (15.8)$$

Using a longer λ can increase an unambiguous range of Z. A longer wavelength can be numerically obtained by using the concept of beat wavelength. For two wavelengths λ_1 and λ_2, the beat wavelength is given by $|\Lambda_{12} = \lambda_1 \lambda_2 / |\lambda_1 - \lambda_2||$. Choosing closer values of λ_1 and

λ_2 can increase the Λ_{12} and thereby increasing the unambiguous range of Z. This is the premise of multiwavelength optical phase unwrapping technique, which is described in detail in the following section.

15.3.1 Two-Wavelength Optical Phase Unwrapping

In two-wavelength optical phase unwrapping, the longer wavelength is obtained by using two shorter wavelengths λ_1 and λ_2 to produce the beat wavelength Λ_{12}. The phase map for Λ_{12} is obtained by subtracting one phase map from the other.

$$\phi_{12} = \begin{cases} \phi_1 - \phi_2 & \text{if } \phi_1 > \phi_2 \\ \phi_1 - \phi_2 + 2\pi & \text{if } \phi_1 < \phi_2 \end{cases}. \qquad (15.9)$$

The phase map ϕ_{12}, called the coarse map, has a surface profile of Z_{12} given by $Z_{12} = \Lambda_{12}\varphi_{12}/2\pi$. However, the phase noise in each single-wavelength phase map is magnified by the same factor as the magnification of the wavelengths. In the two-wavelength optical phase unwrapping method introduced in [28], the phase noise is reduced by using the following steps.

First, using one of the two wavelengths, say λ_1, the quantity of integer multiples of λ_1 present in the range Z_{12} is calculated. In the next step the result is added to the single-wavelength surface profile Z_1. This significantly reduces the phase noise in the coarse map. However, at the boundaries of wavelength intervals λ_1, the noise of the single-wavelength phase map appears as spikes. These are then removed by comparing the result with the coarse map surface profile Z_{12}. If the difference is more than half of λ_1, addition or subtraction of one λ_1 depending on the sign of the difference removes the spikes.

The resultant fine map has a noise level equal to that of single-wavelength surface profile. If a single-wavelength phase map ϕ_m contains a phase noise of $2\pi\varepsilon_m$, the two-wavelength phase unwrapping method works properly for $\varepsilon_m < \lambda_m/4\Lambda_{12}$ [28]. Using a larger beat wavelength therefore reduces the maximum noise limit. For $\lambda_1 = 653.83$ nm and $\lambda_2 = 550.18$ nm, the maximum noise limit is $\varepsilon_m \approx 4.7\%$. Using a larger beat wavelength therefore reduces the maximum noise limit.

15.3.2 Three-Wavelength Optical Phase Unwrapping

In the three-wavelength phase unwrapping method, the beat wavelength is increased without reducing the maximum noise limit by using a third wavelength λ_3. Consider three wavelengths λ_1, λ_2, and λ_3. First, two surface profiles Z_{13} and Z_{23} with beat wavelengths of Λ_{13} and Λ_{23} are produced using two sets of wavelengths: λ_1 and λ_2, and λ_2 and λ_3, respectively. These two surface profiles are then combined to produce a coarse map of coarse maps ϕ_{13-23} with surface profile Z_{13-23} with a beat wavelength of $|\Lambda_{13-23} = \Lambda_{13}\Lambda_{23}/|\Lambda_{13} - \Lambda_{23}||$. Note that $\Lambda_{13-23} = \Lambda_{12}$. The surface profile Z_{13-23} is identical to the surface profile Z_{12}.

The noise reduction procedure is similar to the one used in two-wavelength optical phase unwrapping. First, the quantity of integer multiples of Λ_{13} present in the range Z_{13-23} is calculated. The result $Z(a)$ is given by

$$Z(a) = \text{Int}\left[\frac{Z_{13-23}}{\Lambda_{13}}\right]\Lambda_{13}. \quad (15.10)$$

Then $Z(a)$ is added to the surface profile Z_{13}, which results in $Z(b) = Z(a) + Z_{13}$. $Z(b)$ is then compared with Z_{13-23}. If the difference $Z(c)$ is more than half of Λ_{13}, one Λ_{13} is added or subtracted depending on the sign of the difference, as shown below.

$$Z(d) = \begin{cases} Z(c) + \Lambda_{13} & \text{if } Z(c) > \Lambda_{13}/2 \\ Z(c) & \text{if } -\Lambda_{13}/2 \leq Z(c) \leq \Lambda_{13}/2 \\ Z(c) - \Lambda_{13} & \text{if } Z(c) < -\Lambda_{13}/2 \end{cases}. \quad (15.11)$$

The resulting surface profile $Z(d)$ is called the intermediate fine map and has significantly reduced noise. The remaining noise is further reduced by using a single wavelength, say λ_1. $Z(d)$ is divided into integer multiples of λ_1, resulting in the surface profile $Z(e)$, where

$$Z(e) = \text{Int}\left[\frac{Z(d)}{\lambda_1}\right]\lambda_1. \quad (15.12)$$

Now $Z(e)$ is added to the single-wavelength surface profile Z_1. The result $Z(f) = Z(e) + Z_1$ is then is then compared with Z_1. If the difference is more than half of λ_1, one λ_1 is added or subtracted depending on the sign difference. The resultant final map has a noise level equal to that in the single-wavelength surface profile

Z_1. Therefore, the three-wavelength optical unwrapping method increases the beat wavelength, and as a result, provides a longer range free of 2π ambiguities, without increasing the noise in the final phase map.

Suppose the three chosen wavelengths are $\lambda_1 = 653.83$ nm, $\lambda_2 = 603.48$ nm, and $\lambda_3 = 550.18$ nm. The first two wavelengths give a beat wavelength of $\Lambda_{12} = 10.53$ μm, followed by $\Lambda_{13} = 3.47$ μm and $\Lambda_{23} = 6.2$ μm, respectively. The resultant beat wavelength $\Lambda_{13-23} = \Lambda_{13}\Lambda_{23}/|\Lambda_{13} - \Lambda_{23}| = 7.84$ μm.

The three wavelength phase unwrapping method works when the maximum noise level ε_m in the single-wavelength phase map is the smaller of $\Lambda_{13}/4\ \Lambda_{12} \approx 11.1\%$ or $\lambda_1/4\Lambda_{13} \approx 4.7\%$. Therefore, the three-wavelength phase unwrapping method increases the beat wavelength without magnifying the noise in the final phase map.

15.3.3 Application for Two-Wavelength Optical Phase Unwrapping

The results of two-wavelength optical phase unwrapping using a microelectrode array biosensor are shown in Fig. 15.9. The sensor has 16 gold electrodes deposited on a Pyrex glass substrate, and the center is a 125 μm diameter circle with an approximate thickness of 2 μm. The two wavelengths used here are $\lambda_1 = 653.83$ nm and $\lambda_2 = 550.18$ nm, with a beat wavelength of $\Lambda_{12} = 3.47$ μm. Image area is

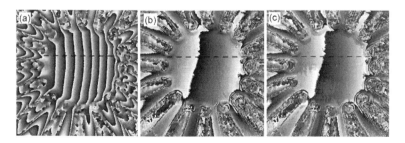

Figure 15.9 Two-wavelength optical phase unwrapping of a MEMS sensor. Image size is 184 μm × 184 μm. (a) Single-wavelength phase map using $\lambda_1 = 653.83$ nm, (b) two-wavelength coarse map produce by $\lambda_1 = 653.83$ nm and $\lambda_2 = 550.18$ nm, and (c) two-wavelength fine map with reduced noise [18].

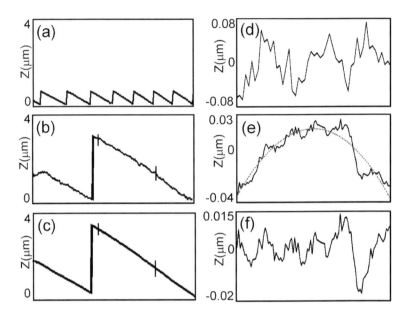

Figure 15.10 Surface profiles of a MEMS biosensor for two-wavelength phase unwrapping. (a) Single-wavelength surface profile, (b) surface profile of a coarse map, (c) surface profile of a fine map with reduced noise, and (d) noise of the coarse map in the region between the two markers in plot (b). RMS noise is 43.27 nm. (e) Noise of fine map in the area shown in (c). The dotted line is the best-fit parabolic curvature and the black solid line is data. (f) Corrected phase noise of the unwrapped phase map after subtracting the curvature of the object. RMS noise is 10.29 nm [18].

184 µm per side. Figure 15.9 (a) shows the single-wavelength phase map with $\lambda_1 = 653.83$ nm. The unwrapped phase image (coarse map) produced by two-wavelength phase unwarping is shown in Fig. 15.9b and the final phase map with reduced noise is shown in Fig. 15.9c.

Figure 15.10 shows cross sections of phase maps along the lines shown in Fig. 15.9 and the phase noise in the selected regions. Figure 15.10a–c shows cross sections of a single-wavelength phase map with $\lambda_1 = 653.83$ nm, a coarse map with $\Lambda_{12} = 3.47$ µm, and fine a map with respectively. For each map, the vertical axis is 4 µm. To compare the phase noise in coarse and fine maps, the root mean square (RMS) value in the region shown was calculated. As shown in

Fig. 15.10d, a coarse map has an RMS noise of 43.27 nm. The reduced RMS phase noise in fine map is shown in Fig. 15.10e. Since the center of the MEMS device has a curvature, a paraboloid is fitted to the data. The dotted line shows the best-fit parabolic curve. After subtracting the curvature from the data, the Fig. 15.10f shows the corrected RMS phase noise of 10.29 nm in the fine map.

15.3.4 Application for Three-Wavelength Optical Phase Unwrapping

The object here is the same microelectrode array biosensor described in Section 1.3.3. In three-wavelength optical phase unwrapping method red ($\lambda_1 = 653.83$ nm), amber ($\lambda_2 = 603.48$ nm) and green ($\lambda_3 = 550.18$ nm) LEDs are used as the three wavelengths. There three wavelengths provide a beat wavelength $\Lambda_{13-23} = 7.84$ μm. Figure 15.11a–c shows the single-wavelength phase map with $\lambda_1 = 653.83$ nm, three-wavelength coarse maps with a 7.84 μm beat wavelength, and the final fine map with reduced noise. Image area is 184 μm per side.

Cross sections and phase noise of the coarse and fine maps are shown in Fig. 15.12. Each cross section is obtained along the lines shown in Fig. 15.11. Figure 15.12a–c shows the surface profiles of a single-wavelength phase map with $\lambda_1 = 653.83$ nm, coarse map and the fine map respectively. Vertical axis for each map is 11 μm.

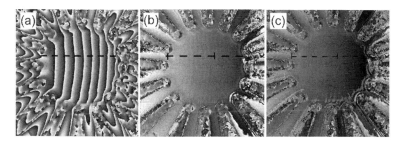

Figure 15.11 Three-wavelength optical phase unwrapping of a MEMS sensor. Image size is 184 μm × 184 μm. (a) Single-wavelength phase map using $\lambda_1 = 653.83$ nm; (b) three-wavelength coarse map with $\Lambda_{13-23} = 7.84$ μm produced by $\lambda_1 = 653.83$ nm, $\lambda_2 = 603.48$ nm, and $\lambda_3 = 550.18$ nm; (c) three-wavelength fine map with reduced noise [18].

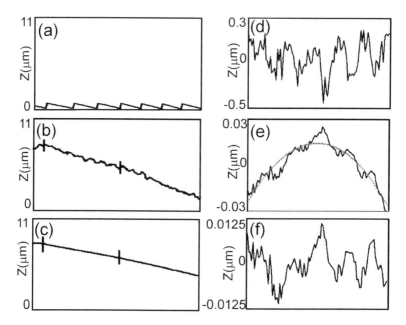

Figure 15.12 Surface profiles for three-wavelength phase unwrapping. (a) Single-wavelength surface profile with $\lambda_1 = 653.83$ nm, (b) surface profile of a coarse map with $\Lambda_{13-23} = 7.84$ μm, (c) surface profile of the final unwrapped phase map with reduced noise, and (d) noise of a coarse map in the area shown in (b). RMS noise is 105.79 nm. (e) Noise of the final unwrapped phase map in the area shown in (c). The dotted line is the best-fit parabolic curvature and the black solid line is data. (f) Final noise of the unwrapped phase map after subtracting the curvature of the object. RMS noise is 4.78 nm [18].

The RMS noise in the selected region of the coarse map is 105.79 nm and is shown in Fig. 15.12d. After following three-wavelength phase unwrapping method, the surface profile of the final fine map with reduced noise is shown in Fig. 15.12e. Because of the curvature of the object surface, a paraboloid is fitted with the final fine map data. The black line is data and the dotted line shows the best-fit parabolic curve. Corrected phase noise in the fine map is 4.78 nm, and is shown in Fig. 15.12f.

The comparison between the two-wavelength phase unwrapping method and the three-wavelength phase unwrapping method shows that the three-wavelength phase unwrapping method results a

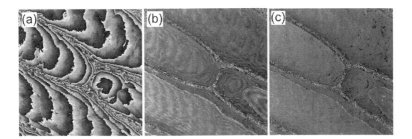

Figure 15.13 Results of three-wavelength optical phase unwrapping of onion cells: (a) single-wavelength phase map with $\lambda_1 = 653.83$ nm, (b) three-wavelength coarse map with $\Lambda_{13-23} = 7.84$ μm beat wavelength, and (c) three-wavelength fine map with reduced noise [18].

larger axial range free of 2π ambiguities, without increasing phase noise. As shown in Sections 15.3.3 and 15.3.4, the two-wavelength phase unwrapping method produced a 3.47 μm unambiguous range with 10.29 nm phase noise, while the three-wavelength phase unwrapping method produced much larger 7.84 μm unambiguous range with smaller 4.78 nm phase noise.

Multiwavelength optical phase unwrapping method can also be used for biological cells. Figure 15.13 shows results of three-wavelength optical phase unwrapping on a phase image of a sample of onion cells. Three wavelengths used here are: red (653.83 nm), amber (603.48 nm), and green (550.18 nm). The beat wavelength is 7.84 μm. Image size is 184 μm × 184 μm. Figure 15.13a shows a single-wavelength phase map with $\lambda_1 = 653.83$ nm. Figure 15.13b,c shows a coarse map with $\Lambda_{13-23} = 7.84$ μm and the final fine map with reduced phase noise. The final fine map clearly shows the cell walls without the 2π ambiguities that would exist in a single-wavelength phase image.

15.4 Full-Color FF-OCT

We demonstrated a method of optical tomography for surface and subsurface imaging of biological tissues, based on the principle of wide field optical coherence tomography and capable of providing full-color 3D views of a tissue structure. Contour or tomographic

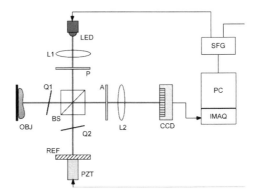

Figure 15.14 Apparatus for color WF-OCT. See text for details [19].

images are obtained with an interferometric imaging system using broadband light sources. The interferometric images are analyzed in the three-color channels and recombined to generate 3D microscopic images of tissue structures with full natural color representation. In contrast to most existing 3D microscopy methods, the presented technique allows monitoring of tissue structures close to its natural color, and can provide critical information for the physiological and pathological applications. Below, we describe the general principles of color 3D microscopy by wide-field optical coherence tomography (WF-OCT) and present some experimental images.

The principle of color 3D microscopy by WF-OCT is described referring to the diagram of apparatus in Fig. 15.14. A high-brightness LED (∼30 lumens) illuminates the Michelson interferometer through a collimating lens L1, a polarizer P, and the broadband PBS. The quarter-wave plates, Q1 and Q2, in the object or reference arm, change the orthogonal polarization states, so all of the reflections from the object or the reference mirror are steered toward the monochrome CCD camera (Sony XC-ST50) through the imaging lens L2. The analyzer A combines the two beams into a common polarization to affect the interference between them. The combination of the polarizer and the analyzer allows continuous adjustment of the relative intensity and the interference contrast between the object and reference beams. The CCD image

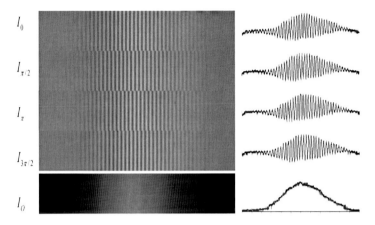

Figure 15.15 Phase shift interference imaging. Four-quadrature-phase interferograms and the extracted interference images are shown, as well as cross-sectional profiles of the interferograms [19].

is acquired by the computer using an image acquisition board (National Instruments IMAQ PCI-1407) and a function generator (SFG) is used to strobe the LED and to dither the reference mirror mounted on a PZT. The object is mounted on a three-axis micrometer translation stage to bring the appropriate object plane into focus.

In Fig. 15.14, we use a plane mirror as the object, tilted by a small angle with respect to the optical axis. The four interference images show that the fringes are shifted by quarter periods. The interference profiles are also plotted in Fig. 15.15. The bottom image and curve show the envelope of the interference profile. The coherence length of the light source can be determined by comparing the width of the interference profile with the fringe periods. They are found to be $\delta_R = 5.8$ µm, $\delta_G = 4.8$ µm, and $\delta_B = 4.0$ µm for the three red, green, and blue LEDs, respectively. The normalized spectra and the interference profiles of the LEDs are shown in Fig. 15.16. The center wavelengths of the LEDs are $\lambda_R = 638.4$ nm, $\lambda_G = 528.5$ nm, and $\lambda_B = 457.4$ nm, while the spectral widths are $\Delta_R = 22.9$ nm, $\Delta_G = 33.2$ nm, and $\Delta_B = 28.6$ nm. The measured coherence lengths δ are consistent with the expected values of $(2 \ln 2/\pi) \lambda^2/\Delta$ within a factor of 1.3.

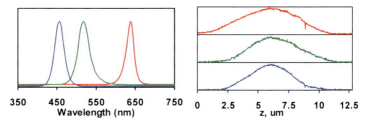

Figure 15.16 (a) The spectra of red, green, and blue LEDs. (b) The interference profiles of the LEDs vs. the axial distance z [19].

15.4.1 Results of Full-Color FF-OCT

Figure 15.17 illustrates the WF-OCT by phase-shifting interferometry using a coin surface as the test object. Figure 15.17a is the direct image of the object in the absence of the reference wave, while Fig. 15.17b is when the reference wave is present. Although the two images are mostly indistinguishable, one can observe fluctuating speckles in portions of the object when the camera image is viewed in real time while the reference mirror is dithered. The coherent portion of the image is extracted as described above to obtain a

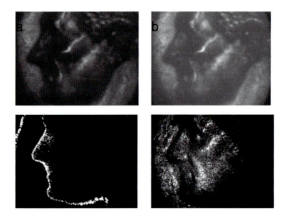

Figure 15.17 Phase shift interference imaging of a coin surface: (a) direct image of the object, (b) image of the object with reference wave, (c) contour image extracted by phase shift interference, and (d) flat view of accumulated contour images. Image volume = 12 mm × 9 mm × 405 μm; voxels = 640 × 480 × 82; voxel volume = 19 μm × 19 μm × 5 μm [19].

contour of the object at a height that corresponds to the reference mirror position, as shown in Fig. 15.17c. We repeat the contour imaging for a number of times (∼40) and average to improve the signal-to-noise ratio (SNR). The 3D imaging is completed by stepping the object's z position over a desired range to obtain a stack of the contour images. The 3D image can then be presented in a number of different ways. In Fig. 15.17d all the contour images are added together resulting in a flat view, where all the image planes are in focus and additionally the overall haze in Fig. 15.17a or 15.17b due to stray reflections from various optical surfaces has been removed.

For generation of color images, the WF-OCT procedure is repeated three times by using red, green, and blue LEDs. Each of these generates a 3D image of the object under the respective color illumination. The three 3D images are then combined as RGB channels to generate the final 3D image with full natural color representation of the volume. In Fig. 15.18, we use a colored coin as the test object. The back side of a dime is painted with red, green, and blue ink and the background is painted white. The first three columns represent separate results obtained using the three LEDs. In each column, the top square is a 7.2 mm × 335 μm xz section view of the 3D image, the middle square is a 7.2 mm ×

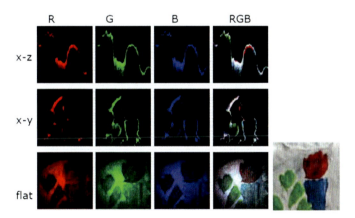

Figure 15.18 Color WF-OCT of a painted coin surface. See text for details [19].

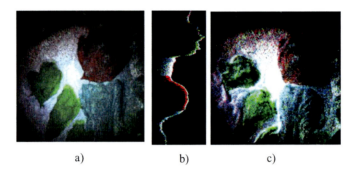

a) b) c)

Figure 15.19 Color WF-OCT 3D of a painted coin surface: (a) *xy* section images, (b) *xz* section images, and (c) 3D perspective views. Image volume = 7.2 mm × 7.2 mm × 335 µm; voxels = 480 × 480 × 67; voxel volume = 15 µm × 15 µm × 5 µm [19].

7.2 mm *xy* section or a contour image, while the bottom square is the flat view for the color channel. The last column is the RGB composite images. In Fig. 15.19, the 3D image is presented in a few different modes. Figure 15.19a is a series of *xy* sections and Fig. 15.19b is a series of *xz* sections. In Fig. 15.19c the reconstructed 3D image is viewed from varying perspective angles. Notice the strong color-independent (white) reflection at the air–paint interface and strong colored reflection from underneath. Penetration depth in some areas is at least about 100 µm. In the *xz* RGB image, the top-right corner of Fig. 15.18, the thin and abrupt layer of red probably accounts for thin layer of paint on the metal surface, whereas there is a pool of white paint in the valley next to it. Also note that the RGB flat image has very respectable color and image quality, compared to the direct photographic image in the lower right corner of Fig. 15.18.

Next we present examples of color 3D microscopy using the three-channel WF-OCT. Figure 15.20 is arranged in the same manner as Fig. 15.18, with the *XZ*- and *XY*-sectional views and the flat view for each of the three color channels and the RGB composite color images, for a 4.7 mm × 4.7 mm × 170 µm volume of a piece of apple skin. Here the penetration depth is at least 150 µm, and it displays details of the top surface and the skin tissue. Because of the large difference in the longitudinal and lateral scales, the minute surface texture is seen greatly exaggerated in the flat views.

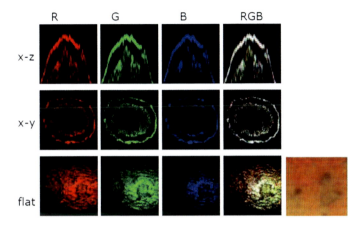

Figure 15.20 Color WF-OCT of a piece of apple skin. See text for details [19].

The cross sectional views clearly show the outer skin layer and a portion of the internal tissue layers. One can also notice the correspondence of the internal structure of a blemish area in the *xz* sections and flat view images. The extra picture in the lower-right corner is a direct photograph of the approximate area of the imaging experiment. Although the color variation of the apple skin is rather subtle and continuous compared to the colored coin experiment, the RGB flat view reproduces the color variation at least approximately. In this data set the colored images are not as one would expect from common macroscopic views of an apple skin. However, under microscope, the coloring of apple skin appears not as a solid layer of red surface, but the red pigments are embedded as numerous particles in otherwise colorless matrix. Still one observes some variation of color that can be attributed to the actual color variation of the surface.

15.5 Discussion

FF-OCT is a noninvasive 3D real-time imaging technique for imaging unprocessed and unstained tissues with micron-level resolution. The success of FF-OCT technologies made OCT probably the most significant development in biomedical imaging in the past decade and an attractive potential diagnostic tool.

The latest applications of FF-OCT in biomedical imaging, with a combination of detection capabilities, controls, and high performance, warrant its capabilities as a powerful clinical and diagnostic instrument.

Examples of latest applications are in the area of skin imaging where the collagen bundles of the dermis and the hypodermis could be identified through architectural and cellular details [31]. Another example is the developing of a rigid needle-like probe using FF-OCT to image biological tissues ex vivo and reveal fine tissue structures [32]. The first use of a commercial prototype of FF-OCT called Light-CT^{TM}, for rapid histology of unprocessed tissue was reported in 2011 [33]. More recently, FF-OCT was demonstrated in identifying and differentiating lung tumors from nonneoplastic lung tissue [34].

When it comes to build an instrument, we must take into consideration various aspects. The instrument must work under a full-field regime to increase the image acquisition time [13–15], use swept laser sources to increase the axial resolution [11, 12], use integrated optics to reduce the instrument size [35], take advantage of using nonlinear optics imaging like multiphotons and second harmonic generation [36, 37], use of other imaging modalities such as spectroscopy and fluorescence [38], and including up-to-date imaging processing on GPU instead of CPU [39, 40].

The penetration depth and the imaging resolution are always very important when deciding which light source to use. Due to the fact that the axial resolution is a function of the full wavelength range (the spectral width) of the light source, broadband femtosecond-pulse lasers with high repetition rates are more suitable for high-resolution imaging; however, these lasers are expensive.

Holography is another optics area where a low-coherence source can be used to image the 3D structure of an object [41] with a significant reduction of the coherent noise. When a low-coherence source is used the fringe visibility requires a more precise optical alignment. Holography has the advantage of recording the whole imaging volume at a time.

In summary, combining all of these aspects together with the costs weighed against each other a new full-field instrument will be added to existing noninvasive imaging modalities in investigating biological samples.

References

1. Michelson, A. A., and Morley, E. W. (1887). On the relative motion of the earth and the luminiferous ether. *Am. J. Sci.*, **34**, p. 11.
2. Huang, D., Swanson, E. A., Lin, C. P., Schuman, J. S., Stinson, W. G., Chang, W., Hee, M. R., Flotte, T., Gregory, K., Puliafito, C. A., et al. (1991). Optical coherence tomography. *Science*, **254**, pp. 1178–1181.
3. Fujimoto, J. G., Brezinski, M. E., Tearney, G. J., Boppart, S. A., Bouma, B. E., Hee, M. R., Southern, J. F., and Swanson, E. A. (1995). Optical biopsy and imaging using optical coherence tomography. *Nat. Med.*, **1**, pp. 970–972.
4. Fercher, A. F., Mengedoht, K., and Werner, W. (1988). Eye-length measurement by interferometry with partially coherent light. *Opt. Lett.*, **13**, pp. 186–188.
5. Wojtkowski, M., Bajraszewski, T., Targowski, P., and Kowalczyk, A. (2003). Real-time in vivo imaging by high-speed spectral optical coherence tomography. *Opt. Lett.*, **28**, pp. 1745–1747.
6. Podoleanu, A., Rogers, J., Jackson, D., and Dunne, S. (2000). Three dimensional OCT images from retina and skin. *Opt. Express*, **7**, pp. 292–298.
7. Rogers, J., Podoleanu, A., Dobre, G., Jackson, D., and Fitzke, F. (2001). Topography and volume measurements of the optic nerve usingen-face optical coherence tomography. *Opt. Express*, **9**, pp. 533–545.
8. Schuman, J. S., Puliafito, C. A., and Fujimoto, J. G. (2004). *Optical Coherence Tomography of Ocular Diseases*. SLACK Thorofare, NJ.
9. Wojtkowski, M., Srinivasan, V., Fujimoto, J. G., Ko, T., Schuman, J. S., Kowalczyk, A., and Duker, J. S. (2005). Three-dimensional retinal imaging with high-speed ultrahigh-resolution optical coherence tomography. *Ophthalmology*, **112**, pp. 1734–1746.
10. Drexler, W., Morgner, U., Ghanta, R. K., Kärtner, F. X., Schuman, J. S., and Fujimoto, J. G. (2001). Ultrahigh-resolution ophthalmic optical coherence tomography. *Nat. Methods*, **7**, pp. 502–507.
11. Srinivasan, V. J., Huber, R., Gorczynska, I., Fujimoto, J. G., Jiang, J. Y., Reisen, P., and Cable, A. E. (2007). High-speed, high-resolution optical coherence tomography retinal imaging with a frequency-swept laser at 850 nm. *Opt. Lett.*, **32**, pp. 361–363.

12. Choma, M. A., Sarunic, M. V., Yang, C., and Izatt, J. A. (2003). Sensitivity advantage of swept source and Fourier domain optical coherence tomography. *Opt. Express*, **11**, pp. 2183–2189.
13. Dubois, A., Vabre, L., Boccara, A. C., and Beaurepaire, E. (2002). High-resolution full-field optical coherence tomography with Linnik microscope. *Appl. Opt.*, **41**, pp. 805–812.
14. Vabre, L., Dubois, A., and Boccara, A. C. (2002). Thermal-light full-field optical coherence tomography. *Opt. Lett.*, **27**, p. 530.
15. Moneron, G., Boccara, A. L., and Dubois, A. (2007). Polarization-sensitive full-field optical coherence tomography. *Opt. Lett.*, **32**, pp. 2058–2060.
16. Potcoava, M. C., and Kim, M. K. (2009). Fingerprint biometry applications of digital holography and low-coherence interferography. *Appl. Opt.*, **48**, pp. H9–H15.
17. Warnasooriya, N., and Kim, M. (2009). Quantitative phase imaging using three-wavelength optical phase unwrapping. *J. Mod. Opt.*, **56**, pp. 85–92.
18. Warnasooriya, N., and Kim, M. K. (2007). LED-based multi-wavelength phase imaging interference microscopy. *Opt. Express*, **15**, pp. 9239–9247.
19. Yu, L. F., and Kim, M. K. (2004). Full-color three-dimensional microscopy by wide-field optical coherence tomography. *Opt. Express*, **12**, pp. 6632–6641.
20. Pedrini, G., and Tiziani, H.J. (2002). Short-coherence digital microscopy by use of a lensless holographic imaging system. *Appl. Opt.*, **41**, pp. 4489–4496.
21. Goodman, J. (2005). *Introduction to Fourier optics*. Roberts.
22. Yu, L., Rao, B., Zhang, J., Su, J., Wang, Q., Guo, S., and Chen, Z. (2007). Improved lateral resolution in optical coherence tomography by digital focusing using two-dimensional numerical diffraction method. *Opt. Express*, **15**, pp. 7634–7641.
23. Servin, M., Marroquin, J. L., Malacara, D., and Cuevas, F. J. (1998). Phase unwrapping with a regularized phase-tracking system. *Appl. Opt.*, **37**, pp. 1917–1923.
24. Ying, L. (2006). *Wiley Encyclopedia of Biomedical Engineering*. John Wiley & Sons.
25. Zebker, H. A., and Lu, Y. (1998). Phase unwrapping algorithms for radar interferometry: residue-cut, least-squares, and synthesis algorithms. *J. Opt. Soc. Am. A*, **15**, pp. 586–598.
26. Cheng, Y. Y., and Wyant, J. C. (1984). Two-wavelength phase shifting interferometry. *Appl. Opt.*, **23**, pp. 4539–4543.

27. Cheng, Y. Y., and Wyant, J. C. (1985). Multiple-wavelength phase-shifting interferometry. *Appl. Opt.*, **24**, pp. 804–807.
28. Gass, J., Dakoff, A., and Kim, M. K. (2003). Phase imaging without 2 pi ambiguity by multiwavelength digital holography. *Opt. Lett.*, **28**, pp. 1141–1143.
29. Tziraki, M., Jones, R., French, P. M. W., Melloch, M. R., and Nolte, D. D. (2000). Photorefractive holography for imaging through turbid media using low coherence light. *Appl. Phys. B*, **70**, pp. 151–154.
30. Repetto, L., Piano, E., and Pontiggia, C. (2004). Lensless digital holographic microscope with light-emitting diode illumination. *Opt. Lett.*, **29**, pp. 1132–1134.
31. Dalimier, E., and Salomon, D. (2012). Full-field optical coherence tomography: a new technology for 3D high-resolution skin imaging. *Dermatology*, **224**, pp. 84–92.
32. Latrive, A., and Boccara, A. C. (2011). In vivo and in situ cellular imaging full-field optical coherence tomography with a rigid endoscopic probe. *Biomed. Opt. Express*, **2**, pp. 2897–2904.
33. Jain, M., Manzoor, M., Shukla, N., Mukherjee, S., and Nadolny, S. (2011). Modified full-field optical coherence tomography: a novel tool for rapid histology of tissues. *J. Pathol. Inform.*, **2**, p. 28.
34. Jain, M., Narula, N., Salamoon, B., Shevchuk, M. M., Aggarwal, A., Altorki, N., Stiles, B., Boccara, C., and Mukherjee, S. (2013). Full-field optical coherence tomography for the analysis of fresh unstained human lobectomy specimens. *J. Pathol. Inform.*, **4**, p. 26.
35. Sun, J., and Xie, H. (2011). MEMS-based endoscopic optical coherence tomography. *Int. J. Opt.*, **2011**, Article ID 825629, 12 pages.
36. Liu, G., and Chen, Z. (2011). Fiber-based combined optical coherence and multiphoton endomicroscopy. *J. Biomed. Opt.*, **16**, pp. 036010–036014.
37. Graf, B. W., and Boppart, S. A. (2012). Multimodal in vivo skin imaging with integrated optical coherence and multiphoton microscopy. *IEEE J. Sel. Top. Quantum Electron.*, **18**, p. 7.
38. Ju, M. J., Lee, S. J., Kim, Y., Shin, J. G., Kim, H. Y., Lim, Y., Yasuno, Y., and Lee, B. H. (2011). Multimodal analysis of pearls and pearl treatments by using optical coherence tomography and fluorescence spectroscopy. *Opt. Express*, **19**, pp. 6420–6432.
39. Fang, L., Li, S., Nie, Q., Izatt, J. A., Toth, C. A., and Farsiu, S. (2012). Sparsity based denoising of spectral domain optical coherence tomography images. *Biomed. Opt. Express*, **3**, pp. 927–942.

40. Huang, Y., Liu, X., and Kang, J. U. (2012). Real-time 3D and 4D Fourier domain Doppler optical coherence tomography based on dual graphics processing units. *Biomed. Opt. Express*, **3**, pp. 2162–2174.
41. Monemhaghdoust, Z., Montfort, F., Emery, Y., Depeursinge, C., and Moser, C. (2011). Dual wavelength full field imaging in low coherence digital holographic microscopy. *Opt. Express*, **19**, pp. 24005–24022.

Chapter 16

Dual-Modality Full-Field Optical Coherence and Fluorescence Sectioning Microscopy: Toward All Optical Digital Pathology on Freshly Excised Tissue

Fabrice Harms

LLTech, Pépinière Paris Santé Cochin, 29 rue du Faubourg Saint Jacques, 75014 Paris, France
Institut Langevin, ESPCI, 1 rue Jussieu, 75005 Paris, France
fharms@lltech.fr

16.1 Introduction: Clinical Context

The localization, assessment, and removal of tumor and/or suspicious tissue constitute today the cornerstone of cancer care. To optimize the diagnosis and resection processes, a precise identification of tumor tissue and an exact resection of tumor boundaries are required. Currently the most precise diagnosis is available through the use of histology slide preparation on excised tissue or biopsies. The process mainly relies on paraffin embedding of the fixed biopsy or excised tumor, on its mechanical slicing with

---

Handbook of Full-Field Optical Coherence Microscopy: Technology and Applications
Edited by Arnaud Dubois
Copyright © 2016 Pan Stanford Publishing Pte. Ltd.
ISBN 978-981-4669-16-0 (Hardcover), 978-981-4669-17-7 (eBook)
www.panstanford.com

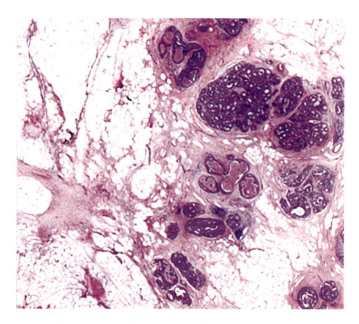

Figure 16.1 Example of a digitized histology slide (H&E staining): breast cancer—ductal carcinoma in situ (DCIS). Hematoxylin appears in purple; eosin in pink. The high density of cell nuclei, visible in purple areas, reveals a concentration of cancerous cells.

typical thicknesses of 4 or 5 µm, then on staining using contrast agents for the enhancement of specific tissue and cell structures, usually hematoxylin for cell nuclei and eosin for supporting tissue (H&E staining) (Fig. 16.1).

Even if this process is accurate and based on decades of experience, it is not necessarily optimized for modern surgery procedures. The slide preparation is indeed based on several mechanical and biochemical steps, which require significant experience, equipment, and time. Depending on hospitals, such a process typically lasts from 1 day to 1 week, which is not on line with the need for a fast diagnosis guiding the surgical removal of tissue, in particular regarding the precise assessment of tumor margins for minimizing resection. Moreover, sampling errors in biopsy procedures, still significant even if guided using modern computed tomography (CT) scans or ultrasonography systems, cannot be assessed using conventional histology without the need for repeating the procedure a couple of days after.

This led to the development of intraoperative diagnosis methods, which currently mainly consist of the use of frozen sections analysis [27, 30, 36] and of modern cytopathology examination (e.g., Pap smears). Frozen sections correspond to fast H&E slide preparation using alternative tissue processing methods: freshly excised tissue is frozen and sliced in a cryostat and then stained using conventional H&E preparation. It allows for a significant acceleration of the procedure, which takes about 20 minutes. However, the use the technique has progressively diminished over years and been restricted to a couple of clinical situations (e.g., Mohs surgery, lymph nodes), since the process is highly operator dependent and shows significant artifacts mainly due to tissue freezing. As a consequence the accuracy of the corresponding diagnosis is often limited (e.g., average 80% on breast tissue). Moreover the intrinsic destructive nature of the technique is not adapted to recent clinical trends: biopsy and surgery procedures are more and more tissue conservative, leading to a reduced average biopsy size, and the development of personalized medicine and molecular analysis requires a minimum amount of tumor tissue, so there is no room for supplementary tissue consumption.

Cytopathology analysis is based on the use of a fine needle for the aspiration of cells in a suspicious tissue site, followed by a smearing of cells across a glass slide, staining, and microscopy analysis. It is often used where conventional biopsy is difficult (e.g., sarcomas) and for thyroid, gynecology, and fluid collection among others. However, it does not provide any information about tissue architecture, which is often required for a precise evaluation of inflammatory conditions. It is also highly operator dependent.

There is consequently a significant room for the development of fast, noninvasive tissue characterization methods at a cellular level, compatible with the requirements of intraoperative procedures—in particular in terms of duration—and allowing for highly accurate cancer diagnosis.

16.2 Optical Coherence Tomography in Pathology: Early Multimodal Approaches and Limitations

16.2.1 Applying OCT to Pathology Assessment of Biological Tissue

The introduction of optical coherence tomography (OCT) in the early 1990s [18] opened a new field of research in biomedical imaging, targeting noninvasive optical imaging of tissue for fast diagnosis. Since OCT can be considered as an equivalent of ultrasonography using optical waves, it provides in depth images of biological tissue with the resolution of optics (typically from 1 to tenths of microns), but with a more limited penetration depth due to the use of higher wavelengths. The use of short coherence length light sources allows producing "optical biopsies," thus defining an optical section in scattering tissue. The introduction of spectral domain optical coherence tomography (SD-OCT) in the late 1990s led to a significant improvement of the technique in terms of imaging speed and sensitivity [10, 12, 40], which led to the development of commercially available clinical devices, mainly in the field of ophthalmology.

The development of medical applications of OCT was indeed mainly driven by ophthalmology [12, 13, 41], with the capability of the technique to visualize the layered structure of the retina as well as microscopic details of the anterior segment. Various other medical fields have been and are currently explored, including dermatology, upper aerodigestive tract, gastrointestinal tract, breast tissue and lymph nodes, and more recently cardiology. New assessment methods based on OCT are now available for clinicians, based on the analysis of tissue morphology from a noninvasive process, for early diagnosis, guidance, or follow-up of pathological conditions (retinal diseases, visualization of atheroma plaques in vessels, etc.).

When considering cancer diagnosis several studies using OCT have shown capabilities to visualize suspicious and/or cancerous lesions, on skin [40], breast [8, 17, 34, 43], gastrointestinal tract [37], and many other tissue types. However, this identification was mainly limited to macroscopic areas, basically groups of tumor cells following invasion, and usually not able to distinguish all microscopic tissue structures, except for large-scale morphologic features such as adipocytes, large vessels, or glands. This restriction was mainly due to two major limitations of OCT:

- In conventional scanning SD-OCT, the transverse resolution is intrinsically limited due to the intrinsic requirement of a large depth of focus in order to avoid axial scanning [12]. The typical achieved resolution is between 5 and 20 µm, so fine tissue structures cannot be resolved. For example, the average size of a cell is 10 µm in diameter, and bundles of collagen fibers exhibit a thickness of a couple of microns.
- The contrast of OCT is mainly linked to the local variation of refractive index within the object. It is usually observed that cell nuclei, which are of particular importance for cancer diagnosis, cannot be distinguished from their surrounding tissue environment (cytoplasm, collagen fibers).

These limitations highlight the need for increasing both the resolution and the specificity of in depth optical imaging of tissue, in particular regarding critical microstructures such as cell nuclei.

16.2.2 High-Resolution OCT for Pathology: Full-Field Optical Coherence Tomography

Considering optical resolution, full-field optical coherence tomography (FF-OCT) has shown significantly superior performance to other OCT techniques [6, 39]. FF-OCT is a time domain, parallel acquisition OCT technique that makes use of relatively large-numerical-aperture (NA) objectives (typically between 0.3 and 0.8) to achieve higher resolution than Fourier domain OCT devices, typically 1 µm in 3D. Indeed, since time domain techniques do not require a large depth of focus for simultaneous in-depth voxel acquisition, large-NA optics can be used. The use of a parallel acquisition method allows for a simplified setup with no scanning arrangement, but also for faster acquisition, depending on the detectors used. Extensive technical details about the technique are given in previous chapters.

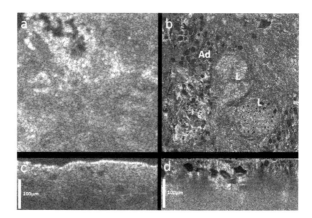

Figure 16.2 Comparative images of a breast tissue sample acquired with swept-source OCT (Thorlabs OCSS1300SS) and FF-OCT (LLTech Light-CT™ scanner). (a and b) En face images (C-scans); (c and d) cross sections (B-scans). FF-OCT images (b–d) reveal superior details, such as lobules (L) and adipocytes (Ad).

Figure 16.2 illustrates this resolution capability, which is of particular importance when considering pathological examination of resected tissue. This resolution feature, combined with other key characteristics, makes FF-OCT the OCT technique that currently shows the best fit with pathology requirements considering the assessment of biological tissue. The following paragraphs develop these features:

- The possibility to use relatively large-NA microscope objectives, typically 0.3 to 0.5, combined with the use of spatially incoherent light sources with a large spectral width such as halogen lamps, provides high-resolution images of biological tissue, with reduced speckle crosstalk.
- Disclosed setups are most of the time based on a Linnik arrangement and exhibit a 3D resolution of ~1 μm both in the transverse and in the axial directions, which represent a factor of 5 to 20 when compared to conventional OCT techniques. Such a resolution is compatible with the observation of tissue microstructures: The typical size of a nucleus being between 5 to 10 μm, FF-OCT shows good

performance where conventional OCT, with a resolution between 5 and 20 μm, fail to resolve cellular structures of interest.
- FF-OCT is based on a full-field acquisition method, without any point-by-point scanning such as in conventional OCT or in confocal microscopy. The signal acquisition is performed through the use of 2D detectors such as charge-coupled device (CCD) or complementary metal-oxide semiconductor (CMOS) cameras. Besides the obvious advantage in terms of simplicity of the setup, it provides en face OCT images (C-scans), in a similar geometry to conventional pathology microscopes.
- The combination of this en face acquisition geometry with transverse movements and correlation of multiple fields of view, usually called *image stitching* or *image mosaicing*, provides large reconstructed fields of view of several millimeters square. This feature makes it possible for the user to digitally zoom in and out on an image, going from a macroscopic to a microscopic view of a sample: It is compatible with the current practice of pathologists, who often switch between several magnifications using multiple objectives on a conventional pathology microscope.

An extensive literature describes various implementations of FF-OCT and corresponding lab setups. Recently, with the availability of a commercial FF-OCT device—the Light-CT™ scanner, LLTech, France (see Fig. 16.3)—several studies have been published on the use of FF-OCT for the intraoperative pathology assessment of biopsied or excised tissue samples, without the need for any sample preparation [20]. Promising results have been reported, in particular in terms of sensitivity, on several tissue types such as breast [2], lung [19], skin [10], and brain [3].

The FF-OCT technique shows several advantages when compared to traditional intraoperative pathology diagnosis methods, such as frozen section analysis (FSA):

- Due to the parallel acquisition geometry, the technique is fast and compatible with clinical requirements in terms of

Figure 16.3 FF-OCT commercial microscope (Light-CT™ scanner, LLTech, France).

time: the typical duration of a FSA being 20 to 30 minutes, a 10 mm × 10 mm FF-OCT image can be obtained within 15 minutes.
- FF-OCT is operator independent, since no preparation, staining, or slicing is required.
- FF-OCT does not induce artifacts due to sample preparation, such as freezing artifacts commonly observed during FSA, which can cause false diagnosis (Fig. 16.4).

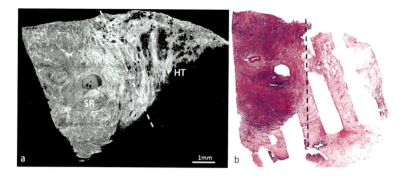

Figure 16.4 Breast tissue sample, including tumor margin, imaged with (a) FF-OCT (Light-CT™ scanner, LLTech, France) on a fresh unprepared resection and (b) digitized frozen section slide after sample preparation.

However, even if these studies have shown good performance for the identification of tumor tissue, limited performance in terms of specificity has been reported. Such a result is also common to other OCT methods. There is consequently some room for new approaches targeting the improvement of optical contrast on tissue microstructures which are commonly observed in pathology as a basis for specific diagnosis, and on which OCT techniques cannot provide a good contrast. An obvious approach consists of the combination of at least two optical imaging methods, each of them providing a specific and complementary contrast.

16.2.3 Multimodal OCT Approaches

The intrinsic contrast of OCT showing limited performance in terms of specific identification of some microstructures or in terms of tumor grade assessment, several multimodal OCT setups have been developed during the last decade, either based on ex vivo microscopy experiments or on in vivo probes, in an attempt to supplement the clinical information extracted from OCT images alone. These setups described the use of OCT in combination with complementary optical contrasts, in an attempt to create fused OCT images including highlights on specific tissue structures, thus potentially improving OCT-based diagnosis. These contrasts include polarization [12], spectroscopy [1], attenuation [32], fluorescence [5], photoacoustic [42], Doppler [23], two-photon [15], and elastography contrasts [33, 35].

These combinations have shown several drawbacks, possibly limiting their performance for tissue diagnosis in the field of pathology:

- Spectroscopic, attenuation, and polarization contrasts are strongly affected by the nature and properties of tissue structures above the imaging depth and require complex computing and/or multiple acquisition to recover the real information coming from the selected plane of interest.
- Doppler and photoacoustic contrasts are usually applied to the visualization of blood flux: it gives information about

neovascularization linked to cancerous areas but not about morphologic structures such as nuclei or metabolism.
- Two-photon contrast, if used as an endogeneous technique, is limited to signals coming from the autofluorescence of a couple of proteins such as elastin, thus limiting the contrast to corresponding fibers. It is not revealed by H&E staining, so it does not give similar information than histology. Also, the safety of illumination—usually using high-power, high-frequency femtolasers—is still debated.
- Elastography is a noninvasive method, already successfully used as a complementary contrast modality in ultrasonography, related to the change in stiffness of tumor invaded volumes. When adapted to OCT, it suffers from a significant decrease of the corresponding resolution and is currently limited to the visualization of macroscopic suspicious areas, not at a cellular level.
- Fluorescence can specifically target structures of particular interest for histology and not revealed by morphology such as cell nuclei, but it requires for this purpose the use of exogeneous contrast agents. Authorized fluorescent dyes are today limited and not yet specific to cancer cells. However, with further development on nontoxic specific dyes, currently under development, this method has a great potential for targeted histology-like diagnosis on tumor tissue.

16.3 Multimodal Full-Field Optical Coherence Tomography for Pathology Applications

16.3.1 *Translating FF-OCT to Pathology Diagnosis*

It has been shown in previous sections that FF-OCT is currently the OCT technique providing the best matching when considering pathology applications, mainly in terms of 3D resolution and acquisition geometry. Figure 16.5 describes a comparative image between histology and FF-OCT: A good correlation is reported here, although the FF-OCT image was acquired before any sample

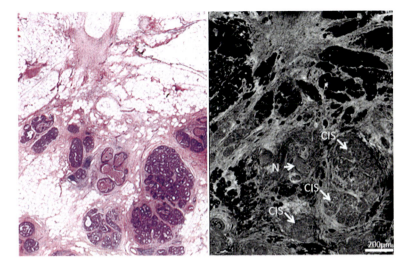

Figure 16.5 Comparative histology/FF-OCT image of a breast tumor tissue sample—ductal carcinoma in situ (DCIS). Left: H&E histology slide. Right: FF-OCT image allowing for the identification of carcinoma nests (CIS), enlarged ducts with necrosis (N). FF-OCT acquisition depth: 30 μm.

preparation on a volume resection from a whole lumpectomy, and such a macroscopic view allows for the identification of most of the structures of breast tissue (adipocytes, collagen fibers, lobules, ducts, etc.).

Published studies targeting the use of FF-OCT for pathology diagnosis have established new reading criteria for clinicians, mainly based on the identification of tissue architecture and microstructures—such as the presence of normal structures or of a stromal reaction—since FF-OCT provides a detailed view of the morphology of the tissue. When considering the requirements for an accurate pathology diagnosis, based on the gold-standard histology, it appears that the precise identification of cell nuclei is of particular importance. The size and shape of a nucleus, the ratio between nucleus and surrounding cytoplasm, the color of a stained nucleus using H&E are typically observed by pathologists and used to provide an accurate diagnosis, in particular in terms of specificity.

The translation FF-OCT to pathology, with an overview of current FF-OCT performance, could as a consequence highly *benefit from the use of a complementary imaging modality providing a specific contrast on cell nuclei.* Fluorescence microscopy, using specific contrast agents, is currently the best candidate for such a combination. Moreover, with current developments of new contrast agents applying for compliance with safety regulations (FDA approval, CE marking) in the near future, such a combined method and corresponding device could accelerate the adoption of FF-OCT in pathology, as a novel artifact-free, noninvasive diagnosis method.

Among fluorescence microscopy setups, one approach is particularly adapted to a combined use with FF-OCT: *structured illumination microscopy (SIM)*. SIM is a full-field optical sectioning method compatible with fluorescence imaging, based on a parallel acquisition geometry using 2D detectors such as cameras [9, 28]. It is an alternative method to conventional confocal microscopy, based on a simplified and cheaper instrumental arrangement since it does not require point by point scanning or the mandatory use of expensive high spatial coherence source such as lasers. Since FF-OCT and SIM share major optical characteristics (full-field acquisition geometry, use of incoherent sources), the combination of the two techniques will be based on a simplified optical arrangement, while still compatible with pathology requirements.

16.3.2 Combined FF-OCT and Fluorescence Sectioning Microscopy

16.3.2.1 Setup

A proposed setup [16] is described in Fig. 16.6. It takes advantage of the full-field image acquisition geometry of FF-OCT and SIM to minimize the noncommon optical path between both imaging techniques and facilitate instrumental integration. Indeed, both techniques are based on the use of a standard microscope configuration, with an additional objective in the reference arm on the case of FF-OCT. The main microscope optical path is then common, and illumination and detection subsystems are modified in order to combine characteristics of both FF-OCT and SIM.

FF-OCT setups have been extensively described in the past [6, 39], and are usually based on imaging interferometers, such as

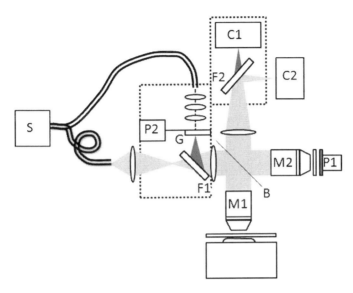

Figure 16.6 Instrumental setup—combined FF-OCT/SIM for fluorescence. S, light source; C1, fluorescence camera; C2, FF-OCT camera; M1, M2, microscope objectives; P1, P2, piezoelectric modulators; F1, F2, filters; G, Ronchi grid; B, beam splitter. Areas in dotted lines correspond to modifications of the FF-OCT setup for the integration of SIM.

Linnik or Mirau microscopes. The proposed setup consists of a Linnik interferometer as the basis of the FF-OCT setup with the addition of SIM illumination and detection subsystems. Subsystems in dotted lines in Fig. 16.6 correspond to modified parts of a conventional Linnik-based FF-OCT setup.

An incoherent broadband light source S, typically a tungsten-halogen or xenon-fibered bulb, is sent to the two illumination subsystems using fiber bundles: a Köhler illumination arrangement in the case of FF-OCT, and a modulated binary 2D transmissive mask in the case of SIM, using a Ronchi grid G (typically 40 lp/mm) and a piezoelectric shifter P2. The two illumination arrangements are combined using filters F1, as the combination of a long pass filter (IR cold mirror; $\lambda > 700$ nm) for FF-OCT imaging and of a dichroic beamsplitter centered at 500 nm (Semrock FITC-Di01) for the selection of fluorescence excitation light (in order to adapt to selected fluorophores, see Section 16.3.3.1, "Sample Selection and Preparation"). The detection subsystem is composed of the FF-OCT camera C2 (Dalsa Pantera 1M60, 1 Mpixels, 60 Hz, 500 ke well depth) plus an additional camera for fluorescence imaging C1 (Andor iXon3 EMCCD, 512 × 512 pixels). Spectral separation between the FF-OCT and the SIM detection channels is achieved using filters F2, as the combination of a dichroic beamsplitter centered at 700nm (IR hot mirror) and of a long pass filter @515 nm (Semrock FITC-LP01). Both cameras having substantially similar pixel size (12 and 13 μm, respectively), a single achromat is used as the tube lens for both channel in order to simplify instrumental integration, and its focal length is selected to fulfill sampling requirements (Nyquist frequency). The Linnik interferometer for FF-OCT is based on a broadband nonpolarizing beamsplitter B such as a pellicle beamsplitter, on 2two identical microscope objectives M1 and M2 (OlympusUPlanAPO × 30 NA 1.05, silicon oil immersion) with a large NA for fluorescence collection, and on a reference mirror (8% reflectivity) modulated using a piezoelectric shifter P1 (Piezomechanik GmbH).

16.3.2.2 Full-field optical coherence tomography: resolution

Details about FF-OCT theory and various instrumental characterizations are extensively described in other chapters and in the literature. The main feature of FF-OCT being the achievable 3D resolution, in particular in the case of pathology applications, we describe here the characterization of the resolution performance of the proposed setup.

The transverse resolution of both FF-OCT and SIM is driven by the diffraction theory, in particular using the Airy disk formula:

$$r = \frac{1.22 \cdot \lambda}{2 \cdot \mathrm{NA}}$$

where r is the Airy disk radius, λ is the central wavelength, and NA is the numerical aperture of the microscope objective. The corresponding transverse resolution is 0.43 μm for FF-OCT and 0.32 μm for SIM, the difference being due to the shift of the central wavelength between both modalities used for spectral separation. These characteristics share the same order of magnitude, so

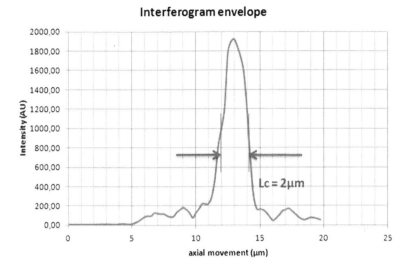

Figure 16.7 Axial envelope of the measured FF-OCT interferogram and corresponding coherence length.

merged SIM and FF-OCT images can be directly visualized without significant processing such as resampling.

As for all OCT techniques, the thickness of the optical section provided by FF-OCT, which corresponds to the axial resolution, is linked to the coherence length of the source. This coherence length is measured by the axial envelope of the interferogram, which is obtained by a fine axial translation of a reflective surface positioned as the sample. Figure 16.7 shows the measured envelope on the experimental setup, so the axial resolution is given by

$$\Delta z = \frac{L_C}{2n},$$

with L_c being the coherence length and n the refractive index of the imaging medium. The measured axial resolution calculated from the coherence length is 0.72 μm.

16.3.2.3 Structured illumination fluorescence microscopy

The simplest implementation of SIM illumination consists of the use of a transmissive binary 2D mask such as a Ronchi or sinusoidal grid

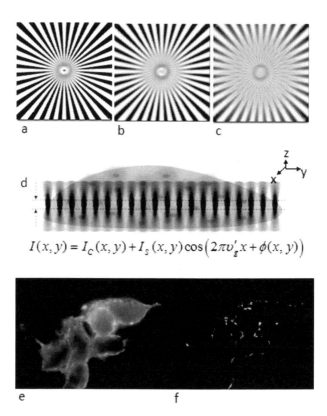

Figure 16.8 Optical sectioning using structured illumination. (a–c) Simulation of the effect of defocus on spatial frequencies demonstrating faster blurring of fine details; (d) schematic representation of a projected 2D mask on a volume sample illustrating the principle of optical sectioning using SIM; (e and f) conventional (e) and SIM (f) fluorescence images of cells with tagged membrane (quantum dots) showing the optical sectioning capability of SIM (grating 40 lp/mm).

conjugated with the imaging plane (focal plane of the microscope objective). The grid is spatially modulated in the plane orthogonal to the optical axis, along the direction orthogonal to the linear pattern. Briefly, as illustrated in Fig. 16.8, the method is based on the faster defocalization of high spatial frequencies, that is, fine details. As a result the imaged mask is only sharp in the depth of focus of the objective. A combination of at least three images (I1,

I2, I3) corresponding to phase shifts of the grid of 0, $2\pi/3$, and $4\pi/3$, that is, nonoverlapping images of the mask, allows for the reconstruction of an optical section, getting rid of parasitic out of focus signal, in particular in the case of fluorescence microscopy. The use of differences of pairs of images remove constant out-of-focus signal and grid patterns, according to the following basic processing [9]:

$$I_S = \sqrt{(I_1 - I_2)^2 + (I_1 - I_3)^2 + (I_2 - I_3)^2}$$

To compare with FF-OCT, we characterize here the axial resolution of the SIM setup (transverse resolution was calculated in the previous section).

The optical section provided by SIM is characterized by the measurement of the so-called *plane spread function* of the SIM setup. The measurement is performed using a homogeneous fluorescent plane (such as a fluorescent layer on top of a microscope glass slide), which is axially scanned through the focal plane of the objective, while performing the reconstruction of the optically sectioned image for each axial position. The average signal is plotted with respect to the axial position of the fluorescent plane. Figure 16.9 shows a measured FWHM of 5.8 µm: This value is slightly different from the theoretical axial resolution of the device, calculated from theoretical equations described in [9], and equal to 4.5 µm for a central wavelength of 550 nm. This difference might be due to translation errors from the manual micrometric linear stage used for displacing the fluorescent plane.

It is noticeable that SIM shows significantly lower sectioning ability when compared to FF-OCT: Axial resolution in the case of SIM is comparable to the depth of focus of the microscope objective, since the technique is basically a confocal—full-field—approach. SIM images are as a consequence expected to appear not as sharp as FF-OCT images. However, the measured axial resolution is comparable to the thickness of a conventional histology slice (typically 4 to 5 µm) obtained using a microtome, so the technique is applicable to pathology applications.

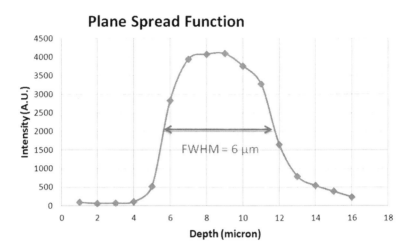

Figure 16.9 Plane spread function measurement.

16.3.3 Multimodal FF-OCT/Fluorescence Images of Healthy and Cancerous Tissue Samples

16.3.3.1 Sample selection and preparation

Skin specimens were selected targeting validation of the multimodal technique, since its tissue architecture—in particular the epidermis and epidermis/dermis junction—is characterized by a dense cell matrix (epidermis) with large, easily recognizable nuclei, without any additional structures, such as collagen, potentially introducing reading artifacts. This tissue type is then fully adapted to the use and validation of SIM as a complementary technique targeting the visualization of nuclei, which are of particular interest in pathology, as previously explained.

Skin specimens were collected from remaining tissue following skin surgery of basal cell carcinoma (BCC), from our clinical partner, and fixed in formalin. This remaining tissue is normally discarded and not used for diagnosis. The Institutional Review Board approval has been obtained for transferring and using the samples. Average size of samples was $5 \times 5 \times 2$ mm.

The staining strategy is focused on the visualization of cell nuclei. A fluorescence dye targeting DNA and enabling deep tissue pene-

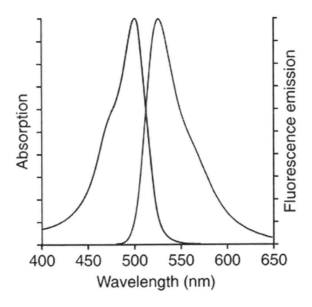

Figure 16.10 Acridine orange absorption/emission spectra.

tration was selected: specimens were stained using acridine orange (invitrogen absorption and emission spectra shown in Fig. 16.10), which shows large quantum efficiency. The staining protocol was duplicated from a previously described protocol used in confocal microscopy [7] for the specific contrast enhancement of nuclei. Each specimen was immersed in a solution of 0.6 mM of acridine orange during 60 s, which was found to be an optimum staining procedure, avoiding artifacts due to nonspecific binding of acridine orange in the dermis with higher concentrations [7]. Each specimen was then rinsed in isotonic saline, and mounted in a specific sample holder. A clean glass cover was mounted on top of the sample and gently pressed in order to provide an even imaging surface.

16.3.3.2 Imaging protocol

FF-OCT and SIM fluorescence images are acquired and reconstructed using standard procedures described in the literature. Each FF-OCT image is reconstructed from the combination of four individual images with an additional path length difference of $\pi/2$

between each of them. Similarly each SIM image is reconstructed from the combination of three images acquired following a lateral shift of the projected Ronchi grid of $2\pi/3$. Published image processing algorithms, such as image normalization and precise calculation of the effective translation using phase correlation, were used to remove or minimize remaining SIM artifacts like residual grid patterns or fluorescence bleaching.

Using the xenon lamp, the integration time was set to 5 ms for the acquisition of FF-OCT images, and 35 to 70 ms for the acquisition of fluorescence SIM images. To obtain a good signal-to-noise ratio, several images were averages for each modality: typically 40 images for FF-OCT and 5 images for SIM. As both types of images can be acquired simultaneously using spectral separation (see Section 16.3.2.1, "Setup"), the average acquisition time for a multimodal image is 1s. This time corresponds to an individual field of view of 245 µm × 245 µm for FF-OCT and about half for SIM. The achievable field of view can easily be enlarged using mosaicing techniques [14].

The transverse image registration between both modalities was calibrated before any acquisition on biological samples, using a reflective structured target (USAF). Measured offsets in both translation and rotation, mainly linked to the relative position of the two cameras conjugated on the sample, were recorded and further applied to the merging of FF-OCT and SIM images.

A pseudocolor (green) was applied to SIM images during the merging process in order to easily visualize the relative contribution of both techniques.

16.3.3.3 Results

We depict here the most representative images, illustrating the capabilities of the combined instrument to visualize both the tissue architecture and cell nuclei, targeting pathology applications.

Figure 16.11 shows FF-OCT images of a large axial resection of a skin biopsy. The sample (16 mm × 7 mm) has been cut from the skin surface (epidermis) to the dermis, and positioned into a sample holder, so all skin microstructures can be visualized in a single image. Figure 16.11a shows a macroscopic FF-OCT image of the specimen, as the result of postacquisition image stitching. It is easily

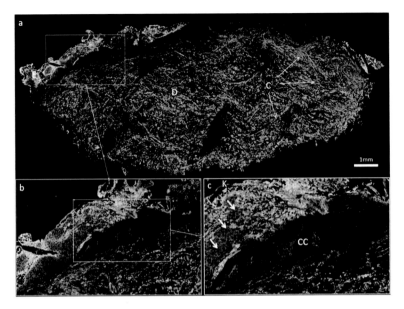

Figure 16.11 FF-OCT images of a skin specimen, including whole view (a) zooms at various levels of the epidermis/dermis junction (b and c). D, dermis; C, collagen bundles; CC, cancer cells; K, keratinocytes; arrows in c, nuclei of cells of stratum spinosum. Acquisition depth: 30 μm.

possible to recognize the macroscopic organization of skin, with the hyperreflective epidermis located at the top left and the dermis (D) constituting the majority of the image, mostly characterized by the wavy aspect of collagen fibers. Figures 16.11b and 16.11c correspond to zooms on the epidermis/dermis junction, where it is possible to visualize the epidermis, the beginning of the dermis, and a dark area at the junction. The epidermis is a favorable layer of the skin for FF-OCT, since cells of the stratum corneum and stratum spinosum are large and easily distinguishable. For example, cell nuclei of the stratum spinosum are clearly visible and dark (arrows in Fig. 16.11c) as well as large, flat keratinocytes (K label in Fig. 16.11c). Such a characteristic justifies the choice of skin as a validation sample type, since it is possible to co-register FF-OCT and fluorescence images and confirm the specific targeting of nuclei using SIM, as will be illustrated by the next figures. The dark area marked in CC in Fig. 16.11c corresponds to cancer cells (BCC in situ).

586 | Dual-Modality Full-Field Optical Coherence and Fluorescence Sectioning Microscopy

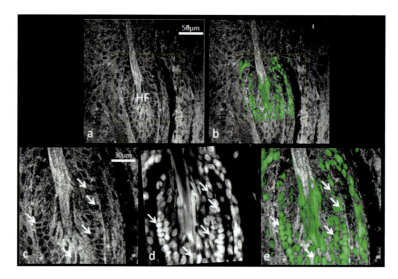

Figure 16.12 Combined FF-OCT/SIM image of a hair follicle (HF). (a) FF-OCT image; (b) merged FF-OCT/SIM image; (c) zoom on FF-OCT image; (d) SIM fluorescence signal of the same zoom area; and (e) merged images. Cell nuclei are marked with arrows on (c), (d), and (e), showing good multimodal correlation. Acquisition depth: 30 µm.

Figure 16.12 shows a first example of a co-registered FF-OCT/SIM fluorescence image of a hair follicle (HF in Fig. 16.12a) located in the dermis, prepared following the previously detailed protocol. Like the epidermis, the hair follicle is a favorable structure since its root is composed of many large cells, which can be visualized by FF-OCT, with nuclei appearing dark (arrows in Fig. 16.12c). When zooming in the area of the root, the FF-OCT image (Fig. 16.12c) and fluorescence image (Fig. 16.12d) show a good correspondence that can be visualized on the merged image (Fig. 16.12e), with arrows designating cell nuclei at the same location on the single modality and merged images. This figure validates both the image registration—and indirectly the calibration of the position of both imaging channels—and also the staining protocol, which makes it possible to target nuclei alone and to collect a proper fluorescence signal using reasonable acquisition times.

When comparing the FF-OCT and SIM images in detail, it is noticeable that some additional nuclei can be visualized in the

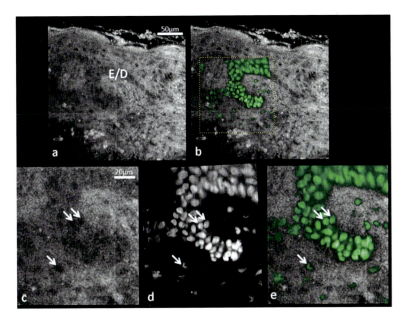

Figure 16.13 Combined FF-OCT/SIM image of the epidermis/dermis junction (E/D). (a) FF-OCT image; (b) merged FF-OCT/fluorescence SIM image; (c) FF-OCT zoom on papilla; (d) SIM fluorescence image of the same zoom area; and (e) merged (c) and (d). Some cell nuclei are indicated by arrows to visualize the correct image registration. Acquisition depth: 30 μm.

fluorescence image of Fig. 16.12d. This is mainly attributed to the significant difference of optical section between both modalities characterized previously (2 μm for FF-OCT and 6 μm for SIM): the SIM channel can collect signal from nuclei that are not included into the corresponding thinner FF-OCT section.

Figure 16.13 gives another example of a merged image, in this case from the epidermis/dermis junction (marked E/D in Fig. 16.13a). The area of a papilla was image in fluorescence and merged with FF-OCT signal (Fig. 16.13b). Details of the papilla are shown in zoomed images (Fig. 16.13c–e), with cell nuclei again indicated by arrows. The same remarks than Fig. 16.12 apply to the example, and it can be noticed here that the merged image brings even better information than for Fig. 16.12, since cell nuclei could not be visualized as clearly as around the follicle root.

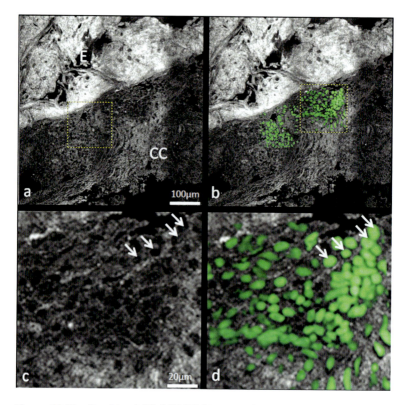

Figure 16.14 Combined FF-OCT/SIM image of basal cell carcinoma (CC) at the epidermis (E)/dermis junction. (a) FF-OCT image; (b) merged FF-OCT/fluorescence SIM image; (c) FF-OCT zoom on cancer cells; and (d) merged image of the same zoom area. Some cell nuclei are indicated by arrows to visualize the correct image registration. Acquisition depth: 30 µm.

As a consequence, the junction line appears clearer when combining both modalities (Fig. 16.12b), and the exact information about the density of cells is far more precisely determined from the merged image.

This capability to significantly improve the assessment of cell density has particular interest in the field of pathology, with many cancers being characterized by a strong aggregation of tumor cells. Figure 16.14 shows a multimodal image of a tumor invasion located at the epidermis/dermis junction (BCC in situ). The high density of tumor cells, which can be assumed when zooming on the FF-OCT image (Fig. 16.14c, with arrows pointing nuclei), is clearly identified on the merged image (Fig. 16.14b,d). Here again the fluorescence image allows to identify more nuclei than the FF-OCT image alone, mainly due to the difference in terms of thickness of the optical section provided by the two techniques.

16.4 Discussion and Conclusion

The proposed setup and corresponding first images demonstrate the feasibility of a multimodal FF-OCT/fluorescence optical sectioning imaging modality, for applications in pathology. The combination of two full-field acquisition methods allows for a simplified instrumental arrangement when compared to other scan-based optical sectioning techniques such as confocal microscopy. FF-OCT provides a fast visualization of tissue morphology at a cellular level, whereas fluorescence SIM supplements this contrast with the visualization of cell nuclei using the specific binding of fluorophores. A couple of alternative setups and corresponding images have been described in the literature [4, 24] and show some minor instrumental differences with the proposed instrument.

The first studies in a clinical context using FF-OCT as a noninvasive diagnosis method have shown good performance, mainly in terms of sensitivity, on various tissue types. In a first study on breast biopsies [2], a mean sensitivity of 94% and a specificity of 75% were obtained. Another preliminary study on prostate biopsies reported an overall agreement of 81% between histology and FF-OCT-based diagnosis. The certain identification of cell nuclei, and in particular of the cell density, was only possible in FF-OCT images alone in favorable cases, for structures such as epidermis or glands. The identification of cell proliferation, the study of the shape of the nucleus, and the assessment of the ratio nucleus/cytoplasm are of particular importance for an accurate pathology diagnosis, in particular targeting high specificity. The proposed multimodal technique provides a new approach to enhance FF-OCT images and has the potential to improve diagnosis performance.

Moreover the combination opens the door to a better matching with histology: The technique provides en face images such as histology slides, and it can be envisaged to deliver false color images with a similar aspect to H&E, with nuclei colored in purple (hematoxylin) and morphology colored in various levels of pink (eosin). Such an approach was already proposed in previous studies using both reflectance and fluorescence confocal microscopy [7] and could dramatically improve the learning curve of pathologists.

The intrinsic nondestructive nature of FF-OCT is another advantage in the field of pathology, when compared to existing intraoperative procedures on biopsies or resection such as ex tempo analysis using frozen sections. Indeed the recent development of molecular analysis in personalized medicine requires a minimum amount of informative tissue. This contradicts the current evolution of biopsy procedures, targeting less removal of tissue, so new methods preserving tissue might be a good alternative to FSA, which is destructive, operator dependent, and time consuming. The proposed method might be an alternative to FSA in the near future, all the more that it has the potential to be faster and is already fully digital.

Currently, the main limitation of this multimodal approach arises from the use of exogenous contrast agents, such as acridine orange in the proposed protocol. A limited number of contrast agents are currently authorized for clinical use, and are not specific to cancer cells. Also, the combined technique cannot anymore be considered noninvasive with use of fluorescent dyes. There are, however, strong efforts currently made to create new optical contrasts agents—mostly having fluorescence properties—with specific binding to cancer cells [21, 26, 38]. Some of these new molecules are currently under clinical evaluation for possible future clinical use. The proposed instrument could help to accelerate the clinical adoption of new diagnosis protocols.

Some recent work demonstrated the feasibility of in vivo imaging using FF-OCT, in particular using new endoscopy instruments [22]. SIM was also demonstrated through an endoscopy arrangement [29]. The combined approach can be adapted in the future to in vivo devices, targeting new clinical applications such as in situ assessment of tumor margins.

References

1. Adie, S. G., Liang, X., Kennedy, B. F., John, R., Sampson, D. D., and Boppart, S. A. (2010). Spectroscopic optical coherence elastography. *Opt. Express*, **18**(25), pp. 25519–25534.
2. Assayag, O., Antoine, M., Sigal-Zafrani, B., Riben, M., Harms, F., Burcheri, A., Grieve, K., Dalimier, E., Le Conte de Poly, B., and Boccara, C. (2014). Large field, high resolution full-field optical coherence tomography: a pre-clinical study of human breast tissue and cancer assessment. *Technol. Cancer Res. Treat.*, **13**, pp. 455–468
3. Assayag, O., Grieve, K., Devaux, B., Harms, F., Pallud, J., Chretien, F., and Varlet, P. (2013). Imaging of non-tumorous and tumorous human brain tissues with full-field optical coherence tomography. *NeuroImage Clin.*, **2**, pp. 549–557.
4. Auksorius, E., Bromberg, Y., Motiejūnaitė, R., Pieretti, A., Liu, L., Coron, E., Aranda, J., Goldstein, A. M., Bouma, B. E., Kazlauskas, A., and Tearney, G. J. (2012). Dual-modality fluorescence and full-field optical coherence microscopy for biomedical imaging applications. *Biomed. Opt. Express*, **3**, pp. 661–666
5. Barton, J. K., Guzman, F., and Tumlinson, A. (2004). Dual modality instrument for simultaneous optical coherence tomography imaging and fluorescence spectroscopy. *J. Biomed. Opt.*, **9**(3), pp. 618–623.
6. Beaurepaire, E., Boccara, A. C., et al. (1998). Full-field optical coherence microscopy. *Opt. Lett.*, **23**, pp. 244–246.
7. Bini, J., Spain, J., Nehal, K., Hazelwood, V., DiMarzio, C., and Rajadhyaksha, M. (2011). Confocal mosaicing of human skin ex vivo: spectral analysis for digital staining to simulate histology-like appearance. *J. Biomed. Opt.*, **16**(7), p. 076008.
8. Boppart, S. A., Luo, W., Marks, D. L., and Singletary, K. W. (2004). Optical coherence tomography: feasibility for basicresearch and image-guided surgery of breast cancer. *Breast Cancer Res. Treat.*, **84**, pp. 85–97.
9. Chasles, F., Dubertret, B., and Boccara, A. C. (2007). Optimization and characterization of a structured illumination microscope. *Opt. Express*, **15**(24), pp. 16130–16140.
10. Choma, M., Sarunic, M., Yang, C., and Izatt, J. (2003). Sensitivity advantage of swept source and Fourier domain optical coherence tomography. *Opt. Express*, **11**(18), pp. 2183–2189.

11. Dalimier, E., and Salomon, D. (2012). Full-field optical coherence tomography: a new technology for 3D high-resolution skin imaging. *Dermatology*, **224**(1), pp. 84–92.
12. Drexler, W., and Fujimoto, J. G. (2009). *Optical Coherence Tomography*, 2nd Ed. Springer International, Switzerland.
13. Drexler, W., Morgner, U., Ghanta, R. K., Kartner, F. X., Schuman, J. S., Fujimoto, J. G., et al. (2001). Ultrahigh-resolution ophthalmic optical coherence tomography. *Nat. Med.*, **7**, pp. 502–507.
14. Garreau, D. S., Patel, Y. G., Li, Y., Aranda, I., Halper, A. C., and Nehal, K. S. (2009). Confocal mosaicing microscopy in skin excisions: a demonstration of rapid surgical pathology. *J. Microsc.*, **233**(1), pp. 149–159.
15. Graf, B. W., and Boppart, S. A. (2012). Multimodal in vivo skin imaging with integrated optical coherence and multiphoton microscopy. *IEEE J. Sel. Top. Quantum Electron.*, **18**(4), pp. 1280–1286.
16. Harms, F., Dalimier, E., Vermeulen, P., Fragola, A., and Boccara, A. C. (2012). Multimodal full-field optical coherence tomography on biological tissue: toward all optical digital pathology. *Proc. SPIE*, **8216**(Multimodal Biomedical Imaging VII), p. 821609.
17. Hsiung, P. L., Phatak, D. R., Chen, Y., Aguirre, A. D., Fujimoto, J. G., and Connolly, J. L. (2007). Benign and malignant lesionsin the human breast depicted with ultrahigh resolution and three-dimensional optical coherencetomography. *Radiology*, **244**, pp. 865–874.
18. Huang, D., Swanson, E. A., Lin, C. P., Schuman, J. S., Stinson, W. G., Chang, W., Hee, M. R., Flotte, T., Gregory, K., Puliafito, C. A., and Fujimoto, J. G. (1991). Optical coherence tomography. *Science*, **254**, pp. 1178–1181.
19. Jain, M., Narula, N., Salamoon, B., Shevchuk, M. M., Aggarwal, A., Altorki, N., Stiles, B., Boccara, C., and Mukherjee, S. (2013). Full-field optical coherence tomography for the analysis of fresh unstained human lobectomy specimens, *J. Pathol. Inform.*, **4**, p. 26.
20. Jain, M., Shukla, N., Manzoor, M., Nadolny, S., and Mukherjee, S. (2011). Modified full field optical coherence tomography: a novel tool for histology of tissues. *J. Pathol. Inform.*, **2**, p. 28.
21. Kim, S., Lim, Y. T., Soltesz, E. G., et al. (2004). Near-infrared fluorescent type II quantum dots for sentinel lymph node mapping. *Nat. Biotechnol.*, **22**, pp. 93–97.
22. Latrive, A., and Boccara, A. C. (2011). In vivo and in situ cellular imaging full-field optical coherence tomography with a rigid endoscopic probe. *Biomed. Opt. Express*, **2**, pp. 2897–2904.

23. Liu, G., Lin, A. J., Tromberg, B. J., and Chen, Z. (2012). A comparison of Doppler optical coherence tomography methods. *Biomed. Opt. Express*, **3**(10), pp. 2669–2680.
24. Makhlouf, H., Perronet, K., Dupuis, G., Lévêque-Fort, S., and Dubois, A. (2012). Simultaneous optically sectioned fluorescence and optical coherence microscopy with full-field illumination. *Opt. Lett.*, **37**, pp. 1613–1615.
25. Masters, B. R., and So, P. T. C. (2001). Confocal microscopy and multiphoton excitation microscopy of human skin in vivo. *Opt. Express*, **8**(1), pp. 2–10.
26. Munro, I., McGinty, J., Galletly, N., et al. (2005). Toward the clinical application of time-domain fluorescence lifetime imaging. *J. Biomed. Opt.*, **10**, p. 051403.
27. Nakazawa, H., Rosen, P., Lane, N., and Lattes, R. (1968). Frozen section experience in 3000 cases: accuracy, limitations, and value in residence training. *Am. J. Clin. Pathol.*, **49**, pp. 41–51.
28. Neil, M. A. A, Juskaitis, R., and Wilson, T. (1997). Method of obtaining optical sectioning by using structured light in a conventional microscope. *Opt. Lett.*, **22**(24), pp. 1905–1907.
29. Bozinovic, N., Ventalon, C., Ford, T., and Mertz, J. (2008). Fluorescence endomicroscopy with structured illumination. *Opt. Express*, **16**, pp. 8016–8025.
30. Olson, T. P., Harter, J., Munoz, A., Mahvi, D. M., and Breslin, T. (2007). Frozen section analysis for intraoperative margin assessment during breast-conserving surgery results in low rates of re-excision and local recurrence. *Ann. Surg. Oncol.*, **14**, pp. 2953–2960.
31. Pierce, M. C., Javier, D. J., and Richards-Kortum, R. (2008). Optical contrast agents and imaging systems for detection and diagnosis of cancer. *Int. J. Cancer*, **123**, pp. 1979–1979.
32. Robbins, P., Saunders, C., Jacques, S. L., Sampson, D. D., McLaughlin, R. A., and Scolaro, L. (2010). Parametric imaging of cancer with optical coherence tomography. *J. Biomed. Opt.*, **15**(4), pp. 046029–046029.
33. Schmitt, J. (1998). OCT elastography: imaging microscopic deformation and strain of tissue. *Opt. Express*, **3**(6), pp. 199–211.
34. Silverstein, M. J., Recht, A., Lagios, M. D., Allred, D. C., Harms, S. E., Holland, R., et al. (2009). Image-detected breast cancer: state-of-the-art diagnosis and treatment. *J. Am. Coll. Surg.*, **209**, pp. 504–520.

35. Sun, C., Standish, B., and Yang, V. X. (2011). Optical coherence elastography: current status and future applications. *J. Biomed. Opt.*, **16**(4), pp. 043001–043001.
36. Taxy, J. (2009). Frozen section and the surgical pathologist: a point of view. *Arch. Pathol. LabMed*, **133**, pp. 1135–1138.
37. Tearney, G. J., Brezinski, M. E., Southern, J. F., Bouma, B. E., Boppart, S. A., and Fujimoto, J. G. (1998). Optical biopsy inhuman pancreatobiliary tissue using optical coherence tomography. *Dig. Dis. Sci.*, **43**, pp. 1193–1199.
38. Troyan, S. L., Kianzad, V., Gibbs-Strauss, S. L., et al. (2009). The FLARE™ intraoperative near-infrared fluorescence imaging system: a first-in-human clinical trial in breast cancer sentinel lymph node mapping. *Ann. Surg. Oncol.*, **16**, pp. 2943–2945.
39. Vabre, L., Dubois, A., and Boccara, A. C. (2002). Thermal-light full-field optical coherence tomography. *Opt. Lett.*, **27**, pp. 530–532.
40. Welzel, J., Lankenau, E., Birngruber, R., and Engelhardt, R. (1997). Optical coherence tomography of the human skin. *J. Am. Acad. Dermatol.*, **37**(6), pp. 958–963.
41. Wojtkowski, M., Leitgeb, R., Kowalczyk, A., Fercher, A. F., and Bajraszewski, T. (2002). In vivo human retinal imaging by Fourier domain optical coherence tomography. *J. Biomed. Opt.*, **7**(3), pp. 457–463.
42. Zhang, E. Z., Povazay, B., Laufer, J., Alex, A., Hofer, B., Pedley, B., and Drexler, W. (2011). Multimodal photoacoustic and optical coherence tomography scanner using an all optical detection scheme for 3D morphological skin imaging. *Biomed. Opt. Express*, **2**(8), pp. 2202–2215.
43. Zhou, C., Cohen, D. W., Wang, Y., Lee, H. C., Mondelblatt, E., Tsai, T. H., Aguirre, A. D., et al. (2010). Integrated optical coherence tomography and microscopy for ex vivo multiscale evaluation of human breast tissues. *Cancer Res.*, **70**, pp. 10071–10079.

PART IV

APPLICATIONS

Chapter 17

Full-Field Optical Coherence Tomography for Rapid Histological Evaluation of ex vivo Tissues

Manu Jain[a] and Sushmita Mukherjee[b]

[a]*Department of Pathology and Laboratory Medicine, Weill Medical College of Cornell University, New York, NY 10065, USA*
[b]*Department of Biochemistry, Weill Medical College of Cornell University, New York, NY 10065, USA*
smukherj@med.cornell.edu

17.1 Introduction

In this chapter, we describe the current status of research in the application of full-field optical coherence tomography (FF-OCT) to histopathological identification and diagnosis of tissues. Specifically, we discuss how FF-OCT can be used as a diagnostic tool in diverse clinical settings for the quick assessment of ex vivo tissues, without the need for tissue processing or staining. We start the chapter with an overview of current intraoperative diagnostic tools, especially frozen section analysis (FSA) and its limitations. We then provide a brief introduction to the basic principles and

---

Handbook of Full-Field Optical Coherence Microscopy: Technology and Applications
Edited by Arnaud Dubois
Copyright © 2016 Pan Stanford Publishing Pte. Ltd.
ISBN 978-981-4669-16-0 (Hardcover), 978-981-4669-17-7 (eBook)
www.panstanford.com

instrumentation of FF-OCT and how it might overcome some of the limitations of FSA. Next, we provide specific examples from published studies from our group, which demonstrate the ability of FF-OCT to recapitulate tissue architecture at histological resolution in fresh (unfixed, unsectioned, unstained) tissues [1–3]. Toward this end, we present a detailed atlas of various tissues from a normal rat model [1]. We also utilized FF-OCT to evaluate spermatogenesis within seminiferous tubules in a busulfan-treated, Sertoli-cell-only rodent model [3]. Additionally, we present results from a study utilizing human lobectomy specimens [2], where we compare FF-OCT images with gold-standard hematoxylin and eosin (H&E)-stained histopathological images generated from the same specimens. Based on the narrative summarized above, we foresee FF-OCT as a potentially powerful tool for rapid assessment of ex vivo tissues that can be applicable in various clinical contexts. Finally we end the chapter with certain limitations of the currently available commercial prototype of FF-OCT and how these limitations might be addressed in the future.

17.2 Current Practice for Intraoperative Diagnosis and Its Limitations

FSA is the mainstay for rapid, intraoperative assessment of tissue, for example, in situations where a surgeon might want to ensure a negative (tumor-free) margin during a solid tumor resection. Additionally, cytology interpretations are sometimes used for giving a quick diagnostic impression [4]. Although frozen sections are mainly done for the assessment of surgical margins intraoperatively, they may sometimes be employed to determine the nature of a lesion or to determine the adequacy of diagnosable tissue obtained (for later confirmatory histopathology or for tissue banking) [5]. Thus, especially in the cases of margin assessments, FSA greatly influences a surgeon's decision during an operative procedure and consequently affects patients' prognosis.

Although FSA is known to have a high overall diagnostic accuracy of 89%–98% [6–8], it has significant limitations. These limitations mainly arise from tissue processing as a result of freezing and sectioning of the tissue [4, 9], potentially causing artifacts. In rare cases, it may physically destroy the tissue. This in turn yields poor-quality histological sections that are difficult to analyze and sometimes leads to false diagnoses. Moreover, if such artifacts occur at the surgical margins, it may not be possible to determine whether the margin is positive or negative for malignancy [10], which in turn has major implications regarding prognosis and follow-up treatment options. Moreover, soft tissues such as brain and breast tissues are difficult to freeze and section and thus are not good candidates for FSA. With increasing demand for molecular analyses for confirmatory diagnoses, tissue banking is becoming an integral part of the histopathology workflow, which requires saving a "representative" part of the tissue for later analysis. However, this poses a problem when the biopsy is small in size and may be completely lost during FSA tissue processing [4]. Lastly, frozen sections result in a procedural delay of around 20 minutes, which lengthens the time the patient stays under general anesthesia, increasing patient morbidity as well as cost of the procedure [11].

Thus, there is a need for novel intraoperative diagnostic tools that can quickly generate histology-quality images from fresh ex vivo tissues (unprocessed, and unstained). One such technique is FF-OCT. FF-OCT has been shown as a potentially powerful tool for performing fast histology on fresh, as well as formalin-fixed ex vivo tissues obtained from both rat [1, 3, 12, 13] and human [2] organs, without introducing any tissue artifacts.

17.3 Basic Principles and Instrumentation of FF-OCT

FF-OCT is based on the principles of white-light interference microscopy. The optical arrangement is generally based on a bulk Michelson interferometer, with identical microscope objectives in both sample and reference arms. This configuration is referred to as the Linnik interferometer [14]. Due to the low temporal coherence of the source, interference occurs only when the optical path lengths of the two interferometer arms are identical within ~ 1 μm. When a biological object is placed under the objective in the sample arm, the light reflected by the reference mirror interferes with the light reflected or backscattered by the sample structures contained in a limited volume. This volume is a slice orthogonal to the objective axis, located at a depth inside the object defined by an optical path length difference of zero. The thickness of the slice is determined by the width of the fringe envelope. The extraction of the signal yields therefore an en face tomographic image. The signal is extracted from the large background of incoherent backscattered light using a phase-shifting method. Two interferometric images are recorded successively with a charge-coupled device (CCD) camera, the phase being changed by 180° in the interferometer between each image. The phase shift is accomplished by a displacement of the reference mirror with a piezoelectric translation stage. FF-OCT uses a simple tungsten halogen lamp as a light source.

The clinical FF-OCT prototype used in our studies (Light-CT^{TM}, LLTech SAS, Paris, France) utilizes two matched 10×/0.3 numerical aperture (NA) water immersion objectives (Olympus America, Center Valley, PA), one to collect reflections and backscattering signals from the specimen and the other to collect reflection signal from a reference mirror. The instrument design is shown in Fig. 17.1. This device uses a spatially and temporally incoherent light source of low power (Quartz-Halogen Schott KL 1500 Compact, Mainz, Germany). Light-CT^{TM} has a higher resolution (1.5 μm transverse and 0.8 μm axial) as compared to the traditional OCT systems [2]. The native field of view is 0.8 mm × 0.8 mm; however, larger fields of view can be acquired by image tiling. The system is also able to collect 3D image stacks through the top ∼60 μm of the specimen (the precise penetration depth varies with specimen type). Imaging of a single plane of 2.72 mm × 2.72 mm field of view takes ∼43 seconds.

17.4 FF-OCT-Generated Histology Atlas of Rat Organs

To establish the ability of FF-OCT in generating histology-quality images from fresh tissues, we harvested various organs from healthy rats and imaged them using Light-CT^{TM}.

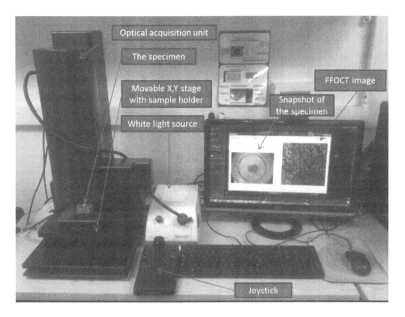

Figure 17.1 Full-field optical coherence tomography instrumentation. A photograph showing the layout of different components of the LLTech light-collisional thick target model system. Reproduced with permission from Ref. [2].

17.4.1 Skin

The normal skin is composed of three layers: (1) superficial epidermis, made of keratinized stratified squamous epithelium; (2) dermis, mainly comprised of collagen fibers, elastin fibers, blood vessels, nerves, and lymphatics, along with various skin appendages (hair follicles, sebaceous glands, and sweat glands); and (3) hypodermis, a subcutaneous layer mainly composed of adipose tissue. There is a distinct dermal-epidermal junction between the epidermis and dermis [15].

In Fig. 17.2A, we present a cross section of rat skin from its flank. Here, one can clearly appreciate the outer two layers of skin, that is, superficial epidermis and dermis, along with the intervening dermal-epidermal junction. Hypodermis was not present in our section. The epidermis appears as a multilayered epithelium, although the various layers of the epidermis are not recognizable at the given resolution. The dermis shows connective tissue (bright signals) along with some sebaceous glands and hair follicles.

17.4.2 Stomach

The normal rat stomach has four layers: (1) the innermost mucosa, consisting of an epithelium lined by gastric pits, lamina propria composed of loose connective tissue and packed with gastric glands, and a thin layer of smooth muscle called the muscularis mucosae; (2) the submucosa, mainly composed of fibrous connective tissue; (3) the muscularis externa (propria), comprised of three layers of smooth muscle bundles (inner oblique, middle circular, and outer longitudinal layers); and (4) the serosa, the outermost connective tissue layer that is continuous with the peritoneum [15].

Figure 17.2B is a cross-sectional FF-OCT image of a rat's stomach showing distinct layers of mucosa with its lining epithelium and gastric pits toward the lumen. Underneath the mucosa, the submucosa is seen with fibrous connective tissue, followed by the muscularis externa (propria) with smooth muscle bundles and the outermost fibrofatty layer of the serosa visible.

17.4.3 Liver

The normal liver parenchyma is mainly composed of cords of hepatocytes radiating from central veins and separated by sinusoidal spaces. A portal triad containing a branch of the hepatic arteriole, portal vein, and bile duct is seen in the liver parenchyma [16].

On FF-OCT, Fig. 17.2C shows the normal architecture of the rat's liver. The liver parenchyma is largely composed of hepatic cords radiating from a central vein. Also identifiable are the components of portal triads, that is, the portal vein (the biggest structure with a lumen), the hepatic arteriole (smaller luminal structures), and the bile duct. The bile duct is distinguishable from other luminal structures, that is, vessels, by a layer of epithelial cells (dull gray signal) lining its luminal border. In addition, the connective tissue capsule (bright signal) is also seen covering the liver surface.

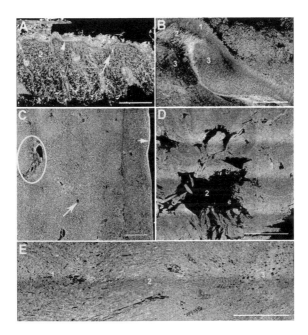

Figure 17.2 (A) Cross section of skin showing superficial epidermis (Region 1), dermis (Region 2), dermal-epidermal junction (arrowhead), superficial part of a hair follicle (arrow). (B) Cross section of stomach showing mucosa with gastric pits (Region 1), submucosa (Region 2), muscularis propria (Region 3), and serosa (region 4). (C) Cross section of liver showing outer capsule (arrowhead), cords of hepatocytes radiating from central veins (arrow), and portal triads (circle). (D) Cross section of heart showing branching muscle fibers of the myocardium (Region 1) and the ventricular chamber (Region 2). (E) Cross section of kidney showing convoluted tubules (Region 1), medullary rays (Region 2), and collecting ducts (Region 3). Scale bar: 1 mm. Reproduced with permission from Ref. [1].

17.4.4 Heart

The rat's heart has two ventricles (right and left) separated by interventricular septum and two atria (right and left). The atrium and ventricle are separated by valves (tricuspid and mitral valves). The heart wall is composed of striated, branching cardiac muscle fibers called myocardium [15].

Figure 17.2D, is a cross-sectional FF-OCT image of a rat's heart showing mostly myocardium with its branching cardiac muscle fibers, surrounding the empty space (signal void area) of a ventricular chamber.

17.4.5 Kidney

The rat's kidney has two distinct components, cortex and medulla. The cortex is characterized by glomeruli and highly specialized convoluted tubules. The medulla is made of straight tubules, running parallel to each other, called medullary rays, which finally drain into the collecting duct system. Collecting ducts are readily recognized in the renal medulla as tubules with relatively larger lumen [17].

Figure 17.2E shows an FF-OCT image of a cross section of a rat's kidney composed of an outer cortical area with convoluted tubules and a medulla with the straight tubules forming medullary rays and finally draining into larger-diameter collecting ducts.

17.4.6 Prostate

The prostate is composed of acinar glands embedded in a fibromuscular stroma. These glands are lined by a double-layered epithelium consisting of the inner secretory and outer basal layers. The prostate gland is enveloped by a fibrous capsule and periprostatic fat [16].

Figure 17.3 is an FF-OCT image of a rat's prostate, where prostatic acini are seen with the surrounding fibromuscular stroma. In addition, on zooming, papillary infoldings of the acini are evident. Also identifiable are the capsule, periprostatic fat, and large blood vessels.

17.4.7 Lung

The normal lung parenchyma is made up of a network of thin-walled alveoli. The rest is branching respiratory airways, starting from biggest diameter bronchi, and then bronchioles leading to alveolar duct, which eventually ends into a blind sac, called the alveolar

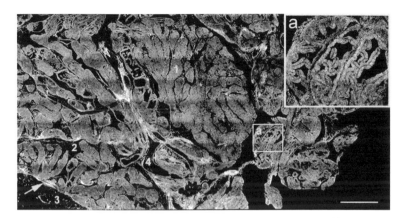

Figure 17.3 Cross section of prostate gland showing prostatic acini (Region 1) and fibromuscular stroma (Region 2), capsule (arrow), periprostatic fat (Region 3), and blood vessels (Region 4). Papillary folds of the acinar glands are shown in Inset (a) (zoomed in relative to the main image). Scale bar: 1.5 mm. Reproduced with permission from Ref. [1].

sac, where major exchange of gases takes place. This branching respiratory airway is accompanied all the way by pulmonary and bronchial arteries. The lung is surrounded by pleura, a thin layer of connective tissue which is lined by mesothelial cells [18].

Figure 17.4 shows an FF-OCT image of the normal architecture of a rat's lung where all the components of the branching airways such as bronchus, bronchiole, alveolar duct, and finally alveoli are seen. The bronchus is lined by multilayered epithelium (dull gray signal) and has cartilage in its wall (condensed bright signal). Pulmonary vessels are also identified and appear distinct from airways by the lack of epithelial lining (dull gray signal) seen in latter. Also identifiable is the lining visceral pleura of the lung.

17.4.8 Urinary Bladder

The urinary bladder is composed of urothelium, lamina propria, muscularis propria, and adventitia or serosa. The normal urothelium is a stratified transitional epithelium, which is three to seven cell layers thick. The underlying lamina propria is made of collagen

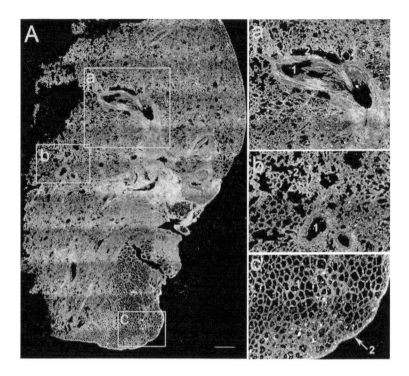

Figure 17.4 Cross section of lung. Inset (a) showing bronchus at the hilum (Region 1) and its accompanying pulmonary vessel (Region 2). Inset (b) shows a bronchiole (Region 1) and an alveolar duct (Region 2). The majority of the lung parenchyma, as seen in Inset (c), is composed of alveoli (Region 1) and surrounded by visceral pleura (Region 2). Scale bar: 0.5 mm. Reproduced with permission from Ref. [1].

bundles and medium-sized blood vessels. The muscularis propria is made of smooth muscle fibers. The serosa, i.e., the outermost layer, is mainly composed of connective tissue [19].

In Fig. 17.5, all layers of the urinary bladder are distinctly visible. The stratified nature of the urothelium (dull gray signal) can also be appreciated. The lamina propria is seen with its connective tissue (bright signals). In addition, muscular bundles of the muscularis propria, along with the underlying serosa (bright signals from connective tissue), are identifiable.

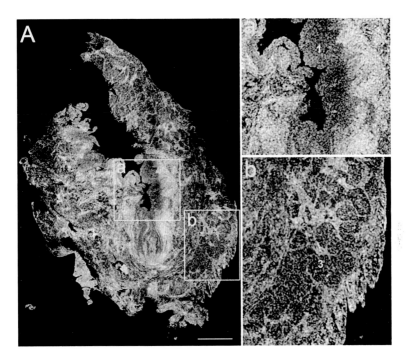

Figure 17.5 Cross section of urinary bladder. Inset (a) shows stratified urothelium (Region 1) and underlying connective tissue of lamina propria (Region 2). Inset (b) shows smooth muscular bundles of muscularis propria (Region 3) and the outermost layer of serosa (Region 4). Scale bar: 0.5 mm. Reproduced with permission from Ref. [1].

17.5 FF-OCT to Identify Spermatogenesis in Rat Testis

Nonobstructive azoospermia (NOA), the lack of sperm in the ejaculate, affects nearly 1% of all men and 10%–15% of infertile men [20]. Microdissection testicular sperm extraction (micro-TESE) coupled with intracytoplasmic sperm injection (ICSI) has replaced conventional testis biopsies for obtaining sperm for in vitro fertilization for men with NOA. However, low-magnification inspection of the tubules with a surgical microscope is insufficient to definitively identify sperm-containing tubules, leading to a successful retrieval in only 40%–60% of men. Furthermore, as this technique requires removal of tissue for assessment, there is a loss

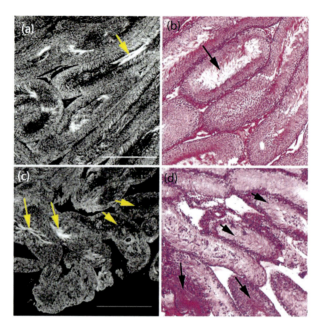

Figure 17.6 Comparative full-field optical coherence tomography and H&E-stained histology images from the testes of normal adult Sprague–Dawley rats and rats exhibiting a Sertoli-cell-only phenotype. (a) Seminiferous tubules in the testis of a normal rat. The tubules are relatively uniform in size and shape (diameter 328 + 11 μm). (b) Same specimen processed and stained for conventional (H&E) histology. Arrows point to the sperm within the tubule lumen. (c) Seminiferous tubules in the testis of a rat treated with busulfan. Tubules, on average, are thinner and show a greater degree of heterogeneity in size and shape (diameter 178 + 35 μm). Only ∼10% of the tubules show normal spermatogenesis as identified by presence of sperm tails (bright white hair-like structures within the lumen; long arrows). The remainder of the tubules showed no sperm within the lumen (short arrows). (d) H&E staining of the same specimen confirms the observations. Long and short arrows point to tubules with and without spermatogenesis, respectively. Field of view in each panel = 1 mm^2. Reproduced with permission from Ref. [3].

of testosterone-producing Leydig cells, which further reduces the serum testosterone levels in infertile men with already significantly lower serum testosterone levels [21]. Lower testosterone levels can result in serious long-term consequences such as osteoporosis, increased insulin resistance, and depression [22]. Sometimes, these procedures can unfortunately result in a permanent decrease in serum testosterone levels [23, 24], requiring long-term testosterone replacement therapy. Thus optimizing the ability to identify sperm-containing tubules before removing the tissue could reduce these potential risks and increase the benefit:risk ratio for men attempting fatherhood by this method.

We have previously shown that FF-OCT can be utilized to evaluate spermatogenesis within seminiferous tubules in a busulfan-treated, Sertoli-cell-only rodent model [3]. Figure 17.6 shows stark differences not only between the size and the shape of the seminiferous tubules upon busulfan treatment but also in the content of its lumen, that is, presence or absence of sperms, specifically sperm tails (bright white hair-like structures) within the lumen. The normal tubules are narrow and elongated with sperms in the lumen. On the other hand, tubules from Sertoli-cell-only rodent model shows a marked heterogeneity in their size and shape (diameter 178 + 35 μm) with only ~10% of the tubules having normal spermatogenesis, with no visible sperms within the lumens of the rest of the tubules.

Thus, FF-OCT has the potential to facilitate real-time visualization of spermatogenesis in humans and aid in micro-TESE for men with infertility. The unique advantage of this technique, in addition to fast speed of image acquisition and lack of tissue processing, is the use of a nonlaser safe light source (150 W halogen lamp) for tissue illumination, which ensures that the sperm extracted for in vitro fertilization are not photodamaged or mutagenized [3].

17.6 FF-OCT for the Analysis of Human Lobectomy Specimens

Lung cancer is the second most common cancer in both men and women. Routinely, lung tumors are detected on chest X-rays and computed tomography scans. Biopsies are then obtained from the suspicious lesions and submitted for histopathological confirmation. Sometimes smaller wedge resections of lung are performed and sent for FSA to confirm negative surgical margins. This is done to avoid

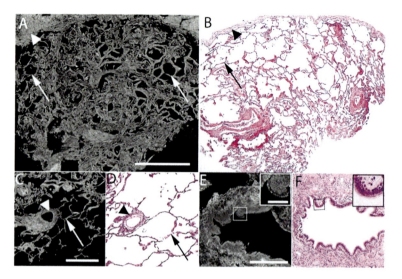

Figure 17.7 Comparative full-field optical coherence tomography and H&E images of nonneoplastic lung. (A, B) Large-field images show lung parenchyma composed of alveoli (signal void areas; arrows) surrounded by pleura (connective tissue-bright signals; arrowheads). Some thickening of the alveolar septa is shown (right arrow). (C, D) Images of blood vessel (arrowheads) and surrounding alveoli (arrows). (E, F) Images of a bronchus, with columnar epithelial lining (box and inset) and underlying connective tissue (connective tissue-bright signal). Scale bars for FF-OCT: (A) 1 mm, (C, E) 0.5 mm, and inset in (E) 0.1 mm. H&E total magnifications: (B) ×40 and (D, F) ×200. Inset in (F) ×2.5 zoom. Reproduced with permission from Ref. [2].

radical lobectomies, especially in patients who have small tumors and those with already compromised lung function [5].

As mentioned earlier, although histopathological analysis and FSA are the gold standard in diagnosis and management of tumors, they are nevertheless fraught with significant limitations delaying pathology feedback, increasing surgical time, with associated increases in morbidity and cost for the patient [11].

In a previously published report [2], our group has shown the ability of FF-OCT in not only recapitulating normal histology of human lung, as shown in Fig. 17.7, but also in identifying and differentiating neoplastic from nonneoplastic tissue. Moreover, we could recognize various histological patterns of tumors in these FF-

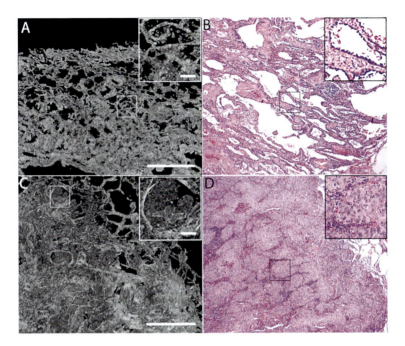

Figure 17.8 Comparative full-field optical coherence tomography and H&E images of neoplastic lung. (A, B) Images of adenocarcinoma of lung with lepidic-predominant pattern. Boxed areas and insets show tumor cells lining the alveolar septa. (C, D) Images of adenocarcinoma of lung with solid-predominant pattern. Boxed areas and insets shows clusters of tumor cells. Scale bars for FF-OCT: (A, C) 1 mm and insets in (A, C) 0.1 mm. H&E total magnifications: (B, D) ×100 and insets in (B, D) ×200. Reproduced with permission from Ref. [2].

OCT images, especially for adenocarcinomas (Fig. 17.8). Specifically, the more differentiated adenocarcinomas with predominant lepidic patterns (cells growing along the alveolar septa) were distinguishable from the invasive tumor with a predominant solid pattern (clusters of cells). The tumor cells appear to have a dull gray signal, similar to other normal cells lining the bronchi (Fig. 17.7e,f) but were distinguishable from normal cells by their size and arrangement. Despite correctly identifying the presence of tumor in all the tumor specimens, a high false positive rate was obtained for nonneoplastic lung tissue. These false positives were especially

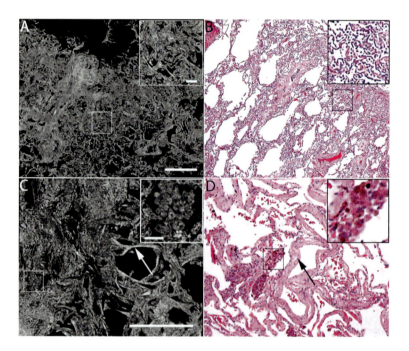

Figure 17.9 Comparative full-field optical coherence tomography and H&E images of nonneoplastic lung with false positive diagnosis. (A, B) Images of nonneoplastic tissue with collapse of normal lung architecture. Boxed areas and insets show dense connective tissue where it is difficult to rule out presence of tumor. (C, D) Images of nonneoplastic tissue showing clusters of the smoker's macrophages (boxed areas and insets) and thickened alveolar septa (arrows). Inset shows bright spots in the cytoplasm of the macrophages, which are likely to be tar. Scale bars for FF-OCT: (A) 1 mm, (C) 0.5 mm, inset in (A) 0.1 mm, and inset in (C) 0.05 mm. H&E total magnifications: (B) ×40 and (D) ×100. Insets in (B, D) ×200. Reproduced with permission from Ref. [2].

encountered in areas with extensive lung collapse or abundant smoker's macrophages, making it difficult to rule out the presence of tumor (Fig. 17.9).

Based on this study we foresee FF-OCT in surgical margin assessments, especially in cases of limited lung resections, for example, in patients with compromised lung function. The fact that FF-OCT does not require any tissue processing and can generate histological-quality images from fresh tissue can help overcome the problem of freezing and sectioning artifacts observed with FSA. In addition FF-OCT can also be used to determine the adequacy of the biopsy material in freshly excised tissue before sending them for histopathology, molecular studies, or banking. Thus it can reduce the number of repeat biopsies and the associated complications, such as bleeding, pneumothorax, and occasional needle track seeding, as well as improve costs and time-to-decision for the patients [25].

17.7 Potential Clinical Applications of FF-OCT

FF-OCT has several advantages over other optical imaging techniques that are being currently explored, such as rapid speed of image acquisition of a relatively large area of the tissue, ability to image both fresh and formalin-fixed specimens, and the use of a simple (non-laser-based) light source that makes it a safe and ideal tool for analyzing tissues. In addition, with a relatively small foot print of the commercial Light-CT^{TM} system and its user-friendly format, it can be installed in either a gross examination room and/or operation suite to get quick histological impressions on freshly excised tissue. Thus, based on the above-mentioned published works [1–3] and its advantages, we envision the use of the commercial Light-CT^{TM} in the following clinical or surgical contexts. Of course, similar applications can be found in a variety of tissue types, in addition to the organ-specific applications mentioned above for human lobectomy [2] and testis specimens [3].

(a) *FF-OCT as an adjunct tool to FSA in intraoperative consultation:* FF-OCT can help in surgical margin assessment. The fact that FF-OCT does not require any tissue processing and can generate histology-quality images from fresh tissue can help overcome the problem of freezing and sectioning artifacts observed with FSA. In addition, since no physical sectioning or processing of the tissue is required, the imaged tissue can be preserved in its entirety for ancillary techniques (immunohistochemistry, genetic analysis etc.) [4]. This will be especially valuable when the section under evaluation is small and might be lost during tissue processing. Furthermore, the ability of FF-OCT to generate

large field images (up to 3 mm^2) at a relatively fast speed (a single plane of 2.72 mm × 2.72 mm can be acquired in ∼43 s) can help speed up intraoperative decision making [1].
(b) *FF-OCT in obtaining diagnostically relevant tissue during biopsy procedure:* By determining the adequacy of the biopsy material in freshly excised tissue before sending them for histopathology, FF-OCT can reduce the number of repeat biopsies and the associated complications, such as bleeding, infection, and occasional needle track seeding, as well as improve costs and time-to-decision for the patients [25].
(c) *FF-OCT in biobanking:* FF-OCT can be used for tissue selection noninvasively in the specimen before cryopreservation [26]. This can be especially used for tumor selection in prostatectomy specimens, where the tumor is not always obvious on gross (naked eye) inspection of the tissue.

17.8 Current Limitations of FF-OCT and Possible Solutions

The major limitation of the current clinical prototype of FF-OCT (the commercial Light-CT^{TM}) is the absence of detailed cellular and nuclear features in most tissues. This is predominantly due to the use of a relative low-NA (0.3) objective in this prototype, along with the presence of speckle artifacts that further reduces effective resolution. However, these limitations and the high false positive rate reported for lung tumors by Jain et al. [2] can be overcome in the future. For example, laboratory-based systems have already been described with much higher resolution (0.7 μm × 0.9 μm axial and lateral resolution, respectively, using a pair of 40×/0.8 NA objectives) [27]. Such systems should be able to provide the necessary resolution, for example, to identify cellular and nuclear features that will allow the distinction between normal and malignant cells.

Similarly, laboratory-based systems using a Xenon flash lamp as the light source have been described [28] where the frame rates can be increased to 10 microsecond/frame. In this system, rather than producing two phase-opposed images by path length modulation and capturing them consecutively on a single CCD camera (like in Light-CT^{TM}), a quarter-wave plate introduced in one of the interferometric arms is used to create a $\lambda/2$ phase retardation between the object and the reference. A polarizing beam splitter then splits the interference signal into two phase-opposed interference images, which are simultaneously captured by two identical CCD cameras. The final tomographic image is then calculated from this pair of interference images.

In addition, several approaches are being considered for ex vivo specimens to increase contrast by using chemical agents that alter the optical properties of the tissue (e.g., acetic acid, aluminum chloride) [29, 30]. Future studies will address the efficacy of such methods and how tissues thus treated behave in terms of subsequent gold-standard histopathology. Some of the criteria for clinical application of such techniques will include ensuring that the treatment is fast, that it penetrates to the depths that FF-OCT can image, and that it does not alter tissue properties that might later be required to generate permanent histopathology sections, carry out ancillary studies, or bank the tissue.

In conclusion, the fast speed of image acquisition of a relatively large area, lack of tissue processing and associated artifacts, and the use of a safe light source (halogen lamp) makes FF-OCT an attractive potential diagnostic tool for quick assessments of freshly excised tissue in multiple clinical settings. Additionally, further improvements made in the future iterations of Light-CT^{TM} (along the lines outlined above), combined with its small footprint and user-friendly interface, and would make this device even more useful to the clinical pathology labs.

References

1. Jain, M., Shukla, N., Manzoor, M., Nadolny, S., and Mukherjee, S. (2011). Modified full-field optical coherence tomography: a novel tool for rapid histology of tissues. *J. Pathol. Inform.*, **2**, p. 28.
2. Jain, M., Narula, N., Salamoon, B., et al. (2013). Full-field optical coherence tomography for the analysis of fresh unstained human lobectomy specimens. *J. Pathol. Inform.*, 4, p. 26.

3. Ramasamy, R., Sterling, J., Manzoor, M., et al. (2012). Full field optical coherence tomography can identify spermatogenesis in a rodent sertoli-cell only model. *J. Pathol. Inform.*, **3**, p. 4.
4. Taxy, J. (2009). Frozen section and the surgical pathologist: a point of view. *Arch. Pathol. Lab. Med.*, **133**, pp. 1135–1138.
5. Sienko, A., Allen, T., Zander, D., and Cagle, P. (2005). Frozen section of lung specimens. *Arch. Pathol. Lab. Med.*, **129**, pp. 1602–1609.
6. Nakazawa, H. (1968). Frozen section experience in 3000 cases: accuracy, limitations, and value in residence training. *Am. J. Clin. Pathol.*, **49**, pp. 41–52.
7. Ishak, K. (1988). Benign tumors and pseudotumors of the liver. *Appl. Pathol.*, **6**, pp. 82–104.
8. Nagasue, N., Akamizu, H., Yukaya, H., and Yuuki, I. (1984). Hepatocellular pseudotumor in the cirrhotic liver. Report of three cases. *Cancer*, **54**, pp. 2487–2494.
9. Weinberg, E., Cox, C., Dupont, E., et al. (2004). Local recurrence in lumpectomy patients after imprint cytology margin evaluation. *Am. J. Surg.*, **188**, pp. 349–354.
10. Lechago, J. (2005). Frozen section examination of liver, gallbladder, and pancreas. *Arch. Pathol. Lab. Med.*, **129**, pp. 1610–1618.
11. McLaughlin, S., Ochoa-Frongia, L., Patil, S., Cody, H., and Sclafani, L. (2008). Influence of frozen-section analysis of sentinel lymph node and lumpectomy margin status on reoperation rates in patients undergoing breast-conservation therapy. *J. Am. Coll. Surg.*, **206**, pp. 76–82.
12. Wang, J., Léger, J. F., Binding, J., Boccara, A. C., Gigan, S., and Bourdieu, L. (2012). Measuring aberrations in the rat brain by coherence-gated wavefront sensing using a Linnik interferometer. *Biomed. Opt. Express*, **3**, pp. 2510–2525.
13. Unglert, C. I., Namati, E., Warger, W. C., et al. (2012). Evaluation of optical reflectance techniques for imaging of alveolar structure. *J. Biomed. Opt.*, **17**, p. 071303.
14. Dubois, A., Vabre, L., Boccara, A., and Beaurepaire, E. (2002). High-resolution full-field optical coherence tomography with a Linnik microscope. *Appl. Opt.*, **41**, pp. 805–812.
15. Wilson, F. (1999). *Histology Image Review*. Mcgraw-Hill.
16. Keer, H. N., Gaylis, F. D., Kozlowski, J. M., et al. (1991). Heterogeneity in plasminogen activator (PA) levels in human prostate cancer cell lines: increased PA activity correlates with biologically aggressive behavior. *Prostate*, **18**, pp. 201–214.

17. Anderson, J. (2007). Surgical anatomy of the retroperitoneum, adrenals, kidneys, and ureters. In: Wein, A. J., ed. *Campbell-Walsh Urology*. Vol. 1, 9th Ed. Saunders Elsevier, Philadelphia, PA.
18. Husain, A. (2009). The lung. In: Kumar, V., ed. *Robbins and Cotran Pathologic Basis of Disease*. 8th Ed. Saunders Elsevier, Philadelphia, PA.
19. Epstein, J., Amin, M., and Reuter, V. (2010). *Biopsy Interpretation of the Bladder*. 2nd Ed. Lippincott Williams & Wilkins, Philadelphia, PA.
20. Thonneau, P., Marchand, S., Tallec, A., et al. (1991). Incidence and main causes of infertility in a resident population (1,850,000) of three French regions (1988-1989). *Hum. Reprod.* **6**, pp. 811–816.
21. Andersson, A. M., Jørgensen, N., Frydelund-Larsen, L., Rajpert-De Meyts, E., and Skakkebaek, N. E. (2004). Impaired Leydig cell function in infertile men: a study of 357 idiopathic infertile men and 318 proven fertile controls. *J. Clin. Endocrinol. Metab.*, **89**, pp. 3161–3167.
22. MacIndoe, J. H. (2003). The challenges of testosterone deficiency. Uncovering the problem, evaluating the role of therapy. *Postgrad. Med.*, **114**, pp. 51–53, 57–58, 61–62.
23. Ishikawa, T., Yamaguchi, K., Chiba, K., Takenaka, A., and Fujisawa, M. (2009). Serum hormones in patients with nonobstructive azoospermia after microdissection testicular sperm extraction. *J. Urol.*, **182**, pp. 1495–1499.
24. Takada, S., Tsujimura, A., Ueda, T., et al. (2008). Androgen decline in patients with nonobstructive azoospemia after microdissection testicular sperm extraction. *Urology*, **72**, pp. 114–118.
25. Smayra, T., Braidy, C., Menassa-Moussa, L., Hlais, S., Haddad-Zebouni, S., and Aoun, N. (2012). Risk factors of pneumothorax and hemorrhage in lung biopsy: a single institution experience. *J. Med. Liban.*, **60**, pp. 4–13.
26. Dalimier, E., and Salomon, D. (2012). Full-field optical coherence tomography: a new technology for 3D high-resolution skin imaging. *Dermatology*, **224**, pp. 84–92.
27. Grieve, K., Dubois, A., Moneron, G., Guiot, E., and Boccara, C. (2005). Full-field OCT: ex vivo and in vivo biological imaging applications. In V. V. Tuchin, ed. *Coherence Domain Optical Methods and Optical Coherence Tomography in Biomedicine*, pp. 31–38.
28. Dubois, A., Moneron, G., Grieve, K., and Boccara, A. (2004). Three-dimensional cellular-level imaging using full-field optical coherence tomography. *Phys. Med. Biol.*, **49**, pp. 1227–1234.

29. Scope, A., Mahmood, U., Gareau, D. et al. (2010). In vivo reflectance confocal microscopy of shave biopsy wounds: feasibility of intraoperative mapping of cancer margins. *Br. J. Dermatol.*, **163**, pp. 1218–1228.
30. Rajadhyaksha, M., Menaker, G., Flotte, T., Dwyer, P. J., and González, S. (2001). Confocal examination of nonmelanoma cancers in thick skin excisions to potentially guide mohs micrographic surgery without frozen histopathology. *J. Invest. Dermatol.*, **117**, pp. 1137–1143.

Chapter 18

FF-OCT Imaging: A Tool for Human Breast and Brain Tissue Characterization

Osnath Assayag
ESPCI ParisTech/Institut Langevin, 1 rue Jussieu, 75005 France
osnath.assayag@gmail.com

18.1 Introduction

Histological and immunohistochemical analyses of surgery samples remain the current gold standard method used to analyze tumorous tissue due to advantages of subcellular level resolution and high contrast. However, these methods require lengthy (12 to 72 h), complex multiple steps, and use of carcinogenic chemical products that would not be technically possible intra-operatively. In addition, the number of histological slides that can be reviewed and analyzed by a pathologist is limited, and it defines the number and size of sampled locations on the tumor, or the surrounding tissue.

In current practice for surgical interventions to treat solid tumors, intra-operative diagnosis and assessment of the surgical margins are critical to ensure that the whole tumor has been excised. These information are largely provided by frozen sections.

Handbook of Full-Field Optical Coherence Microscopy: Technology and Applications
Edited by Arnaud Dubois
Copyright © 2016 Pan Stanford Publishing Pte. Ltd.
ISBN 978-981-4669-16-0 (Hardcover), 978-981-4669-17-7 (eBook)
www.panstanford.com

The frozen section procedure, also called cryosection, consists in freezing small samples of tissue from the defect cavity after removal of the gross tumor. The samples are sent to the pathologists that will perform fast histological slides. The slides quality is lower than the quality of formalin fixed paraffin embedded tissue processing but may still be performed to assess surgical margins, in an effort to minimize the need for a second surgical procedure. Unfortunately, frozen section analysis reportedly only has an overall sensitivity of 73% [Olson et al. (2007)] and is associated with several drawbacks: it is time consuming (20–30 minutes) [McLaughlin et al. (2008)], highly operator dependent, destructive to the sample as part of the tissue is lost during slide preparation, allows for only limited sampling, and introduces compression and freezing artifacts [Ishak (1988); Nagasue et al. (1984); Sienko et al. (2005); Taxy (2009)] that significantly hamper interpretation. Alternative techniques for immediate assessment of specimen margins, such as touch prep cytology [Valdes et al. (2007)] or intraoperative radiography [Goldfeder et al. (2006)], are also fraught with disadvantages, including poor sensitivity, limited spatial resolution, and added time.

Consequently, improvements in the ability to perform accurate intra-operative margin assessment using a subsurface optical microscopic biopsy technique is attractive, particularly if it is easy to perform, operator independent, nondestructive, fast and provides high enough spatial resolution to mimic traditional histopathological analysis.

In this chapter we will show how full-field optical coherence tomography (FF-OCT) can be used to image human breast and brain tissue and assess how these images can help the pathologist in his intraoperative diagnosis.

18.2 Breast Tissue Features Recognition and Pathological Modifications

Even though breast cancer mortality has steadily decreased since 1990, with a decrease of 3.2% per year in women younger than 50 and of 2% per year for women 50 and older [Facts (2011)] thanks to a more systematic and early screening, the re-excision rate is typically 20%, and may be as high as 40% [Cabioglu et al. (2007); Fleming et al. (2004); Gonzalez-Angulo et al. (2007); Ishak (1988); Papa et al. (1999); Swanson et al. (2002); Willner et al. (1997)] and is mainly due to a poor evaluation of the tumor margins. Breast has been therefore the first organ we have studied. Moreover, it is a relatively differentiated organ and a large amount of tissue can be made available; indeed when a mastectomy is performed a substantial quantity of tissue is not necessary for diagnosis.

All the images shown in this section have been taken on freshly excised breast tumor specimens from mastectomy and lumpectomy from the Tenon Hospital (Paris, France) and the Institut Curie (Paris, France).

Written consent was obtained prior to imaging according to the standard procedure in Tenon Hospital (Paris) for each patient undergoing biopsy or surgical resection.

All FF-OCT images were taken on ~ 2 cm^2 breast specimen, which is comparable to the typical size of the samples used for frozen section analysis. All FF-OCT images shown were taken at 20 μm below the surface of the breast specimen.

The FF-OCT images have been taken using the Light-CT™ scanner, which is an FF-OCT based imaging system commercialized by the LLTech Company in Paris, France.

The safety of the FF-OCT imaging procedure has been checked on 4 samples: the value of the different evaluation markers (HES staining and immunohistochemistry testing) on the tumorous specimens imaged with FF-OCT were not significantly different from those obtained on matched control specimens from similar regions of the specimen that had not been not imaged.

The initial step of the study consisted in identifying the main architectural features of human breast tissue on FF-OCT images by comparing them with standard histopathology slides prepared from the same specimens.

The morphologic features of the major components of breast tissue in the FF-OCT images could be identified, including fibroadipose tissue, epithelial ducts, vasculature and fibrous tissue (Fig. 18.1). For example, it was found that the galactophorous ducts, when cut tangentially, are recognizable by light-gray color, which we hypothesize is generated by their highly scattering thick elastic

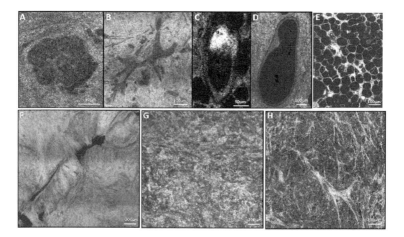

Figure 18.1 Breast tissue basic structures. Lobule (A), galactophorous duct (B), cross section of a galactophorous duct with calcifications (C), vessel (D), adipocytes (E), scar fibrous tissue (F), normal fibrous tissue (G), and fibrous tissue surrounding carcinomatous cells in tumorous stroma (H).

membranes (Fig. 18.1C). In contrast, the same glands, when cut longitudinally, exhibit a characteristic epithelial layer of varying thickness that appears dark gray (Fig. 18.1B).

Intraluminal secretions are sometimes identified. The lobules (Fig. 18.1A) are identified as dark-gray, granular, rounded structures. The vessels (Fig.18.1D), when cut tangentially, do not have the thick epithelial layer of the galactophorous ducts, but are also associated with the presence of an elastic membrane. The adipocytes (Fig. 18.1E) are hyposcattering, and appear as black rounded structures. Their membranes scatter more and appear gray. The characteristic honeycomb configuration of the fat cells is easily identified. The healthy fibrous tissue appears grainy with medium backscattering signal (Fig. 18.1G). The fibrous tissue in the stroma is made of hyperscattering trabeculae (Fig. 18.1H). The scar fibrous tissue is made of thick and large trabeculae, less scattering than the stroma (Fig. 18.1F). Calcifications appear white (Fig. 18.1C) due to their very high backscattering levels. The high resolution of the FF-OCT technique makes the very thin membranes of the vessels and adipocytes visible (Fig. 18.1D,E). Lobules, ducts and adipocytes are features that are clearly distinguished in both FF-OCT images and histology. FF-OCT images presented an advantage in being able to distinguish the granular fibrous appearance of normal tissue and the fine trabeculae of malignant tissue, whereas these details are not well visualized in histology with HES staining and require use of special trichome stains.

A drawback of the FF-OCT images is the lack of visibility of individual nuclei as the backscattering signal of the nuclei is low in breast tissues. Indeed, pathologists found it misleading that necrosis was not always well visualized with FF-OCT. The overall signal amplitude, or whiteness, of the tissue therefore had to be taken into account over and above detection of necrosis as the predominant indication of malignancy. Overall tissue architecture and form was sometimes better visualized with FF-OCT than in histology in cases where histology slices had been taken too close to the edge of the block. In comparison with studies on similar tissues comparing histology with conventional OCT [Hsiung et al. (2007)], FF-OCT performs better than conventional OCT, particularly on distinguishing the smallest features, due to its higher resolution. Figure 18.2 shows normal breast tissue of a postmenopausal woman. Characteristic structures of breast tissue such as the lobules, galactophorous ducts, adipocytes and a normal fibrous tissue are identifiable. Some entanglement of the fibrous tissue and adipose tissue is noted.

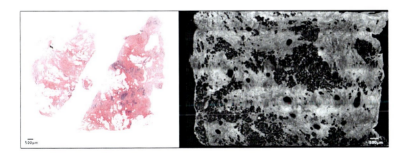

Figure 18.2 Healthy breast tissue specimen. Postmenopausal woman. Whole slide image of the conventional gold standard histology preparation (A) and the corresponding FF-OCT image (B). Characteristic structures of breast tissue such as the lobules, galactophorous ducts, adipocytes, and normal fibrous tissue are seen.

The fibrous tissue appears grainy with a medium back scattering level. Ducts are cut longitudinally; thus they look dark gray. A digital zooming on the image reveals acini in the lobules.

18.3 Distinction between Benign/Normal and Malignant Tissue

After gaining an understanding of the appearance of normal morphological features of human breast tissue on FF-OCT images, it is important to discern what aberrations of these morphological features would allow characterization of the tissue as benign/normal or malignant.

18.3.1 *Malignant Tissue Characterization*

The presence of an invasive tumor results in a modification of the breast tissue aspect and by the appearance of a stellate or nodular structure. Figures 18.3 and 18.4 show two invasive tumor, in both

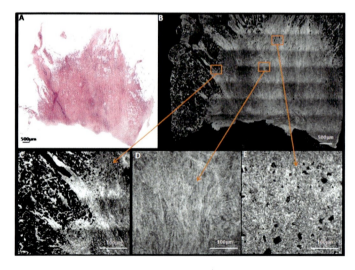

Figure 18.3 Invasive adenocarcinoma: stellate tumor (A and B). Invaded adipose tissue (C), tumorous fibrous tissue (D), and small-size adipocytes (E).

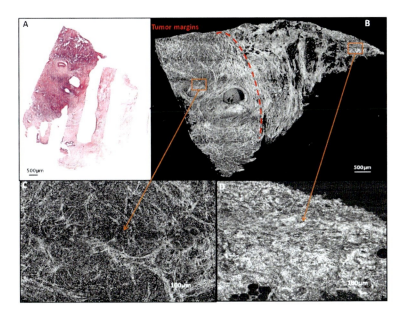

Figure 18.4 Invasive adenocarcinoma: nodular tumor (A and B). Fibrous tissue surrounding carcinomatous cells (C) and normal fibrous tissue (D).

cases fibrous and adipose tissue are separated. In both lesions, the trabeculae of the highly scattering tumor-associated fibrous tissue are observed (Figs. 18.3D and 18.4C). This is in contrast to the normal fibrous tissue that appears grainy and produces less scattering. Furthermore, tumor-associated adipocytes are smaller than those outside of the tumor (Fig. 18.3E). In the stellate tumor, the fibrous tissue is seen invading the adipose tissue (Fig. 18.3C). In the nodular tumor (Fig. 18.4), foci of carcinoma cells appear as gray zones surrounded by the highly scattering tumor-associated fibrous tissue trabeculae (Fig. 18.4C). The different aspect of fibrous tissue allows to assess tumor margins.

Figure 18.5 shows an invasive ductal carcinoma (IDC) with an associated ductal carcinoma in situ (DCIS) component. Enlarged lobules and ducts filled with DCIS (Fig. 18.5C,D) are clearly visible on the image. In addition, the invasive component is characterized by the presence of highly scattering fibrous tissue (Fig. 18.5B) and foci of darker-gray carcinoma cells (Fig. 18.5D).

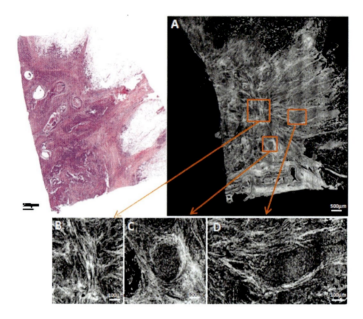

Figure 18.5 Ductal invasive adenocarcinoma with ductal carcinoma in situ (DCIS) component. Histology and FF-OCT images with digital zooms. Enlarged lobules and ducts filled with DCIS (C and D). Highly scattering fibrous tissue (B) and foci of darker-gray carcinoma cells (D).

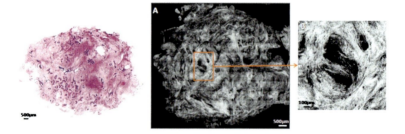

Figure 18.6 Fibroadenoma with characteristic abnormal enlarged ducts (B).

18.3.2 Benign Lesions Identification

Benign lesions can also be distinguished on FF-OCT images. Figure 18.6 shows a fibroadenoma where the enlarged ductules (Fig. 18.6B), characteristic of the lesion, are easily recognizable.

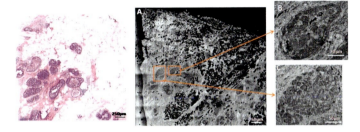

Figure 18.7 Ductal carcinoma in situ (DCIS) with an abnormal duct (B) and an enlarged lobule (C).

A ductal carcinoma in situ (DCIS), defined by the presence of enlarged abnormal ducts (Fig. 18.7B) is represented Fig. 18.7. An abnormal enlarged lobule, indicating an additional lesion called lobular carcinoma in situ (LCIS), can be seen in Fig. 18.7C.

18.4 Breast Tissue Classification Using FF-OCT

The analysis produced a distinct set of criteria that need to be satisfied for accurate discrimination between benign/normal and malignant tissue, along with a recommended workflow that should be followed. The workflow and criteria are presented as a flowchart (Fig. 18.8), and briefly outlined below:

(1) Assess the FF-OCT images following the criteria used for low magnification assessment of standard histology slides. These would include identifying features such as the presence or absence of stellate lesions, and the entanglement of the adipose and fibrous tissues.
(2) If there is no evidence of tumors at low magnification, it is necessary to verify the presence of intact normal breast tissue structures such as lobules, galactophorous ducts and vessels at high magnification.
(3) If there is no obvious tumor and no evidence of intact normal structures, the following criteria for reading the FF-OCT images must be used:

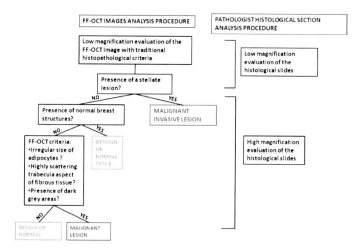

Figure 18.8 Diagnosis decision tree for FF-OCT images of human breast tissue.

(a) Assess the shrinkage of the adipose tissue, which is characterized by the presence of adipocytes of different sizes at the periphery of an invasive carcinoma. In most cases, at the interface of the lesion, fat cells appear smaller and less rounded.

(b) Search for the presence of gray zones, typically surrounded by white fibrous structures (corresponding to foci of carcinoma surrounded by fibrous tissue).

(c) Assess the color and morphology of this fibrous tissue. The appearance of the fibrous tissue is different depending on the type of collagen present. In FF-OCT images, the fibrous tissue of malignant tumor-associated stroma is very white, while it is grayer for scar-associated fibrous tissue or normal/benign breast fibrous tissue. Furthermore, the fibrous tissue architecture differs, usually appearing as thin trabeculae in malignant tumor-associated stroma, while appearing as thick trabeculae in tissue associated with reactive/reparative change (i.e., scar tissue) (Fig. 18.1F). Trabeculae are not seen in normal or benign fibrous tissue (Fig. 18.1F–H).

18.4.1 Diagnostic Accuracy of FF-OCT Images

In order to evaluate the efficiency of FF-OCT as a diagnostic tool for breast tumor, it is necessary to perform a so called specificity and sensitivity study. The sensitivity describes the capability of a diagnostic test to give a positive result when a lesion is present. It is opposed to specificity that describes the capability of a test to give a negative result when there is no lesion.

Two breast pathologists (Dr Martine Antoine from Tenon Hospital, Paris and Dr Brigitte Sigal from Institut Curie, Paris) were asked to classify breast specimen as malignant or benign/normal by only observing FF-OCT images with no access to histological slides nor any patient information. 74 specimens were analyzed by Dr Antoine and 77 specimen by Dr Sigal. The analysis of 74 samples yielded 28 true positives, 34 true negatives, 3 false negatives and 9 false positives, giving a sensitivity of 90% and a specificity of 79%. The analysis of 77 samples yielded 31 true positives, 33 true negatives, 2 false negatives and 11 false positives giving a sensitivity of 94% and a specificity of 75%.

Sensitivity values obtained with FF-OCT are higher than those obtained with cryosection analysis that are about 73%, as mentioned earlier. However, typical specificity values of cryosection analysis are much higher, ranging around 99%. The low value of FF-OCT specificity is explained by the large amount of false positive obtained for both pathologists. These misleading results are due, in most cases, to lobular lesion. Indeed, pathologists could hardly define if the lobules were abnormally large to be healthy.

False positive results were often associated with adenofibromas. Adenofibromas are liable to cause confusion as the deformation of the mammary tissue architecture associated with this benign lesion can give the impression of malignancy.

18.5 FF-OCT Imaging of Healthy and Tumorous Human Brain Parenchyma

Primary central nervous system (CNS) tumors represent a heterogeneous group of tumors with benign, malignant and slow-growing evolution. Overall survival from brain tumors depends on the complete resection of the tumor mass, as identified through postoperative imaging, associated with updated adjuvant radiation therapy and chemotherapy regimen for malignant tumors. For low-grade tumors located close to eloquent brain areas, a maximally safe resection that spares functional tissue warrants the current use of intraoperative techniques that guide a more complete tumor resection.

Human brain tissue can hardly be frozen therefore frozen section procedure for intra operative procedure is rarely performed. Instead, cytological smear tests are used, however tissue architecture information is thereby lost and the analysis is carried out on only a limited area of the sample (1 mm × 1 mm). Reviewing the state of the art, a need is identified for a quick and reliable method of providing the neurosurgeon with architectural and cellular information without the need for injection or oral intake of exogenous markers in order to guide the neurosurgeon and optimize surgical resections.

In the CNS, published studies that evaluate OCT [Bizheva et al. (2005); Böhringer et al. (2006, 2009); Boppart et al. (2004)]; using time domain (TD) or spectral domain (SD) OCT systems had insufficient resolution (10 to 15 µm axial) for visualization of fine morphological details. A study of 9 patients with gliomas carried out using a TD-OCT system led to classification of the samples as malignant versus benign [Böhringer et al. (2009)]. However, the differentiation of tissues was achieved by considering the relative attenuation of the signal returning from the tumorous zones in relation to that returning from healthy zones. The classification was not possible by real recognition of CNS microscopic structures. Another study showed images of brain microstructures obtained with an OCT system equipped with an ultrafast laser that offered axial and lateral resolution of 1.3 µm and 3 µm respectively [Bizheva et al. (2005)]. In this way, it was possible to differentiate malignant from healthy tissue by the presence of blood vessels, microcalcifications and cysts in the tumorous tissue. However the images obtained were small (2 mm × 1 mm), captured on fixed tissue only and required use of an expensive large laser thereby limiting the possibility for clinical implementation. Other studies have focused on animal brain. In rat brain in vivo, it has been shown that optical coherence microscopy (OCM) can reveal neuronal cell bodies and myelin fibers [Srinivasan et al. (2012)], while FF-OCT can also reveal myelin fibers [Ben et al. (2011)], and movement of red blood cells in vessels [Binding et al. (2011)]. In this section, we present the first study that analyzes nontumorous and tumorous human brain tissue samples using FF-OCT.

18.5.1 *Human Brain Parenchyma Morphological Structure Recognition*

Figure 18.9 shows the different structures of a human brain parenchyma.

The cortex and the white matter are clearly distinguished from one another. Indeed, a sub population of neuronal cell bodies (Fig. 18.9B,C) as well as myelinated axon bundles leading to the white matter could be recognized (Fig. 18.9D,E). Neuronal cell bodies appear as dark triangles (Fig. 18.9C) in relation to the bright surrounding myelinated environment. The FF-OCT signal is produced by backscattered photons from tissues of differing refractive indices. The number of photons backscattered from the nuclei in neurons appears to be too few to produce a signal that allows their differentiation from the cytoplasm, and therefore the whole of the cell body (nucleus plus cytoplasm) appears dark. Myelinated axons are numerous, well discernible as small fascicles and appear as bright white lines (Fig. 18.9E). As the cortex does not contain many myelinated axons, it appears dark gray. Brain vasculature is visible (Fig. 18.9F,G), and small vessels are distinguished by a thin collagen membrane that appears light gray. The different regions of the human hippocampal formation are easily recognizable (Fig. 18.10). Indeed, CA1 field and its stratum radiatum, CA4 field, the hippocampal fissure, the dentate gyrus, and the alveus are easily distinguishable. Other structures become visible by zooming in digitally on the FF-OCT image. The large pyramidal neurons of the CA4 field (Fig. 18.10B) and the granule cells that constitute the stratum granulosum of the dentate gyrus are visible, as black triangles and as small round dots, respectively (Fig. 18.10D). In the normal cerebellum, the lamellar

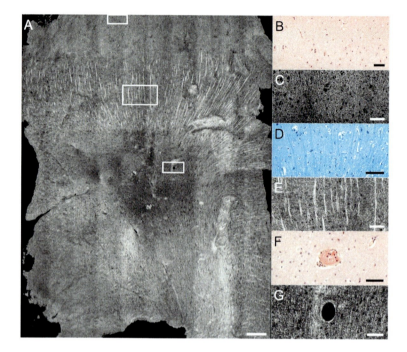

Figure 18.9 Cortex is distinguished from white matter. (A) Cortex appears gray. (B and C) Neuronal cell bodies (arrows), (D and E) myelinated axon bundles (arrow) leading to white matter, and (F and G) vasculature (arrow). (B and F) Hemalun and phloxin stainings and (D) Luxol blue staining. Rectangles indicate locations of zooms. Scale bars show 500 μm (A), 50 μm (B, C, F, and G), and 80 μm (D and E).

or foliar pattern of alternating cortex and central white matter is easily observed (Fig. 18.11A). By digital zooming, Purkinje and granular neurons also appear as black triangles or dots, respectively (Fig. 18.11C), and myelinated axons are visible as bright white lines (Fig. 18.11E).

18.5.2 *Benign Lesions Identification*

Meningiomas are a diverse set of tumors arising from the meninges, they are the most common benign brain tumors. The classic morphological features of a meningioma are visible on the FF-OCT

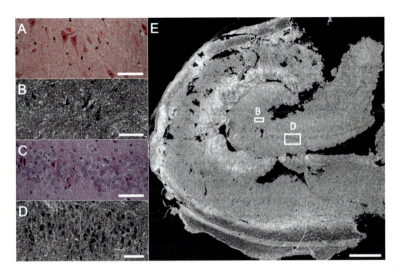

Figure 18.10 Hippocampus. CA1 field and stratum radiatum, CA4 field, the hippocampal fissure, the dentate gyrus, and the alveus are distinguished. (A and B). Pyramidal neurons (arrows) of CA4 and (C and D) granular cells constitute the stratum granulosum of the dentate gyrus. (A and C) Hemalun and phloxin stainings. Rectangles indicate locations of zooms. Scale bars show 40 μm (A and B), 80 μm (C and D), and 900 μm (E). (B, C, F, and G) and 80 μm (D and E).

image: large lobules of tumorous cells appear light gray (Fig. 18.12A), demarcated by collagen-rich bundles (Fig. 18.12B) that are highly scattering and appear a brilliant white in the FF-OCT images. The classic concentric tumorous cell clusters (whorls) are also very clearly distinguished (Fig. 18.12D). In addition the presence of numerous cell whorls with central calcifications (psammoma bodies) is revealed (Fig. 18.12F). Collagen balls appear bright white on the FF-OCT image (Fig. 18.12H). As the collagen balls progressively calcify, they are consumed by the black of the calcified area, generating a target-like image (Fig. 18.12H). Calcifications appear black in FF-OCT as they are crystalline and so allow no penetration of photons to their interior.

Another benign lesion is the choroid plexus papilloma that appears as an irregular coalescence of multiple papillas composed of elongated fibrovascular axes covered by a single layer of choroid

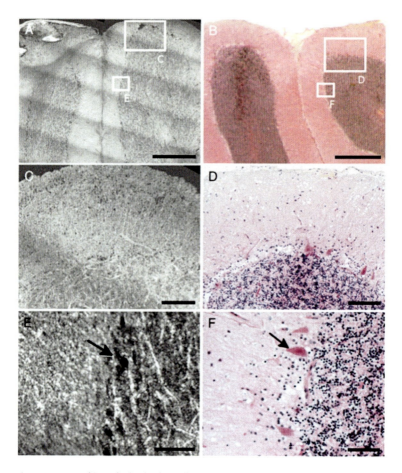

Figure 18.11 (A and B) The lamellar or foliar pattern of alternating cortex and central white matter. (C, D, E, and F) Zooms show cerebellar cortex and granular layer and (E and F) Purkinje (arrow) and granular neurons are distinguished as black triangles or dots, respectively, and myelinated axons as bright white lines. (B, D, and F) Hemalun and phloxin stainings. Rectangles indicate locations of zooms. Scale bars show 800 μm (A and B), 350 μm (C and D), and 100 μm (E and F).

glial cells (Fig. 18.13). By zooming in on an edematous papilla, the axis appears as a black structure covered by a regular light-gray line (Fig. 18.13B). If the papilla central axis is hemorrhagic, the fine regular single layer is not distinguishable (Fig. 18.13C).

Figure 18.12 (A) Meningioma psammoma. (B and C) Collagen bundles, (D and E) whorls, (F and G) calcifications, and (H and I) collagen balls. (C, E, G, and I) Hemalun and phloxin stainings. Rectangles indicate locations of zooms. Scale bars show 500 μm (A), 50 μm (B and C), and 10 μm (D, E, F, G, H, and I).

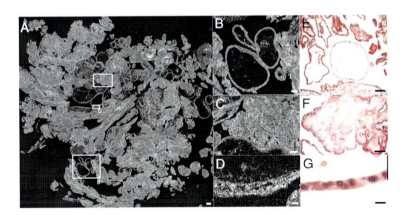

Figure 18.13 (A) Papilloma cauliflower-like aspect. (B and E) Empty papilla, (C and D) blood-filled papilla, and (D and G) single layer of plexus cells. (E, F, and G) Hemalun and phloxin stainings. Rectangles indicate locations of zooms. Scale bars show 150 μm (A), 50 μm (B, C, E, and F), and 20 μm (D and G).

Additional digital zooming in on the image reveals cellular level information, and some nuclei of plexus choroid cells can be recognized.

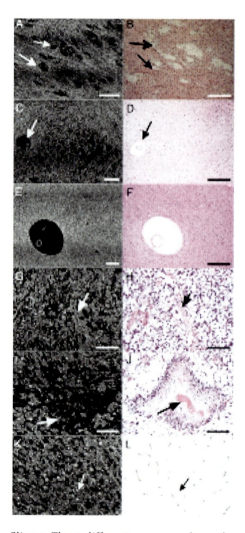

Figure 18.14 Glioma. Three different cases are shown here (A and B; C and F; and G and L). (A and B) Microcysts (arrows) in an oligo-astrocytoma grade 2, (C and D) microcystic areas and Virchow-Robin space (arrows) in an astrocytoma grade 2, (E and F) enlarged Virchow-Robin spaces in an astrocytoma grade 2, (G and H) microvessels (arrow) and tumorous glial cells in an oligoastrocytoma grade 3, and (I and J) pseudopalisading necrosis in an oligo-astrocytoma grade 3. (K and L) Vasculature (arrows) in an oligo-astrocytoma grade 3 (B, D, F, H, and J). Scale bars show 250 μm (A and B), 100 μm (C and F), 20 μm (G and H), and 10 μm (I–L).

However, cellular atypia and mitosis are not visible. These represent key diagnosis criteria used to differentiate choroid plexus papilloma (grade I) from atypical plexus papilloma (grade II).

18.5.3 Malignant Brain Lesion Imaging

Contrary to the choroid plexus papillomas that have a very distinctive architecture in histology (cauliflower-like aspect), very easily recognized in the FF-OCT images (Fig. 18.13A to G), diffusely infiltrating gliomas do not present a specific tumor architecture (Fig. 18.8) as they diffusely permeate the normal brain architecture. Hence, the tumorous glial cells are largely dispersed through a nearly normal brain parenchyma (Fig. 18.14E). The presence of infiltrating tumorous glial cells attested by high magnification histological observation (irregular atypical cell nuclei compared to normal oligodendrocytes) is not detectable with the current generation of FF-OCT devices, as FF-OCT cannot reliably distinguish the individual cell nuclei due to lack of contrast (as opposed to lack of resolution). In our experience, diffuse low-grade gliomas (less than 20% of tumor cell density) are mistaken for normal brain tissue on FF-OCT images. However, in high-grade gliomas (Fig. 18.14G,K), the infiltration of the tumor has occurred to such an extent that the normal parenchyma architecture is lost. This architectural change is easily observed in FF-OCT and is successfully identified as high-grade glioma, even though the individual glial cell nuclei are not distinguished. Moreover, some details characteristic of an invasive lesion are visible. Thus, necrosis appears as dark diamond shaped (Fig. 18.14I) area and white powdery substance in center of dark space (Fig. 18.14J, white arrow) is lysed cells (necrotic cells/centers). Dark arrow on histology shows a vessel. It is worth noticing that all the structures mentioned are visible on a single FF-OCT image while in histology an additional coloration is required to visualize vasculature (Hemalun and phloxin stainings and CD34 immunostaining, Fig. 18.14L).

18.6 Discussion and Conclusion

We have presented the results of two preclinical studies evaluating FF-OCT imaging as a tool for tumor and normal tissue characterization in the case of breast and brain tumors.

In its current stage of development, FF-OCT is of interest for several use cases. First, it represents a paradigm shift in the handling of immediate assessment of resected tissue. Indeed it allows for the first time, the acquisition, in a nondestructive manner, of tissue images over an adequately large area, while producing images of sufficiently high resolution, thereby allowing pathologists to mimic routine histopathology diagnostic workflow by zooming in and out on the images.

The tissue sample does not require preparation and image acquisition is rapid therefore FF-OCT imaging appears promising as an intraoperative tool to help neurosurgeons that could assess the margins directly in the surgical unit during surgery.

The imaging system is compact, it can be placed in the operating room and a simple halogen lamp provides sufficient resolution (of the order of 1m lateral and axial) for microstrutures distinguish brain tissue and breast tissue.

In the case of breast lesions, we established an "alphabet" allowing to recognize breast tissue basic microstructures. Consequently we have developed a set of initial criteria and a recommended workflow for interpreting FF-OCT images to accurately differentiate whether the alterations observed are benign/normal or malignant. The specificity and sensitivity values calculated for our method are close to the ones obtained for current techniques usually used to interpret fresh tissue samples from lumpectomy. In the case of brain tissue, we showed that it was possible to distinguish basic structures of the brain parenchyma and to recognize the presence of lesions when the latter leads to a modification of the general architecture of the tissue. Nevertheless, different types of meningiomas could be identified on the images thus FF-OCT may serve as an intraoperative tool, in addition to frozen section analysis, to refine differential diagnosis between pathological entities with different prognoses and surgical managements. However, malignant lesions like gliomas could be detected on FF-OCT images only if the carcinomatous cell density is greater than around 20% (i.e., the point at which the effect on the architecture becomes noticeable). The FF-OCT technique is therefore not currently suitable for the evaluation of low tumorous infiltration or tumorous margins of brain tumors. One of the current limitations of the technique is the difficulty in estimating the size and shape and the nuclei/cytoplasm ratio nucleus of the cells present. This prevents accurate classification of lesions as well as their grade.

To respond more precisely to surgical needs, it would be preferable to provide in vivo imaging using a surgical probe. Development of FF-OCT to allow in vivo imaging is underway, and first steps include increasing camera acquisition speed. First results of in vivo rat brain imaging have been achieved with an FF-OCT prototype setup, and show real-time visualization of myelin fibers and movement of red blood cells in vessels [Ben et al. (2011); Binding et al. (2011)]. Integration of the FF-OCT system into a surgical probe is currently underway and preliminary images of skin and breast tissue have been captured with a rigid probe FF-OCT prototype [Latrive and Boccara (2011)].

To improve the specificity and sensitivity of the technique, new agents contrast should be added. Recently, elastography has been added to the FF-OCT imaging modality. The supplementary elastographic contrast has been evaluated on human breast tissue specimens [Nahas et al. (2013)].

References

Ben, A. et al. (2011). Single myelin fiber imaging in living rodents without labeling by deep optical coherence microscopy. *J. Biomed. Opt.*, **16**, p. 11.

Binding, J., Ben Arous, J., Léger, J.-F., Gigan, S., Boccara, C., and Bourdieu, L. (2011). Brain refractive index measured in vivo with high-NA defocus-corrected full-field OCT and consequences for two-photon microscopy. *Opt. Express*, **19**, 6, pp. 4833–4847.

Bizheva, K., Unterhuber, A., Hermann, B., Povazay, B., Sattmann, H., Fercher, A. F., Drexler, W., Preusser, M., Budka, H., Stingl, A., and Le, T. (2005). Imaging ex vivo healthy and pathological human brain tissue with ultrahigh-resolution optical coherence tomography. *J. Biomed. Opt.*, **10**, 1, p. 11006.

Böhringer, H. J., Boller, D., Leppert, J., Knopp, U., Lankenau, E., Reusche, E., Hüttmann, G., and Giese, A. (2006). Time-domain and spectral-domain optical coherence tomography in the analysis of brain tumor tissue. *Laser Surg. Med.*, **38**, 6, pp. 588–597.

Böhringer, H. J., Lankenau, E., Stellmacher, F., Reusche, E., Hüttmann, G., and Giese, A. (2009). Imaging of human brain tumor tissue by near-infrared laser coherence tomography. *Acta Neurochir.*, **151**, 5, pp. 507–517,

Boppart, S. A., Luo, W., Marks, D. L., and Singletary, K. W. (2004). Optical coherence tomography: feasibility for basic research and image-guided surgery of breast cancer. *Breast Cancer Res. Treat.*, **84**, 2, pp. 85–97.

Cabioglu, N., Hunt, K., Sahin, A., Kuerer, H., Babiera, G., Singletary, S., Whitman, G., Ross, M., Ames, F., Feig, B., Buchholz, T., and Meric-Bernstam, F. (2007). Role for Intraoperative Margin Assessment in Patients Undergoing Breast-Conserving Surgery, *Ann. Surg. Oncol.*, **14**, 4, pp. 1458–1471.

Facts, C. (2011). Cancer Facts & Figures.

Fleming, F. J., Hill, A. D. K., Mc Dermott, E. W., O'Doherty, A., O'Higgins, N. J., and Quinn, C. M. (2004). Intraoperative margin assessment and re-excision rate in breast conserving surgery. *Eur. J. Surg. Oncol.*, **30**, 3, pp. 233–237.

Goldfeder, S., Davis, D., and Cullinan, J. (2006). Breast specimen radiography: can it predict margin status of excised breast carcinoma? *Acad. Radiol.*, **13**, 12, pp. 1453–1459.

Gonzalez-Angulo, A. M., Morales-Vasquez, F., and Hortobagyi, G. N. (2007). Overview of resistance to systemic therapy in patients with breast cancer. *Adv. Exp. Med. Biol.*, **608**, pp. 1–22.

Hsiung, P.-L., Phatak, D. R., Chen, Y., Aguirre, A. D., Fujimoto, J. G., and Connolly, J. L. (2007). Benign and malignant lesions in the human breast depicted with ultrahigh resolution and three-dimensional optical coherence tomography. *Radiology*, **244**, 3, pp. 865–874.

Ishak (1988). Benign tumors and pseudotumors of the liver. *Appl. Pathol.*, **6**, 2, pp. 82–104.

Latrive, A., and Boccara, A. C. (2011). Full-field optical coherence tomography with a rigid endoscopic probe. *Biomed. Opt. Express*, **2**, 10, pp. 2897–2904.

McLaughlin, S. A., Ochoa-Frongia, L. M., Patil, S. M., Cody, H. S., and Sclafani, L. M. (2008). Influence of frozen-section analysis of sentinel lymph node and lumpectomy margin status on reoperation rates in patients undergoing breast-conservation therapy. *J. Am. Coll. Surgeons,* **206**, 1, pp. 76–82.

Nagasue, N., Akamizu, H., Yukaya, H., and Yuuki, I. (1984). Hepatocellular pseudotumor in the cirrhotic liver: report of three cases. *Cancer,* **54**, 11, pp. 2487–2494.

Nahas, A., Bauer, M., Roux, S., and Boccara, A. C. (2013). 3D static elastography at the micrometer scale using full field OCT. *Biomed. Opt. Express,* **4**, 10, pp. 2138–2149.

Olson, T. P., Harter, J., Muñoz, A., Mahvi, D. M., and Breslin, T. (2007). Frozen section analysis for intraoperative margin assessment during breast-conserving surgery results in low rates of re-excision and local recurrence. *Ann. Surg. Oncol.,* **14**, 10, pp. 2953–2960.

Papa, M. Z., Zippel, D., Koller, M., Klein, E., Chetrit, A., and Ari, G. B. (1999). Positive margins of breast biopsy: is reexcision always necessary? *J. Surg. Oncol.,* **70**, 3, pp. 167–171.

Sienko, A., Allen, T. C., Zander, D. S., and Cagle, P. T. (2005). Frozen section of lung specimens. *Arch. Pathol. Lab. Med.,* **129**, 12, pp. 1602–1609.

Srinivasan, V. J., Radhakrishnan, H., Jiang, J. Y., Barry, S., and Cable, A. E. (2012). Optical coherence microscopy for deep tissue imaging of the cerebral cortex with intrinsic contrast. *Opt. Express,* **20**, 3, pp. 2220–2239.

Swanson, G. P., Rynearson, K., and Symmonds, R. (2002). Significance of margins of excision on breast cancer recurrence. *Am. J. Clin. Oncol.,* **25**, 5, pp. 438–441.

Taxy, J. B. (2009). Frozen section and the surgical pathologist: a point of view. *Arch. Pathol. Lab. Med.,* **133**, 7, pp. 1135–1138.

Valdes, E. K., Boolbol, S. K., Cohen, J.-M., and Feldman, S. M. (2007). Intraoperative touch preparation cytology; does it have a role in re-excision lumpectomy? *Ann. Surg. Oncol.,* **14**, 3, pp. 1045–1050.

Willner, J., Kiricuta, I. C., and Kölbl, O. (1997). Locoregional recurrence of breast cancer following mastectomy: always a fatal event? Results of univariate and multivariate analysis. *Int. J. Radiat. Oncol.,* **37**, 4, pp. 853–863.

Chapter 19

Full-Field Optical Coherence Microscopy in Ophthalmology

G. Latour,[a,b] K. Grieve,[c] G. Georges,[d] L. Siozade,[d] M. Paques,[c] V. Borderie,[c] L. Hoffart,[e] and C. Deumié[d]

[a]*Université Paris-Sud, Laboratoire Imagerie et Modélisation en Neurobiologie et Cancérologie, CNRS, Université Paris-Saclay, Orsay, France*
[b]*Ecole Polytechnique, Laboratoire d'Optique et Biosciences, CNRS, INSERM, Université Paris-Saclay, Palaiseau, France*
[c]*Institut de la Vision, UPMC Univ Paris 06, INSERM, CNRS, Paris, France*
[d]*Aix-Marseille Université, CNRS, Centrale Marseille, Institut Fresnel UMR 7249, 13013 Marseille, France*
[e]*Aix-Marseille Université, Hôpital de la Timone, Service d'Ophtalmologie, Marseille, France*
gael.latour@u-psud.fr, katharine.grieve@inserm.fr, vincent.borderie@upmc.fr, michel.paques@gmail.com, gaelle.georges@fresnel.fr, laure.siozade@fresnel.fr, carole.deumie@fresnel.fr, louis.hoffart@ap-hm.fr

19.1 Introduction

The eye is a sophisticated imaging device. It forms real images of real objects that compose the outer scene on the light-sensitive surface of the retina. Moreover, to obtain a sharp image this system must be stigmatic and the outer scene must be focused at the fixed plane where the photoreceptors are located. The visual acuity, whatever

---

Handbook of Full-Field Optical Coherence Microscopy: Technology and Applications
Edited by Arnaud Dubois
Copyright © 2016 Pan Stanford Publishing Pte. Ltd.
ISBN 978-981-4669-16-0 (Hardcover), 978-981-4669-17-7 (eBook)
www.panstanford.com

the position of the observed object, is ensured thanks to the process of accommodation, that is, by the adaptation of the optical power processed by the lens. From an optical point of view, the anterior segment of the eye is mainly composed of the cornea and the lens. The cornea serves three main functions: protection from the outer environment, transmission of light (light must enter the eyeball and reach the retina to be detected by the photoreceptors) and refraction of light (the cornea is responsible for two-thirds of the eye's optical power).

Nowadays, optical coherence tomography (OCT) is considered to be an indispensable diagnostic instrument in ophthalmology. Commercial OCT systems are based on Fourier domain optical coherence tomography (FD-OCT) technology using spectral detection or a swept source (therefore also known as spectral OCT or swept-source OCT; in this chapter we maintain the term "FD-OCT"). OCT is mainly used for retinal imaging for diagnosis and the follow-up of retinal diseases but its field of applications in ophthalmology is growing with the development of the technology (higher speed, better spatial resolution), especially with swept-source OCT. However, the spatial resolution reached by commercial OCT devices does not allow the characterization of the cellular organization and the follow-up of subcellular modifications in the case of pathology.

Full-field optical coherence microscopy (FF-OCM) appears to present a promising alternative by providing images with en face and cross-sectional micrometer-scale resolutions. Based on the same optical principle of low-coherence interferometry, FF-OCM is a 3D imaging technique that offers histological resolution on intact ex vivo tissues but without the time-consuming processing steps of conventional histology. In ophthalmology, FF-OCM provides morphological characterization of the ocular tissues (cornea, lens and retina). The technique shows great potential for imaging of pathological corneas and corneal graft evaluation. Coupling FF-OCM to other imaging modalities opens up new perspectives.

The majority of the images shown in this chapter have been captured using laboratory prototypes, while the study concerning corneal pathology was carried out using a commercial device (LLTech, France). All setups are similar in configuration and performance. The typical spatial resolution is around 1 μm in all three dimensions.

19.2 Cornea

19.2.1 Morphology and Characterization

19.2.1.1 Description of the tissue

The cornea and sclera form the external envelope of the ocular globe. On the optical axis of the eye, the cornea and lens are the two main optical refractive elements of the eye and constitute the anterior segment (Fig. 19.1a). The cornea is characterized by two key optical functions: to transmit and to focus light on the retina. The reasons for its transparency in the visible range (more than 90% in the 400–1300 nm range [53]) are not completely understood. The optical properties have been widely studied and the absence of blood vessels associated with nanoscale architecture probably plays an important role in the optical transparency of the tissue [7, 27, 33, 39, 53, 55]. From a macroscopic point of view, the cornea is a curved shape disc with a mean horizontal diameter of 11.4 mm, slightly

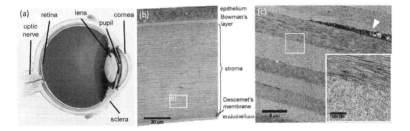

Figure 19.1 Description of the eye and the cornea. (a) Eye diagram showing the main optical components of the ocular tissue (adapted from the National Eye Institute, National Institutes of Health (NEI/NIH). (b) Histological toluidine blue–stained cross-section of the cornea observed in conventional white-light microscopy with the typical area (white square) observed in transmission electron microscopy (TEM). (c) TEM of the posterior stroma showing the stacked organization of the collagen lamellae with a keratocyte (arrow) between two lamellae. Each lamella is composed of nanometric collagen fibrils (inset).

greater than the mean vertical diameter, with a central thickness of around 500 μm and a peripheral thickness of around 700 μm under physiological conditions.

At the microscopic scale, the corneal tissue is arranged in five morphological layers, each contributing to the corneal integrity and the macroscopic properties [13, 47] (see stained histological section in Fig. 19.1b):

- The epithelium is the outer part of the cornea. It is composed of five to seven cell layers and acts as a barrier to foreign bodies, especially water and pathogens, thanks to an impervious layer at the surface. It plays an important role in controlling corneal hydration. Epithelial cells are continuously regenerated.
- Bowman's layer is the intermediate layer between the basement membrane of the epithelium and the anterior part of the stroma. This acellular region is composed of tightly packed, randomly oriented collagen fibrils.
- The stroma represents 90% of the corneal thickness. It is composed of water (around 78% in physiological conditions), collagen and proteoglycans. More than 200 stacked collagen lamellae can be found in the center of human cornea and this number increases in the periphery to reach 500 lamellae. The organization of the lamellae is not homogeneous in depth. In the anterior part, lamellae are interwoven and some of them are anchored to Bowman's layer, whereas in the mid and posterior part, they are parallel to each other [42] (Fig. 19.1c). These lamellae are arranged in various orientations [41, 56, 68]. This complex spatial organization may be responsible for both the shape and strength of the tissue. Each lamella is composed of collagen fibrils of uniform diameter around 30–35 nm with a regular interfibrillar distance (inset in Fig. 19.1c). Keratocytes are specialized fibroblasts responsible for the synthesis of collagen in case of injury but also for the components of the ground substance (mainly proteoglycans and glycoproteins), and for the continuous renewal of collagen fibrils and extracellular matrix. They are flat in

shape (approximately 1 μm thick) but extend up to 50 μm laterally in several directions giving the cells a stellate appearance [43]. They appear as dark forms between collagen lamellae in stained histological cross-sections (Fig. 19.1b).
- Descemet's membrane, the basement membrane adjacent to the endothelium, is an acellular region.
- The endothelium, composed of a monolayer of endothelial cells, plays a key role as barrier and pump that maintain the cornea in a physiological hydration state. The endothelial cells prevent the cornea from swelling, which would cause opacity of the tissue, by controlling the corneal hydration through an active pump mechanism. The endothelial cells do not regenerate and the endothelial cell density, along with their shape and size, is a physiological parameter of corneal health monitored by ophthalmologists. The endothelial cell characteristics are also a key deciding factor for the selection of human graft corneas in eye banks. A healthy subject has a cell density of 2500–3000 cells/mm^2 and this density progressively decreases with age. When the endothelial cell density decreases below 400–500 cells/mm^2, the endothelial function is impaired and the cornea becomes edematous.

Corneal tissue is mainly composed of water (around 78%). The degree of hydration and therefore the refractive index depends on depth, with the posterior cornea being more hydrated than the anterior (Table 19.1). The mean refractive index of the entire cornea is considered to be 1.376.

Table 19.1 Refractive indices of different corneal layers [41, 51]

Cornea layer	Refractive index
Tear film	1.336
Epithelium	1.401
Bowman's layer	1.380
Stroma: anterior part	1.380
Stroma: posterior part	1.373

19.2.1.2 Conventional corneal imaging techniques

19.2.1.2.1 Ex vivo imaging techniques The gold standard for tissue characterization at the cellular scale is histology. The first steps of the histology process involve fixing and embedding the fresh tissue. The block is then oriented along cross-sections and sliced in sections (typically from 0.5 to 2 micrometers thick). Finally, the sections are stained in order to obtain high-contrast images of the tissues observed in conventional white-light microscopy. The staining principally enables the observation of cell nuclei (Fig. 19.1b): epithelial cells in the anterior cornea, keratocytes between the collagen lamellae in the stroma and endothelial cells in the posterior cornea. Histology is widely used to characterize tissue physiology and its evolution with pathology at the cellular scale (see Section 19.2.3.2. for the observation of pathological tissues).

Conventional white-light microscopic observation of histological sections can be coupled to electron microscopy in order to achieve subcellular imaging. After fixation and inclusion, ultrathin sectioning (a few hundred nanometers thick) is performed followed by staining with heavy metals to create contrast between different structures. In the stroma, the stacked organization of the lamellae and individual collagen fibrils that compose lamellae are visible in transmission electron microscopy (TEM) (Fig. 19.1c) [8]. The orientation of the lamellae can be studied by scanning electron microscopy (SEM) [56]. High-quality and high-resolution electron microscopy images provide additional information to histology.

However, both stained histology and electron microscopy require complex and time-consuming tissue preparation. These techniques only provide 2D images and are limited to ex vivo ocular tissues. Moreover, the stromal hydration rate is usually dramatically modified by the fixation process. However, they give key information about the cellular organization and the ultrastructure of the tissue.

For in vivo diagnosis or ex vivo 3D characterization, optical techniques provide quick and reliable architectural and cellular information.

19.2.1.2.2 In vivo imaging techniques Ophthalmologists have a wide choice of observation techniques depending on the properties they want to observe and analyze. The refractive properties can be analyzed by measuring the shape and radius of the corneal interfaces. This is the aim of keratometry measurements or specular corneal topography but also of OCT. Measuring corneal thickness is also of great interest in order to characterize swelling abnormalities or keratoconus (pathology leading to modification of cornea shape and thickness, more details are given in paragraph 19.2.3.2). This measurement can be achieved with ultrasound pachymetry, OCT or other corneal elevation topography (based for example on Scheimpflug photography). Finally, the transparency has also to be characterized for clinical applications. Usually this is achieved by qualitative observation of the light scattered by the tissue under slit illumination. These macroscopic measurements can be compared to microscopic information captured with in vivo microscopes. Some optical techniques widely used for imaging the cornea are described in the following paragraphs.

The slit lamp is the key tool of ophthalmic examination. It consists of a high-intensity light source that is focused into the eye to produce a bright slit image that can be adjusted in width and height. The ophthalmologist observes the magnified view of the eye structures in order to perform the anatomical diagnosis. However, the spatial resolution is not sufficient for characterization of the cellular organization of the ocular tissues.

In vivo confocal reflectance microscopy is a laser scanning imaging technique that provides en face optical sections. A pinhole in front of the detector performs a spatial selection of the light coming from the focal plane by limiting the detection of light scattered or reflected from structures outside of this plane. The laser is scanned in order to acquire a 2D image; therefore, any movement during the acquisition blurs the final image. 3D information is available by scanning the microscope objective in depth, focusing the beam deeper and deeper [52]. The spatial resolution is around 1–2 μm laterally and 8–20 μm axially for the commercial devices in ophthalmology. The contrast in confocal reflectance imaging is related to the spatial variation of the refractive index within the tissue, and therefore the use of contrast agents is not necessary. Confocal reflectance microscopy is becoming widely used in ophthalmology, particularly for monitoring purposes before and

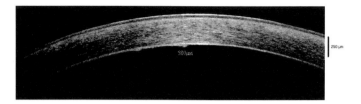

Figure 19.2 OCT image of cornea (acquired with an Optovue system). Scale bar: 250 μm.

after transplant and refractive surgery. Epithelial and endothelial cells, nerve plexus and keratocytes are clearly visualized, thereby allowing characterization of the physiology and pathology of the cornea [15, 25, 52].

19.2.1.2.3 Corneal imaging with OCT Nowadays the gold-standard for in vivo retinal imaging is FD-OCT, as it enables noncontact high-resolution 3D imaging without the use of contrast agents. OCT signals are based on the optical reflectance and scattering properties of the tissue. Characterization of the anterior segment achieved with FD-OCT [29] is also now routine in the clinic (Fig. 19.2), and is useful for mapping of the corneal thickness, early detection of keratoconus (creation of pachymetry and epithelial thickness maps), detection and diagnosis of pathology, flap thickness measurements in LASIK surgery, measurement of the iridocorneal angle (between the cornea and the iris) for glaucoma management and characterization of intracorneal and intraocular implants [28]. However, despite recent advances in swept-source OCT technology, histology-like cellular and subcellular characterization of the corneal tissue is not possible due to a lack of spatial resolution.

19.2.2 FF-OCM Imaging of the Healthy Cornea

FF-OCM, based on the same optical principle of low-coherence interferometry as that used by OCT, is an attractive alternative to achieve 3D imaging of ex vivo tissues with subcellular resolution. The technical description of the scientific principle of FF-OCM and the optical setup used to perform the FF-OCM images that appear in this chapter can be found in Chapter 1. The images produced are based on intrinsic reflectance and scattering signals and the tissue is observed without any preparation or staining. To avoid high reflection from the first optical interface (the epithelium), and also to maintain hydration of the tissue, the cornea is immersed in a liquid and water immersion objectives are used.

The strength of FF-OCM lies in the characterization of ex vivo unstained tissue. En face images are recorded on intact fresh or fixed tissue. From the 3D collected data, cross-sectional reconstruction can be performed numerically (see Fig. 19.3a for the terms employed for the orientations of the sections). The axial scanning steps must small enough (no larger than 1 µm) to be limited by the cross-sectional resolution (around 1 µm) and not by the optical axial sectioning. This enables an cross-sectional reconstruction of the ex vivo cornea over the entire thickness of the tissue. The

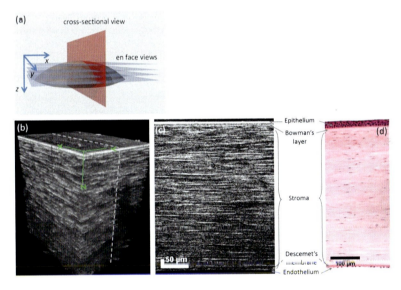

Figure 19.3 Histological and imaging cross-sections of a healthy human cornea. (a) Schematic diagram of the cornea, labeling the terms employed in this chapter to describe the orientations of the sections. (b) 3D FF-OCM imaging of an ex vivo human cornea. (c) Cross-sectional reconstruction from the initial 3D tomographic image (along the dashed line in (b)) compared to (d) stained histological cross-section observed with a conventional white-light microscope. Scale bar: 100 µm. Adapted from Ref. [32].

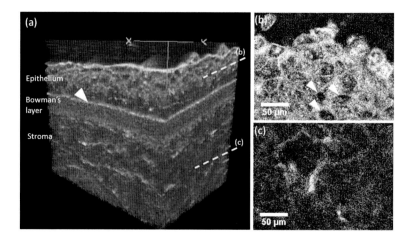

Figure 19.4 FF-OCM images of the anterior part of the human cornea. (a) 3D imaging of the cornea (250 × 190 × 150 μm³ stack with 0.5 μm axial steps) with the epithelium, the basement membrane (white arrow), Bowman's layer, and the stroma. En face images of (b) the epithelium with the hyporeflective corneal epithelial bullae (white arrows) and (c) the anterior stroma. Adapted from Ref. [32].

different morphological layers of the cornea can be identified with a reliable correspondence with histological cross-section (Fig. 19.3) [2, 23, 24, 32]. The high penetration depth and imaging quality is mainly due to the transparency of the cornea in the near infrared and the low dispersion mismatch between the two interferometer arms.

Representative images of the anterior part of the cornea are presented in Fig. 19.4. Epithelial cells are clearly observed with high signals from cell-cell junctions and nuclear boundaries (Fig. 19.4b). At the bottom of the epithelium, the basement membrane is clearly visible with a hyper reflective signal (arrow in Fig. 19.4a). Although collagen lamellae are the main component of the stroma, the typical collagen fibril diameter is well below the optical resolution meaning that the lamellae structure is not visible in OCT. However, the optical inhomogeneities within the focal volume combine to give a low contribution to backscattered intensity, so that the detected signal is averaged to a uniform gray. The anterior stroma is characterized by a high density of keratocytes. On progressing in depth toward the posterior stroma, keratocyte size increases while their density decreases. Their characteristic stellate shape can be recognized throughout. En face OCT images allow characterization of the size and density of the cells, as well as their distribution (parallel to each other and between the collagen lamellae) and shape (elongated in the en face direction (Fig. 19.4c). The stacked organization that appears in cross-sectional reconstruction (Fig. 19.3c) is due to the signals from the keratocytes positioned between collagen lamellae.

In the posterior part of the cornea, Descemet's membrane is recognized as a hyporeflective layer between the stroma and the hyperreflective endothelial layer. The endothelial cells are recognized by their hexagonal shape and arrangement in a regular honeycomb mosaic as shown in Fig. 19.5. These cells are of regular size, shape and distribution in the normal cornea. From these images, quantitation of the endothelial cell density could be performed. However, due to the micrometer-scale cross-sectional resolution and to the fact that the endothelium is a single cell layer, only small regions of endothelium are visible on a single en face image due to the natural shape of the cornea and to the folds in the posterior part of ex vivo tissue. Imaging the endothelium is similarly challenging in confocal microscopy, though due to the lower axial resolution (i.e., increased depth of field) images are less affected by the thinness of the layer.

FF-OCM is particularly well suited to imaging of the subcellular organization of the human cornea. Images produced are close to traditional histological sections observed in conventional microscopy but without needing to perform the multiple tissue processing of steps of fixing, embedding, slicing, and staining.

19.2.3 Pathological Corneas

19.2.3.1 Monitoring the evolution of edema

FF-OCM can image corneal edema, and its subcellular 3D resolution allows imaging of the evolution of the tissue as a function of edema development. The possibility of viewing both en face and cross-sections is useful for the characterization of the evolution of the stromal microstructure.

652 | Full-Field Optical Coherence Microscopy in Ophthalmology

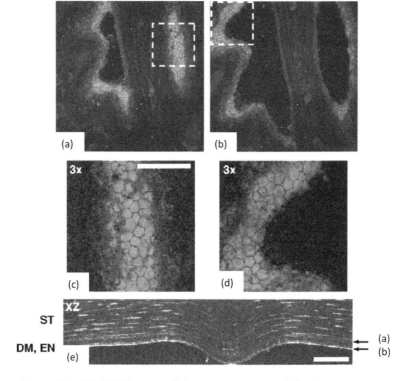

Figure 19.5 FF-OCM images of the posterior part of the human cornea. (a, b) En face images of the endothelium at the depths indicated in (e) the cross-section with the stroma (ST), Descemet's membrane (DM), and the endothelium (EN). (c, d) Zoomed regions of interest (dashed boxes) with the endothelial cell mosaic. Scale bar: 100 µm. Reprinted from Ref. [2].

During edema development, structural disturbances occur that increase light scattering and lead to the opacification of the cornea. In physiological conditions, edema may occur as the result of a wound, an endothelial dysfunction, pathology, or indeed during conservation in a tissue bank. Due to an excessive level of hydration, the tissue tends to swell. The accumulation of fluid causes local ruptures in the cell layers of the stroma and epithelium ("lakes" [21]) and the local-scale order that ensures transparency breaks down. These fluid-filled lakes can be nanometric to micrometric in scale. Multiwavelength modeling and experimental studies highlight different behaviors depending on the degree of edema [7, 17, 53]. In the healthy physiological cornea, the wavelength dependence of the scattered light's intensity reveals a major contribution from the fibrillar structures (i.e., Rayleigh scattering by nanostructures). On the other hand, when the cornea swells, the signal is mainly composed of scattering from larger structures (Mie scattering). In addition, other studies discovered an increase in scattering from keratocytes, due to a deficiency in the proteins which normally ensure the homogenization of the local refractive index [30, 43]. Several mechanisms of transparency loss have been studied in the literature, but the whole phenomenon is complex and the contribution of each scatterer is still unclear.

The relationship between the degree of hydration and the scattering in the reflected half-space (i.e., the only half-space accessible in vivo) has been characterized on ex vivo human corneas [10, 32]. Figure 19.6 shows the scattering intensity as a function of the scattering angle (in the incidence plane) for human corneal grafts with controlled and measured thickness. These results highlight that the scattering intensity increases with the thickness and the degree of edema. In particular, the angle-resolved information enables the surface contribution (low scattering angle) to be dissociated from bulk scattering (higher scattering angle) [33].

The relationship between the corneal microstructure resolved by FF-OCM and the scattering level of the tissue has been probed using an electromagnetic model [10]. The cross-sectional images are used as a refractive index mapping of the corneal bulk. The image stack is used as the basis for the creation of a tissue model for electromagnetic calculations in order to determine the scattered intensity at the microstructural scale. Figure 19.7 shows the agreement between measured backscattering intensity (total integrated scattering [TIS]) and the numerical results from FF-OCM cross-sectional images which demonstrate dilation of the tissue with edema development.

In addition to scattering properties of the tissue, FF-OCM images can help to track structural modifications of the tissue as edema evolves. Ex vivo corneas were imaged at different hydration states, which therefore had various thicknesses. Of the five layers that

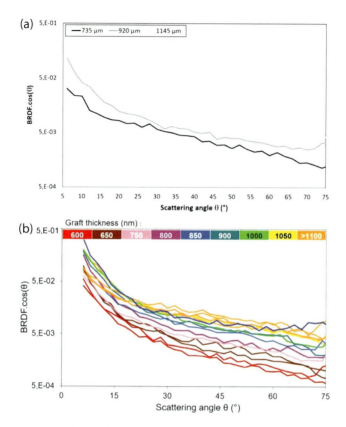

Figure 19.6 Angle-resolved scattered intensity measured in the reflected half-space for corneas in different edematous states. (a) Intensity levels are plotted for a single cornea at different hydration states after swelling and de-swelling protocols and (b) for different graft corneas with various thicknesses.

constitute the cornea, the stroma is the most affected by edema. Micrometer-scale modifications visible in FF-OCM could partially explain the transparency loss. The increase in thickness causes opacity due to increased light scattering, and a decrease in overall refractive index [16, 40]. In FF-OCM images, epithelium (Fig. 19.8a,b) and Descemet's membrane appear to be unaffected whereas the posterior part of the stroma is greatly modified by swelling (Fig. 19.8c,d). Keratocytes (recognized by their hyperreflective signal in physiological corneas) generally become less visible with

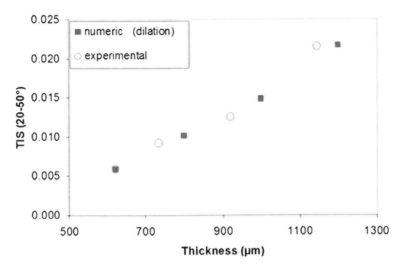

Figure 19.7 Comparison between the total integrated scattering (TIS) measured and simulated with parameters extracted from FF-OCM images.

edema, and lakes (recognized by their hyporeflective signal) appear between collagen lamellae. The link between structural modification and transparency loss is illustrated by viewing a paper grid through the cornea in the two different hydration states (Fig. 19.8e,f).

Figure 19.9 focuses on the evolution of the stroma during corneal swelling. En face views at different depths (Fig. 19.9c) reveal the different behavior of the anterior and posterior stroma. It appears that swelling mainly occurs along the posterior-anterior axis, with swelling decreasing from posterior to anterior. This phenomenon has already been shown in a previous study using electron microscopy [42] and is confirmed on whole unfixed corneas with FF-OCM. As shown in Figs. 19.8a and 19.8b, comparing anterior stroma of healthy and edematous corneal grafts, en face views at 70 μm below the epithelial layer are similar (Fig. 19.9c, images $z1_a$ and $z1_b$), whereas differences are observed in the posterior part where liquid infiltrations appear as darker areas in Figs. 19.9b and 19.9c (images $z2_b$ and $z3_b$). This behavior may be linked to the fact that the anterior stroma is more rigid and cohesive with interwoven lamellae than the posterior stroma [42].

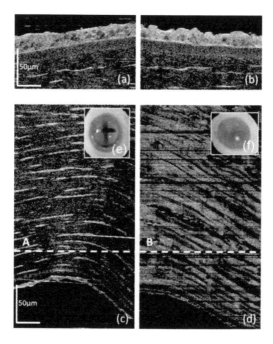

Figure 19.8 FF-OCM cross-sections of the same cornea at two different hydration states: (a, b) epithelium and anterior stroma and (c, d) posterior stroma. The measured thicknesses are 650 μm (a, c) and 1200 μm (b, d). Inset (e) and (f) Transparency loss of the tissue: macroscopic photographs of a grid as viewed through a corneoscleral button.

FF-OCM imaging leads us to conclude that for corneal thicknesses greater than 800 μm, lakes appear in the posterior part of the stroma, and these micrometric structures can partly explain the loss of transparency. However, the size and density of these lakes varied from one cornea to another (13 corneas studied). Moreover, in addition to the long, thin infiltrations due to lamellar separation, the presence of shorter and thicker lakes sparsely distributed in the stroma is noted. Generally, two lake distributions were visible depending on the graft studied: wide lamellar interspaces (over 100 μm in a plane parallel to the surface of the cornea) at high density (Fig. 19.10b), or smaller lakes (up to 50 μm) sparsely distributed (Fig. 19.10a). Each graft preferentially presented one of

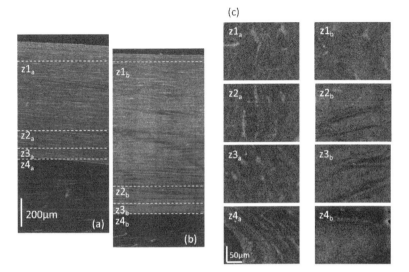

Figure 19.9 (a, b) FF-OCM cross-section from the same cornea at two swelling steps. (c) En face images at different thicknesses indicated on (a) and (b) by dotted lines: $z1_a$ and $z1_b$ at 70 μm beneath the epithelial layer, $z2_a$ and $z2_b$ at 200 μm above the endothelial layer (EL), $z3_a$ and $z3_b$ at 70 μm above the EL, and $z4_a$ and $z4_b$ at the EL. Scale bars: (a, b) 200 μm and (c) 50 μm.

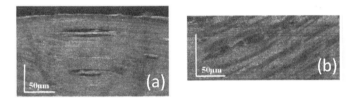

Figure 19.10 Lake details: (a) short and thick lakes and (b) elongated liquid infiltrations within the corneal stroma. Scale bar: 50 μm. Reprinted from Ref. [10].

these two distributions and no correlation with the thickness could be established as yet.

FF-OCM is a powerful tool for noninvasive characterization of both morphological changes in tissue and the hydration process, in particular the progression of corneal edema where FF-OCM enables the follow-up of modifications with micrometer-scale resolution.

19.2.3.2 Imaging of common pathologies

In addition to edema, specific lesions of corneal layers can be identified in a number of corneal diseases using FF-OCM.

One study [14] found that the pathologies of the cornea that have the highest incidence of transplant were keratoconus, bullous keratopathy, infectious keratitis and trauma, in that order. In addition, regraft was a major reason for transplant, but information on the first graft indication is not necessarily known. A second study [11] found the most common indications for transplant were keratoconus (29%), bullous keratopathy (21%), and other diagnosis (32%), including Fuchs' endothelial dystrophy (15%) and stromal dystrophies (3%) such as lattice corneal dystrophy.

An FF-OCM imaging study was performed on surgical specimens of 5 diseased corneas (keratoconus, bullous keratopathy, Fuchs' dystrophy, stromal scar after infectious keratitis, and lattice corneal dystrophy) [19]. They were obtained at the time of keratoplasty and stored in a de-swelling medium containing Dextran that results in corneas of physiological thickness. They were imaged using a FF-OCM commercial device (LLTech, France) following surgery, immersed six hours in a de-swelling medium, and then fixed in formaldehyde (10%) for histopathology.

Histological sections were cut 4 µm thick, stained with hematoxylin-eosin-safran (HES) or Periodic Acid Schiff (PAS), and observed with a conventional white-light microscope.

For comparison, in vivo FD-OCT images were taken on the same corneas before keratoplasty with an RTVue (Optovue, Inc, Fremont, CA) with a corneal adapter. This device provides corneal images with an cross-sectional resolution of 5 µm.

Keratoconus

Keratoconus is a progressive disease in which the cornea thins and becomes more steeply curved, creating either mild or severe astigmatism and myopia. Keratoconus may also cause acute swelling

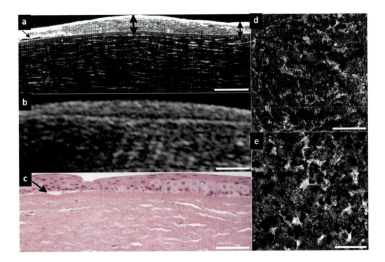

Figure 19.11 Keratoconus. (a) FF-OCM cross-section in the ex vivo cornea. (b) Portion of an FD-OCT image on the same eye in vivo before keratoplasty. The improvement in resolution between (a) and (b) can be appreciated. An epithelial scar can be seen to the far left in (a) and (c) (arrow). (c) Histology of the same sample. (d) En face view in the upper stroma of the keratoconus specimen and (e) en face view in the upper stroma of a normal cornea for comparison. Scale bars: (a–c) 100 μm and (d, e) 200 μm. Reprinted from Ref. [19].

leading to tears in Descemet's membrane and scarring of the cornea and vision loss. It can be an inherited condition, or be caused by trauma or disease.

In a corneal specimen tissue obtained from a keratoconus patient who underwent deep anterior lamellar keratoplasty, irregularities of the epithelium and Bowman's layer thickness were easily identified in FF-OCM images (Fig. 19.11). Stroma was also affected, with an alteration of the normal organization of collagen lamellae that had an undulated pattern. The irregular thickness of the Bowman's layer and the undulated arrangement of both the collagen lamellae and keratocyte layers are revealed in the cross-sectional view (Fig. 19.11a). Measurements on this image reveal a variability of Bowman's layer thickness from edges to center (7 μm to 12 μm) in comparison with Bowman's layer thickness in a normal cornea (10 μm to 12 μm, as measured on a normal cornea). Epithelial thickness is also variable from edges to center in the keratoconus cornea: (39 μm to 45 μm) in comparison with epithelial thickness in normal cornea (37 μm to 40 μm). Locations of maximum and minimum thickness measurements are indicated by double-headed arrows in Fig. 19.11a. In the anterior stroma of the keratoconus patient, keratocytes are numerous, small and arranged according to the undulated pattern of the layers (Fig. 19.11d). For comparison, in the upper stroma of a normal cornea, keratocytes are more sparse and larger and arranged in flat layers (Fig. 19.11e).

Bullous keratopathy

Bullous keratopathy is the consequence of long standing stromal and epithelial edema due to any cause. The epithelium separates from Bowman's layer beneath due to an inflow of fluid. Thickening of the underlying basement membrane is common. Ingrowths of collagen may occur between the epithelium and Bowman's layer. It is difficult to diagnose stromal edema at an early stage; however, bullous keratopathy can be otherwise recognized by increased thickness of cornea and spaces between stromal lamellae. The endothelial layer may be extremely thin or absent, while Descemet's membrane shows thickening.

Intraepithelial edema and subepithelial fibrosis were clearly imaged in cross-sections and en face sections of a bullous keratopathy specimen with FF-OCM (Fig. 19.12). The epithelial thickness is normal where subepithelial fibrosis is moderate and decreases where subepithelial fibrosis is prominent (Fig. 19.12b). Bowman's layer is clearly seen under the area of subepithelial fibrosis. Intraepithelial edema appears as dark spaces in the en face views of the epithelial layer (Fig. 19.12d), while the region of subepithelial fibrosis appears bright white (Fig. 19.12e).

Fuchs' dystrophy

Fuchs' dystrophy is an common age-related condition which may be inherited. It occurs in the endothelial layer. Normally, the endothelial cells prevent excess fluid from accumulating and so help the cornea maintain its transparency. In Fuchs' dystrophy, deteriorated

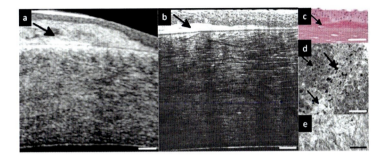

Figure 19.12 Bullous keratopathy. (a) Portion of an FD-OCT image of the cornea in vivo before keratoplasty. (b) Corresponding FF-OCM cross-section of this area ex vivo. Arrows in (a), (b), and (c) indicate the area of subepithelial fibrosis. (c) Corresponding histology. (d) En face view in the upper epithelium, where the arrows indicate intraepithelial edema that is seen as dark spaces (top arrow), gray wing cell layer (middle arrow), and bright surface epithelial cells (lower arrow). (e) En face view located in the zone of subepithelial fibrosis, which appears as a bright white area. Scale bars: (a, b) 200 µm and (c–e) 100 µm. Reprinted from Ref. [19].

endothelial cells die and allow fluid to build up in the cornea causing swelling and loss of corneal transparency.

In a Descemet's membrane specimen obtained from a Fuchs' dystrophy patient who underwent endothelial keratoplasty, the cross-sectional FF-OCM view showed excrescences of Descemet's membrane (guttae) that are typical of Fuchs' dystrophy patients (Fig. 19.13b), while en face pictures also showed the typical dark areas corresponding to guttae (Fig. 19.13d). Abnormal collagen deposits and guttae beneath Descemet's membrane are visible in Fig. 19.13b and Fig. 19.13d. Thickness measurements on Fig. 19.13b (Fuchs' specimen) and Fig. 19.13c (normal cornea specimen) reveal the thickening associated with Fuchs' dystrophy: Descemet's membrane plus endothelial thickness is 20 µm on average compared to 14 µm in the normal cornea.

Stromal scar following infectious keratitis

Keratitis is an inflammation of the cornea that sometimes occurs with infection after bacteria or fungi enter the cornea, for example after injury or more commonly from wearing contact lenses.

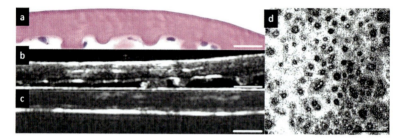

Figure 19.13 Fuchs' dystrophy. (a) Histology of Descemet's membrane with Fuchs' dystrophy. (b) FF-OCM cross-sectional and (d) en face view, respectively, in the same sample. For comparison, (c) shows a cross-section in a normal cornea. Scale bars: (a–c) 50 μm and (d) 100 μm. Reprinted from Ref. [19].

In a corneal specimen tissue obtained from a patient who underwent deep anterior lamellar keratoplasty as a surgical treatment for stromal scar after infectious keratitis, the consequences of infectious keratitis were well identified on cross-sections and en face FF-OCM pictures as highly reflective structures corresponding to the corneal opacity that progressively decreased from the upper to the lower stroma (Fig. 19.14). The scar region is visible as a hyperreflective bright white region in the cross-section view (Fig. 19.14c) and en face views (Fig. 19.14e,f). Bowman's layer is absent (Fig. 19.14b,c).

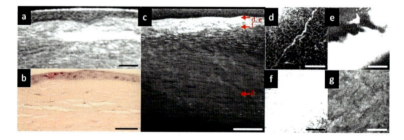

Figure 19.14 Stromal scar after infectious keratitis. (a) Portion of an FF-OCT image on the in vivo eye before keratoplasty, (b) corresponding histology, and (c) corresponding FF-OCM cross-section in the ex vivo cornea. (d–g) En face views at the depths indicated by the arrows in (c); (d) and (e) are separated by only a few microns and so share an arrow. Scale bars: (a–c) 200 μm, (d) 50 μm, and (e–g) 100 μm. Adapted from Ref. [19].

Figure 19.14d shows a subepithelial nerve in proximity to Bowman's layer. The scar extends from the subepithelial region (Fig. 19.14e) into the stroma (Fig. 19.14f). Lower stroma (Fig. 19.14g) is also affected and shows an abnormal lack of keratocytes and a dense gray aspect.

Lattice corneal dystrophy

Lattice corneal dystrophy is a rare condition that is characterized by accumulation of abnormal protein fibers throughout the middle and anterior stroma.

From a lattice corneal dystrophy patient who underwent deep anterior lamellar keratoplasty, corneal opacities were well identified in both cross-sectional and en face FF-OCM pictures as hyperreflective structures predominant in the anterior stroma (Fig. 19.15). The high density of keratocytes in the upper stroma can be seen in cross-sectional (Fig. 19.15b) and en face (Fig. 19.15d) views. The filament deposits below the epithelium can be seen in Fig. 19.15d.

The higher resolution of FF-OCM images was evident when compared with FD-OCT in all surgical specimens (compare, respectively, Figs. 19.11a and 19.11b, Figs. 19.12a and 19.12b, Figs. 19.14a and 19.14c, and Figs. 19.15a and 19.15b). The corneal pathology observed in the FF-OCM images of surgical specimens that had been

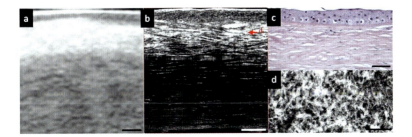

Figure 19.15 Lattice corneal dystrophy. (a) Portion of an FD-OCT image on the in vivo eye before keratoplasty and (b) corresponding FF-OCM cross-section in the ex vivo cornea. (c) Corresponding histology. (d) En face view in the upper stroma, where the depth of this section is indicated by the arrow in (b). Scale bar: 200 μm. Adapted from Ref. [19].

imaged after surgery with no fixation and no staining was confirmed by conventional histology after fixation and appropriate staining.

For ex vivo imaging of corneal pathology, FF-OCM has advantages over both conventional OCT and confocal microscopy. Cross-sectional and en face resolutions are, respectively, 1 µm and 1 µm in FF-OCM as compared to approximately 5 µm and 15 µm in conventional FD-OCT. Confocal microscopy imaging properties match FF-OCM's en face resolution of 1 µm, but cross-sectional resolution is limited to 5–20 µm, so cross-sectional views constructed from 3D confocal data sets appear blurred in comparison to FF-OCT cross-sections [52].

19.2.4 Characterization of Laser Ablations

FF-OCM can be used to observe local defects in the cornea. This is particularly interesting in the context of laser ablations. Indeed, lasers are routinely used nowadays in corneal surgery, especially for refractive surgery. The use of lasers rather than mechanical cutting has broadened the eligible indications and reduced potential microbial contamination. Very regular and precise cutting over a large area reduces the risk of side effects such as astigmatism or optical aberrations (responsible for night glare), while preserving the corneal epithelium, leading to faster postoperative recovery. Moreover, the mechanism of interaction between the femtosecond laser and the corneal tissue has been widely studied and the parameters for cutting are well known (i.e., irradiance, radiance exposure, incident wavelength, damage threshold) [37, 63, 64]. It is now easy to adjust the laser settings to cut to a specific depth without affecting the surrounding tissue.

This technique has more recently been applied to corneal transplants. In this case, the flexibility and the precision of the femtosecond lasers allows custom cuts, which permits a perfect match between the donor and the recipient corneas. However, if laser incisions are relatively well controlled in healthy corneas, in pathological tissues this is more problematic. Indeed, indications for transplant are principally highly degraded optical properties and in these conditions, tissue is often clouded. This is mainly due to an increase in the scattering level that modifies the interaction mechanisms with the laser. Currently, the optical properties of the tissue are not taken into account. Surgeons adapt the technique case by case, by supplementing with mechanical cutting, by increasing the energy delivered by the laser or by repeating the laser cutting procedure twice. Visualization of the laser incisions could help to optimize the laser parameters and also better understand the laser–tissue interaction.

Feasibility studies carried out in vitro require cross-sectional histology [59] or electron microscopy for the ultrastructure characterization [1]. Nuzzo et al. [48] present a study showing the effect of laser fluence on the tissue. Figure 19.16 shows a stained histological section of a cornea with laser cutting and images obtained with TEM along this cut. The authors conclude that corneal trephination performed by the femtosecond laser preserves the ultrastructure of the disrupted collagen fibers. In pathological corneas (edematous), a layer of cellular and collagenic debris, thicker in the anterior stroma

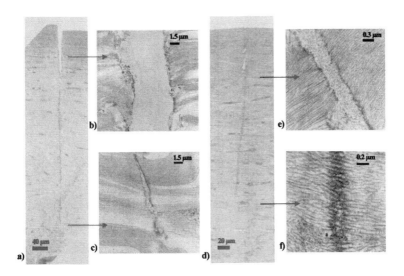

Figure 19.16 (a) and (d) Stained histological sections of laser incisions in a human edematous cornea performed at NA = 0.3 and NA = 0.5 and constant energies of 500 and 170 nJ, respectively. (b), (c), (e) and (f) TEM micrographs of the collagen (b) and (e) in the anterior stroma and (c) and (f) at the end of the incision. Adapted from Ref. [67].

and thinner in the posterior stroma, runs along the edges of the incision that was performed at a constant laser energy density.

To overcome the artifacts caused by tissue preparation in histology, some studies use high-resolution FD-OCT imaging of laser ablations. Brown et al. [9] used an FD-OCT device to compare the laser incisions with mechanical cutting. FD-OCT shows the presence of cavitation bubbles caused by the laser-tissue interaction and the presence of debris in the vicinity of the incision. FF-OCM offers an alternative to histology and FD-OCT. FF-OCM imaging enables visualization of the effect of laser incisions performed at energies close to or below the damage threshold. [32].

Lamellar laser incisions (parallel to the surface) were made with a commercial surgical laser in the anterior stroma of a human corneal graft to a depth of 160 μm. A few minutes after the incision, the graft was imaged by a commercial FD-OCT device (Fig. 19.17a) and then 36 h after cutting by FF-OCM (Fig. 19.17b). Hyperreflective areas in FF-OCM images correspond to the laser damage within the stroma. It was also noted that the tissue near the incision is disturbed over an area only a few micrometers thick and that the incision border shows a higher signal, indicating collagen hyperdensification, matching the delamination effect highlighted in the TEM images shown in Fig. 19.16. The remaining tissue presents no apparent change. It is also possible to highlight the presence of cavitation bubbles a few tens of micrometers in diameter (which are smaller than those observed immediately after laser cutting in FD-

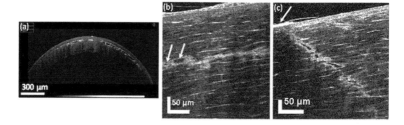

Figure 19.17 (a, b) Cross sections of a lamellar laser incision in the same human cornea obtained (a) with commercial FD-OCT and (b) with FF-OCM. (c) FF-OCM cross-section of a perforating laser incision in the human cornea. Scale bars: (a) 300 μm and (b, c) 50 μm. Adapted from Ref. [32].

OCT (Fig. 19.17a) due to the resorption that occurs a few hours after cutting).

A perforating incision (perpendicular to the surface) was also imaged in the same way (Fig. 19.17c). FF-OCM cross-sections show a smooth incision of good quality over the entire depth. As above, the disruption of the collagen structure near the incision and the presence of cavitation bubbles a few tens of micrometers in diameter are highlighted. On the surface, there is an increase in the FF-OCM signal which may indicate the interaction between the femtosecond laser and the epithelial cells.

These results illustrate the interest of FF-OCM as a complement to or a replacement for histology for imaging laser ablations, where the delicate preparation of the damaged sample for histology could induce measurement artifacts.

19.2.5 *Perspectives*

19.2.5.1 Eye banking

As no tissue preparation, modification or staining of any kind is required for FF-OCM imaging, it appears suitable for the evaluation of human donor corneas in tissue banks. In current practice in eye banks, aside from an excellent analysis of the corneal endothelium thanks to conventional white-light microscopy, the rest of the cornea is grossly observed in light microscopy. Fine details may escape the checklist and thus compromise the outcome of the corneal transplant. Donor corneas that have been assessed by light microscopy are stored with no digital image record of their structure. FF-OCM imaging automatically creates a digital image catalog of the corneas that have been assessed. Furthermore, in current practice, clinical history taking of the donor might be incomplete and lack important details like history of refractive surgery or infectious keratitis.

Using FF-OCM, it is possible to identify the different corneal structures in normal and diseased corneas [19]. For instance, clear imaging of Bowman's layer enables detection of irregularities, breaks, scars or absence of this layer that are features of keratoconus, laser in situ keratomileusis, scars after infectious keratitis, or photorefractive keratectomy. These four corneal conditions, which are contraindications for using donor tissue for penetrating or anterior lamellar keratoplasty, are currently difficult to detect in eye banks when the donor history is not precisely known. The fine assessment of the lamellar stroma, in terms of organization, number, thickness, and reflectivity of the collagen lamellae and keratocyte density, provides a new means to assess the condition of the stroma in donor corneas and, potentially, its optical properties. This should be of major interest for selection of donor tissue for penetrating or anterior lamellar keratoplasty.

Descemet's membrane can be precisely assessed with FF-OCM with regard to its thickness and structure. This information is certainly complementary to that provided by light microscopy in the selection of donor tissue for penetrating or endothelial keratoplasty. Precise assessment of Descemet's membrane thickness could also be useful for selection of donor corneal tissue for Descemet's membrane endothelial keratoplasty where thin Descemet's membrane may be more difficult to separate from the stroma.

Finally, previous corneal laser surgery is also a contraindication for donor tissue. However, wound detection within the stroma is currently challenging with conventional microscopes used in eye banks. FF-OCM imaging could help to detect the wound signature. Polarization-sensitive OCT (see Chapter 13) could determine local defects by highlighting local variation of birefringence induced by laser ablation and collagen disruption and remodeling.

Preoperative evaluation of corneal grafts is currently very limited, and the extent to which their histological characteristics influence the long-term outcome of the graft (except for endothelial counts) remains unknown. FF-OCM imaging of donor corneas would allow better quality control of corneas destined for graft, as well as better matching of each specific cornea to each particular patient. In addition, the acceptance rate for donor corneas for graft may increase as FF-OCM can detect the layer and region of any defect, allowing graft of the remaining healthy corneal layers rather than rejection of the entire button. Applying FD-OCT to imaging of donor corneas demonstrated the improved localization of corneal opacities using the higher definition of OCT as compared to traditional slit-lamp examination [5] that led to improved decision making on transplant suitability of the donor corneas. FF-OCM with its higher resolution should further improve on the level of detail in determining opacities in the cornea.

19.2.5.2 Characterization of artificial corneas

Corneal surgery has been a pioneering area of bioengineering, with the development of cell cultures of the ocular surface and the development of artificial corneas. FF-OCM could be used to evaluate the optical characteristics of bioengineered tissue, namely cultured limbal cells and artificial cornea [12, 20]. Specifically, due to the fact that it is noninvasive, FF-OCM could be used to characterize the cell colonization of artificial corneas at multiple time points in living tissue. It could also be used to assess suitability of the artificial cornea for transplantation.

19.3 Lens

19.3.1 *Anatomy*

The crystalline lens, which sits behind the cornea at the back of the anterior chamber, provides one third of the eye's optical power. It has a capsule, an epithelium and in the interior it contains fibers a few microns in diameter, arranged in concentric onion-like layers (Fig. 19.18). It contains no nerves, blood vessels or connective tissue. This structure of interleaved fibers allows it to remain flexible, and the crystalline proteins that form the fibers maintain the lens transparency. Disorganization of the fiber structure leads to opacity. It has a variable refractive index and can vary its power by changing form, thanks to ciliary muscles that deform the lens in order to change its focal length. This process is accommodation.

19.3.2 *Imaging the Lens*

The lens can be imaged in vivo by both ultrasound and magnetic resonance imaging (MRI), giving information about the thickness of the lens at relatively low resolution. Recently, custom OCT imaging of the in vivo lens was achieved in order to obtain information

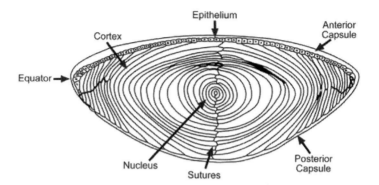

Figure 19.18 Schematic of crystalline lens anatomical structure (figure reproduced with permission of Dr Joan Roberts from http://photobiology.info/Roberts.html).

on its form and optical properties [49, 50]. Ex vivo images obtained by electron microscopy reveal the organization of the lens fibers; however, to obtain this type of image, the lens must be sliced [61].

19.3.3 FF-OCM Lens Imaging

Using FF-OCM, it was possible to resolve the fiber structure of the whole ex vivo rat lens in 3D [24]. As disorganizations in this fiber structure may be responsible for clouding, the information on the 3D fiber structure that can be obtained by FF-OCM imaging could contribute to understanding of cataractogenesis.

In the lens (Fig. 19.19), the capsule, the epithelial cell layer, and the cortical fibers can be discerned with FF-OCM. The interface between the lens capsule and the lens epithelial cells appears highly reflective. The micrometer-scale 3D resolution of FF-OCM offers the most precise view of the intact lens to date, allowing identification of individual lens fibers.

19.3.4 Conclusion and Perspectives

Study of the lens in humans and animal models using FF-OCM associated with image processing techniques could reveal information about the 3D orientations of the lens fibers and the

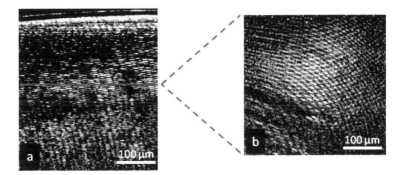

Figure 19.19 Ex vivo rat lens imaging. (a) Cross section reconstruction of the lens anterior cortex. Note the high reflectivity of the lens capsule, the visibility of the lens epithelial cell layer between the two anterior hyperreflective lines, and the underlying lens fibers. The regular array and the curved shape of the lens fibers are clearly revealed. (b) En face view inside the anterior cortex, the depth location of which is indicated by the line on the cross-section.

role of their organization in the transparency of the lens. This may bring new insights into the process of cataractogenesis. The noninvasiveness and high resolution of the imaging technique make it particularly well suited in the investigation of the fragile lens and its intricate structure.

19.4 Retina

19.4.1 Anatomy

The retina, a fine, vascularized membrane, covers the surface of the choroid anteriorly. Stray light is absorbed by the pigmented epithelium, which is situated between the choroid and the photoreceptors of the retina. When light enters the eye it is focused on the retina, where it is absorbed by photoreceptors and converted into action potential by several layers of nerve cells, the last of which are the photoreceptors (Fig. 19.20). The visual stimuli, now encoded as neuronal input, undergo complex processing by several cell types and are transmitted via the optic nerve to the brain. Stray light is absorbed by the pigmented epithelium, which is situated

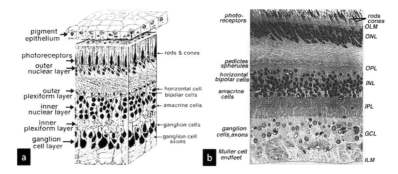

Figure 19.20 (a) Schematic of retinal layers and (b) light micrograph of a histological slice of the retina (figure reproduced with permission of Dr Bryan Jones from http://webvision.med.utah.edu/).

immediately below the photoreceptors. The organization of the cell layers in the retina permits a pretreatment of the signal at the retinal level before it is transmitted to the brain.

19.4.2 Retinal Imaging Techniques

Direct images of the retina are obtained using a traditional ophthalmoscope, or with a fundus camera or slit lamp. The traditional ophthalmoscope sends its image directly into the eye of the operator, while the modern ophthalmoscope will display the retinal image on screen. The image of the fundus is in the en face orientation.

The scanning laser ophthalmoscope (SLO) is nowadays a commonly used instrument for fundus imaging. The SLO scans a laser beam over the retinal surface and collects the reflected light in order to construct the fundus image point by point (Fig. 19.21a). A confocal pinhole placed in the detection channel provides optical sectioning ability of a few hundred microns per depth slice. The typical en face resolution is 5 µm.

OCT has become the leading technique for optical imaging of the retina since its introduction in the late nineties. Its resolution, penetration depth, and native cross-sectional orientation make it particularly interesting for the visualization of the retinal layers (Fig. 19.21b). Current FD-OCT systems on the market achieve cross

Figure 19.21 (a) SLO fundus image (courtesy of Timothy J. Bennett, Penn State Hershey Eye Center) and (b) OCT retinal image of drusen: accumulations of extracellular material that can be an indication of age-related macular degeneration (courtesy of James D. Strong, Penn State Hershey Eye Center).

sectional resolution of 5 μm and en face resolution of the order of 10 μm, to depths of 2 mm. Volumes are scanned at rapid rates thanks to spectral domain detection schemes or swept-source technology [18]. Recently it has become common to combine OCT with a SLO in a single instrument so that the location of the OCT scan can be navigated within the SLO field.

19.4.3 FF-OCM Retinal Imaging

FF-OCM with its resolution of 1 μm in three dimensions can provide images of the ex vivo retina that reveal individual cells. Noninvasive cellular imaging of retinal tissues is of clinical relevance to assess the content of surgically removed tissues, and is of interest for research purposes since histology of these delicate tissues is challenging. In addition, high-resolution FF-OCM imaging of ex vivo tissues may provide insight into the signals captured in vivo by FD-OCT systems.

FF-OCM of the retina was first performed on rodent and porcine retinal tissues using a prototype instrument in 2004 [24]. All animal manipulation was in accordance with the "ARVO Statement for the Use of Animals in Ophthalmic and Vision Research." Pigmented and albino rats were deeply anesthetized by pentobarbital and killed by head dislocation. The porcine eyes were obtained from a local abattoir. Eyes were preserved in phosphate-buffered saline (PBS) or paraformaldehyde during transfer to the laboratory, dissected with small scissors and forceps, and put in a container filled with PBS.

A stack of tomographic images were acquired at successive depths, typically in 1 μm steps. Postacquisition volumetric reconstruction then allowed navigation within the image stack in any direction, to allow examination of a given area of interest.

Images of the retina are shown in Fig. 19.22. All images were obtained from isolated retinas. In the rat retina, the nerve fiber

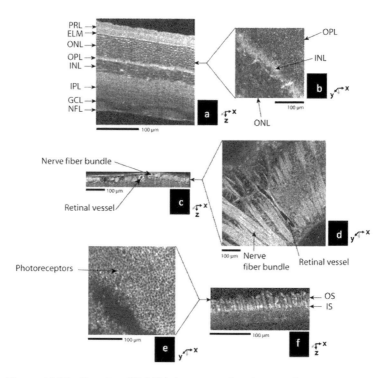

Figure 19.22 Ex vivo FF-OCM imaging of rodent and porcine retina samples. (a) Reconstructed cross-sections from stacks of en face images in the rat retina. (b) En face image showing the plexiform and nuclear layers. (c, d) Cross-sectional and en face views, respectively, of the nerve fiber bundles in the rat retina. (e, f) En face and cross-sectional views, respectively, of pig photoreceptors. Arrows on cross-sections indicate the depths of the corresponding en face images. PRL, photoreceptor layer; ELM, external limiting membrane; ONL, outer nuclear layer; OPL, outer plexiform layer; INL, inner nuclear layer; IPL, inner plexiform layer; GCL, ganglion cell layer; NFL, nerve fiber layer; OS, outer segment; IS, inner segment. Scale bar: 100 μm. Adapted from Ref. [24].

bundles (Fig. 19.22d) and, within the porcine retina, individual outer segments (Fig. 19.22e,f) are visible. The size of the outer segments suggests that these are cones. Conversely, in the rod-dominant retina of rats, no separation between outer segments of photoreceptors was visible. This was thought to be because the diameter of rods in rats is comparable to or smaller than the FF-OCM system resolution capability. In reconstructed cross-sections of the rat retina (Fig. 19.22a,b), the nerve fiber layer, the three nuclear layers, and the plexiform layers are clearly recognizable.

The penetration of FF-OCM was sufficient to image an arteriovenous crossing (Fig. 19.23a–e). In the optic nerve, the convergence of axons and vessels is visible (Fig. 19.23f–i).

The 3D data set at 1 µm resolution in all three dimensions offers the possibility of viewing both en face and cross-sectional images to provide complementary information. Indeed, the en face image can reveal structures that pass unseen in the cross-section and vice versa. For instance, the nerve fiber layer is most clearly visible in en face images, while determination of the thickness of retinal layers appears best performed on cross-sectional images. Volumetric reconstruction allows further exploitation of the available information through the analysis of a given area of interest from multiple angles.

Preliminary analysis of signal levels arising from different tissues shows that overall, fibrillar structures give a high signal, for example within the nerve fiber layer. This is similar to what was found in lens fibers. Note also that the membranes, such as Bruch's membrane, and also the interface between cells and extracellular milieu give high signal levels.

19.4.4 Perspectives in Retinal Imaging

In vivo FF-OCM retinal imaging remains challenging due to the Linnik configuration of the FF-OCM setup and its requirement for identical reference and sample arms: if the in vivo retina is the sample, then the reference arm requires an optical path that is almost identical to the (patient dependent) anterior segment in front of the reference mirror. Adding adaptive optics to the imaging

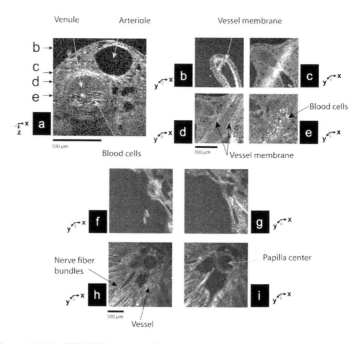

Figure 19.23 FF-OCM images of an arteriovenous crossing in the porcine retina (a–e) and of the disc in the rodent retina (f–i). (a) Cross-sectional reconstruction; (b–e) en face images. Arrows in the cross-section indicate the respective depths of the en face images. Successive en face images through the arteriovenous crossing reveal the relationship between the artery on top and the vein below. The vein is filled with erythrocytes, while the artery is empty. (f–i) En face images of the disc in a rat retina. Note the convergence of axons toward the disc and the presence of radiating vessels. The disc appears devoid of reflective structures in the deeper scans, probably because of the perpendicular disposition of the fibers and/or a masking effect of overlying structures. Scale bar: 100 μm. Reprinted from Ref. [24].

setup in order to correct aberrations induced by the passage through the anterior segment could provide a solution.

At present, while limited to ex vivo imaging, capturing high resolution images of the ex vivo retina is of interest for two main reasons: (1) clinically, for intraoperative verification of tissues removed during surgery and (2) for research, where high-resolution analysis of a retinal sample using FF-OCM may aid understanding of features of in vivo OCT images obtained with a regular resolution clinical instrument.

The retina is a particularly difficult tissue to examine by histology. The retina is composed mostly of tightly packed neuronal and glial cells, but the absence of myelin makes it very fragile. The retina notably retracts with fixation hence there is unavoidable tissue distortion during histological processing. Finally, human retinal tissues alter quickly postmortem. In vivo imaging techniques (such as conventional OCT), although giving access to retinal architecture [62], still lack cellular resolution. Hence, analysis of the 3D arrangement of retinal and vitreal structures remains a tedious task.

19.5 Future Developments in Ophthalmology

19.5.1 *In vivo FF-OCM Imaging*

In vivo imaging of the rat cornea was demonstrated in 2005 on a FF-OCM setup that used a rapid camera [22]. En face images revealed some cellular features in the corneal layers, in a similar manner to confocal imaging. However, the potential for image averaging was reduced due to movement artifacts, so that images were noisier than on ex vivo samples, and cross-sectional reconstruction was not possible as the 3D stacks could not be captured in a fast enough timeframe. With recent improvements in CMOS camera technology, including greater full-well depth (so reducing noise) and increase in capture speed, the possibility of in vivo corneal imaging with FF-OCM with images comparable to those captured in ex vivo tissues is approaching (see Chapters 6 and 7 on high-speed FF-OCM). Recent studies have demonstrated the ability to perform in vivo imaging on skin [36].

19.5.2 *Coupling with Other Modalities: A Way to Increase Information*

FF-OCM for ophthalmic imaging has been widely investigated in this chapter. The contrast of this reflectance technique, as in the case of confocal reflectance microscopy, stems from reflected and backscattered signals. As FF-OCM signals are due to spatial variation of the refractive index, no specific information related to the biochemical composition of the probed structures is observed on the images. FF-OCM imaging enables the architecture of the cells in the different parts of the eye to be highlighted (morphology, density, and nuclear boundaries), but contrast cannot be selectively enhanced for specific features with FF-OCM alone. Coupling FF-OCM to other types of microscopy can contribute additional contrast to structures of interest.

Fluorescence FF-OCM

Combining FF-OCM with fluorescence provides the possibility to add functional information to the morphological images provided by FF-OCM [6, 66]. Contrast agents can be used to label cells or blood vessels. Some studies have shown the interest of this combination for retinal imaging to image blood vessels in addition to the anatomical images obtained with OCT [4, 57].

Multiphoton FF-OCM

The stroma is mainly composed of stacked collagen lamellae and the diameter of the collagen fibrils is well below the optical resolution (around 35 nm). These structures are not visible in OCT. Another complementary emerging technique in the field of biomedical imaging is multiphoton microscopy (MPM). Signals are based on the nonlinear interaction of the laser beam with the tissue providing noninvasive 3D micrometer-scale imaging, as in FF-OCM. A key advantage of MPM is the different endogenous contrast mechanisms. Figure 19.24 shows multiphoton imaging of unstained ex vivo intact human cornea. Two-photon excited fluorescence (2PEF) signals can be generated from endogenous fluorophores mainly located in the cell cytoplasm: epithelial and endothelial cells and keratocytes (Fig. 19.24a,d) [3, 38, 54]. Second harmonic generation (SHG) is a second-order nonlinear process. Second harmonic signals are emitted by noncentrosymmetric organization such as fibrillar collagen within biological tissues. Strong SHG

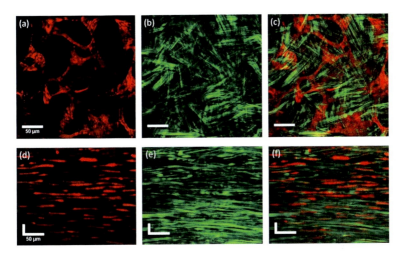

Figure 19.24 Ex vivo multiphoton microscopy imaging of intact unstained human cornea. Images in false color: two-photon excited fluorescence (2PEF) in red and second harmonic generation (SHG) in green. (a–c) Transverse sections (also called en face views in this chapter) at 200 μm deep within the cornea. (d–f) Cross sections (numerical reconstruction) along the depth of the anterior part of the cornea (from the surface to 200 μm deep). Keratocytes exhibit 2PEF signals (in red) and collagen lamellae appear with striated features that traduce the orientation of the lamellae in forward SHG signals (in green). (c) and (f): merged 2PEF and SHG images. Scale bar: 50 μm.

signals are emitted within the stroma providing contrasted images with striated features related to the orientation of the collagen fibrils that compose the lamellae (Fig. 19.24b,e) [3, 26, 44, 65]. Key information about the 3D organization of the collagen lamellae can be obtained from these SHG images or by using polarization-sensitive SHG signals [34, 68]. Finally, third harmonic generation (THG) is a signature of optical heterogeneities and THG signals are detected at lamellar interfaces giving information similar to OCT [3].

Multiphoton microscopy seems complementary to other conventional imaging techniques used in ophthalmology due to its capacity to image fibrillar collagen [35]. First demonstrations of in vivo multiphoton imaging of the cornea [34, 60] and of the retina

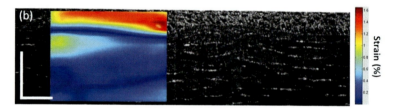

Figure 19.25 Static elastography combined with FF-OCM. (a) FF-OCM imaging of an ex vivo porcine cornea and (b) strain map (as a percentage) superimposed on the cross-section after compression of the tissue. Scale bar: 100 μm. Adapted from Ref. [45].

[58] have been performed. Moreover, some attempts at combining OCT with multiphoton microscopy have been performed in order to benefit from the complementary information of the two techniques [4, 31].

Full-Field Optical Coherence Elastography

The cornea contributes approximately two thirds of the optical power of the eye. Any modification in the shape or mechanical strength of the tissue induces a loss of vision. Mechanical studies are generally performed to follow the deformation of the tissue and to determine its mechanical characteristics at the scale of the tissue. However, these macroscopic properties are probably linked to the microscopic organization [68]. Micrometer-scale imaging via FF-OCM combined with determination of local mechanical properties via elastography should provide new information about the tissue, as was recently demonstrated using full-field optical coherence elastography (Fig. 19.25) [45, 46].

Acknowledgments

GL gratefully aknowledges Michèle Savoldelli (Department of Ophthalmology, AP-HP Hôtel-Dieu, Paris, France) for preparing corneal tissue and providing histologic section and TEM imaging, Banque Française des Yeux (Paris, France) for providing human cornea buttons for scientific use, Amir Nahas (Institut Langevin, ESPCI, Paris, France) for his contribution on FF-OCM elastography, and Marie-Claire Schanne-Klein for stimulating discussions. KG, MP, and VB thank Wajdene Ghouali, Salima Bellefqih, Otman Sandali, and Laurent Laroche for their work on pathological cornea study; the EFS Cornea Bank, Hôpital Saint Antoine (Paris) and Hôpital des Quinze-Vingts (Paris, France) for their contribution of samples; LLTech (Paris, France) for the use of its FF-OCM device, which was provided free of charge; and Manuel Simonutti, Jose Sahel, Jean-François Le Gargasson, Arnaud Dubois, and Claude Boccara for their work on retinal tissue imaging and in vivo FF-OCM studies. GG, LS, LH, and CD acknowledge the French Blood Center of Alpes-Mediterranée (France) and particularly Dr. Y. Nouaille de Gorce for providing samples of human graft cornea and Olivier Casadessus for experimental and theoretical help.

References

1. Abahussin, M., Hayes, S., Edelhauser, H., Dawson, D. G., and Meek, K. M. (2013). A microscopy study of the structural features of post-lasik human corneas. *PLoS One*, **8**(5), p. e63268.
2. Akiba, M., Maeda, N., Yumikake, K., Soma, T., Nishida, K., Tano, Y., and Chan, K. P. (2007). Ultrahigh-resolution imaging of human donor cornea using full-field optical coherence tomography. *J. Biomed. Opt.*, **12**(4), p. 041202.
3. Aptel, F., Olivier, N., Deniset-Besseau, A., Legeais, J.-M., Plamann, K., Schanne-Klein, M.-C., and Beaurepaire, E. (2010). Multimodal nonlinear imaging of the human cornea. *Invest. Ophthalmol. Vis. Sci.*, **51**(5), pp. 2459–2465.
4. Auksorius, E., Bromberg, Y., Motiejunaite, R., Pieretti, A., Liu, L., Coron, E., Aranda, J., Goldstein, A. M., Bouma, B. E., Kazlauskas, A., and Tearney,

G. J. (2012). Dual-modality fluorescence and full-field optical coherence microscopy for biomedical imaging applications. *Biomed. Opt. Express*, **3**(3), pp. 661–666.

5. Bald, M. R., Stoeger, C., Galloway, J., Tang, M., Holiman, J., and Huang, D. (2013). Use of fourier-domain optical coherence tomography to evaluate anterior stromal opacities in donor corneas. *J. Ophthalmol.*, **2013**, Article ID 397680, 5 pages.

6. Beaurepaire, E., Moreaux, L., Amblard, F., and Mertz, J. (1999). Combined scanning optical coherence and two-photon-excited fluorescence microscopy. *Opt. Lett.*, **24**(14), pp. 969–971.

7. Benedek, G. B. (1971). Theory of transparency of the eye. *Appl. Opt.*, **10**(3), pp. 459–473.

8. Beuerman, R. W., and Pedroza, L. (1996). Ultrastructure of the human cornea. *Microsc. Res. Techniq.*, **33**(4), pp. 320–335.

9. Brown, J. S., Wang, D., Li, X., Baluyot, F., Iliakis, B., Lindquist, T., Shirakawa, R., Shen, T., and Li, X. (2008). In situ ultrahigh-resolution optical coherence tomography characterization of eye bank corneal tissue processed for lamellar keratoplasty. *Cornea*, **27**(7), pp. 802–810.

10. Casadessus, O., Georges, G., Siozade-Lamoine, L., Deumié, C., and Hoffart, L. (2012). Light scattering from edematous human corneal grafts' microstructure: experimental study and electromagnetic modelization. *Biomed. Opt. Express*, **3**(8), pp. 1793–1810.

11. Claesson, M., Armitage, W. J., Fagerholm, P., and Stenevi, U. (2002). Visual outcome in corneal grafts: a preliminary analysis of the swedish corneal transplant register. *Brit. J. Ophthalmol.*, **86**(2), pp. 174–180.

12. De Sa Peixoto, P., Deniset-Besseau, A., Schmutz, M., Anglo, A., Illoul, C., Schanne-Klein, M.-C., and Mosser, G. (2013). Achievement of cornea-like organizations in dense collagen 1 solutions: clues to the physicochemistry of cornea morphogenesis. *Soft Matter*, **9**, pp. 11241–11248.

13. DelMonte, D. W., and Kim, T. (2011). Anatomy and physiology of the cornea. *J. Cataract. Refract. Surg.*, **37**(3), pp. 588–598.

14. Edwards, M., Clover, G. M., Brookes, N., Pendergrast, D., Chaulk, J., and McGhee, C. N. (2002). Indications for corneal transplantation in New Zealand: 1991–1999. *Cornea*, **21**(2), pp. 152–155.

15. Erie, J. C., McLaren, J. W., and Patel, S. V. (2009). Confocal microscopy in ophthalmology. *Am. J. Ophthalmol.*, **148**(5), pp. 639–646.

16. Farrell, R. A. (1994). Corneal transparency. In *Principles and Practice of Ophthalmology: Basic Sciences*. WB Saunder, Philadelphia.

17. Farrell, R. A., McCally, R. L., and Tatham, P. E. R. (1973). Wavelength dependencies of light scattering in normal and cold swollen rabbit corneas and their structural implications. *J. Physiol.*, **233**(3), pp. 589–612.
18. Geitzenauer, W., Hitzenberger, C. K., and Schmidt-Erfurth, U. M. (2011). Retinal optical coherence tomography: past, present and future perspectives. *Brit. J. Ophthalmol.*, **95**(2), pp. 171–177.
19. Ghouali, W., Grieve, K., Bellefqih, S., Sandali, O., Harms, F., Laroche, L., Paques, M., and Borderie, V. (2015). Full-field optical coherence tomography of human donor and pathological corneas. *Curr. Eye Res.*, **40**(5), pp. 526–534.
20. Ghoubay-Benallaoua, D., Sandali, O., Goldschmidt, P., and Borderie, V. (2013). Kinetics of expansion of human limbal epithelial progenitor cells in primary culture of explants without feeders. *PLoS One*, **8**(12), p. e81965.
21. Goldman, J. N., Benedek, G. B., Dohlman, C. H., and Kravitt, B. (1968). Structural alterations affecting transparency in swollen human corneas. *Invest. Ophthalmol. Vis. Sci.*, **7**(5), pp. 501–519.
22. Grieve, K., Dubois, A., Simonutti, M., Paques, M., Sahel, J., Le Gargasson, J.-F., and Boccara, C. (2005). In vivo anterior segment imaging in the rat eye with high speed white light full-field optical coherence tomography. *Opt. Express*, **13**(16), pp. 6286–6295.
23. Grieve, K., Moneron, G., Dubois, A., Le Gargasson, J.-F., and Boccara, C. (2005). Ultrahigh resolution ex-vivo ocular imaging using ultrashort acquisition time en face optical coherence tomography. *J. Opt. A*, **7**(8), pp. 368–373.
24. Grieve, K., Paques, M., Dubois, A., Sahel, J., Boccara, C., and Le Gargasson, J.-F. (2004). Ocular tissue imaging using ultrahigh-resolution, full-field optical coherence tomography. *Invest. Ophthalmol. Vis. Sci.*, **45**(11), pp. 4126–4131.
25. Guthoff, R. F., Zhivov, A., and Stachs, O. (2009). In vivo confocal microscopy, an inner vision of the cornea: a major review. *Clin. Exp. Ophthalmol.*, **37**(1), pp. 100–117.
26. Han, M., Giese, G., and Bille, J. (2005). Second harmonic generation imaging of collagen fibrils in cornea and sclera. *Opt. Express*, **13**(15), pp. 5791–5797.
27. Hart, R. W., and Farrell, R. (1969). Light scattering in the cornea. *J. Opt. Soc. Am.*, **59**(6), pp. 766–774.

28. Huang, D., Li, Y., and Tang, M. (2008). Anterior eye imaging with optical coherence tomography. In *Optical Coherence Tomography*. Springer, Berlin, Heidelberg, pp. 961–981.
29. Izatt, J. A., Hee, M. R., Swanson, E. A., Lin, C. P., Huang, D., Schuman, J. S., Puliafito, C. A., and Fujimoto, J. G. (1994). Micrometer-scale resolution imaging of the anterior eye in vivo with optical coherence tomography. *Arch. Ophthalmol.*, **112**(12), pp. 1584–1589.
30. Jester, J., Moller-Pedersen, T., Huang, J., Sax, C., Kays, W., Cavangh, H., Petroll, W., and Piatigorsky, J. (1999). The cellular basis of corneal transparency: evidence for "corneal crystallins". *J. Cell Sci.*, **112**(5), pp. 613–622.
31. Lai, T., and Tang, S. (2014). Cornea characterization using a combined multiphoton microscopy and optical coherence tomography system. *Biomed. Opt. Express*, **5**(5), pp. 1494–1511.
32. Latour, G., Georges, G., Siozade-Lamoine, L., Deumie, C., Conrath, J., and Hoffart, L. (2010). Human graft cornea and laser incisions imaging with micrometer scale resolution full-field optical coherence tomography. *J. Biomed. Opt.*, **15**(5), p. 056006.
33. Latour, G., Georges, G., Siozade-Lamoine, L., Deumie, C., Conrath, J., and Hoffart, L. (2010). Light scattering from human corneal grafts: bulk and surface contribution. *J. Appl. Phys.*, **108**(5), p. 053104.
34. Latour, G., Gusachenko, I., Kowalczuk, L., Lamarre, I., and Schanne-Klein, M.-C. (2012). In vivo structural imaging of the cornea by polarization-resolved second harmonic microscopy. *Biomed. Opt. Express*, **3**(1), pp. 1–15.
35. Latour, G., Kowalczuk, L., Savoldelli, M., Bourges, J.-L., Plamann, K., Behar-Cohen, F., and Schanne-Klein, M.-C. (2012). Hyperglycemia-induced abnormalities in rat and human corneas: the potential of second harmonic generation microscopy. *PLoS One*, **7**(11), p. e48388.
36. Latrive, A., and Boccara, A. C. (2011). In vivo and in situ cellular imaging full-field optical coherence tomography with a rigid endoscopic probe. *Biomed. Opt. Express*, **2**(10), pp. 2897–2904.
37. Lubatschowski, H. (2005). *Laser in Medicine: Maser-Tissue Interaction and Applications*. John Wiley & Sons.
38. Lyubovitsky, J. G., Spencer, J. A., Krasieva, T. B., Andersen, B., and Tromberg, B. J. (2006). Imaging corneal pathology in a transgenic mouse model using nonlinear microscopy. *J. Biomed. Opt.*, **11**(1), p. 014013.
39. Maurice, D. M. (1957). The structure and transparency of the cornea. *J. Physiol.*, **136**, pp. 263–286.

40. McCally, R. L., and Farrell, R. A. (1990). Light scattering from cornea and corneal transparency. In *Noninvasive Diagnostic Techniques in Ophthalmology*. Springer Verlag, New York, pp. 189–210.
41. Meek, K. M., and Boote, C. (2009). The use of x-ray scattering techniques to quantify the orientation and distribution of collagen in the corneal stroma. *Prog. Retin. Eye Res.*, **28**(5), pp. 369–392.
42. Müller, L. J., Pels, E., and Vrensen, G. F. J. M. (2001). The specific architecture of the anterior stroma accounts for maintenance of corneal curvature. *Brit. J. Ophthalmol.*, **85**(4), pp. 437–443.
43. Møller-Pedersen, T. (2004). Keratocyte reflectivity and corneal haze. *Exp. Eye Res.*, **78**(3), pp. 553–560.
44. Morishige, N., Wahlert, A. J., Kenney, M. C., Brown, D. J., Kawamoto, K., Chikama, T.-I., Nishida, T., and Jester, J. V. (2007). Second-harmonic imaging microscopy of normal human and keratoconus cornea. *Invest. Ophthalmol. Vis. Sci.*, **48**(3), pp. 1087–1094.
45. Nahas, A., Bauer, M., Roux, S., and Boccara, A. C. (2013). 3D static elastography at the micrometer scale using full field OCT. *Biomed. Opt. Express*, **4**(10), pp. 2138–2149.
46. Nahas, A., Tanter, M., Nguyen, T.-M., Chassot, J.-M., Fink, M., and Boccara, C. (2013). From supersonic shear wave imaging to full-field optical coherence shear wave elastography. *J. Biomed. Opt.*, **18**(12), p. 121514.
47. Nishida, T. (2005). Cornea and sclera: anatomy and physiology. In *Cornea*, 2nd Ed. Elsevier.
48. Nuzzo, V., Aptel, F., Savoldelli, M., Plamann, K., Peyrot, D., Deloison, F., Donate, D., and Legeais, J.-M. (2009). Histologic and ultrastructural characterization of corneal femtosecond laser trephination. *Cornea*, **28**(8), pp. 908–913.
49. Ortiz, S., Perez-Merino, P., Duran, S., Velasco-Ocana, M., Birkenfeld, J., de Castro, A., Jimenez-Alfaro, I., and Marcos, S. (2013). Full OCT anterior segment biometry: an application in cataract surgery. *Biomed. Opt. Express*, **4**(3), pp. 387–396.
50. Ortiz, S., Perez-Merino, P., Gambra, E., de Castro, A., and Marcos, S. (2012). In vivo human crystalline lens topography. *Biomed. Opt. Express*, **3**(10), pp. 2471–2488.
51. Patel, S., Marshall, J., and Fitzke, F. W. (1995). Refractive index of the human corneal epithelium and stroma. *J. Refract. Surg.*, **11**(2), pp. 100–141.
52. Petroll, W. M., Weaver, M., Vaidya, S., McCulley, J., and Cavanagh, H. D. (2013). Quantitative 3-dimensional corneal imaging in vivo using a modified HRT-RCM confocal microscope. *Cornea*, **32**, pp. 36–43.

53. Peyrot, D. A., Aptel, F., Crotti, C., Deloison, F., Lemaire, S., Marciano, T., Bancelin, S., Alahyane, F., Kowalczul, L., Salvodelli, M., Legeais, J.-M., and Plamann, K. (2010). Effect of incident light wavelength and corneal edema on light scattering and penetration: laboratory study of human corneas. *J. Refract. Surg.*, **26**, pp. 786–795.

54. Piston, D. W., Masters, B. R., and Webb, W. W. (1995). Three-dimensionally resolved nad(p)h cellular metabolic redox imaging of the in situ cornea with two-photon excitation laser scanning microscopy. *J. Microsc.*, **178**(1), pp. 20–27.

55. Plamann, K., Aptel, F., Arnold, C. L., Courjaud, A., Crotti, C., Deloison, F., Druon, F., Georges, P., Hanna, M., Legeais, J.-M., Morin, F., Mottay, E., Nuzzo, V., Peyrot, D. A., and Savoldelli, M. (2010). Ultrashort pulse laser surgery of the cornea and the sclera. *J. Opt.*, **12**(8), p. 084002.

56. Radner, W., Zehetmayer, M., Aufreiter, R., and Mallinger, M. (1998). Interlacing and cross-angle distribution of collagen lamellae in the human cornea. *Cornea*, **17**(5), pp. 537–543.

57. Rosen, R. B., Hathaway, M., Rogers, J., Pedro, J., Garcia, P., Laissue, P., Dobre, G. M., and Podoleanu, A. G. (2009). Multidimensional en-face OCT imaging of the retina. *Opt. Express*, **17**(5), pp. 4112–4133.

58. Sharma, R., Yin, L., Geng, Y., Merigan, W. H., Palczewska, G., Palczewski, K., Williams, D. R., and Hunter, J. J. (2013). In vivo two-photon imaging of the mouse retina. *Biomed. Opt. Express*, **4**(8), pp. 1285–1293.

59. Soong, H. K., Mian, S., Abbasi, O., and Juhasz, T. (2005). Femtosecond laser-assisted posterior lamellar keratoplasty: initial studies of surgical technique in eye bank eyes. *Ophthalmology*, **112**(1), pp. 44–49.

60. Steven, P., Bock, F., Hüttmann, G., and Cursiefen, C. (2011). Intravital two-photon microscopy of immune cell dynamics in corneal lymphatic vessels. *PLoS One*, **6**(10), p. e26253.

61. Taylor, V. L., Al-Ghoul, K. J., Lane, C. W., Davis, V. A., Kuszak, J. R., and Costello, M. J. (1996). Morphology of the normal human lens. *Invest. Ophthalmol. Vis. Sci.*, **37**, pp. 1396–1410.

62. Tick, S., Rossant, F., Ghorbel, I., Gaudric, A., Sahel, J.-A., Chaumet-Riffaud, P., and Paques, M. (2011). Foveal shape and structure in a normal population. *Invest. Ophthalmol. Vis. Sci.*, **52**(8), pp. 5105–5110.

63. Vogel, A., Noack, J., Hüttman, G., and Paltauf, G. (2005). Mechanisms of femtosecond laser nanosurgery of cells and tissues. *Appl. Phys. B*, **81**(8), pp. 1015–1047.

64. Vogel, A., and Venugopalan, V. (2003). Mechanisms of pulsed laser ablation of biological tissues. *Chem. Rev.*, **103**(2), pp. 577–644.
65. Yeh, A. T., Nassif, N., Zoumi, A., and Tromberg, B. J. (2002). Selective corneal imaging using combined second-harmonic generation and two-photon excited fluorescence. *Opt. Lett.*, **27**(23), pp. 2082–2084.
66. Yoo, H., Kim, J. W., Shishkov, M., Namati, E., Morse, T., Shubochkin, R., McCarthy, J. R., Ntziachristos, V., Bouma, B. E., Jaffer, F. A., and Tearney, G. J. (2011). Intra-arterial catheter for simultaneous microstructural and molecular imaging in vivo. *Nat. Med.*, **17**, pp. 1680–1684.
67. Nuzzo, V., Plamann, K., Savoldelli, M., Merano, M., Donate, D., Albert, O., Gardeazabal Rodirguez, P. F., Mourou, G., and Legeais, J. M. (2007). In situ monitoring of second-harmonic generation in human corneas to compensate for femtosecond laser pulse attenuation in keratoplasty. *J. Biomed. Opt.*, **12**(6), p. 064032.
68. Benoit, A., Latour, G., Schanne-Klein, M.-C., and Allain J.-M. (2016). Simultaneous microstructural and mechanical characterization of human corneas at increasing pressure. *J. Mech. Behav. Biomed. Mater.*, **60**, pp. 93–105.

Chapter 20

Investigation of Spindle Structure and Embryo Development for Preimplantation Genetic Diagnosis by Subcellular Live Imaging with FF-OCT

Ping Xue and Jing-gao Zheng

Department of Physics and State Key Lab of Low-Dimensional Quantum Physics, Tsinghua University and Collaborative Innovation Center of Quantum Matter, Beijing 100084, China
xuep@tsinghua.edu.cn

This chapter firstly gives a brief introduction to the technique of full-field optical coherence microscopy/tomography (FF-OCM/FF-OCT), which is one extension of optical coherence tomography (OCT). The principle of FF-OCT and its performances are briefly discussed. We also demonstrate some application examples in embryology and developmental biology, which are of high value for preimplantation genetic diagnosis.

20.1 Introduction

Optical coherence tomography (OCT) [1, 2] is an attractive imaging technique for medical and biological applications. It enables in vivo and noninvasive imaging of the biological samples without any need of labeling or contact with the samples. Also, OCT can image the samples at a penetration depth of several millimeters. The low-level signals reflected from the tissues and cells can be enhanced by coherently mixing them with a strong reference signal. Moreover, using a low-coherence-length source, OCT has the potential of achieving high depth resolution in the order of several microns or even a subcellular scale.

The first-generation time domain OCT (TD-OCT) system is based on a point-by-point scanning approach. In order to build a 3D image it needs a 3D scanning system and thus takes long time. Recently, Fourier domain OCT (FD-OCT) [3, 4] is getting more and more attention because of its higher speed and sensitivity. FD-OCT can provide depth resolved information with no need of the mechanical axial scanning and mainly divided into two types. One is spectral domain OCT (SD-OCT) that is based on a parallel spectrometer with a charge-coupled device (CCD) to measure the longitudinal cross-sectional image with the single-shot exposure [5, 6]. The other one is swept-source OCT (SS-OCT). With SS-OCT, the depth information can be also extracted by sweeping the source spectrum with a tunable laser [7].

In order to increase the imaging speed and reduce the mechanical complexity of lateral scanning system, the concept of full-field OCT (FF-OCT) [8] has been introduced. Without any 2D lateral scanning, FF-OCT can provide 2D transverse cross-sectional (or en face) OCT image by illuminating the entire field of view and collecting the backscattered light with a 2D CCD or a complementary metal-oxide semiconductor (CMOS) camera. Furthermore, the light source of FF-OCT can just be a thermal light source, which has a wide spectral bandwidth that leads to an axial resolution of $\sim$1 µm [9]. Also, the average optical power density on the samples is only of 10^{-1} W/cm^2, about 7 to 10 orders of magnitude lower than confocal imaging [10]. Therefore, the in vivo sample receives much less optical dose, which is favorable for long-time and live observation. On the other hand, as FF-OCT is compatible with conventional microscope, it could be well integrated with a commercial microscope and is then very easy and convenient to operate. It is also able to achieve very high transverse resolution

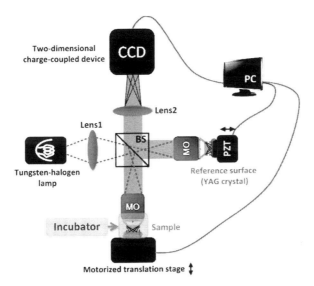

Figure 20.1 Schematic diagram of FF-OCT based on the Linnik interference microscope. BS, beam splitter; MO, microscope objectives; PZT, piezoelectric transducer; PC, personal computer.

with the use of high-numerical-aperture (NA) water immersion microscope objectives. Therefore, superior to other OCT systems, FF-OCT is capable of imaging with high resolution both in transverse and longitudinal directions. These unique characteristics make FF-OCT a powerful imaging technique in biomedical studies, especially in embryology and developmental biology [11–15].

20.2 Experimental Design

The schematic of FF-OCT setup is shown in Fig. 20.1 [16]. It is based on the low-coherence Linnik interference microscope geometry with a kind of thermal light source called tungsten-halogen lamp in a Köhler illumination system, illuminating the entire sample field of view uniformly with an optical power density of 10^{-1} W/cm^2 and it is significantly smaller than the optical power density (10^6–10^9 W/cm^2) for fluorescence excitation in confocal and multiphoton excitation laser scanning microscopy [10].

More specifically, the Köhler illumination system mainly consists of the Lens 1 and microscope objectives. The central wavelength of the light source was 600 nm and the spectral bandwidth was 180 nm. Because of the low temporal coherence of this tungsten-halogen lamp and on the basis of the coherence gating, the effective interference occurs only when the difference of optical path lengths of the two interferometer arms is within 0.9 μm in air (0.7 μm in water), and this is the very axial (depth) resolution of our FF-OCT. To image more details of the samples (i.e., to increase the transverse resolution of the system), two identical high-NA microscope objectives (Nikon 20×, water immersion, 0.5 NA) are employed in both arms, achieving a transverse resolution of 0.7 μm. Light is split into the reference and sample arms by a broadband beam splitter and by use of a fixed achromatic lens with 300 mm focal length, the interferograms resulted from the backreflected light of both arms are digitized by a 2D silicon CCD camera array (IMPERX, MDC-1004, 12-bit, 1004 × 1004 pixels, 48Hz). The backreflected signals of the samples from a certain depth can be extracted by use of a specific phase-shifting method. Phase shifting is accomplished by a piezoelectric transducer (PZT), which attached to the reference mirror (in Fig. 20.1, the reference mirror is a polished YAG crystal). The PZT creates a sinusoidal phase modulation at the frequency which is chosen to be exactly one half or a quarter of the image acquisition rate of a CCD to extract en face tomographic images of the sample from every consecutive two or four interferograms, respectively [14–17]. Using associated data processing, each en face tomographic image is then calculated. A high-precision motorized translation stage (minimum step = 0.3 μm) is placed in the sample arm, moving the sample in the axial direction to get a 3D volumetric image. Modulation acquisition synchronizing, sample scanning, image acquisition, calculation and display are all implemented by a personal computer with a C^{++} program.

For the observation and dynamic study of the live samples, there should be an incubator equipped on the translation stage of the sample arm or an incubator that encloses the whole setup, stably providing appropriate temperature and humidity to keep the samples alive as long as possible.

20.3 Extraction of the Signals

20.3.1 *Phase Shifting*

As mentioned above, to extract the signals of the sample from the interference fringes (interferograms), some phase-shifting methods have been developed in FF-OCT [15–19]. In general, there are two kinds of phase-shifting manners, one is discrete phase shifting and the other one is continuous phase shifting.

In the discrete manner, the phase is changed discretely between each intensity measurement. Because the data acquisition should be done after shifting the phase, these two procedures cannot operate at the same time, leading to the fact that obtaining one en face image is a little time consuming. Therefore, this manner is often limited to the static study. While in the continuous manner, the phase is continuously changed between each intensity measurement and at the same time the data acquisition is performed. Therefore, in contrast to the discrete manner, the imaging rate can be much higher. As a result, some dynamic investigations of biological samples are able to be carried out.

There are two kinds of continuous phase-shifting methods that are most widely used in the technique of FF-OCT. One is called the four-phase-shifting method and it uses four consecutive 2D interferograms to extract an en face images at a certain depth of the samples [14]. The other one is called the two-phase-shifting method and it uses only two consecutive 2D interferograms, but it can also realize the same function as the former one [18, 19]. Though the imaging speed of the two-phase-shifting method is two times as fast as four-phase shifting, one drawback of this method is that the interference fringes cannot be perfectly eliminated. However, when imaging biological samples, the fringes are usually not visible in the images. One possible explanation for this is that the sizes of the microstructures in most biological samples are only several times of the wavelength of the light source and the various small particles inside the samples are not uniformly distributed [17, 18]. More details about this effect will be discussed in the next section.

20.3.2 Extract the Signals with Two-Phase Shifting

Here, we demonstrate the detailed procedures of signal processing with the two-phase-shifting method [18, 19]. In this method, the PZT creates a sinusoidal phase modulation at the frequency which is chosen to be exactly one half of the image acquisition rate of a CCD to extract en face tomographic images of the sample from every consecutive two interferograms. As the reference mirror is attached to the PZT, it oscillates at the same frequency.

The intensities of backreflected light from both sample and reference arms detected by the each pixel of the 2D CCD camera can be written as

$$I(x, y) = I_r(x, y) + I_s(x, y) + 2\sqrt{I_r(x, y) I_{s_c}(x, y)} \cos(\Delta\phi(x, y))$$
$$= I_r(x, y) + I_{(s+\text{setup})_{\text{inc}}}(x, y) + I_{s_c}(x, y)$$
$$+ 2 I_0 r s_c \cos(\Delta\phi(x, y)), \quad (20.1)$$

where I_s and I_r are the intensities of backreflected light from sample and reference arms, while I_{Sc}, $I_{(S+\text{Setup})\text{inc}}$, and I_0 represent the proportion of light reflected from the sample that contributes to the effective interference fringes, the noninterfering proportion of light reflected from the sample and other parts of the setup, half the intensity of the light source, respectively. r and S_c are the reflectivities of the reference mirror and the sample that interferes, while $\Delta\phi$ is the optical phase difference of the two arms. The parameters x and y are the coordinates of each pixel of the CCD camera. If the chip of the camera consists of $M \times M$ pixels, then

$$\begin{aligned} x &= 1, 2, \ldots M, \\ y &= 1, 2, \ldots M. \end{aligned} \quad (20.2)$$

Because the optical path length of the reference arm is modulated by the PZT in sinusoidal mode, Eq. 20.1 can be rewritten as

$$I(x, y, t) = I_r(x, y, t) + I_{(s+\text{setup})_{\text{inc}}}(x, y, t) + I_{s_c}(x, y, t)$$
$$+ 2 I_0 r s_c \cos\left[\Delta\phi(x, y) + \psi \sin(2\pi f t + \theta)\right], \quad (20.3)$$

where ψ and f are the amplitude and the frequency of the sinusoidal phase modulation and θ represents the phase at the time origin.

The CCD camera acquires a pair of images per modulation period $T(1/f)$. A number, N, of interferograms can be accumulated to obtain high-signal-to-noise-ratio (SNR) images. In doing so, the integrated intensities detected by each pixel of the CCD over half of the phase modulation period T can be written as

$$E_m(x, y, t) = \eta N \int_{\frac{(m-1)T}{2}}^{\frac{mT}{2}} I_m(x, y, t) dt, \quad (20.4)$$
$$m = 1, 2$$

where η is the quantum efficiency of the CCD sensor.

Because when the intensity of the light is strong enough, the shot noise resulting from the fundamental photon noise dominates among different kinds of noises, that is, electrical noise, thermal noise, etc., also because the photon noise is Poisson distributed, that is, the shot noise variance is equal to the number of photons received by each pixel of the CCD camera, the illumination intensity is adjusted such that the CCD pixel wells are close to the saturation limit to maximize the SNR of the system when imaging the samples. Assuming that the CCD operates close to saturation at shot noise limited, the number of photoelectrons produced by each CCD pixel is close to the full-well capacity of each pixel which is denoted as ξ_{sat}. Since for common biological samples: $r \gg s_c$, we can yield the following relation according to Eq. (1.4):

$$\xi_{sat} \approx \frac{\eta I_0 T}{4}(r + (s + \text{setup})_{\text{inc}}). \quad (20.5)$$

We now have three useful equations: 20.3–20.5. The next step is the most crucial one that we can obtain S_c, which contains the sample information by extracting the AC term or separating the DC and AC terms. Using three equations above, we obtain the final extracting equation as follows:

$$(E_1 - E_2)^2 = 8(N\xi_{sat})^2 \frac{r}{(r + (s + setup)_{\text{inc}})^2} S_c \cos^2 \Delta\phi$$
$$\propto S_c \cos^2 \Delta\phi. \quad (20.6)$$

Then we can calculate S_c at each couple of coordinate x and y according to Eq. 20.6 to get a 2D en face tomographic images at a certain depth of samples.

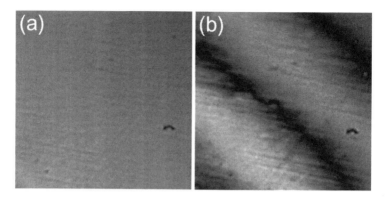

Figure 20.2 (a) An FF-OCT image of a flat surface area of one CaF_2 crystal imaged with the four-phase-shifting method. (b) The same area with the two-phase-shifting method.

20.3.3 Comparison between Two and Four-Phase Shifting

As we can notice from Eq. 20.6, when using the method of two-phase shifting, there is always the term of $\cos^2 \Delta\phi$ adhered to the term of S_c. Therefore, as we mentioned in the previous section, the interference fringes cannot be perfectly eliminated.

One example for this is shown in Fig. 20.2 where the two-phase-shifting method is compared to the four-phase-shifting one. Figure 20.2a shows an FF-OCT image of a flat surface area of a CaF_2 crystal with the four-phase-shifting method, and we cannot see any interference fringes in the figure, while Fig. 20.2b shows the same area with the two-phase-shifting method, and the fringes are clearly seen in this figure.

However, when imaging the biological samples, the fringes are usually not visible in the images because of the sizes and distributions of the microstructures various small particles in most biological samples. Corresponding comparison of biological images between two- and four- phase shifting are shown in Fig. 20.3. The sample used here is a zygote-stage embryo. Figure 20.3a shows an en face tomographic image of the zygote with the four-phase-shifting method, while Fig. 20.3b shows the en face image at the same depth with the two-phase-shifting method. In contrast to the two images in

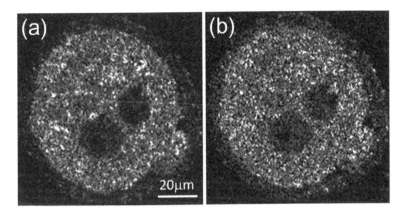

Figure 20.3 (a) An FF-OCT tomographic image of a zygote-stage embryo (biological sample) imaged with the four-phase-shifting method. (b) The same tomographic image with the two-phase-shifting method.

Fig. 20.2, the two in Fig. 20.3a,b are almost the same with each other, and the interference fringes are not visible in Fig. 20.3b.

20.3.4 3D Reconstruction Image Display

To get the whole information of the sample, a stack of tomographic images should be acquired. In order to realize this, the sample is moved step by step in the axial direction by a motorized translation stage (shown in Fig. 20.1). Once a 3D dataset is recorded, sections with arbitrary orientation can be extracted and volume-rendering images can be computed. Three-dimensional reconstruction and the segmentation can be performed with commercial software platform 3Dmed 3.0 and Amira 5.2.0.

On the other hand, to enhance the visual quality in 3D, the raw images from the FF-OCT should be improved. We propose an image processing pipeline to achieve this goal. Firstly, the contrast is increased as the dynamical range of the original images is low. Secondly, due to the resolution reducing with depth increasing, a linearly decreasing gamma is used as penetrating in depth. Finally, the images are normalized to cover the full gray-scale range. In addition, registration of neighboring slices is necessary because of the random movement of the sample.

20.4 System Characteristics

20.4.1 Axial Resolution

In contrast to conventional and confocal microscopy, OCT is capable of achieving very high axial image resolutions independent of focusing conditions. In other words, the axial and transverse resolutions are independent of each other [1]. Here, we focus the discussion on the related characteristics of FF-OCT system.

Like other kinds of OCT systems, the axial resolution of FF-OCT system is also determined by the coherence length of the light source rather than the depth of field (DOF) as in microscopy. The light source widely used in FF-OCT is the thermal light source (such as tungsten halogen lamp in Fig. 20.1), which has the property of both low temporal coherence and also low spatial coherence as well [11–19]. Therefore, unlike using the laser systems, such as ultrashort femtosecond lasers, there is little cross-talk between the adjacent pixels in the images. Furthermore, the spectrum of a thermal light source is very smooth and does not contain any spikes or emission lines that could cause side lobes in the coherence function and create artifacts in the images.

The well-known expression for the axial resolution Δz of an OCT system is

$$\Delta z = \frac{2 \ln 2}{\pi} \frac{\lambda_0^2}{\Delta \lambda} \approx 0.44 \frac{\lambda_0^2}{\Delta \lambda}, \quad (20.7)$$

where λ_0 is the central wavelength of the effective spectrum of the system and $\Delta \lambda$ represents full-width at half-maximum (FWHM) of the effective spectrum. The effective spectrum of the system is related to the spectrum of the light source, the spectral transmission of the system optical components and the spectral response of the CCD camera. Because there is little absorption by the optical components in most cases, the FWHM is mainly determined by the spectral response of the cameras. As we can see from Eq. 20.7, Δz is inversely proportional to the FWHM of the effective spectrum, so higher axial resolution can be obtained by increasing the FWHM or decreasing the central wavelength.

Sometimes, when imaging the biological samples whose primary constituent is water, water immersion microscope objectives are often utilized in both arms of the interferometer to optimize the dispersion match that affects the axial resolution. In these cases, Eq. 20.7 should be modified as

$$\Delta z = \frac{2 \ln 2}{\pi} \frac{\lambda_0^2}{n \Delta \lambda} \approx 0.44 \frac{\lambda_0^2}{n \Delta \lambda}, \quad (20.8)$$

where n is the refractive index of water (immersion media). Assuming one typical FF-OCT system whose λ_0 is 600 nm, $\Delta \lambda$ is 180 nm and n is 1.33 (water), we can calculate the axial resolution with Eq. 20.8 that Δz is 0.7 μm (0.9 μm in air).

20.4.2 Transverse Resolution

Unlike the axial resolution, the derivation of transverse resolution of the OCT systems including FF-OCT is the same as that of the conventional or confocal microscopy. Therefore, the transverse resolution is determined by the focused transversal spot size.

Assuming the optical system to be diffraction limited, the well-known expression for the transverse resolution Δx of an OCT system is

$$\Delta x = 0.61 \frac{\lambda_0}{\text{NA}}, \quad (20.9)$$

where NA is the numerical aperture of the objectives in the sample arm. Therefore, the higher the NA we use, the better the transverse resolution we can get.

In the technique of FF-OCT, high-NA microscope objectives, such as water immersion microscope objectives, are often used to achieve a transverse resolution of the order of 1 μm [16–21], which is 10–20 times higher than those of the techniques of other kinds of the OCT systems, such as TD-OCT and FD-OCT where relatively low-NA objectives are used for optimizing the imaging depths [22–27] (see the next section). For example, assuming one typical FF-OCT system that utilizes water immersion microscope objectives with an NA of 0.5 and whose λ_0 is 600 nm, then according to Eq. 20.9, we can calculate the transverse resolution that Δx is 0.7 μm. The high transverse resolution is the most significant advantage for FF-OCT over other OCT systems.

There is another useful expression for the transverse resolution Δx which is determined by the focal length of the lens and the spot size on the objective lens [1], as shown in Eq. 20.10:

$$\Delta x = \frac{4\lambda_0}{\pi}\frac{f}{d}, \qquad (20.10)$$

where f is the focal length and d represents the spot size on the objective lens. Considering the immersion media, Eq. 20.10 should be modified as

$$\Delta x = \frac{4\lambda_0}{\pi}\frac{f}{nd}, \qquad (20.11)$$

where n is the is the refractive index of the immersion media. Equations 20.9 and 20.11 can be derived from each other using the following relation:

$$\text{NA} = n\sin\left(\frac{d}{2f}\right) \approx \frac{nd}{2f}. \qquad (20.12)$$

20.4.3 Depth of Focus/Field

The depth of focus or field (DOF) is defined as twice the Rayleigh length $2z_R$ and it is also determined by the NA as the following relation:

$$\text{DOF} = 2z_R = \frac{\pi \Delta x^2}{2\lambda} \propto \frac{1}{\text{NA}^2}, \qquad (20.13)$$

where Δx is the transverse resolution of the OCT system. Therefore, although increasing the NA of the objective can increase the transverse resolution, it decreases the DOF. This is the reason why both TD-OCT and FD-OCT, which produce cross-sectional images, often utilize low-NA lenses instead of high-NA lens [22–27]. Otherwise, one can obtain only a little axial information of the samples. For these reasons, the transverse resolutions of these techniques are often 10–20 times lower than those of conventional and confocal microscopy.

However, the scanning mode of FF-OCT is much different from other OCT systems, as the sample arm is directly scanned rather than the reference arm. Therefore, in general, there is no relationship between the NA of the objective and the DOF or imaging depth of FF-OCT. The only factors that affect the imaging depth of the FF-OCT system are the wavelength of the effective spectrum, the scattering property of the sample itself and the sensitivity of the

FF-OCT system. Usually, the imaging depth of FF-OCT can be several hundred microns. If FF-OCT system uses a near-infrared light source, such as a source with a central wavelength of 1300 nm, the imaging depth can be up to 1 mm [17].

In fact, with no need of replacing the low-high-NA lens with high-NA ones, that is, without sacrificing the DOF, a few methods have been developed to improve the transverse resolution in TD-OCT and FD-OCT, such as dynamic focusing, de-convolution, zone-focusing, and image fusion techniques [28–31].

20.4.4 Sensitivity

The sensitivity is defined as the minimum detectable reflected optical power compared to a perfect reflector. Because the backreflected signal decays with the increase of the depth, the sensitivity determines the imaging depth, that is, the maximum depth at which imaging is also possible. In addition, the imaging speed can be increased by improving the system sensitivity, as the integration time for acquiring enough signals can be reduced. Often, the SNR and the dynamic range are used interchangeably with sensitivity. We present here the detection sensitivity of FF-OCT system [1, 14, 18].

Assuming the noise of each pixel of the CCD camera is v, the number of photoelectrons produced at each pixel of the CCD can be written as $E + v$. Therefore, the extracted actual signal of sample is modified to (see Eq. 20.6)

$$\text{Signal} = ((E_1 + v_1) - (E_2 + v_2))^2, \quad (20.14)$$

where the noise v_m is of the following properties:

$$\begin{cases} \langle v_m \rangle = 0, \, m = 1, 2 \\ \langle v_m v_n \rangle = \begin{cases} \sigma^2, & m = n \\ 0, & m \neq n \end{cases} \end{cases}, \quad (20.15)$$

where the angle brackets $\langle \rangle$ represent a time average and σ^2 is the variance of the shot noise.

As mentioned before, to obtain the images of the highest SNR in FF-OCT, the illumination intensity is usually adjusted so that the CCD pixel wells are close to the saturation limit. Also, according to the fact that the shot noise variance is equal to the number of photons received by each pixel of the CCD camera, we can write the following relation:

$$\sigma^2 = N\xi_{sat}, \qquad (20.16)$$

where N is the number of the accumulated interferograms and ξ_{sat} is full-well capacity of the CCD camera.

The noise level then can be calculated when there is no interference. By averaging Eq. 20.14 when $E_1 = E_2$ (corresponding to the case of no interference), we obtain

$$\text{Noise} = 2N\xi_{sat}. \qquad (20.17)$$

As the minimum detectable sample signal is equal to the noise level, we yield:

$$\text{Signal}_{min} = \text{Noise} = 2N\xi_{sat}. \qquad (20.18)$$

Then using Eq. 20.6 and $\langle \cos^2 \Delta\phi \rangle = 0.5$, we obtain the sensitivity of the FF-OCT system:

$$\text{Sensitivity} = \frac{1}{S_{c_{min}}} = \frac{2rN\xi_{sat}}{(r+(s+\text{setup})_{inc})^2}. \qquad (20.19)$$

Also, using Eq. 20.19, the dynamic range of the FF-OCT system can be calculated. For any 3D biological samples, there is always the following relation:

$$S_{c_{max}} \leqslant S_{inc}. \qquad (20.20)$$

Therefore, the dynamic range of the FF-OCT system is:

$$\text{Dynamic Range} = \frac{S_{c_{max}}}{S_{c_{min}}} \leqslant \frac{2rN\xi_{sat}S_{inc}}{(r+(s+\text{setup})_{inc})^2}. \qquad (20.21)$$

As we can see from Eq. 20.19 or 20.21, the sensitivity strongly depends on the full-well capacity ξ_{sat} of the CCD camera and the number of the accumulated interferograms N. Therefore, to optimize the sensitivity or dynamic range, one often selects a CCD camera with large full-well capacity and accumulate many interferograms at the same depth.

For full-well capacity ξ_{sat}, its typical value of a commercially available CCD is usually of the order of $10^4 - 10^5$, and the value can be increased further by the method of pixel binning in which the charges generated from adjacent several pixels are summed up. For example, our CCD camera (IMPERX, MDC-1004, 48Hz), when using

2×2 pixel binning, has a full-well capacity ξ_{sat} of $4 \times 42{,}000 = 168{,}000$.

For the number of the accumulated interferograms, theoretically as we can see from Eq. 20.19, the sensitivity increases as the accumulating number increases. However, at the same time, the imaging time increases as well, so the accumulating number should not be too large. Often, the number is adjusted according to the imaging rate of the CCD camera and the system sensitivity so that the time needed for one en face image is in the order of 1 s. But of course, the extracted image should have enough SNR to do the relevant studies. If the image quality is not good enough, one has to increase the number even sacrificing the imaging time. Therefore, although FF-OCT allows the extraction of the signal simultaneously over millions of pixels at video rates or even faster, the actual imaging time of each en face point of the samples with FF-OCT is equal to the imaging time of the whole image (as mentioned above: in the order of 1 s), which is several orders of magnitude longer than that in scanning OCT. For these reasons, FF-OCT is more favorable for imaging the stable and immobile samples; otherwise, relatively fast motion of the samples during the image acquisition time may blur the interference signal and in consequence, the image contrast vanishes or cause corresponding image distortions. So compared to the TD-OCT and FD-OCT, whose acquisition time is only of the order of 1 μs or even less at each image point [32–35], FF-OCT shows a notable drawback of low imaging speed.

In addition, as Eq. 20.19 shows, the proportion of light that does not interfere, $(s + setup)_{inc}$ should be as low as possible and hence all the optical elements of the FF-OCT system are often antireflection coated to minimize un-interfering reflections. The strong specular reflection on the surface of the samples can be reduced significantly by using water immersion microscope objectives. As a result, the main contribution to $(s + setup)_{inc}$ is the light backreflected from the biological structures located outside the coherence volume and collected by the microscope objective, s_{inc}.

The detection sensitivity is also associated with the reference mirror reflectivity, r. By simple mathematic derivation to Eq. 20.21, we can see that when the value of r is equal to $(s + setup)_{inc}$, the detectable sample signal s_c is minimized, while the sensitivity is maximized. Therefore, the reference mirror reflectivity r is chosen to be close to this optimum value. Typically when imaging the biological samples, r is of the order of 10^{-2}. Then, the maximized sensitivity can be written as

$$\text{Sensitivity}_{\max} = \frac{N\xi_{\text{sat}}}{2r}. \tag{20.22}$$

For example, for a typical FF-OCT system (the accumulating number $N \approx 100$, the full-well capacity $\xi_{\text{sat}} \approx 10^5$, the reference mirror reflectivity $r \approx 10^{-2}$), the sensitivity is $\sim$90 dB with the en face imaging time of the order of 1 s.

20.5 Applications in Embryology and Developmental Biology

In recent years there are increasing applications of optical imaging for developmental biology study. This is basically due to the two crucial requirements for the techniques: one is safety for long time observation and the other is high resolution for acquisition of detail information.

In fact, the observation of preimplantation development of the embryo increasingly demands for new optical methodology that is capable of in real time and noninvasive 3D imaging with high resolution. However, the widely used optical microscopy can only implement en face imaging and does not have tomographic capability. Confocal microscopy and multiphoton excitation microscopy do have the capability of 3D imaging but the tissue sample needs to be biochemically preprocessed and labeled by dye molecules or fluorophores. Also, for fluorescence detection based imaging technique laser beam is scanned and focused on the sample to excite the dye molecule. Along the beam path, especially at the focus the sample experiences very strong optical intensity or power and suffers from photon bleaching and other side effects.

In comparison, the FF-OCT requires no need of sample preprocessing as it measures the backscattering other than fluorescence. FF-OCT uses thermal light source for plane illumination and hence has much less light damage to living cells. Furthermore, FF-OCT has faster imaging rate, higher depth resolution and penetration. So as a nonlabeled and noninvasive technique, FF-OCT is a potential breakthrough for the observation of preimplantation development of the embryo.

Actually, OCT has already been successfully demonstrated for high resolution imaging in many animal models for developmental biology study including *Xenopus laevis* (South African clawed frog) [36], *Brachydanio rerio* (zebrafish) [37], *Rana pipiens* (American leopard frog) [10], embryonic avian heart [38, 39], mouse [39], or adult *Drosophila melanogaster* [40]. But the observation of preimplantation development of the embryo of mammals [17, 41, 42] is still challenging. The weak backscattered morphological signals and the small scale of the embryo as well as the incubation requirement make it very hard to implement long time and clear observation of embryo preimplantation development.

Recently, our group have developed an FF-OCT system and successfully applied the system to study the mouse oocytes before fertilization and embryos during the first 3 to 4 days of embryonic life both statically and dynamically [16, 20, 21]. During the research, useful morphological structure information was obtained with FF-OCT. We shall demonstrate the details in the next two sections: one is about the structures of spindles and the other is about early patterning and polarity of the mouse embryos.

20.5.1 *Imaging the Structures of Spindles*

The spindles are intracellular structures and composed of microtubules. By controlling the chromosomal movements, the spindles play key roles in proper chromosome alignment and segregation during meiosis and mitosis. Defects in either meiotic spindle of the oocyte or mitotic spindle of the zygote can cause chromosome misalignments, resulting in subsequent aneuploidy, infertility, spontaneous abortions, and congenital malformations [43, 44]. It has been reported that the appearance of the normal meiotic spindle at metaphase II (MII) is suggested to be a good indicator of oocyte quality. Other groups also made various attempts for a better and further understanding of the relationship between the presence/location/abnormalities of the spindles and various assisted reproduction technology (ART)-related outcomes [45–47]. Studying spindles in living cells has become an important issue. To visualize and evaluate the structures of spindles, a lot of spindle imaging systems have been utilized and developed. Phase contrast microscopy is capable of creating contrast based on optical interference effects, but the spindle cannot be visualized due to the lack of enough contrast [48]. Confocal microscopy and transmission electron microscopy can image the detailed structure of the spindles, but these methods use fixed or dye labeled samples, providing limited value for dynamic and subsequent developmental studies [45]. To image the spindles noninvasively, polarized light microscopy, which is based on the birefringence of the spindles, is used. Nonetheless, orientation-dependent and nonquantitative characteristics limit its value in further analysis. In response to the clinical and research need, a promising technology called Polscope is developed [49]. With the Polscope, spindles can be visualized noninvasively regardless of their orientation. However, it has been reported that sometimes the spindles became visible only after manual rotation with micropipette even when using the Polscope [50]. Furthermore, because the Polscope measures the retardance, which is accumulative effect of optical property, it fails to achieve 3D morphology of the samples. Until now, a method which simultaneously allows noninvasive, perfectly orientation-independent and 3D imaging of the spindles in living cells has never been demonstrated.

Here, we demonstrate the successful application of FF-OCT for noninvasive 3D morphological imaging of the meiotic and mitotic spindles in living oocytes and zygotes [16]. Also, by use of postprocessing of the 3D dataset obtained with FF-OCT, useful morphological and spatial parameters of the spindles are accurately quantified.

As the mouse is an excellent model, wild-type ICR/CD1 mice (Laboratory Animal Facility, Tsinghua University, PRC) were used. To collect in vivo matured oocytes that arrested at the second meiotic metaphase (MII-arrested oocytes), females (8-week-old) were superovulated by injecting 5 IU of pregnant mare's serum gonadotrophin (Bo'en Pharmaceutical Ltd., Chifeng, PRC). Forty-eight hours later, 5 IU of human chorionic gonadotropin (Livzon

Pharmaceutical Group Inc., Zhuhai, PRC) was injected and then the oocytes were obtained from the oviducts 17 h later. Cumulus cells were removed by brief incubation in 0.05% hyaluronidase (Sigma, Louis, MO; cat. no. H3506) at 37°C for 5 min. On the other hand, to collect zygotes at the mitotic metaphase and telophase, females were mated overnight with males and examined for vaginal plugs. The presence of a plug was taken as embryonic day 0.5 or E0.5 for short. The zygotes at the mitotic metaphase and telophase obtained at the period between late E0.5 and early E1.5. For live imaging, the oocytes and the zygotes were then immediately placed into a droplet of human tubal fluid (HTF) medium with 20% FBS (Hyclone, USA) covered with mineral oil (M4080, Sigma, USA) in our incubator. About 80 and 70 optical slices were needed to scan the entire oocytes and zygotes, respectively, using the axial step of 1.2 µm. The imaging rate was 1.25 s per optical slice.

The MII-arrested oocyte has two notable morphological characteristics. One is that, the meiotic spindle appears in the outermost layer of the cytoplasm of the oocyte and orients with its long axis parallel to the cortical membrane. The other one is that, the chromosomes are controlled by the spindle and aligned in the midregion of the spindle. Figure 20.4a shows the 3D cross-sectional view of a typical MII-arrested oocyte with FF-OCT. Besides the zona pellucida (ZP), the first polar body (1PB) and the cytoplasm of the oocyte, a relatively dark and long area, which extends along the cortical membrane, is also visible with enough contrast against the oocyte cytoplasm. To see the area more clearly, image cropping and color inversion were applied to the 3D dataset obtained with FF-OCT. Figure 20.4b shows the 3D reconstruction of this dark area after these image processings. Therefore, the cropped dark area appears bright in the figure. By the shape and orientation, the area was ascertained to be the spindle. To confirm this further, we employed the immunofluorescence technique to label the spindle (SP) and chromosomes (CH) of the same oocyte and scanned it with two-photon laser scanning microscopy (TP-LSM). Figure 20.4c shows the false-color 3D reconstruction of the quantum dot 585-labeled spindle and Hoechst 33342-stained chromosomes. Detailed structures of the spindle can be seen and the chromosomes in the midregion are clearly visible as well. Then using the 3D dataset

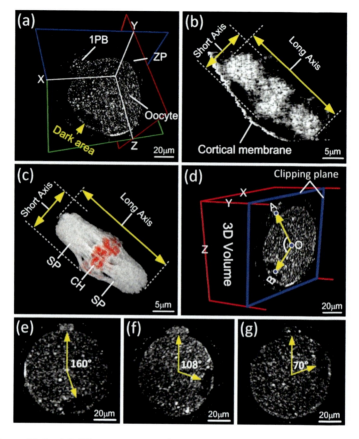

Figure 20.4 (a) 3D cross-sectional view of a typical MII-arrested oocyte using FF-OCT. (b) 3D reconstruction of the dark area in (a) after cropping and color inversion. (c) 3D reconstructions of the spindle and the chromosomes of the same oocyte using TP-LSM. (d) A cross-sectional view where the centers of the oocyte, 1PB, and the spindle can be seen at the same time using the 3D dataset obtained with FF-OCT. (e–g) Other oocytes with different angles of meiotic spindle deviation from the 1PBs.

obtained with FF-OCT and TP-LSM, we measured the short and long axes of the bright area in Fig. 20.4b and the spindle in Fig. 20.4c. As a result, the lengths of short and long axes were 11.7 µm and 26.6 µm with TP-LSM. While with FF-OCT, the lengths were 11.2 µm and 27.0 µm, consistent with the TP-LSM's measurements with the accuracy of FF-OCT resolution. Therefore, we affirm that the dark area in the FF-OCT images is the spindle, reflecting the microtubules of the spindle are of lower scattering coefficient than the cytoplasm. On the other hand, because the nucleus was less scattering than the cytoplasm [20], the chromosomes that aligned in the metaphase plate also contributed to the low brightness of the midregion of the spindle in Fig. 20.4a.

In addition, when using the Polscope, to measure the angle of meiotic spindle deviation from the 1PB, the oocyte has to be manually rotated, not only until both of the 1PB and the spindle can be clearly viewed in the same observation plane, but also until the spindle appears in the outermost layer of the cytoplasm. However, no matter how the 1PB and the spindle orient in the observation plane, just by postprocessing, the angle can be precisely measured using the 3D dataset obtained with FF-OCT. To measure the angle, 3D geometry centers of the cytoplasm, the 1PB, and the spindle of the oocyte were calculated. Points O, A, and B in Fig. 20.4d are the corresponding 3D geometry centers. With these three centers, the angle was quantified to be 101°. Because the 1PB and the spindle were not oriented in the same XY plane as shown in Fig. 20.4a, when scanning in the Z direction with FF-OCT, the two structures cannot appear in the same observation plane. However, by clipping the 3D volume of the oocyte with a plane which contains these three points, the 1PB and the spindle can also be simultaneously seen in a cross-sectional view, as shown in Fig. 20.4d. Figure 20.4e–g shows other oocytes with different angles of meiotic spindle deviation from the 1PBs. The angles were measured to be 160°, 108°, and 70°, respectively. Accurate measurement of the angle in this way avoids the subjectivities of observers and is more reliable. Therefore, FF-OCT may provide new insight into the relationship between the angle and subsequent developmental competence.

Also, to demonstrate the capabilities of FF-OCT to image the mitotic spindle, the zygotes at the metaphase and telophase were imaged. Figure 20.5a shows the cross-sectional view of a typical zygote at metaphase. In Fig. 20.5a, the spindle can be clearly seen in the middle of the zygote. The short and long axes of this spindle were quantified to be 23.9 µm and 43.2 µm, respectively. Figure 20.5b shows the cross-sectional view of a typical zygote at telophase. The spindle has been divided into two parts and the cleavage furrow has

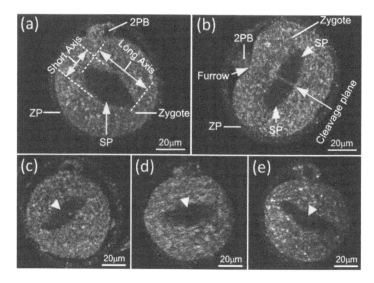

Figure 20.5 Mitotic spindle imaging of the zygotes (a) at metaphase and (b) at telophase. (c–e) Other zygotes with mitotic spindles (arrow).

been established. The bright area between the separated spindles would be the first cleavage plane of the zygote. The interesting structures of the spindles both in Figs. 20.5a and 20.5b obtained by FF-OCT were in good accord with the biological characteristics at mitotic metaphase and telophase. Figure 20.5c–e shows other zygotes with mitotic spindles (arrow). Therefore, with FF-OCT, 3D morphological images of the spindles can be achieved in living cells without any fixation and staining both at meiosis and mitosis.

Our results suggested that the technique of FF-OCT might be valuable in ART and many other spindle related studies.

20.5.2 Imaging Early Patterning and Polarity

Early patterning and polarity of mammalian embryos are fundamental and important in developmental biology. The early mammalian embryos have been thought to be highly regulative and many embryologists assume that there is no prepatterning during early development [51–53]. However, recent studies have proposed that factors in early patterning such as pronuclei, zona pellucida (ZP), second polar body (2PB), and the first and second two cleavages, are associated with the polarity axis in several ways [54–59], but some of these findings are not consistent with each other. Whether these factors play decisive roles in the establishment of the polarity has yet to be defined.

The controversies not only arise from the remarkable regulating capacity of embryos, but also from experimental methods used in the related investigations. Most of the studies involved dye-labeling techniques. Though some groups claimed that their labeling did not affect the normal development [54, 56, 59], this does not exclude the possibilities of latent disturbances to the normal processes. On the other hand, most of the current findings are based on 2D imaging methods, such as conventional microscope and differential interference contrast microscopy [55–58]. In such investigations, spatial localizations of related factors can only be qualitatively analyzed but not correctly quantified. Also, most embryos are not well oriented [54]; their relative positions are often a puzzle to observers. To solve the problem, a few groups utilized 3D imaging techniques such as confocal laser scanning microscopy and multiphoton excitation laser scanning microscopy [55, 59, 60], but they all required sample preprocessing and dye labeling. Furthermore, when using these systems, optical damage to the sample is not ignorable, as the optical power density is as high as 10^6 to 10^9 W/cm^2 [10]. Therefore, label-free 3D imaging techniques with low optical dose are in high demand in the study of early patterning and polarity.

20.5.2.1 Static studies of early patterning and polarity

Here we demonstrate the application of FF-OCT to image early patterning and polarity of preimplantation mouse live embryos. We shall show the label-free 3D subcellular structural images at various typical preimplantation stages, including zygote, two-cell, four-cell, and blastocyst stages [20].

Also, as the mouse is an excellent model to investigate early patterning and polarity of mammalian development, wild-type eight-week-old female ICR/CD1 mice (Laboratory Animal Facility, Tsinghua University, China) were used in our studies. Females were mated overnight with males and examined for vaginal plugs. The presence of a plug was taken as E0.5. Zygote, two-cell, four-cell, and blastocyst stage embryos were obtained at E0.5, early E1.5, late E1.5, and E3.5, respectively. Embryos were imaged in a droplet of human tubal fluid medium with 20% FBS (Hyclone, USA). For each embryo, about 70 optical slices were needed with a step of 1.2 µm. Using two-phase-shifting modulation and averaging 30 images at the same depth, the total acquisition time for each embryo was about 90 s, with an imaging rate of 24 fps.

Figure 20.6a shows 3D reconstruction of a zygote-stage (E0.5) embryo, and the cross-sectional view in Fig. 20.6b. Besides ZP, zygote, and 2PB, three relatively dark areas (DA1, DA2, and DA3) are also clearly visible. Two of them lie in the zygote and the third in the 2PB. Three-dimensional reconstruction of three dark areas segmented from the image stack is shown in Fig. 20.6c. By the size, shape, and location, the dark areas were ascertained to be the nuclei. To confirm this, DNAs of the same embryo were stained with Hoechst 33342 and scanned with TP-LSM. Figure 20.6d shows 3D reconstruction of three nuclei, from left to right, three nuclei are the nucleus of the 2PB (2PBN), female pronucleus (FPN), and male pronucleus (MPN), respectively [61]. Owing to TP-LSM's capability of 3D quantitative measurement, a triangle, whose vertices are fluorescence centers of three nuclei, was characterized. The lengths of three sides were 19.1, 44.4 and 26.0 µm, respectively, shown in Fig. 20.6d.

In comparison, a similar triangle was also measured with FF-OCT. Its vertices were three dark areas in FF-OCT image, shown in Fig. 20.6c. The corresponding side lengths of the triangle were measured to be 19.1, 44.0, and 25.7 µm, coincident with the TP-LSM measurements with an accuracy of FF-OCT resolution. Furthermore, the FF-OCT image gave essentially the same volumes of three dark areas as those of corresponding nuclei in the TP-LSM image. Therefore, we may affirm that the dark areas in the FF-OCT image are no other than the nuclei. On the other hand, chromatin is structurally loose during interphase, and its size is far below the wavelength of light. Therefore, the nucleus, whose main part is chromatin, has a low level of the scattering signal and shows relatively dark in the FF-OCT image. As different organelles have

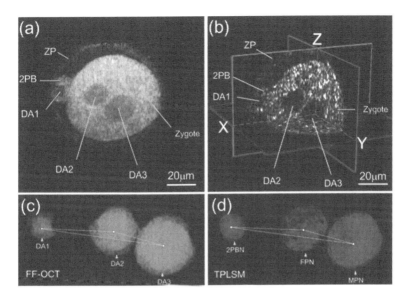

Figure 20.6 (a) 3D reconstruction of a zygote-stage embryo with FF-OCT. (b) Cross-sectional view of the same embryo. ZP, zygote, 2PB and three dark areas (DA1, DA2, and DA3) are clearly distinguishable. (c) 3D reconstruction of three dark areas. Three vertices of the triangle are 3D geometry centers of three dark areas and corresponding three lengths are 19.1, 44.0, and 25.7 μm. (d) 3D reconstruction of three nuclei (2PBN, FPN, and MPN) labeled with Hoechst 33342. Three vertices of the triangle are 3D fluorescence centers of the nuclei and three lengths are 19.1, 44.4, and 26.0 μm.

different scattering coefficients and thus give different gray-level signals, FF-OCT is also a functional imaging technique capable of distinguishing functional areas in a cell without any dye labeling.

Figure 20.7a shows a two-cell-stage (early E1.5) embryo imaged by a conventional 2D microscope whose first cleavage plane, 2PB, and blastomere 2 (BM2) do not clearly appear in the observation plane. However, FF-OCT can give a 3D digital dataset of an embryo. By some quantitative rotational operation to the 3D dataset, one can readily observe all the major structural information of two-cell embryo. The same embryo is shown in Fig. 20.7b with its ZP, 2PB, BM1, BM2, and the first cleavage plane clearly displayed. Therefore, no matter how the embryo orients in the observation plane, by FF-OCT imaging and image processing, one can make full use of all the

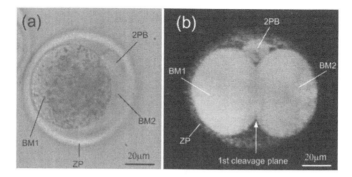

Figure 20.7 (a) 2D image of a two-cell-stage embryo with conventional 2D microscope. The first cleavage plane and BM2 do not clearly appear in the observation plane. (b) 3D reconstruction of the same embryo after rotational operation to the 3D dataset. Major structures are all clearly revealed.

sample embryos, without any exclusion of sample embryo due to improper orientation.

Three-dimensional reconstruction of a four-cell-stage (late E1.5) embryo is shown in Figs. 20.8a and 20.8b. There are a total of six planes where cells contact with each other in this tetrahedral-shaped embryo. All six planes are able to be revealed at different angles of view, and one pair of them is shown in Figs. 20.8a and

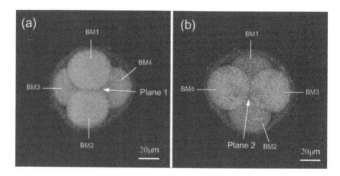

Figure 20.8 (a) 3D reconstruction of a four-cell-stage embryo. Plane 1 where BM1 and BM2 contact is observable. (b) At the opposite angle of view, Plane 2, where BM3 and BM4 contact is observable. The two planes are nearly perpendicular to each other.

Applications in Embryology and Developmental Biology | 715

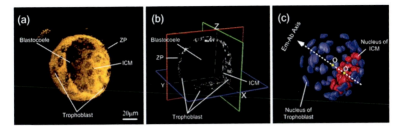

Figure 20.9 (a) 3D reconstruction of a blastocyst. (b) Cross-sectional view of the same blastocyst. Blastocoele, trophoblast, and ICM are clearly visible. (c) 3D reconstruction of the segmented nuclei of this blastocyst. Point O is the geometry center of the whole blastocyst and Point O' is that of the ICM. The Em-Ab axis is defined by the line which contains the two points O and O'.

20.8d. The angle between these two planes is measured to be 87°, which is significant for normal embryonic development, because second two divisions have already decided the fates of blastomeres and the angle may be one of the reasons for this [57–59]. Other parameters such as spatial locations, relative distances, and volumes of four blastomeres can also be precisely measured using the 3D dataset obtained with FF-OCT.

Three-dimensional reconstruction of a blastocyst stage (E3.5) embryo is shown in Fig. 20.9a, and the cross-sectional view in Fig. 20.9b. Besides blastocoele, the structure of inner cell mass (ICM), which is biologically defined as the mass of cells inside a blastocyst, and trophoblast, biologically defined as the cells forming the outer layer of a blastocyst, are also clearly visible. The segmented nuclei of ICM and trophoblast are shown in Fig. 20.9c. The spatial locations of cells can be measured by calculating every center of the dark areas. The embryonic-abembryonic axis in Fig. 20.9c is then characterized by the line that contains the geometry centers of the whole blastocyst (marked with O) and inner cells (O'). Incorrect positioning of polarity axis due to the limitation of the traditional measurement may be one of the possible causes for the controversies in understanding early development, while accurate measurement of polarity axis with FF-OCT in this way is more reliable and may be the key to clarify them.

Our results suggest that FF-OCT offers a powerful methodology for the study of early patterning, polarity formation, and many other preimplantation-related researches in the mammalian developmental biology.

20.5.2.2 Dynamic studies of early patterning and polarity

For further analysis, we extended our work and made some successful attempts to study the dynamics of developmental processes during mouse preimplantation embryonic lives [21]. We focused our imaging on the event of the first cleavage to investigate 3D spatial morphogenetic relationship between the 2PB and the first cleavage plane. Coupled with 3D quantitative study, we showed that only 25% of the predicted first cleavage planes, as defined by the apposing plane of two pronuclei passed through the 2PB. Also, only 27% of the real cleavage planes passed through the 2PB. These results suggest that the 2PB is not a convincing spatial cue for the event of the first cleavage.

Zygote-stage embryos whose two pronuclei apposed closely were obtained on late E0.5. Embryos were then imaged in a droplet of HTF medium with 20% FBS (Hyclone, USA) covered with mineral oil (M4080, Sigma, USA). As the embryos were sensitive to the prolonged and frequent light exposure [60], we only imaged the 3D structures of the 2PB and two pronuclei rather than the entire embryos. The time needed to image all the 3D structures of 2PB, two pronuclei of each embryo ranged from 20 to 60 s, depending on their spatial orientations. Also, embryos were imaged only twice, before and after the first cleavages to minimize the optical dose. With this kind of time-lapse imaging, the normal development was ensured and the 3D structural images were sufficient to reveal important 3D spatial relations between the 2PB and the first cleavage plane.

First, according to the angular classification methods used in Ref. [54] and on the basis of the conclusions that the first cleavage planes can be defined (or predicted) by the two apposing pronuclei [54], we classified embryos into two categories without excluding any embryo from analysis: Type A, in which the predicted cleavage plane (PCP) occurs within 30° of the 2PB; type B, between 30° and 90°. According to Ref. [54], the PCPs that belonged to type A were

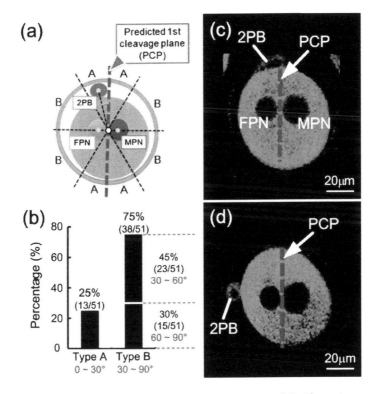

Figure 20.10 (a) The schematic model of types A and B. The points are the 3D geometry centers of the 2PB, female pronucleus (FPN), and male pronucleus (MPN). The point in the center is the middle point of the 3D geometry centers of two pronuclei. (b) Percentages of each type. (c) 3D reconstruction of a typical embryo of type A. (d) 3D reconstruction of a typical embryo of type B.

considered to pass through the 2PB and the PCPs which belonged to type B were considered not to pass through the 2PB. The schematic model, the percentages of each type and 3D reconstructions of two typical embryos of each type are shown in Fig. 20.10. For visualization of the two pronuclei, a part of the zygotes were removed with clipping planes. Based on our previous work [20], the dark areas in Figs. 20.10c and 20.10d were the areas of nuclei: the biggest one is the MPN; the one of middle size is FPN and the smallest is the 2PB's. In Fig. 20.10a, the points are the 3D geometry

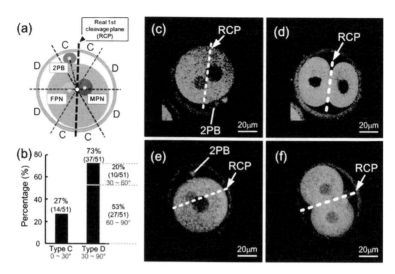

Figure 20.11 (a) The schematic model of types C and D. (b) Percentages of each type. (c, d) 3D reconstruction of a typical embryo of type C before and after the first cleavage. (e, f) 3D reconstruction of a typical embryo of type D before and after the first cleavage.

centers of the 2PB, FPN and MPN. The point in the center is the middle point of the 3D geometry centers of two pronuclei. The PCP which was concluded to be perpendicular to apposing direction of two pronuclei [54], and the angles were measured with these points. Only 25% (13 out of 51, type A) of the PCPs were consistent with the assertion that the first cleavage plane coincides with the A-V axis whose animal pole is assumed to be marked with the 2PB, while 75% for type B (38 out of 51: 23 between 30° and 60°; 15 between 60° and 90°) were not, as shown in Fig. 20.10b. Our results suggest that the PCP does not pass through the 2PB. Instead, it tends to occur in a random way, but a little more frequently between 30° and 60°.

For further analysis, to address whether the real first cleavage plane (RCP) passes through the 2PB, the embryos were kept incubating in the incubator and imaged again after their first cleavages. To determine the occurrences of the first cleavages, we used the conventional 2D imaging every 1 h. The embryos were imaged for the second time right after the determination. For our samples, the average time interval between the first FF-OCT imaging and the occurrences of the first cleavage was 8 h. Then we quantified every RCP using the 3D datasets obtained with the second imaging at two-cell stage by FF-OCT. Because the two areas of nuclei of each future blastomere move apart during division in the direction that is perpendicular to the first cleavage plane, the 3D RCP was characterized by 3D geometry centers of two nuclei of each two-cell stage blastomere. Also, we classified the embryos into two categories: type C, in which the RCP occurs within 30° of the original 2PB at the zygote stage, but not the one which has moved to another position at two-cell stage; type D, between 30° and 90°. According to Ref. [54], the RCPs which belonged to type C were considered to pass through the 2PB and the RCPs which belonged to type D were considered not to pass through the 2PB. Figure 20.11a shows the schematic model and Fig. 20.11b shows the percentages of each type. Three-dimensional reconstructions of a typical embryo of type C before and after its first cleavage are shown in Figs. 20.11c and 20.11d, while Figs. 20.11e and 20.11f show this for type D. For visualization of the two pronuclei and the nuclei of the blastomeres, a part of the embryos were removed with clipping planes. Only 27% (14 out of 51, type C) of the RCPs were formed within 30° of the original 2PB, whereas 73% (37 out of 51, type D: 10 between 30° and 60°; 27 between 60° and 90°) were not, as shown in Fig. 20.11b. These results suggest that the RCP does not pass through the 2PB and does not strongly depend on the 2PB. On the other hand, if there is still assumed to be a predetermined A-V axis and the first cleavage plane is assumed to be meridional with reference to this axis, the 2PB is not a good spatial cue that marks the animal pole. At the opposite point of view, if the 2PB does mark animal pole, only a little proportion of the RCPs would align with the A-V axis. Instead, the RCPs tend to occur perpendicular (between 60° and 90°) to the A-V axis more frequently.

Our studies demonstrate the feasibility of FF-OCT in providing new insights and potential breakthrough to the controversial issues of early patterning and polarity in mammalian developmental biology.

20.6 Conclusion

As an extension of OCT, FF-OCT is an emerging technology for submicron-scale label-free 3D imaging. FF-OCT detects backscattered light rather than fluorescence, requiring no labeling nor preprocessing. Equipped with a broadband light source and high-NA microscope objectives, FF-OCT is capable of achieving 3D images with subcellular resolution in both longitudinal and transverse directions. Furthermore, the entire field of view is uniformly illuminated with the average optical power density much lower than confocal imaging. Therefore, the sample receives much less optical dose, which is favorable for long-time and live observation. These characteristics make FF-OCT a powerful imaging methodology in various biomedical studies, especially in embryology and developmental biology.

References

1. Drexler, W., and Fujimoto, J. G., eds. (2007). *Optical Coherence Tomography: Technology and Applications.* Series Title: *Biological and Medical Physics, Biomedical Engineering.* Springer, Berlin, Heidelberg, New York.
2. Huang, D., Swanson, E. A., Lin, C. P., Schuman, J. S., Stinson, W. G., Chang, W., Hee, M. R. Flotte, T., Gregory, K., Puliafito, C. A., and Fujimoto, J. G. (1991). Optical coherence tomography. *Science,* **254**, pp. 1178–1180.
3. Leitgeb, R., Hitzenberger, C. K., and Fercher, A. F. (2003). Performance of fourier domain vs. time domain optical coherence tomography. *Opt. Express,* **11**, pp. 889–894.
4. Nassif, N., Cense, B., Park, B. H., Park, B. H., Yun, S. H., Chen, T. C., Bouma, B. E., Tearney, G. J., and de Boer, J. F. (2004). In vivo human retinal imaging by ultrahigh-speed spectral domain optical coherence tomography. *Opt. Lett.,* **29**, pp. 480–482.
5. Grajciar, B., Pircher, M., Fercher, A., and Leitgeb, R. (2005). Parallel Fourier domain optical coherence tomography for in vivo measurement of the human eye. *Opt. Express,* **13**, pp. 1131–1137.
6. Nakamura, Y., Makita, S., Yamanari, M., Itoh, M., Yatagai, T., and Yasuno, Y. (2007). High-speed three-dimensional human retinal imaging by line-

field spectral domain optical coherence tomography. *Opt. Express*, **15**, pp. 7103–7116.

7. Huber, R., Wojtkowski, M., and Fujimoto, J. G. (2006). Fourier domain mode locking (FDML): a new laser operating regime and applications for optical coherence tomography. *Opt. Express*, **14**, pp. 3225–3237.

8. Beaurepaire, E., Boccara, A. C., Lebec, M., Blanchot, L., and Saint-Jalmes, H. (1998). Full-field optical coherence microscopy. *Opt. Lett.*, **23**, pp. 244–246.

9. Vabre, L., Dubois, A., and Boccara, A. C. (2002). Thermal-light full-field optical coherence tomography. *Opt. Lett.*, **27**, pp. 530–532.

10. Göppert-Mayer, M. (1931). Über Elementarakte mit zwei Quantensprüngen. *Ann. Phys.*, **401**, pp. 273–294.

11. Grieve, K., Dubois, A., Simonutti, M., Paques, M., Sahel, J., Gargasson, J. F. L., and Boccara, C. (2005). In vivo anterior segment imaging in the rat eye with high speed white light full-field optical coherence tomography. *Opt. Express*, **13**, pp. 6286–6295.

12. Grieve, K., Paques, M., Dubois, A, Sahel, J., Boccara, C., and Gargasson, J. F. L. (2004). Ocular tissue imaging using ultrahigh-resolution, full field optical coherence tomography. *Invest. Ophthalmol. Vis. Sci.*, **45**, pp. 4126–4131.

13. Perea-Gomez, A., Camus, A., Moreau, A., Grieve, K., Moneron, G., Dubois, A., Cibert, C., and Collignon, J. (2004). Initiation of gastrulation in the mouse embryo is preceded by an apparent shift in the orientation of the anterior-posterior axis. *Curr. Biol.*, **14**, pp. 197–207.

14. Dubois, A., Grieve, K., Moneron, G., Lecaque, R., Vabre, L., and Boccara, C. (2004). Ultrahigh-resolution full-field optical coherence tomography. *Appl. Opt.*, **43**, pp. 2874–2883.

15. Dubois, A. (2004). Effects of phase change on reflection in phase-measuring interference microscopy. *Appl. Opt.*, **43**, pp. 1503–1507.

16. Zheng, J. G., Huo, T. C., Tian, N., Chen, T. Y., Wang, C. M., Zhang, N., Zhao, F Y., Lu, D. Y., Chen, D. Y., Ma, W. Y., Sun, J. L., and Xue, P. (2013). Noninvasive three-dimensional live imaging methodology for the spindles at meiosis and mitosis. *J. Biomed. Opt.*, **18**, p. 050505.

17. Sacchet, D., Moreau, J., Georges, P., and Dubois, A. (2008). Simultaneous dual-band ultra-high resolution full-field optical coherence tomography. *Opt. Express*, **16**, pp. 19434–19446.

18. Dubois, A., Moneron, G., and Boccara, C. (2006). Thermal-light full-field optical coherence tomography in the 1.2 µm wavelength region. *Opt. Commun.*, **266**, pp. 738–743.

19. Na, J. H., Choi, W. J., Choi, E. S., Ryu, S. Y., and Lee, B. H. (2008). Image restoration method based on Hilbert transform for full-field optical coherence tomography. *Appl. Opt.*, **47**, pp. 459–466.
20. Zheng, J. G., Lu, D. Y., Chen, T. Y., Wang, C. M., Tian, N., Zhao, F. Y., Huo, T. C., Zhang, N., Chen, D. Y., Ma, W. Y., Sun, J. L., and Xue, P. (2012). Label-free subcellular 3D live imaging of preimplantation mouse embryos with full-field optical coherence tomography. *J. Biomed. Opt.*, **17**, p. 070503.
21. Zheng, J. G., Huo, T. C., Chen, T. Y., Wang, C. M., Zhang, N., Tian, N., Zhao, F. Y., Lu, D. Y., Chen, D. Y., Ma, W. Y., Sun, J. L., and Xue, P. (2013). Understanding three-dimensional spatial relationship between the mouse second polar body and first cleavage plane with full-field optical coherence tomography. *J. Biomed. Opt.*, **18**, p. 010503.
22. Fercher, A. F., Hitzenberger, C. K., Drexler, W., Kamp, G., and Sattmann, H. (1993). In vivo optical coherence tomography. *Am. J. Ophthalmol.*, **116**, pp. 113–114.
23. Schmitt, J. M., Yadlowsky, M., and Bonner, R. F. (1995). Subsurface imaging of living skin with optical coherence tomography. *Dermatology*, **191**, pp. 93–98.
24. Fujimoto, J. G., Brezinski, M. E., Tearney, G. J., Boppart, S. A., Bouma, B. E., Hee, M. R., Southern, J. F., and Swanson, E. A. (1995). Optical biopsy and imaging using optical coherence tomography. *Nat. Med.*, **1**, pp. 970–972.
25. Golubovic, B., Bouma, B. E., Tearney, G. J., and Fujimoto, J. G. (1997). Optical frequency-domain reflectometry using rapid wavelength tuning of a Cr^{4+}: forsterite laser. *Opt. Lett.*, **22**, pp. 1704–1706.
26. Yun, S. H., Tearney, G. J., Bouma, B. E., Park, B. H., and de Boer, J. F. (2003). High-speed spectral-domain optical coherence tomography at 1.3 μm wavelength. *Opt. Express*, **11**, pp. 3598–3604.
27. Oh, W. Y., Yun, S. H., Tearney, G. J., and Bouma, B. E. (2005). 115 kHz tuning repetition rate ultrahigh-speed wavelength-swept semiconductor laser. *Opt. Lett.*, **30**, pp. 3159–3162.
28. Qi, B., Himmer, A. P., Gordon, L. M., Yang, X. D. V., Dickensheets, L. D., and Vitkin, I. A. (2004). Dynamic focus control in high-speed optical coherence tomography based on a microelectromechanical mirror. *Opt. Commun.*, **232**, pp. 123–128.
29. Yasuno, Y., Sugisaka, J., Sando, Y., Nakamura, Y., Makita, S., Itoh, M., and Yatagai, T. (2006). Non-iterativenumerical method for laterally superresolving Fourier domain optical coherence tomography. *Opt. Express*, **14**, pp. 1006–1020.

30. Yu, L. F., Rao, B., Zhang, J., Su, J., Wang, Q., Guo, S., and Chen, Z. (2007). Improved lateral resolution in optical coherence tomography by digital focusing using twodimensional numerical diffraction method. *Opt. Express*, **15**, pp. 7634–7641.

31. Drexler, W., Morgner, U., Kartner, F. X., Pitris, C., Boppart, S. A., Li, X. D., Ippen, E. P., and Fujimoto, J. G. (1999). In vitro ultrahigh-resolution opticalcoherence tomography. *Opt. Lett.*, **24**, pp. 1221–1223.

32. Huber, R., Adler, D. C., and Fujimoto, J. G. (2006). Buffered Fourier domain mode locking: unidirectional swept laser sources for optical coherence tomography imaging at 370,000 lines/s. *Opt. Lett.*, **31**, pp. 2975–2977.

33. Moon, S., and Kim, D. Y. (2006). Ultra-high-speed optical coherence tomography with a stretched pulse supercontinuum source. *Opt. Express*, **14**, pp. 11575–11584.

34. Choi, D., Hiro-Oka, H., Furukawa, H., Yoshimura, R., Nakanishi, M., Shimizu, K., and Ohbayashi, K. (2008). Fourier domain optical coherence tomography using optical demultiplexers imaging at 60,000,000 lines/s. *Opt. Lett.*, **33**, pp. 1318–1320.

35. Wieser, W., Biedermann, B. R., Klein, T., Eigenwillig, C. M., and Huber, R. (2010). Multi-Megahertz OCT: high quality 3D imaging at 20 million A-scans and 4.5 GVoxels per second. *Opt. Express*, **2**, pp. 14687–14704.

36. Yang, V. X. D., Gordon, M., Seng-Yu, E., Lo, S., Qi, B., Pekar, J., Mok, A., Wilson, B., and Vitkin, I. (2003). High speed, wide velocity dynamic range Dopler optical coherence tomography: imaging in vivo cardiac dynamics of Xenopus laevis. *Opt. Express*, **11**, pp. 1650–1658.

37. Boppart, S. A., Brezinski, M. E., Bouma, B. E., Tearney, G. J., and Fujimoto, J. G. (1996). Investigation of developing embryonic morphology using optical coherence tomography. *Dev. Biol.*, **177**, pp. 54–64.

38. Jenkins, M. W., Adler, D. C., Gargesha, M., Huber, R., Rothenberg, F., Belding, J., Watanabe, M., Wilson, D. L., Fujimoto, J. C., and Rollins, A. M. (2007). Ultrahigh-speed optical coherence tomography imaging and visualization of the embryonic avian heart using a buffered Fourier domain mode locked laser. *Opt. Express*, **15**, pp. 6251–6267.

39. Jenkins, M. W., Rothenberg, F., Roy, D., Nikolski, V. P., Hu, Z., Watanabe, M., Wilson, D. L., Efimov, I. R., and Rollins, A. M. (2006). 4D embryonic cardiography using gated optical coherence tomography. *Opt. Express*, **14**, pp. 736–748.

40. Choma, M. A., Izatt, S. D., Robert, M. D., Wessells, J., Bodmer, R., and Izatt, J. A. (2006). In vivo imaging of the adult Drosophila melanogaster heart

with real-time optical coherence tomography. *Circulation*, **114**, pp. 35–36.

41. Honda, Y. K., Tanikawa, H. D., Fukuda, J., Kawamura, K. H., Sato, N. K., Sato, T. H., Shimizu, Y. S., Kodama, H., and Tanaka, T. N. (2005). Expression of Smac/DIABLO in mouse preimplantation embryos and its correlation to apoptosis and fragmentation. *Mol. Hum. Reprod.*, **11**, pp. 183–188.
42. Oh, W. Y., Bouma, B. E., Iftimia, N., Yun, S. H., Yelin, R., and Tearney, G. J. (2006). Ultrahigh-resolution full-field optical coherence microscopy using InGaAs camera. *Opt. Express*, **14**, pp. 726–735.
43. Mantikou, E., Wong, K. M., Repping, S., and Mastenbroek, S. (2012). Molecular origin of mitotic aneuploidies in preimplantation embryos. *Biochim. Biophys. Acta*, **1822**, pp. 1921–1930.
44. Eichenlaub-Ritter, U., Shen, Y., and Tinneberg, H. R. (2002). Manipulation of the oocyte: possible damage to the spindle apparatus. *Reprod. Biomed. Online*, **5**, pp. 117–124.
45. Borini, A., Lagalla, C., Cattoli, M., Sereni, E., Sciajno, R., Flamigni, C., and Coticchio, G. (2005). Predictive factors for embryo implantation potential. *Reprod. Biomed. Online*, **10**, pp. 653–668.
46. Petersen, C. G., Oliveira, J. B. A., Mauri, A. L., Massaro, F. C., Baruffi, R. L. R., Pontes, A., and Franco Jr, J. G. (2009). Relationship between visualization of meiotic spindle in human oocytes and ICSI outcomes: a meta-analysis. *Reprod. Biomed. Online*, **18**, pp. 235–243.
47. Gu, Y. F., Lin, G., Lu, C. F., and Lu, G. X. (2009). Analysis of the first mitotic spindles in human in vitro fertilized tripronuclear zygotes after pronuclear removal. *Reprod. Biomed. Online*, **19**, pp. 745–754.
48. Oldenbourg, R. (1999). Polarized light microscopy of spindles. *Method. Cell. Biol.*, **61**, pp. 175–208.
49. Liu, L., Oldenbourg, R., Trimarchi, J. R., and Keefe, D. L. (2000). A reliable, noninvasive technique for spindle imaging and enucleation of mammalian oocytes. *Nat. Biotechnol.*, **18**, pp. 223–225.
50. Rienzi, L., Ubaldi, F., Martinez, F., Iacobelli, M., Minasi, M. G., Ferrero, S., Tesarik, J., and Greco, E. (2003). Relationship between meiotic spindle location with regard to the polar body position and oocyte developmental potential after ICSI. *Hum. Reprod.*, **18**, pp. 1289–1293.
51. Tarkowski, A. K. (1959). Experiments on the development of isolated blastomeres of mouse eggs. *Nature*, **184**, pp. 1286–1287.
52. Tsunoda, Y., and McLaren, A. (1983). Effect of various procedures on the viability of mouse embryos containing half the normal number of blastomeres. *J. Reprod. Fertil.*, **69**, pp. 315–322.

53. Ciemerych, M. A., Mesnard, D., and Zernicka-Goetz, M. (2000). Animal and vegetal poles of the mouse egg predict the polarity of the embryonic axis, yet are nonessential for development. *Development*, **127**, pp. 3467–3474.
54. Hiiragi, T., and Solter, D. (2004). First cleavage plane of the mouse egg is not predetermined but defined by the topology of the two apposing pronuclei. *Nature*, **430**, pp. 360–364.
55. Kurotaki, Y., Hatta, K., Nakao, K., Nabeshima, Y., and Fujimori, T. (2007). Blastocyst axis is specified independently of early cell lineage but aligns with the ZP shape. *Science*, **316**, pp. 719–723.
56. Gray, D., Plusa, B., Piotrowska, K., Na, J., Tom, B., Glover, D. M., and Zernicka-Goetz, M. (2004). First cleavage of the mouse embryos responds to change in egg shape at fertilization. *Curr. Biol.*, **14**, pp. 397–405.
57. Piotrowska, K., and Zernicka-Goetz, M. (2001). Role for sperm in spatial patterning of early mouse embryos. *Nature*, **409**, pp. 517–521.
58. Gardner, R. L. (2002). Experimental analysis of second cleavage in the mouse. *Hum. Reprod.*, **17**, pp. 3178–3189.
59. Piotrowska, K., Wianny, F., Pedersen, R. A., and Zernicka-Goetz, M. (2001). Blastomeres arising from the first cleavage division have distinguishable fates in normal mouse development. *Development*, **128**, pp. 3739–3748.
60. McDole, K., Xiong, Y., Iglesias, P. A., and Zheng, Y. (2011). Lineage mapping the pre-implantation mouse embryo by two-photon microscopy, new insights into the segregation of cell fates. *Dev. Biol.*, **355**, pp. 239–249.
61. Bouniol-Baly, C., Nguyen, E., Besombes, D., and Debey, P. (1997). Dynamic organization of DNA replication in one-cell mouse embryos: relationship to transcriptional activation. *Exp. Cell Res.*, **236**, pp. 201–211.

Chapter 21

FF-OCT for Nondestructive Material Characterization and Evaluation

David Stifter

Christian Doppler Laboratory for Microscopic and Spectroscopic Material Characterization (CDL-MS-MACH), Center for Surface and Nanoanalytics (ZONA), Johannes Kepler University Linz, Altenberger Str. 69, A-4040 Linz, Austria
david.stifter@jku.at

Applications of full-field optical coherence tomography (FF-OCT) outside the biomedical field are the topic of this chapter. Since its introduction and based on the original motivation of the underlying white-light interferometry technique in microelectronic inspection tasks, FF-OCT is continuously gaining impetus for the characterization of materials and for the evaluation of technical structures in a nondestructive way. Examples, from the measurement of surface topography and of layer thicknesses to a full three-dimensional evaluation of internal structures within scattering media, are given together with novel and advanced concepts for functional imaging.

21.1 Introduction

Accessing the third dimension was a long sought-after goal in microscopy and till today worldwide efforts to introduce optimized

Handbook of Full-Field Optical Coherence Microscopy: Technology and Applications
Edited by Arnaud Dubois
Copyright © 2016 Pan Stanford Publishing Pte. Ltd.
ISBN 978-981-4669-16-0 (Hardcover), 978-981-4669-17-7 (eBook)
www.panstanford.com

technical solutions and new concepts for achieving this goal have not ceased. Such a novel instrumental solution was presented in 1987 for the inspection and metrology of microelectronic structures, as an alternative to scanning electron microscopy and the then already known confocal microscopy technique. This method was called coherence probe microscopy (CPM) and was based on a Linnik interferometer illuminated by white light and on an area detector for delivering full-field (FF) images at defined depth positions [1], several years before optical coherence tomography (OCT) was introduced for 3D biomedical imaging [2]. Supported by the increasing success of OCT for accessing the internal structure of scattering biomaterials, the original concept of CPM—and of white-light interferometry (WLI) in general—was equally optimized for imaging within turbid media, finally leading to full-field optical coherence microscopy (FF-OCM) [3] and FF-OCT [4]. As a consequence, the majority of applications for conventional, that is, raster scanning, OCT and also for FF-OCT can be found nowadays in the area of life science and in biomedical research and diagnostics. However, the original driving forces for the development of low-coherence interferometry (LCI) and WLI imaging techniques, namely metrology, inspection, and nondestructive testing/evaluation (NDT/NDE) of technical structures and materials, are gaining ground, as summarized for OCT in [5] and as described in detail for FF-OCT in this chapter.

The chapter is organized as follows. At first different FF-OCT techniques and measurement schemes used for the to-date reported nonbiomedical applications are shortly revisited, introducing also advanced imaging concepts, which allow the modification of contrast or which can supply complimentary information to the structural data usually delivered. In the main part of this chapter, a collection of already published application ideas and material characterization studies outside the biomedical field is given, starting with surface profilometry and topography measurements, going through thin-film metrology and finally showing promising results for structural investigations of the interior of semitransparent and scattering materials. Functional material imaging, achieved by making use of the polarization state of the backreflected and backscattered light or by evaluating the phase for precise displacement measurements, is finally presented. At the end a brief outlook to anticipated technical developments and promising future evaluation and characterization tasks for FF-OCT is given.

21.2 Methods

The majority of the FF-OCT systems used for nondestructive evaluation and material characterization was and is realized in the time domain (TD) configuration with a broadband light source, a Linnik interferometer, and an area camera as the main components, exactly reflecting the original instrumental concept of CPM [1]. For the measurement in CPM, the sample is moved in the z direction and the interference signal captured for each detector pixel during a total depth scan. Since the original use was the detection of the surface position and the location of underlying interfaces, the envelope of the interference amplitude was evaluated, but it was already stated that the determination of the phase of the coherence function can be used as well, allowing extremely precise surface metrology and film thickness monitoring. This procedure became then standard for the so-called scanning WLI (SWLI) systems [6], which generally also take advantage of a more stable, common path interferometer setup by using a Mirau objective [7], and which have been used till now for surface topography measurements reaching subnanometer precision. However, especially for high-numerical-aperture (NA) systems the beam splitter can cause strong aberrations in the Mirau microscope. Furthermore, for the imaging of the interior of scattering samples it is advisable that the coherence gate and the focus can be independently adjusted to match the position of both within the plane of interest [4]. These facts are again in favor for the Linnik configuration, currently widely used for FF-OCT imaging.

Besides using FF-OCT in the TD, scanning the wavelength is an alternative approach, as introduced in [8] for real-time surface shape measurements. Wavelength scanning interferometry (WLSI, not to be mistaken with the above-mentioned SWLI concept) has the advantage that no movable parts are needed for depth scanning together with an increased signal-to-noise ratio, as demonstrated in Ref. [9] with an FF swept-source (SS) phase microscope, inspired by the developments in Fourier domain (FD) OCT [10]. Nevertheless, the fact that the whole depth range is acquired at once in FF-SS-OCT does not allow a dynamic adjustment of the focus position to the imaging depth. Consequently, this technique is only suitable for the detection of the surface topography, for the investigation of very thin specimens and of well-defined layers with medium or low NA objectives.

For FF-SS-OCT a Fourier transform of the signal of each pixel for a full wavelength sweep has to be usually performed to obtain depth resolved data. In contrast, in the TD the signal has to be de-modulated for each pixel, which led to the development of smart-pixel detector arrays, consisting of individual Si photodiodes coupled to complementary-metal-oxide semiconductor (CMOS) electronic circuits for signal amplification and de-modulation. With such a system dynamic processes were demonstrated to be followed in 3D with video rate [11]. Acquisition and de-modulation of frames with frame rates of several kHz are feasible for such parallel OCT (pOCT) systems, however, providing only a much reduced number of pixels. If dynamic processes have to be followed very fast in high detail or if extended 3D volumes have to be rapidly scanned with high lateral resolution, then the only option is to employ CMOS cameras, which provide—as compared to conventional charge-coupled device (CCD) detectors—increased frame rates together with a higher dynamic range. Usually, for dynamic imaging two consecutive frames with a defined phase shift have to be captured (dual-shot acquisition), which requires that the sample is quasi-stationary between these two frames. Alternatively, with single-shot techniques these two frames are taken simultaneously, further reducing the restrictions to the maximum sample movement allowed. Conventionally, for obtaining the final image the squared difference of the two frames is taken [12]. This procedure is working well for samples with less defined structures, as found in scattering biological tissue. In contrast, for technical samples, like multilayer structures with smooth surfaces and interfaces, de-modulation artifacts would be introduced in the final image. Such artifacts can be minimized by advanced 2D de-modulation algorithms, as, for example, shown for dynamic imaging of technical samples by using the 2D Riesz transform (RT) for de-modulation [13]. Finally, one should also keep in mind that in such a high-speed configuration TD-FF-OCT is outperforming the fastest raster scanning SS-OCT systems nowadays available: the acquisition of en face images with Megapixel size and (ultra)high isotropic resolution (i.e., the lateral resolution is comparable to the longitudinal one) would require laser sweep rates around 300 MHz, a figure which is orders of magnitude higher than the sweep rate of the fastest Fourier domain mode-locked (FDML) lasers currently available [14].

In addition to structural data further information on complementary sample properties can be required, which leads to modified contrasting and functional imaging concepts for FF-OCT. Due to the fact that specular reflexes of smooth surfaces can easily saturate the 2D detector, dark-field (DF) imaging would be of advantage in order to detect weakly reflecting structures within such samples. DF imaging was already demonstrated to be helpful when detecting small point defects in highly reflecting multilayer components [15]. In this case the sample was tilted by a large angle so that the specular reflex of the surface was not transmitted by the objective. However, it is not always possible to tilt the sample of interest far enough, especially if features in an extended depth region should be imaged and if the working distance between sample and objective is small. In this case, a manipulation of the image in the Fourier plane (FP) is the better choice, as already routinely used in conventional 2D microscopy. DF imaging and additional contrast modes have therefore been introduced for FF-OCT in [16] exploiting Fourier plane filtering (FPF), as depicted in Fig. 21.1 for a planar test specimen. In this case even a reflective spatial light modulator (SLM) was used in the Fourier plane of the microscope to permit a flexible choice of the contrast mode. Depending upon the displayed pattern on the SLM different portions of the wave field in the FP are altered in phase and transmitted to the image plane on the area detector. Normal bright-field (BF) imaging is possible when the SLM is just used as a normal mirror, as depicted in Fig. 21.1a. By displaying a phase grating on the SLM, selected areas can be diffracted away. This is visible, for example, in Fig. 21.1b for the DF mode, where the center region of the SLM is exhibiting such a grating, which deflects the zero order beam so that only higher Fourier components arising from edges and small features can reach the detector. Contrary to

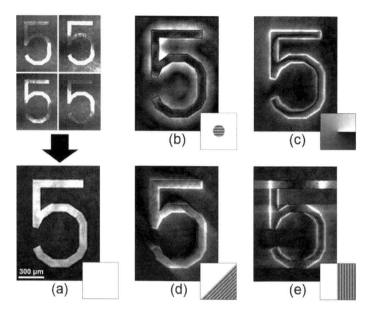

Figure 21.1 Interferometric imaging with FPF exemplified on a planar test feature, with indicated SLM filter functions (gray scale proportional to the introduced phase shift) in the bottom right corners of each image: (a) BF, (b) DF, (c) SP, and SC contrast under (d) 45° and (e) 90° orientation. In (a), a set of raw, phase-shifted interference images is given. Figure from Ref. [16], with kind permission of OSA, © 2010.

the DF mode, which discards part of the light from the sample, spiral phase (SP) contrast redistributes the total intensity to the edges of the structure, as seen in Fig. 21.1c, leading to a much brighter and sharper edge contrast. Also directional features can be enhanced, using a mode analogous to Schlieren contrast (SC), depicted in Fig. 21.1d,e.

Modifying the contrast in 3D imaging by FPF is then shown in Fig. 21.2 for the SP mode. Especially for layered technical samples, enhanced contrast and complementary information can easily be obtained, changing on the fly the displayed pattern on the SLM. Additional quantitative information on the optical thickness of sample features can be derived by evaluating the number of turns and the handedness of the small spirals appearing in the SP mode, as clearly depicted in the inset of Fig. 21.2b.

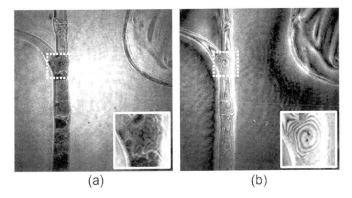

(a) (b)

Figure 21.2 FPF imaging demonstrated on a glass slide with partly opaque features: FF-OCT images in BF mode (a) and with SP filter (b), including enlarged views of the marked area in the insets. Figure taken from Ref. [16], with kind permission of OSA, © 2010.

Additional contrast and information is also obtained, when taking the polarization state into account. Polarization-sensitive (PS) OCT provides in this way access to the phase retardation, additionally to the conventional reflectivity image, as shown in Ref. [17] for PS-FF-OCT performed on biological tissue. For technical applications PS imaging can deliver crucial information on the internal birefringence distribution, which reflects sample anisotropies and the internal stress state, as summarized in [5] for conventional PS-OCT imaging of polymer samples and as shown later for PS-FF-OCT in this chapter.

Functional contrast can be also obtained, when linked to the deformation of the sample. A strain-birefringence response of a sample can be exploited by PS-FF-OCT. However, with FF low-coherence-speckle interferometry (LCSI), which is a derivative of conventional electronic speckle interferometry (ESPI) by using a low coherence light source, the displacement of internal structures can directly be accessed [18]. Since this approach uses a variable path length of the reference beam it is working in the TD. In a similar way, depth-resolved FF displacement measurements have been realized in the FD with wavelength scanning speckle pattern interferometry [19].

21.3 Applications

After the short overview given above on the measurement concepts introduced to date, and which were used or exhibit a promising potential for nondestructive material characterization and structure evaluation, the focus is put in this section on the different types of applications. These are ranging from surface profilometry over thin-film metrology to the structural investigation of features within scattering technical materials or pieces of cultural heritage, leading finally to dynamic and functional material imaging. In addition, advanced image analysis and data processing can provide further crucial information on the investigated specimens.

21.3.1 *Surface Metrology*

Application of FF-OCT for surface metrology tasks is directly linked to the original measurement concepts of CPM and SWLI, as laid out in the above section. Based on the mentioned CPM concept [1], soon a fully automated inspection system for line width measurements of resist lines for the semiconductor industry was presented in 1988 [20]. Interestingly, already several potential future applications were anticipated in this context, including film thickness measurements, acquisition of the 3D structure, characterization of contact holes and trenches and the use of CPM for overlay registration to align wafers. In the following years, a number of instrumental optimizations and variations were presented, including the so-called coherence scanning microscope in Michelson interferometer configuration [21], the above-mentioned Mirau correlation microscope [7] and an interferometric profiler for rough surfaces with adapted envelope de-modulation [22]. These concepts have in common that mostly thermal light sources are used for the reduction of speckle-induced artifacts and that the phase front coming from the sample has to match the one of the reference wave. In such a way a correlation effect, or a probing of the coherence position, takes place, which leads to an additional sectioning effect to that one arising from the use of the low coherence source [23], in contrast to conventional OCT with low NA objectives. In fact high axial resolution can be achieved with rather narrowband sources and high NA optics, as analyzed in detail in [24]. All the techniques mentioned before form the basis for modern SWLI devices [6] for surface topography measurements, and also for thin-film metrology, as laid out later. Finally, an interesting modification employing rather low NA optics for measuring the shape of rough surfaces of industrial objects shall be mentioned, which was already presented in 1992: with the so called coherence radar a complete bore hole, being more than 2 cm deep and only 5 mm wide, was shown to be fully imaged and evaluated [25].

Summarizing, in the context of surface metrology, the distinction between FF-OCT and SWLI is blurred. Some further applications worth being mentioned include mainly topography measurements for the semiconductor and electronics industry, like evaluation of features on wafers and integrated circuits (ICs) or the characterization of microelectromechanical systems (MEMS). In this context reflectivity and topography images were obtained in real time with an interference microscope setup [26], which was used later for FF-OCT imaging [4]. IC structures were also evaluated with an LCI microscope using geometric phase shifting [27], with conventional FF-OCT [28] or with FF-SS-OCT using a superluminescence diode (SLD) in combination with an acousto-optic tunable filter to obtain OCT images and phase profiles [29]. The latter setup was equally used to derive phase maps of microlens arrays, which were not placed in the usual sample position, but in the beam path before the beam splitter [30]. Furthermore, the evaluation of the topography of MEMS by SWLI in a static way was demonstrated in, for example, Ref. [31]. Later on, static and dynamic characterization of MEMS and micro-opto-electromechanical structures (MOEMS) followed [32]. Also, the focus is put on improvements of the measurement concepts and especially on automatization and integration to generate industrial applicable systems and complete metrology solutions; recently, a SWLI system with long working distance and high lateral resolution by using near-ultraviolet (UV) light-emitting diodes (LEDs) and its integration into a nanomeasuring machine was demonstrated [33].

In the case of pOCT and FF-OCT with digital signal processing (DSP) cameras, solder point arrays were shown to be rapidly inspected [34], 3D machine vision was realized for the evaluation

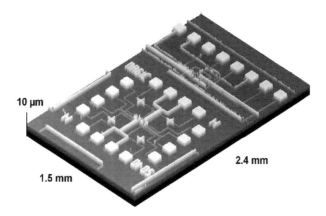

Figure 21.3 3D real-time image of a CMOS magnetic probe being translated sideways at 20 µm/s, measured at a 3D image rate of 1.9 i/s with a ×5 Michelson objective (image size = 640 × 1024 pixels, axial scan step = 40 nm over a range of 5 µm). Figure from Ref. [37], with kind permission of SPIE, © 2011.

of the surface topography [35] and new CMOS lock-in sensors were developed with frame rates up to 1 million/s (at a reduced pixel number of 300 × 300), as, for example, demonstrated for roughness and shape measurements on machined metallic surfaces [36]. High-speed measurements were also achieved by using conventional CMOS cameras with DSP performed on, for example, field programmable gate arrays (FPGA), so 4D online acquisition of moving parts can now be realized, as demonstrated in Fig. 21.3 for a microstructured surface moving at a speed of 20 µm/s [37].

The measurements of the topography in the latter example could successfully be performed through a transparent cover layer. With FF-OCT it was equally shown that the surface of wet pads, used in chemical mechanical polishing of wafers, can be evaluated even when the pads are covered by the polishing fluid and diluted slurry [38]. In a similar sense, a feasibility study was presented to demonstrate the inspection of the inner surface of sawing channels in a Si crystal from the outside using an LCI system, which operates at a wavelength around 1300 nm and uses an InGaAs camera for imaging through the Si material [39].

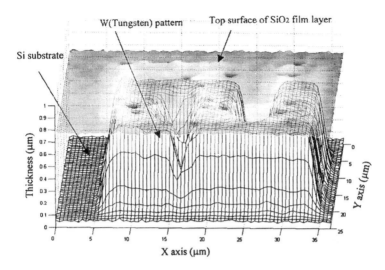

Figure 21.4 3D thickness profile of a SiO_2 film on a patterned wafer. Figure from Ref. [40], with kind permission of OSA, © 1999.

21.3.2 Layer Thickness Determination

The examples of application presented above with respect to the characterization of covered surfaces naturally leads to the evaluation of film thicknesses. In the context of SWLI, this technique could be optimized to measure not only opaque surfaces but also transparent films on MEMS and MOEMS, as successfully shown in Ref. [32]. However, for precise thin-film metrology multiple reflections and their interference can represent a challenge, which could be overcome by applying extensive analysis of the measured data in the FD. An example is shown in Fig. 21.1 with the volumetric film profile of a SiO_2 layer covering a metallic structure, which was deposited on a Si wafer [40]. Multiple interference effects and their influence on layer thickness measurements were also analyzed with a combined FF-OCT and FD-OCT system in [41]. Furthermore, if high-NA objectives are used for FF-OCT in combination with narrowband sources, the longitudinal spatial coherence can be much shorter than the temporal coherence length, making such an approach promising for the thickness determination of thin multilayer structures [24].

Not only rather planar surfaces and layers, like micromembranes for MEMS and multilayers on Si wafers, have been evaluated by FF-OCT [42], but also curved ones, as demonstrated for thickness and refractive index measurements of roundly shaped epoxy droplets [43]. As a last example, the determination of the coating thickness of spherical pharmaceutical pellets, as described in Ref. [44], shall be mentioned.

21.3.3 3D Structural and Functional Evaluation

The evaluation of the internal structure of turbid and scattering materials is more in line with the original motivation of FF-OCT (and of OCT imaging in general), in contrast to the examples given above. In this context, a highly scattering material is paper. Its properties are crucially determined by the underlying fiber and void microstructure, which was analyzed by FF-OCT in Ref. [45]. Due to the limited penetration depth of 10 µm, anisotropic low density structures (reflecting the fiber direction within the paper) could only be observed in the subsurface region. Better penetration was achieved with FF-OCT in the Mirau configuration for the evaluation of flamed and painted wood, with the results and insights being used for studying the varnish and wood structure of an 18th century violin [46] and with exemplary images depicted in Fig. 21.5.

A direct comparison with optical microscopy, as shown in Fig. 21.5, demonstrates that several layers of wood cells can be imaged by FF-OCT, even when a thin varnish layer or ground layer, including a small number of filler particles, is applied to the wooden surface. Pigmented varnish layers on wood were also studied with a similar setup using different spectral regions and showing that the detection of the included pigments is wavelength dependent [47]. In addition, Fourier transform analysis of the interferograms obtained from the pigments can deliver spectral information, as shown with spectroscopic FF-OCT experiments in Ref. [48].

Varnish is usually also applied on artistic paintings. Especially for restorers and art historians it is essential to learn more about the surface and subsurface structure of such varnish layers and the underlying paint, without destroying unique specimens of cultural heritage. Consequently, similarly to conventional raster scanning

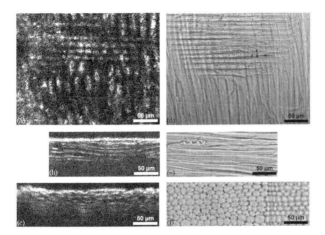

Figure 21.5 Tomographic images (left) compared to optical microscopic images (right) of flamed maple wood according to the (a) and (d) radial plane of the wood structure, (b) and (e) tangential plane, and (c) and (f) transverse plane. Figure from Ref. [46], with kind permission of OSA, © 2009.

OCT [49], also FF-OCT was used for the evaluation of art objects. Besides a study in which Kalman filtering of the signal in combination with sub-Nyquist sampling for high-speed evaluation of such layered varnish structures is presented [50], a comprehensive overview and comparison between different nondestructive optical methods, including FF-OCT, is given in a recent review [51]. As an example, the structure of a gap under the air-varnish interface, of a pigment cluster and of a crack in the varnish and paint layer of a historical painting, which suffered under some questionable restoration procedures, is shown in Fig. 21.6. With FF-OCT it is now possible to accurately determine the varnish layer thickness, to image individual pigments and pigment structures, to identify overpainted regions and also to study additional aspects, like the leveling of the varnish and its influence on the gloss, as presented in Ref. [52].

Similar as for wood, which is a natural composite, nondestructive evaluation of technical composite materials with FF-OCT is a promising task. In this context, tomographic reconstructions and phase maps of a polymer/E-glass composite with FF-SS-OCT were

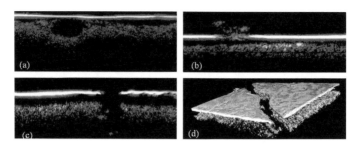

Figure 21.6 Transverse cross sections (250 μm wide and 50 μm deep) obtained with FF-OCT: (a) a gap under the air-varnish interface (47 μm width, 25 μm depth), (b) a curious pigment cluster over the varnish, (c) a varnish crack (32 μm depth), and (d) a varnish crack shown in 3D. Figure taken from Ref. [51], © 2011 Elsevier Masson SAS, all rights reserved.

presented in Ref. [53]. Especially for such semitransparent fiber-reinforced polymer (FRP) composites a reasonable penetration depth within the material and a sufficiently high resolution to observe individual fibers and embedded particles can be reached, as exemplified in Ref. [54]. In this study tomograms from micro-X-ray computed tomography (μ-XCT) were directly compared with the results obtained by a TD-FF-OCT with a compact subnanosecond supercontinuum light source for imaging with ultrahigh axial resolution. An example of such a direct comparison obtained from the same sample region of a FRP is depicted in Fig. 21.7. As shown, FF-OCT delivers the correct dimensions of the fiber pieces, as well as their location and orientation within the material. In addition, FF-OCT is also sensible for inclusions and point defects in the material, which cannot be detected by μ-XCT.

Data sets obtained by FF-OCT, such as in Fig. 21.7, do not only provide the basis for a qualitative structural evaluation in a nondestructive way (one should keep in mind that extended samples usually have to be cut in small pieces for μ-XCT measurements), but can be used, in combination with advanced image and data processing algorithms, to derive quantitative and complimentary information, as presented in Ref. [54] and with an example given in Fig. 21.8.

In this example the obtained FF-OCT data is of sufficient high quality so that a separation of the observed point defects

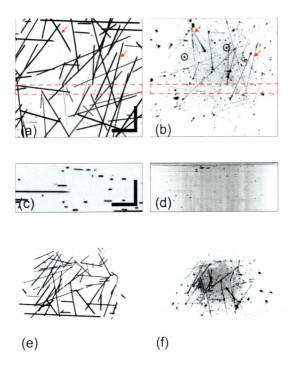

Figure 21.7 Comparison of the complementary information content in micro-X-ray computed tomography (μ-XCT) and FF-OCT imaging, exemplified on a short fiber filled polypropylene compound: left: μ-XCT images; right: FF-OCT images represented as (a, b) en face scan; (c, d) cross-sectional scan at positions between the red lines; (e, f) 3D volume rendering. The good correlation of fibers is indicated by arrows, whereas spherical defects within the material are marked by circles. Horizontal and vertical scale bars: 200 μm. Figure from Ref. [54], with kind permission of Elsevier, © 2012.

and the elongated fiber features is possible by using advanced mathematical routines, for example, based on 2D or 3D shearlet analysis. In a similar way, the orientation of the individual fiber pieces can be mathematically extracted, as shown in Fig. 21.8d [54]. However, for such evaluations, the data has to be preprocessed using advanced image enhancement algorithms for de-noising and background subtraction [55]. Nevertheless, already in an optical way, as introduced in Section 21.2, image and contrast enhancement can be obtained: defects in a filled polymer matrix could be detected by using a spiral phase filter in the Fourier plane of a

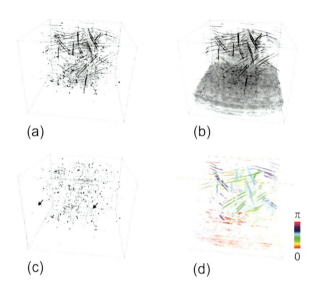

Figure 21.8 Illustration for object separation applying a 2D shearlet approach. 3D volume rendering of the SF-PP compound: (a) preprocessed FF-OCT image stack, (b) separated structures containing elongated fibers, and (c) structures containing spherical defects and microcracks. Partial mismatches may occur for slanted fibers, as indicated in (c). Illustration of local orientation estimation of fiber structures based on the separated objects in the SF-PP compound (d). The different orientations are color encoded. Imaging volume: $1 \times 1 \times 0.3$ mm^3. Figure taken and adapted from Ref. [54], with kind permission of Elsevier, © 2012.

Mach–Zehnder FF-OCT [56], or small scattering defects in optical multilayer components were observed by DF imaging [15].

In case of many technical materials their mechanical properties and their behavior, when subjected to mechanical loading, is of interest. For the FRP composites introduced above it is now feasible to directly monitor the changes of the internal structure during mechanical testing, without affecting the testing routine. As an example, dynamic monitoring of a fracture front within a FRP during tensile mechanical loading was carried out with high-speed FF-OCT in [13]. By advanced image processing methods, complementary information can additionally be obtained, leading toward functional material imaging, besides the conventional structural evaluation. In this context, the displacement field and material flow during the above-mentioned tensile test of a FRP specimen was evaluated using digital speckle correlation, as exemplarily shown in Fig. 21.9. Conventionally, in tensile testing only the surface of the tensile test specimen is monitored by cameras for FF strain analysis and digital image correlation (DIC) [57]. Now with FF-OCT, subsurface planes can equally be evaluated [58].

Another established method beside DIC for displacement measurements is electronic speckle interferometry. If this technique is used in combination with a broadband light source, depth-resolved FF displacement measurements can be carried out. This LCSI method was demonstrated to provide additional useful information on the displacement fields of internal interfaces, namely below the paint of the famous Chinese terracotta warriors [18, 59] or for the detection of interfacial instabilities in adhesive bonded joints [60].

Besides displacement fields, the distribution of the birefringence within a material, either caused by internal stress fields or caused by structural sample anisotropies, can serve as additional source of information, leading to PS-FF-OCT.

PS-FF-OCT, using a single camera and polarization modulation, was shown to detect differences in the phase retardation in resist layers deposited on top of IC structures [61]. In contrast, the approach described in Refs. [58, 62] is relying on a fixed polarization state of the illuminating beam, with two synchronized cameras for the acquisition of the two orthogonal polarization directions. With this system, microcrystallites within polypropylene sheets [58] or the distribution of strain-induced birefringence within photoresist molds for microgearwheels [62] could be visualized. As an example, in Fig. 21.10 the reflectivity image, the phase retardation map and the orientation map of the optical axis, all simultaneously acquired, are depicted for such mold structures [63].

At the end of this section, the retrieval of (mostly) deliberately placed information by FF-OCT is recalled, summarizing applications with respect to security and data storage, like the depth resolved readout of information written in 3D sol-gel materials by a nonlinear absorption process [64]. The retrieval of information, selectively stored on the individual interfaces of a multilayer carrier was the subject of another study [65]. The 2D information (in the form of images, text, and fingerprints) on the smooth carrier layers is

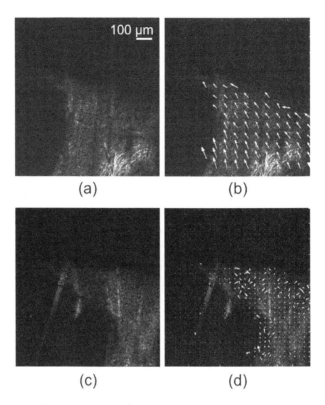

Figure 21.9 Demonstration of FF-OCT imaging for monitoring dynamical testing of a glass fiber-reinforced polymer sample under increasing load: FF-OCT reflectivity (en face) images at two different time states during tensile testing (a, c); corresponding displacement fields determined by digital speckle correlation (b, d). The arrows indicate the local displacement at the two different time states during the process. A rather laminar behavior (b) changes to a turbulent-like behavior with progressing fracture (d) of the strained polymer material. Field of view: 1×1.2 mm^2. Figure taken and adapted from Ref. [58], with kind permission of SPIE, © 2013.

affected by severe artifacts arising from multiple reflections and their interference. Therefore, special algorithms were presented to minimize the parasitic fringe patterns obscuring the information. With respect to fingerprints, also forensic applications have been introduced to date: on weakly reflecting surfaces, SS-FF-OCT was shown to enhance the detection of fingerprints [66]. In addition,

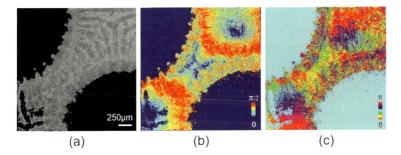

Figure 21.10 PS-FF-OCM images from the bottom of a photoresist mold structure for microgearwheels: (a) reflectivity, (b) phase retardation, and (c) orientation of the optical axis.

security features, like multilayer holograms on banknotes, can be evaluated by FF-OCT and used for counterfeit detection and authenticity checks [67]. Finally, the retrieval of optical information hidden behind a scattering medium with FF-OCT and by applying a contrast discrimination algorithm was demonstrated in [68]. For practical use, a simple portable and hand-operated FF-OCT device for the readout of such hidden 2D information patterns was developed [69], further highlighting the broad field of applicability of the FF-OCT principle.

21.4 Conclusions and Outlook

As detailed in this chapter, the applications of FF-OCT outside the field of life sciences and biomedical diagnostics are manifold, especially for nondestructive material characterization and evaluation. This is at first due to the fact, that the origin of this technique, namely CPM, was developed for surface topography and structural investigations in the semiconductor industry. Secondly, the (ultra)high axial and lateral resolution provides accurate information, delivering directly 2D en face images, to which microscopists are generally used from conventional optical microscopy and from which the 3D structural data can easily be reconstructed. Thirdly, there is a variety of (technical) material systems, which exhibit features on the micron scale within transparent or translucent matrices, all calling for the high isotropic resolution of current FF-OCT systems.

In order to render FF-OCT even wider applicable, novel instrumental developments, the implementation of complementary concepts and the optimization of system designs will be required. This will include at first devices with higher acquisition speed and with real-time imaging and evaluation capabilities. The selective enhancement of contrast, based on the introduced contrast modification by flexible SLM filters, delivers further added value: with novel steerable and multiscale wavelet filter patterns advanced contrast with respect to certain structural characteristics of embedded features according to their size, their orientation or their specific shape can be provided [70, 71]. Advanced evaluation and characterization can also be realized by registering the true color of a sample. This imaging concept was already realized by sequential FF-OCT scans with a red, a green, and a blue LED [72]. The further development and optimization of spectroscopic FF-OCT are aiming in the same direction, that is, to obtain additional information from the spectral response of the sample of interest [73]. An extension of the spectral range is equally desirable, since the penetration depth significantly increases for longer wavelengths, as shown in Ref. [74] for FF-OCT imaging at a center wavelength of 1200 nm, performed on photolithography masks with an InGaAs camera. In this context, FF surface profilometry of rough surfaces with a bolometer camera at a center wavelength of 10 μm should be mentioned [75], a concept which might be expanded to spectroscopic FF-OCT approaches in the midinfrared spectral range. Finally, further useful complementary information and functional material contrast is expected to arise from extended concepts, as shown for PS imaging to detect regions of increased internal strain. Displacement measurements by LCSI and equally by 3D elastography with FF-OCT, recently introduced in Ref. [76] for obtaining virtual palpation maps of biological tissue, are further promising concepts for advanced characterization, delivering strain information and elastic contrast within technical materials.

Considering the results achieved to date in FF-OCT imaging and anticipating novel emerging optical measurement concepts for biology, medicine and material research, which can be combined in a multifunctional way with the principle of FF-OCT for multimodal characterization, FF-OCT is very likely to hold in the near future a prominent position for the evaluation of materials on a microscopic scale.

Acknowledgments

The author thanks B. Heise for helpful discussions and comments on the manuscript. Financial support by the Austrian Federal Ministry of Science, Research and Economy and the National Foundation for Research, Technology and Development is gratefully acknowledged.

References

1. Davidson, M., Kaufman, K., Mazor, I., and Cohen, F. (1987). An application of interference microscopy to integrated circuit inspection and metrology. *Proc. SPIE*, **775**, pp. 223–247.
2. Huang, D., Swanson, E. A., Lin, C. P., Schuman, J. S., Stinson, W. G., Chnag, W., Hee, M. R., Flotte, T., Gregory, K., Puliafito, C. A., and Fujimoto, J. G. (1991). Optical coherence tomography. *Science*, **254**, pp. 1178–1181.
3. Beaurepaire, E., Boccara, A. C., Lebec, M., Blanchot, L., and Saint-Jalmes, H. (1998). Full-field optical coherence microscopy. *Opt. Lett.*, **23**, pp. 244–246.
4. Dubois, A., Vabre, L., Boccara, A.-C., and Beaurepaire, E. (2002). High-resolution full-field optical coherence tomography with a Linnik microscope. *Appl. Opt.*, **41**, pp. 805–812.
5. Stifter, D. (2007). Beyond biomedicine: a review of alternative applications and developments for optical coherence tomography. *Appl. Phys. B*, **88**, pp. 337–357.
6. Deck, L., and Groot, P. D. (1994). High-speed noncontact profiler based on scanning white-light interferometry. *Appl. Opt.*, **33**, pp. 7334–7338.
7. Kino, G. S., and Chim, S. S. C. (1990). Mirau correlation microscope. *Appl. Opt.*, **29**, pp. 3775–3783.
8. Kuwamura, S., and Yamaguchi, I. (1997). Wavelength scanning profilometry for real-time surface shape measurement. *Appl. Opt.*, **36**, pp. 4473–4482.

9. Sarunic, M. V., Weinberg, S., and Izatt, J. A. (2006). Full-field swept-source phase microscopy. *Opt. Lett*, **31**, pp. 1462–1464.
10. Choma, M. A., Sarunic, M. V., Yang, C., and Izatt, J. (2003). Sensitivity advantage of swept source and Fourier domain optical coherence tomography. *Opt. Express*, **11**, pp. 2183–2189.
11. Laubscher, M., Ducros, M., Karamata, B., Lasser, T., and Salathé, R. (2002). Video-rate three-dimensional optical coherence tomography. *Opt. Express*, **10**, pp. 429–435.
12. Grieve, K., Dubois, A., Simonutti, M., Paques, M., Sahel, J., Le Gargasson, J.-F., and Boccara, C. (2005). In vivo anterior segment imaging in the rat eye with high speed white light full-field optical coherence tomography. *Opt. Express*, **13**, pp. 6286–6295.
13. Schausberger, S. E., Heise, B., Bernstein, S., and Stifter, D. (2012). Full-field optical coherence microscopy with Riesz transform-based demodulation for dynamic imaging. *Opt. Lett.*, **37**, pp. 4937–4939.
14. Klein, T., Wieser, W., Reznicek, L., Neubauer, A., Kampik, A., and Huber, R. (2013). Multi-MHz retinal OCT. *Biomed. Opt. Express*, **4**, pp. 1890–1908.
15. Vabre, L., Loriette, V., Dubois, A., Moreau, J., and Boccara, A. C. (2002). Imagery of local defects in multilayer components by short coherence length interferometry. *Opt. Lett.*, **27**, pp. 1899–1901.
16. Schausberger, S. E., Heise, B., Maurer, C., Bernet, S., Ritsch-Marte, M., and Stifter, D. (2010). Flexible contrast for low-coherence interference microscopy by Fourier-plane filtering with a spatial light modulator. *Opt. Lett.*, **35**, pp. 4154–4156.
17. Moneron, G., Boccara, A. C., and Dubois, A. (2007). Polarization-sensitive full-field optical coherence tomography. *Opt. Lett.*, **32**, pp. 2058–2060.
18. Gülker, G., Hinsch, K. D., and Kraft, A. (2001). Deformation montoring on ancient terracotta warriors by microscopic TV-holography. *Opt. Las. Eng.*, **36**, pp. 501–512.
19. Ruiz, P. D., Huntley, J. M., and Wildman, R. D. (2005). Depth-resolved whole-field displacement measurement by wavelength-scanning electronic speckle pattern interferometry. *Appl. Opt.*, **44**, pp. 3945–3953.
20. Davidson, M., Kaufman, K., and Mazor, I. (1988). First results of a product utilizing coherence probe imaging for wafer inspection. *Proc. SPIE*, **921**, pp. 100–114.
21. Lee, B. S., and Strand, T. C. (1990). Profilometry with a coherence scanning microscope. *Appl. Opt.*, **29**, pp. 3784–3788.

22. Caber, P. J. (1993). Interferometric profiler for rough surfaces. *Appl. Opt.*, **32**, pp. 3438–3441.
23. Sheppard, C. J. R., Roy, M., and Sharma, M. D. (2004). Image formation in low-coherence and confocal interference microscopes. *Appl. Opt.*, **43**, pp. 1493–1502.
24. Safrani, A., and Abdulhalim, I. (2011). Spatial coherence effect on layer thickness determination in narrowband full-field optical coherence tomography. *Appl. Opt.*, **50**, pp. 3021–3027.
25. Dresel, T., Häusler, G., and Venzke, H. (1992). Three-dimensional sensing of rough surfaces by coherence radar. *Appl. Opt.*, **31**, pp. 919–925.
26. Dubois, A., and Boccara, A. C. (1999). Real-time reflectivity and topography imagery of depth-resolved microscopic surfaces. *Opt. Lett.*, **24**, pp. 309–311.
27. Roy, M., Svahn, P., Cherel, L., and Sheppard, C. J. R. (2002). Geometric phase-shifting for low-coherence interference microscopy. *Opt. Las. Eng.*, **37**, pp. 631–641.
28. Choi, W. J., Na, J., Ryu, S. Y., and Lee, B. H. (2007). Realization of 3-D topographic and tomographic images with ultrahigh-resolution full-field optical coherence tomography. *J. Opt. Soc. Korea*, **11**, pp. 18–25.
29. Anna, T., Shakher, C., and Mehta, D. S. (2009). Simultaneous tomography and topography of silicon integrated circuits using full-field swept-source optical coherence tomography. *J. Opt. A: Pure Appl. Opt.*, **11**, p. 045501 (9pp).
30. Anna, T., Shakher, C., and Mehta, D. S. (2010). Three-dimensional shape measurement of micro-lens arrays using full-field swept-source optical coherence tomography. *Opt. Las. Eng.*, **48**, pp. 1145–1151.
31. Mahony, C. O., Hill, M., Brunet, M., Duane, R., and Mathewson, A. (2003). Characterization of micromechanical structures using white-light interferometry. *Meas. Sci. Technol.*, **14**, pp 1807–1814.
32. Grigg, D., Felkel, E., Roth, J., Lega, X. C. D., Deck, L., Groot, P. D., and Corporation, Z. (2004). Static and dynamic characterization of MEMS and MOEMS devices using optical interference microscopy. *Proc. SPIE*, **5455**, p. 55 (7pp).
33. Niehues, J., Lehmann, P., and Xie, W. (2012). Low coherent Linnik interferometer optimized for use in nano-measuring machines. *Meas. Sci. Technol.*, **23**, pp. 125002 (1–9).
34. Beer, S., and Seitz, P. (2005). A smart pixel array with massively parallel signal processing for real-time optical coherence tomography perform-

ing close to the physical limits. *IEEE Conf. Proc. Res. Microelectron. Electron.*, **2**, pp. 135–138.
35. Egan, P., Lakestani, F., Whelan, M. P., and Connelly, M. J. (2005). Three-dimensional machine vision utilising optical coherence tomography with a direct read-out CMOS camera. *Proc. SPIE*, **5856**, pp. 427–436.
36. Lambelet, P. (2011). Parallel optical coherence tomography (pOCT) for industrial 3D inspection. *Proc. SPIE*, **8082**, pp. 80820X (1–12).
37. Montgomery, P., Anstotz, F., and Montagna, J. (2011). High speed, on-line 4D microscopy using continuously scanning white light interferometry with a high speed camera and real-time FPGA image processing. *Proc. SPIE*, **8082**, pp. 808210 (1–9).
38. Choi, W. J., Jung, S. P., Shin, J. G., Yang, D., and Lee, B. H. (2011). Characterization of wet pad surface in chemical mechanical polishing (CMP) process with full-field optical coherence tomography (FF-OCT). *Opt. Express*, **19**, pp. 13343–13350.
39. Gastinger, K., Johnsen, L., Simonsen, O., and Aksnes, A. (2011). Inspection of processes during silicon wafer sawing using low coherence interferometry in the near infrared wavelength region. *Proc. SPIE*, **8082**, pp. 80820U (1–9).
40. Kim, S.-W., and Kim, G.-H. (1999). Thickness-profile measurement of transparent thin-film layers by white-light scanning interferometry. *Appl. Opt.*, **38**, pp. 5968–5973.
41. Abdulhalim, I., and Dadon, R. (2009). Multiple interference and spatial frequencies' effect on the application of frequency-domain optical coherence tomography to thin films' metrology. *Meas. Sci. Technol.*, **20**, p. 015108.
42. Tomczewski, S., Pakuła, A., and Sałbut, L. (2012). Simultaneous shape measurement and layers detection by means of low coherence interferometry and optical coherence tomography. *Phot. Lett. Poland*, **4**, pp. 57–59.
43. Na, J., Choi, W. J., Choi, H. Y., Ryu, S. Y., Choi, E. S., and Lee, B. H. (2009). Thickness and index measurements of a transparent specimen by full-field optical coherence microscopy. *Proc. SPIE*, **7184**, pp. 71841B (1–7).
44. Li, C., Zeitler, J. A., Dong, Y., and Shen, Y.-C. (2014). Non-destructive evaluation of polymer coating structures on pharmaceutical pellets using full-field optical coherence tomography. *J. Pharm. Sci.*, **103**, pp. 161–166.
45. Alarousu, E., Gurov, I., Kalinina, N., Karpets, A., Margariants, N., Myllyla, R., Prykari, T., and Vorobeva, E. (2007). Full-field high-resolving optical

coherence tomography system for evaluating paper materials. *Proc. SPIE*, **7022**, pp. 702212 (1–7).

46. Latour, G., Echard, J.-P., Soulier, B., Emond, I., Vaiedelich, S., and Elias, M. (2009). Structural and optical properties of wood and wood finishes studied using optical coherence tomography: application to an 18th century Italian violin. *Appl. Opt.*, **48**, pp. 6485–6491.

47. Latour, G., Georges, G., Siozade, L., Deumie, C., and Echard, J.-P. (2009). Study of varnish layers with optical coherence tomography in both visible and infrared domains. *Proc. SPIE*, **7391**, pp. 73910J (1–7).

48. Latour, G., Moreau, J., Elias, M., and Frigerio, J.-M. (2007). Optical coherence tomography: non-destructive imaging and spectral information of pigments. *Proc. SPIE*, **6618**, pp. 661806 (1–9).

49. Targowski, P., and Iwanicka, M. (2011). Optical coherence tomography: its role in the non-invasive structural examination and conservation of cultural heritage objects: a review. *Appl. Phys. A*, **106**, pp. 265–277.

50. Gurov, I., Karpets, A., Margariants, N., and Vorobeva, E. (2007). Full-field high-speed optical coherence tomography system for evaluating multilayer and random tissues. *Proc. SPIE*, **6618**, pp. 661807 (1–8).

51. Elias, M., Mas, N., and Cotte, P. (2011). Review of several optical non-destructive analyses of an easel painting. Complementarity and crosschecking of the results. *J. Cultural Heritage*, **12**, pp. 335–345.

52. Elias, M., Magnain, C., and Frigerio, J. M. (2010). Contribution of surface state characterization to studies of works of art. *Appl. Opt.*, **49**, pp. 2151–2160.

53. Srivastava, V., Anna, T., Sudan, M., and Mehta, D. S. (2012). Tomographic and volumetric reconstruction of composite materials using full-field swept-source optical coherence tomography. *Meas. Sci. Technol.*, **23**, pp. 055203 (1–10).

54. Heise, B., Schausberger, S. E., Häuser, S., Plank, B., Salaberger, D., Leiss-Holzinger, E., and Stifter, D. (2012). Full-field optical coherence microscopy with a sub-nanosecond supercontinuum light source for material research. *Opt. Fiber Technol.*, **18**, pp. 403–410.

55. Schlager, V., Schausberger, S. E., Stifter, D., and Heise, B. (2011). Coherence probe microscopy imaging and analysis for fiber-reinforced polymers. *LNCS*, **6688**, pp. 424–434.

56. Schausberger, S. E., Heise, B., Maurer, C., Bernet, S., Ritsch-Marte, M., and Stifter, D. (2011). Contrast modification for ultra-high resolution low-coherence interference microscopy by Fourier-plane filtering. *Proc. SPIE*, **8091**, pp. 809109 (1–6).

57. Leiss-Holzinger, E., Cakmak, D. U., Heise, B., Bouchot, J.-L., Klement, E. P., Leitner, M., Stifter, D., and Major, Z. (2012). Evaluation of structural change and local strain distribution in polymers comparatively imaged by FFSA and OCT techniques. *Express Polym. Lett.*, **6**, pp. 249–256.
58. Heise, B., Schausberger, S. E., Buchroithner, B., Aigner, M., Milosavljevic, I., Hierzenberger, P., Bernstein, S., Häuser, S., Jesacher, A., Bernet, S., Ritsch-Marte, M., and Stifter, D. (2013). Full field optical coherence microscopy for material testing: contrast enhancement and dynamic process monitoring. *Proc. SPIE*, **8792**, pp. 87921A (1–6).
59. Hinsch, K. D., Gülker, G., and Helmers, H. (2007). Checkup for aging artwork: optical tools to monitor mechanical behaviour. *Opt. Las. Eng.*, **45**, pp. 578–588.
60. Gastinger, K., Gülker, G., Hinsch, K. D., Pedersen, H. M., Stoeren, T., and Winther, S. (2004). Low coherence speckle interferometry (LCSI) for detection of interfacial instabilities in adhesive bonded joints. *Proc. SPIE*, **5532**, pp. 256–267.
61. Park, K. S., Choi, W. J., Eom, T. J., Kim, J. H., Woo, D. H., and Lee, B. H. (2013). Polarization sensitive full-field optical coherence tomography based on bi-stable polarizaton switch. *Proc. SPIE*, **8589**, pp. 85891D (1–6).
62. Heise, B., Schausberger, S., and Stifter, D. (2013). Chapter 8, Full field optical coherence microscopy: imaging and image processing for micro-material research applications. In *Optical Coherence Tomography*, M. Kawasaki, ed. (InTech, Rijeka), pp. 139–162.
63. Heise, B., Buchroithner, B., Schausberger, S. E., Hierzenberger, P., Eder, G., and Stifter, D. (2014). Simultaneous detection of optical retardation and axis orientation by polarization-sensitive full-field optical coherence microscopy for material testing. *Laser Phys. Lett.*, **11**, p. 055602.
64. Reyes-Esqueda, J., Vabre, L., Lecaque, R., Ramaz, F., Forget, B. C., Dubois, A., Briat, B., Boccara, C., Roger, G., Canva, M., Levy, Y., Chaput, F., and Boilot, J.-P. (2003). Optical 3D-storage in sol–gel materials with a reading by optical coherence tomography-technique. *Opt. Commun.*, **220**, pp. 59–66.
65. Chang, S., Cai, X., and Flueraru, C. (2006). Image enhancement for multilayer information retrieval by using full-field optical coherence tomography. *Appl. Opt.*, **45**, pp. 5967–5975.
66. Dubey, S. K., Anna, T., Shakher, C., and Mehta, D. S. (2007). Fingerprint detection using full-field swept-source optical coherence tomography. *Appl. Phys. Lett.*, **91**, pp. 181106 (1–3).

67. Choi, W.-J., Min, G.-H., Lee, B.-H., Eom, J.-H., and Kim, J.-W. (2010). counterfeit detection using characterization of safety feature on banknote with full-field optical coherence tomography. *J. Opt. Soc. Korea*, **14**, pp. 316–320.
68. Hayasaki, Y., Matsuba, Y., Nagaoka, A., Yamamoto, H., and Nishida, N. (2004). Hiding an image with a light-scattering medium and use of a contrast-discrimination method for readout. *Appl. Opt.*, **43**, pp. 1552–1558.
69. Otaka, M., Yamamoto, H., and Hayasaki, Y. (2006). Manually operated low-coherence interferometer for optical information hiding. *Opt. Express*, **14**, pp. 9421–9429.
70. Heise, B., Schausberger, S. E., Maurer, C., Ritsch-Marte, M., Bernet, S., and Stifter, D. (2012). Enhancing of structures in coherence probe microscopy imaging. *Proc. SPIE*, **8335**, pp. 83350G (1–7).
71. Heise, B. (2012). Coherence probe microscopy. *Imaging Microsc.*, **14**, pp. 29–32.
72. Yu, L., and Kim, M. K. (2004). Full-color three-dimensional microscopy by wide-field optical coherence tomography. *Opt. Express*, **12**, pp. 6632–6641.
73. Dubois, A., Moreau, J., and Boccara, C. (2008). Spectroscopic ultrahigh-resolution full-field optical coherence microscopy. *Opt. Express*, **16**, pp. 17082–17091.
74. Sacchet, D., Moreau, J., Georges, P., and Dubois, A. (2008). Simultaneous dual-band ultra-high resolution full-field optical coherence tomography. *Opt. Express*, **16**, pp. 19434–19446.
75. Groot, P. D., Biegen, J., Clark, J., Lega, X. C. D., and Grigg, D. (2002). Optical interferometry for measurement of the geometric dimensions of industrial parts. *Appl. Opt.*, **41**, pp. 3853–3860.
76. Nahas, A., Bauer, M., Roux, S., and Boccara, A. C. (2013). 3D static elastography at the micrometer scale using Full Field OCT. *Biomed. Opt. Express*, **4**, pp. 2138–2149.

Index

aberrated image field 314
absolute value 72, 77–79, 81, 83, 208, 338
absorbing samples 531
accumulating number 703–704
ACF, *see* autocorrelation function 159–160
achromatic doublet lenses 491–492, 506, 520
acquired images 472, 501, 506, 508
acquisition depth 585–588
adipose tissue 599, 621, 623, 626
AFM, *see* atomic force microscopy
AMF, *see* amplitude modulation function
amplitude images 384–385, 451, 541–542
amplitude modulation function (AMF) 431, 434–437
analytical models 174
analytic signal 283
angles of meiotic spindle deviation 708–709
angular frequency spectrum 370, 373–375
angular spectra 55, 72–73, 75, 82, 84
angular spectrum algorithm 439
angular spectrum gating 76–77, 82, 84
angular spectrum method 306, 440–441, 445

anisotropy parameters 154, 158, 166
anterior lamellar keratoplasty 659, 662–663, 668
aperture planes 58, 66, 105–106
apple skin 558–559
area camera 4–5, 16, 40, 729
artificial corneas 669
atomic force microscopy (AFM) 395
autocorrelation function (ACF) 66, 94, 147, 156, 186–188, 362–364, 366
auxiliary laser interferometer 256–257
averaging demodulated signals 149–150
axial
  direction 4, 7, 30, 358, 361, 379, 382–383, 480, 487, 520, 570, 692, 697
  position 269, 401–402, 581
  range 432, 439, 441, 450, 454, 459, 535
  response 25–26, 28, 376
  sampling step 416, 418
  scan 13, 118, 209, 258
  structure 312

background component 187–188, 204, 440
background noise 23

backscattered light 231, 358, 468, 479, 483, 488, 490, 519, 531, 690, 720, 728
backscattered sample field 139, 144
backscattering object (BO) 136
backscattering structures 18, 31
backscattering zone (BZ) 142
ballistic light 140, 144–146, 149–150, 158–161, 163, 177
beam splitter (BS) 5–6, 57, 232, 245–246, 257, 273, 276–278, 280, 361, 470–473, 499, 507, 520, 544–545, 691
beat wavelength 432, 435, 454, 546–549, 551, 553
Beer's law 161–163, 483, 531
BF, see bright field
biological images 696
biological tissues 329, 342–343, 346, 349, 351, 468–469, 479, 484, 494, 519, 524, 568, 570, 730, 733
biomedical applications 2, 510, 533
biomedical field 32, 727–728
biomedical imaging 93, 357, 560, 568, 678, 728
birefringent sample 473, 476
black triangles 629–630, 632
BO, see backscattering object
Bowman's layer thickness 659
brain tissue 618, 636
breast cancer 566
breast tissue
    classification 625-627
    features recognition 618–622
    imaging 34–35
    sample 570, 572
bright field (BF) 731–732
bright signals 600, 604
broadband light source 5, 40, 175, 187, 273, 360, 362, 368, 521, 536, 554, 720, 729, 743

BS, see beam splitter
B-scan images 542
BZ, see backscattering zone

camera image 280, 556
camera pixel wells 20, 23
cancerous tissue samples 582
CCD, see charge-coupled device
CCD
    exposure time 244
    output signal 250–251
cellular-level resolution imaging 4, 32
central wavelength 140, 147, 159, 252, 309, 377, 578, 581, 692, 698, 701
channeled spectrum 328
charge-coupled device (CCD) 12, 133, 236, 243–244, 247–249, 275–276, 430, 437–438, 469–470, 480, 521, 536–537, 545–546, 690, 694–695
choroidal surface 450–451
chromium-coated glass surface 441
circular imaging 79–80
clinical applications 31, 33, 35, 324, 347, 352, 590, 613, 647
CMOS, see complementary metal-oxide semiconductor
CMOS magnetic probe being 416–418, 736
coherence function 9, 60, 70, 72, 77, 80, 83–84, 95–96, 98–99, 102, 309–310, 362, 386, 400, 698
coherence length 8, 98–100, 148, 152, 187–188, 358, 361, 365–366, 373–374, 382, 446, 448, 534–536, 538, 579
coherence plane 13–14, 343, 346

Index | 757

coherence probe microscopy (CPM) 92, 271, 396, 728–729, 734, 745
coherence region 93, 105, 107, 119
coherence scanning interferometry (CSI) 396
coherence signal 53–55, 71, 75–76, 80, 84
coherence time 364, 373–374
coherence volume 143, 474, 483, 703
coherent light sources 7, 11, 41, 133, 137–138, 152, 544
coherent noise 140, 142, 309, 535, 560
collagen fibrils 644, 678–679
collagen lamellae 643, 645–646, 650–651, 655, 659, 668, 679
combined intensity/spectroscopy FF-OCM image 486
comparative full-field optical coherence tomography 606, 608–610
comparative images 570, 574
complementary metal-oxide semiconductor (CMOS) 8, 252, 271, 304, 348, 396, 470, 571, 690, 730
complex amplitude 61–62, 65, 184–185, 385, 540
complex object images 445
confocal configuration 154, 161–163, 169–170
confocal imaging process 143
confocal microscopes 2, 93, 95, 107, 121
confocal spatial filtering 131–132, 141, 143, 175
connective tissue bright signals 608
continuous fringe scanning 393, 397, 407
continuous signal sampling 197

contrast-enhanced imaging 92, 469
contrast modes 731
conventional demodulation technique 209–210
conventional FF-OCM 5, 16, 30
conventional OCT
  systems 359–360, 363, 365, 368
corneal
  opacities 662–663, 668
  pathology 642, 663–664
  specimen tissue 659, 662
  tissue 644–645, 648, 664
  transparency 661
correct image registration 587–588
correlation function 95, 97–98
corresponding
  FF-OCT image 621
  temporal coherence functions 8
  unwrapped phase maps 384
coupled interferometers 327, 329, 331, 333, 351
CPM, *see* coherence probe microscopy
cross-sectional FF-OCT image 600, 602
cross-sectional reconstruction 344, 649, 651, 676–677
cross-talk signals 338–339
CSI, *see* coherence scanning interferometry
cutoff frequency 196, 200 201, 203

dark field (DF) 731–732
data acquisition time 259–260
DC, *see* direct current
defocused onion 541–543
degree of correlation 135, 137, 140, 172
demodulated FF-OCM image 291

demodulated signals 153–156, 168
depth-dependent sample dispersion 311
depth information 16, 304, 314, 447, 535, 690
Descemet's membrane (DM) 645, 651–652, 654, 660–662, 668
detection sensitivity 1, 4, 7, 18, 22, 24, 30, 247, 493, 502–503, 508, 701, 703
developmental biology 37, 41, 132, 689, 691, 704–705, 707, 709–711, 713, 715, 717, 719–720
DF, see dark field
DHM, see digital holographic microscopy
DIC, see digital image correlation
difference image 279, 285, 289
digital holographic microscopy (DHM) 294, 396
digital image correlation (DIC) 743
digital interference holography (DIH) 304, 431–439, 441–451, 453, 455–457, 459
digital signal processing (DSP) 228, 403, 735–736
DIH, see digital interference holography
direct current (DC) 99, 250, 309, 376, 695
discretization step 192, 197, 199, 205
dispersion mismatch 28, 476, 493, 500–501, 523
displacement fields 742–743
displacement measurements 729, 733, 743, 746
divided interferograms 273–274
DM, see Descemet's membrane
donor corneas 667–669

Doppler frequency 147–148, 243, 245, 247, 249–251, 253–255, 257–259
DSP, see digital signal processing
dual-band FF-OCM 491–492, 496–498

five-step adaptive (FSA) 397, 410–412, 423, 571–572, 590, 595–597, 607–608, 611
flat mirror surface 258–259
flat surface area 696
flexible endoscope 326, 341, 343, 345–346
fluorescence imaging modality 507–508
fluorescence signal 509–510, 586
four-channel polarization phase stepper (FC-PPS) 234, 236
Fourier domain optical coherence tomography (FD-OCT) 98, 188–193, 358–359, 430, 448, 534, 642, 648, 663, 666–667, 690, 699–701, 703
Fourier plane filtering (FPF) 277, 731–732
Fourier transform pairs 363, 368, 435–437
four-phase-shifting method 693, 696–697
FPF, see Fourier plane filtering
FPGA, see field-programmable gate array
fringe image acquisition rate 415, 417, 420
fringe parameters 196, 203–205, 211
fringe signal downsampling 197
FRP, see fiber-reinforced polymer
FSA, see five-step adaptive
full-field configuration 162–163
full-field image acquisition geometry 576

full-field imaging techniques 280, 294
full-field optical coherence elastography 680
full-field optical coherence microscopy (FF-OCM) 1–2, 6–8, 32–36, 53–82, 84, 183–184, 467–468, 478–484, 504–511, 641–642, 648–658, 666–670, 672–678
full-field optical coherence tomography (FF-OCT) 223–242, 393–398, 519–522, 569–572, 574–579, 584–590, 595–598, 607–613, 689–693, 697–701, 703–713, 727–729, 734–740
functional material imaging 728, 734, 742

Gaussian function 7, 21, 26–27, 362, 436–437
geometric phase shifting (GPS) 271–272
GPS, see geometric phase shifting
GPU, see graphics processing unit
graphics processing unit (GPU) 212, 228, 396, 403, 560
GRIN lenses 325–326, 347, 350

hair follicle (HF) 586, 599–601
heterodyne detection schemes 247–248, 277
heterodyne signal detection 249, 252
HF, see hair follicle
high axial resolution 29, 101, 103, 366, 396, 448, 734
high image acquisition rates 13, 134
high-NA microscope objectives 376–377, 699, 720

high-resolution depth imaging 119
high-resolution images 376, 570
high transverse resolution 54, 699
histology-quality images 597, 611
holographic image acquisition 450, 452
human lobectomy specimens 596, 607, 609
human optic nerve sample 453–454

IDC, see invasive ductal carcinoma
illumination aperture function 69, 74–76, 84
image accumulation 30, 247, 458, 481
  acquisition board 546, 554
  acquisition techniques 242
  surface areas 450, 452
  fusion techniques 701
  mapping spectrometry 279
  pixelation effect 334
  processing algorithms 584
  processing techniques 670
  reconstruction technique 285
  restoration technique 285
image processing unit (IPU) 409–410
imaging aperture functions 74–75, 84
  axial resolution 101, 482, 498
  biomedical tissue 533
  endoscope 347
  fiber bundles 337, 340
  interferometer 328–329, 331, 333–334, 345–347, 351, 576
  penetration depth 2, 36, 468, 484, 494, 504
  spatial resolution 7, 41
immersion microscope objectives 15

incoherent light source 134, 141, 143, 172, 351, 360, 367, 570, 598
incorporating sample objectives 131
interferometric images 480–481
information capacity 183, 189, 192–193, 217
inner sample structures 82
integrating buckets 17–18, 21–22
intensity-based tomographic images 481
intensity measurements 489, 693
interference component 243, 249, 252, 261
  images 223–224, 232, 234–236, 242, 245, 254, 555, 613
  microscopes 5, 13, 26, 29, 56, 58, 84, 398, 403, 480
  pattern 99, 303, 312, 358, 381–383, 520
  process 138, 143–144
  signal intensity 228, 247, 253
interferogram width 105, 107–110, 119
interferometer source arm 165
interferometric images 4, 16–17, 24, 31, 107, 267, 269–270, 359, 383, 480–481, 491–492, 498, 500, 554, 598
  imaging system 536, 553
  signal amplitude 16, 31
invasive ductal carcinoma (IDC) 247, 250–251, 623
inverse synthetic aperture microscopy (ISAM) 314
IPU, see image processing unit
ISAM, see inverse synthetic aperture microscopy

Kalman filtering method 208–211

Lambert–Beer's law 161–162, 164, 177
large-field images 608
large reference signal component 13, 133–134
large spectral bandwidth 366, 373
laser diode (LD) 256–257
laser-tissue interaction 665–666
lateral spatial coherence function 94, 370, 372
lattice corneal dystrophy 658, 663
LCI, see low-coherence interferometric
LCIM, see low-coherence interference microscopy
LCSI, see low-coherence speckle interferometry
LD, see laser diode
LED, see light-emitting diode
LF, see line field
light-emitting diode (LED) 9–10, 41, 107, 351, 360, 498, 534, 544, 551, 555–557, 735
light microscopy 667–668
light source wavelength number 188
line field (LF) 303
line-scan camera 431, 446, 448, 540–541
Linnik imaging interferometer 343–344
local amplitude 269, 280–281, 284
longitudinal coherence length 119, 365, 374–375
longitudinal cross-sectional images 134, 690
longitudinal spatial coherence (LSC) 93–97, 99, 103, 105–107, 110, 119, 245–246, 360, 368–369, 373, 375, 379–382, 385–386
low-coherence fringe signals 194–195

Index | 761

low-coherence interference microscopy (LCIM) 93
low-coherence interferometric (LCI) 268, 271, 358, 360, 728
low-coherence interferometry 3–4, 54, 184, 205, 533, 536–538, 540, 642, 648, 728
low-coherence speckle interferometry (LCSI) 733, 746
LSC, *see* longitudinal spatial coherence

magnetic resonance imaging (MRI) 669
malignant tissue 622–623, 625
mammalian developmental biology 37, 716, 719
masking frequencies 199–200
material characterization 39, 469, 479, 511, 727, 729
maximum intensity profiles 155–156, 159–160
MCF, *see* mutual coherent function
measured axial resolution 579, 581
measured sample field 175
measurement concepts 734–735
meiotic spindle deviation 708–709
merged FF-OCT/fluorescence SIM image 587–588
MF, *see* multimode fiber
Michelson processing interferometer 342–343
microscopic surface roughness 393, 395–396
Mirau interferometric objectives 380–382
modulation function 250, 404
monitoring dynamics 289, 293–294, 742

monochromatic light source 360, 368–369, 371, 373, 375, 377, 386, 520
morphological images 511, 678, 710
MRI, *see* magnetic resonance imaging
MSL, *see* multiply scattered light
multilayered model samples 63
multilayered scattering reference mirror 120
multimode fiber (MF) 9, 165–167, 172, 343–344
multiphoton microscopy 678–680
multiple phase-shifted interferometic images 506
multiply scattered light (MSL) 131–132, 135–137, 139–145, 150, 159, 161, 163, 171–173
mutual coherent function (MCF) 141, 175
mutual spectral density signal 76–78

narrow-band illumination 118–120
NA, *see* numerical aperture
nerve fiber layer (NFL) 453, 674–675
NFL, *see* nerve fiber layer
nonpolarizing beam splitter (NPBS) 231, 234, 239–240, 274–275
NPBS, *see* nonpolarizing beam splitter
numbered tomographic images 247
numerical aperture (NA) 54, 164

objective aperture 58, 73, 84, 103, 165

object wave distributions 442–443
OCM, see optical coherence microscopy
OCT, see optical coherence tomography
OCT interference signal 142, 256–258
OCT interferometer 225, 228, 249, 256–257
OCT measurements 153, 163, 247
OCT models 140–141, 144, 174
OCT signal intensities 254–255, 259
OCT signals 134–135, 140, 153, 157–158, 162–163, 167, 208, 226–227, 250–251, 259–260, 311, 327, 330, 333, 335
OCT systems 93, 176, 184, 189, 194–195, 251, 255, 279–280, 358–360, 365–366, 524, 531, 628, 691, 698–700
OCT techniques 16, 93, 207, 230, 271, 367, 534–535, 569–570, 573–574, 579
OD, see optical density
online measurement system 40, 424
OPD, see optical path difference
OPL, see optical path length
optical coherence microscopy (OCM) 4, 40, 54, 85, 92–93, 184, 211, 218, 326, 367, 629
optical coherence tomography (OCT) 3–4, 41, 91–94, 142–144, 146–150, 172–176, 223–242, 308–314, 325–327, 357–360, 467–469, 533–536, 568–569
optical density (OD) 154–155, 158–162, 164, 168, 170, 491
optical field propagation 60–61
optical imaging techniques 3, 323, 534, 611

optical microscopic images 227, 739
optical path difference (OPD) 185, 187, 213, 247, 358, 361, 365, 375, 379–381, 521, 523, 528
optical path length (OPL) 13, 101, 139–140, 152, 158, 225, 269–270, 500–501, 597–598, 674, 692, 694
optical phase unwrapping 536, 544, 547–549, 551, 553
optical transmission functions 61–62, 66
optimal phase modulation parameters 229–230

PA, see plasminogen activator
painted coin surface 557–558
parabolic approximation 59, 61, 70, 74, 82
parallel layers 81, 83
path length differences 93, 100, 102, 174, 328–329, 331, 333, 339, 341, 344–345, 583
pathology applications 574–575, 577–579, 581, 583, 585, 587
PBS, see polarizing beam splitter
PC, see personal computer
PCP, see predicted cleavage plane
peak fringe scanning microscopy (PFSM) 397, 410–411, 418, 423
penetration depth 2, 22, 289, 326, 497, 531, 558, 560, 598, 672, 690, 740, 746
personal computer (PC) 152–153, 378, 409, 412, 691–692
PFSM, see peak fringe scanning microscopy
phase-corrected positive image 448–449
phase distribution 431

phase-opposed interference images 613
phase-quadrature interference images 230, 232–233, 245
phase-shifted interference images 224, 732
phase-shifted interferometric images 5, 16, 270, 492, 499, 506
phase-shifting algorithms 16
phase-shifting interferometry 15–17, 21, 224, 536, 544, 556
phase shift interference imaging 555–556
phase shift interferometry (PSI) 102, 120, 224
phase-stepped interferometric images 269
phase-stepping algorithms 17
phase-stepping approaches 281
piezoelectric transducer 5, 16, 227, 253, 270, 471, 492, 499, 507, 520, 523, 545, 691–692
pixelation removal 335–337
plasminogen activator (PA) 136, 206, 598
point-spread function (PSF) 29, 76, 78–80, 436, 508
polarization sensitive (PS) 142, 145, 344, 468–471, 473, 475, 477–479, 487, 490, 510, 668, 679, 733
polarization-sensitive optical coherence tomography (PS-OCT) 469
polarizing beam splitter (PBS) 235, 240, 252–253, 272, 275–276, 438, 470–471, 477, 486, 544–545, 613, 673
power density spectrum 481
predicted cleavage plane (PCP) 716–718
preimplantation development 704–705

processed images 335, 422
processing interferometer 328–330, 334, 339, 345, 347
processing OCT signals 189, 208, 210
propagating modes 172, 337, 340, 378
PS, *see* polarization sensitive
pseudothermal light source 172–173, 374–375
PSF, *see* point-spread function
PSI, *see* phase shift interferometry
PS-OCT, *see* polarization-sensitive optical coherence tomography

quadrature functions 225, 229
quarter-wavelength intervals 537, 545–546
quasi-homogeneous optical field 372
quasi-monochromatic illumination 76–78, 82

random phase drifts 246–247, 272
random phasor sum 148–150, 176
RBCs, *see* red blood cells
reconstructed spectrum 526–528
reconstructed volume 441, 450–454
recurrence algorithms 203, 205, 207, 212, 214, 217
red blood cells (RBCs) 382, 384, 457, 629, 637
reduced image acquisition number 291
reference and sample fields 97, 139–141, 145, 174–175
reference mirror displacement 156, 158
reference mirror position 137, 139, 145, 147, 434, 557

reference mirror reflectivity 24, 703–704
reference wave field 71, 269–270, 278–279
representative images 584, 650
resolution test target (RTT) 279, 316
retinal imaging 305, 313, 318, 642, 648, 675, 678
retinal imaging retinal sample objects 434
Riesz transform (RT) 31, 283–284, 730
RT, see Riesz transform
RTT, see resolution test target

sampled intensity values 56
sampled interference components 254, 257
sampled OCT signal 211
sampled signal spectrum 197
sample
  field distribution 150
  field experiences interference 95
  image plane 306
  mirror position 157, 177
  preparation 571–572
  scanning 358, 692
  structures 262, 308–309, 598
  surface topography measurements 16
scan interferogram 109–111
scanning fluorescence microscopy techniques 505
scanning laser ophthalmoscope (SLO) 672–673
scanning plane 442, 444–445
scattering materials 295, 728, 738
scattering solution 150, 153–158, 161, 163, 166–170, 176

schematic diagram 185, 360–361, 369, 377–378, 649
SD-OCT, see spectral domain optical coherence tomography
SDPM, see spectral domain phase microscopy
second harmonic generation (SHG) 560, 678–679
sectioned fluorescence images 509
sectioning effect 71–75, 77, 734
SF, see spatial filter
SHG, see second harmonic generation
signal
  contrast 15, 226, 501
  enhancement 161, 163
  envelopes 199, 209
  processing algorithms 217, 281
  processing techniques 269, 280
signal-to-background-noise ratio 23
SII, see spatially incoherent illumination
silicon camera 12, 492–493, 495
SIM, see structured illumination microscopy
SIM fluorescence images 583, 587
simple imaging interferometer 342, 345–346
simultaneous dual-band FF-OCM imaging 495–496
single camera 252, 273, 277, 470, 498, 505, 743
single CCD camera 230, 232, 235, 237, 241, 248, 255, 261, 274
single-shot FF-OCT 241, 267–268, 276–277, 279, 285, 289–290, 292–296
single-shot techniques 269, 271, 273, 275, 277, 279, 294, 730

single-wavelength phase map 547, 549–551, 553
sinusoidal phase modulation 17, 19, 21–22, 228, 271, 471, 474, 492, 692, 694
SLM, *see* spatial light modulator
SLO, *see* scanning laser ophthalmoscope
source autocorrelation function 137, 139–140, 147, 159–160, 172, 174–175
spatial coherence properties 136, 138, 141–142
spatial filter (SF) 152, 165, 167, 360, 438
spatial light modulator (SLM) 273, 294, 296, 731–732
spatial light modulators 296
spatially incoherent illumination (SII) 132–133, 135, 142, 164–171
spatial phase-shifting method 234
spatial power spectral density 372, 375
spectral domain optical coherence tomography (SD-OCT) 358, 534–535, 568, 690
spectral domain phase microscopy (SDPM) 431
spectral fringes 212–213, 215–216
spectral information 447, 480, 482, 526–527, 738
spectral interference pattern 307–308, 312
spectral power distribution 7–8, 10, 26, 479
spectral range 98, 107, 212, 328, 479, 498, 500, 746
spectral resolution 309, 446, 481–482, 498, 505, 527
spectral width 9, 26, 28, 109, 309, 312, 363, 498, 555, 560

spectroscopic FF-OCM 468, 479, 481–483, 485–487, 489–490, 510
spectroscopic measurements 479, 481–482, 484
spectrum sampling step 482, 527
spiral phase (SP) 707, 732
squared downsampled signal 200
SS-OCT, *see* swept-source optical coherence tomography
stacked interferograms 379
structural sample anisotropies 469, 743
structured illumination microscopy (SIM) 430, 506, 508, 576–578, 580–582, 584–585, 587, 590
sub-Nyquist sampling 210, 739
subsurface reflection 202–203
surface blood vessels 289–290
surface profiles 381, 546–548, 550–552
surface roughness 39, 341, 394, 396, 399–400, 420, 423
surgical specimens 658, 663
swept-source optical coherence tomography (SS-OCT) 192, 358, 459, 534–535, 690

talk-generated noise 135–137, 140–141, 161, 164, 169
talk noise contribution 135, 140–141, 171
talk suppression 164, 169, 171–173
TC, *see* temporal coherence
TD, *see* time domain
TD-OCT, *see* time domain optical coherence tomography
TD-OCT and FD-OCT 188, 190–191, 699–701, 703

technical samples 304, 316, 730
temporal coherence (TC) 92–96, 101, 105, 107–109, 111, 116–117, 386
temporal coherence functions 7, 9–10
temporal coherence length 9, 26–27, 110, 360, 368, 375, 386, 737
temporal frequency spectrum 358, 362, 365, 369, 373–374
temporal phase-shifting techniques 268, 270
temporal spectrum 54, 68, 71, 73–76, 81, 84
theoretical values 163, 254, 502, 508–509
thermal light sources 7, 132–134, 165, 170–172, 690–691, 698, 704, 734
thin-film metrology 728, 734–735, 737
three-dimensional imaging 3, 30, 132
three-wavelength phase unwrapping method 548–549, 552–553
time domain (TD) 69, 98, 136, 187, 213, 242, 303, 359, 430, 534–535, 569, 628, 729–730, 733
time domain optical coherence tomography (TD-OCT) 99, 187–188, 191–194, 312, 359, 430, 469, 534–535, 690
tissue architecture 567, 575, 582, 584, 621
tissue processing 2, 32, 595–596, 607, 610–611, 613
tomographic image and spectroscopic information 529
total coherence function 69, 71

TP-LSM, see two-photon laser scanning microscopy
translation stage (TS) 152–153, 165, 167, 245, 520, 692
transmission functions 62, 68, 71
transparent layers 92, 394, 415, 417, 422
transversal spatial spectrum 59, 68, 73, 75
transverse resolution 3, 53, 84, 133, 359, 365, 367–368, 495, 497, 502, 569, 578, 581, 692, 698–701
TS, see translation stage
two-phase-shifting method 693–694, 696–697
two-photon laser scanning microscopy (TP-LSM) 707–708, 712

ultrahigh resolution 458

van Cittert–Zernike theorem 94–96, 368–369, 371–373

water immersion microscope objectives 118, 480, 491, 506, 691, 698–699, 703
wavelength range 191, 333, 490, 500, 535, 560
wavelength scanning 304, 534
WF, see wide field
WF-OCT, see wide-field optical coherence tomography
white light interferometry (WLI) 92, 99, 396, 404, 429, 728
white light scanning interferometry (WLSI) 396–397, 729

wide field (WF) 184, 230, 232, 235, 246, 253, 257, 260, 270, 313, 359, 377, 506
wide-field depth-resolved imaging 430
wide-field optical coherence tomography (WF-OCT) 230, 232, 554, 556
WLI, *see* white light interferometry
WLSI, *see* white light scanning interferometry
Wollaston prism 232, 234, 239–241, 273, 275–276

xenon arc lamps 9, 343, 345, 347
xz B-scan images 541

YZ images 423, 478
y-z planes 441, 454

zona pellucida (ZP) 707, 710, 712–713
ZP, *see* zona pellucida
zygote-stage embryos 696–697, 713, 716